I0787948

Gallium Nitride and Silicon Carbide Power Technologies 3

Editors:

K. Shenai
Argonne National Laboratory
Argonne, Illinois, USA

M. Dudley
Stony Brook University
Stony Brook, New York, USA

M. Bakowski
Acreo – Sweden
Kista, Sweden

N. Ohtani
Kwansei Gakuin University
Sanda, Hyogo, Japan

Sponsoring Divisions:

 **Electronics and Photonics**

 **Dielectric Science & Technology**

Published by
The Electrochemical Society
65 South Main Street, Building D
Pennington, NJ 08534-2839, USA
tel   609 737 1902
fax 609 737 2743
www.electrochem.org

ecstransactions ™

Vol. 58, No. 4

Copyright 2013 by The Electrochemical Society.
All rights reserved.

This book has been registered with Copyright Clearance Center.
For further information, please contact the Copyright Clearance Center,
Salem, Massachusetts.

Published by:

The Electrochemical Society
65 South Main Street
Pennington, New Jersey 08534-2839, USA

Telephone 609.737.1902
Fax 609.737.2743
e-mail: ecs@electrochem.org
Web: www.electrochem.org

ISSN 1938-6737 (online)
ISSN 1938-5862 (print)
ISSN 2151-2051 (cd-rom)

ISBN 978-1-62332-095-9 (Hardcover)
ISBN 978-1-60768-449-7 (PDF)

Printed in the United States of America.

Preface

This issue of *ECS Transactions* is a collection of papers presented at the Symposium on *Gallium Nitride (GaN) and Silicon Carbide (SiC) Power Technologies 3,* held in San Francisco, CA from October 28-31, 2013 as part of the ECS Fall 2013 Meeting. The First Symposium in this series was held in Boston, MA as an integral part of the ECS Fall 2011 Meeting; the Second Symposium was held in Honolulu, Hawaii as part of the ECS Fall 2012 Meeting; both events were indeed grand successes. This new Symposium was founded with a goal to encourage and promote vigorous scientific and technical interactions on a wide range of wide bandgap (WBG) power technology topics from materials to systems. Such an interaction is essential in order to reduce the overall cost and improve performance, and develop power systems that last in the field for a prescribed period of time. Furthermore, as power electronics in the coming years is expected to benefit from a complex interplay of silicon, GaN and SiC technologies it is of outmost importance to understand development, merits and limitations of all three technologies.

The Third Symposium in San Francisco, CA promises even more excitement and enthusiasm as more than 60 papers on related "hot" topics will be presented by authors from all over the world. The Symposium begins with a Plenary Session and is followed by several technical sessions focused on key topics pertaining to materials growth, device design and fabrication, device modeling and characterization, circuit design and assessment, and reliability evaluation. Two Panel Sessions dealing with power module requirements and manufacturing will be conducted with panelists drawn from experts in the field. Details of these Panel Sessions are as follows.

Panel Session #1

"Wide Band Gap Power Module Requirements for Automotive and Grid Integration Applications"

Tuesday, October 29, 2013

4:30 pm – 6:30 pm

Room Continental 9, Tower 3, Ballroom Level, Hilton San Francisco

Chairs:

K. Shenai, Argonne National Laboratory

Bess Ng, NextEnergy

Panelists:

Carlos Cortes, BTCPower

Brad Palmer, Cummins

Anil Paryani, BTCPower

Andy Pinkos, Magna

Satoshi Tanimoto, Nissan Motor

Panel Session #2

"Manufacturing Challenges in Wide Bandgap Power Modules"

Wednesday, October 30, 2013

5 pm – 6:30 pm

Room Continental 9, Tower 3, Ballroom Level, Hilton San Francisco

Chairs:

Mietek Bakowski, Acreo AB

Aris Christou, University of Maryland – College Park

Panelists:

John Kajs, SAIC

Ty McNutt, APEI, Inc.

Takashi Nakamura, Rohm Co.

Krishna Shenai, Argonne National Laboratory

An important feature of this year's Symposium is the introduction of two special sessions completely devoted to power electronics education where university and industry experts will discuss on how to revamp the power electronics curriculum in order to address the daunting challenges of the WBG manufacturing industry.

In order to rapidly promote the Symposium, and accelerate development and manufacturing of wide bandgap power technologies world-wide, an International Advisory Committee (IAC) was set up when the Symposium was first started in 2011. The IAC consists of renowned experts in the field, and is divided into three regions: Americas, Europe, and Far East. The IAC is chaired by K. Shenai, founder and lead organizer of the Symposium. The IAC members include:

IAC Chair:

K. Shenai, Argonne National Laboratory

IAC Members from Americas:

R. Adapa, EPRI, Palo Alto, CA

A. Lidow, Efficient Power Conversion Corp., El Segundo, CA

R. Ram, Massachusetts Institute of Technology, Cambridge, MA

C. Scozzie, Army Research Labs, Adelphi, MD

K. Zia, Aixtron, Sunnyvale, CA

IAC Members from Europe:

U. Grossner, ABB Corporate Research, Baden, Switzerland

M. Leszczynski, Unipress, Warsaw, Poland

L. Lorenz, ECPE, Neubiberg, Germany

M. Saggio, ST Microelectronics, Catania, Sicily

A. Scuderi, ST Microelectronics, Italy

H-J. Wurfl, Ferdinand-Braun-Institut, Berlin, Germany

IAC Members from Far East:

K. Hamada, Toyota Motor Corporation, Japan

H. Matsunami, Emeritus Professor, Kyoto University, Japan

H. Ohashi, National Institute of Advanced Industrial Science and Technology, Japan

T. Oomori, Mitsubishi Electric Corporation, Japan

D. Ueda, Panasonic Corporation, Japan

Bo Zhang, University of Electronic Science and Technology, China

For the first time, this year's Symposium is financially sponsored by the US National Science Foundation (NSF) and IEEE Power Electronics Society (PELS) in addition to ECS DS&T and EPD divisions. We thank all sponsors for their generous financial support.

There is a great deal of interest in developing GaN and SiC material and device technologies for power switching and power amplifier applications. We hope that the *ECS Symposium on GaN and SiC Power Technologies 3* will act as a catalyst to rapidly develop and commercialize these strategic technologies.

K. Shenai, M. Bakowski, M. Dudley, and N. Ohtani

Editors

ECS Transactions, Volume 58, Issue 4
Gallium Nitride and Silicon Carbide Power Technologies 3

Table of Contents

Preface   *iii*

Chapter 1
Plenary Session

A Thermodynamic Interpretation of PVT Growth of Single Crystal SiC Material and   3
Challenges in Reducing Dislocations
  T. Fujimoto, M. Katsuno, H. Tsuge, S. Sato, S. Ushio, K. Tani, H. Yashiro,
  H. Hirano, T. Yano

Unexpected Sources of Basal Plane Dislocations in 4H-SiC Epitaxy   9
  R. E. Stahlbush, N. A. Mahadik

Correlation between Defects and Electrical Characteristics/Reliability Analyzed by   17
Integrated Evaluation Platform for SiC
  M. Kitabatake

Growth of High-Quality GaN Template from Nanometer-Size Lattice Channels by   25
Hydride Vapor Phase Epitaxy
  A. Usui, H. Goto, T. Matsueda, H. Sunakawa, T. Nakagawa, A. Okada, J. Mizuno,
  A. A. Yamaguchi, H. Shinohara, H. Goto

Packaging Techniques for Compact SiC Power Modules Operable in an Extended Tj   33
Range
  S. Tanimoto, K. Watanabe, H. Tanisawa, K. Matsui, S. Sato

Chapter 2
SiC MOS Power Devices

Channel Transport in 4H-SiC MOSFETs: A Brief Review   51
  S. Dhar

Properties of Al-SiO$_2$-SiC(3C) Structures with Thermally Grown and PECVD 61
Deposited SiO$_2$ Layers
  H. M. Przewlocki, T. Gutt, K. Piskorski, P. Borowicz, M. Bakowski

Influence of Ion Implantation in SiC on the Channel Mobility in Lateral N-Channel 71
MOSFETs
  C. Strenger, V. Uhnevionak, A. Burenkov, A. Bauer, P. Pichler, T. Erlbacher,
  H. Ryssel, L. Frey

On the Temperature Dependence of the Hall Factor in n-Channel 4H-SiC MOSFETs 81
  V. Uhnevionak, A. Burenkov, C. Strenger, A. Bauer, P. Pichler

Key Reliability Issues for SiC Power MOSFETs 87
  A. Lelis, D. Habersat, R. Green, E. Mooro

Chapter 3
SiC Epitaxy

Comparison of SiC Epitaxial Growth from Dichlorosilane and Tetrafluorosilane 97
Precursors
  H. Song, T. Rana, M. V. S. Chandrashekhar, S. U. Omar, T. S. Sudarshan

Reducing the Wafer Off Angle for 4H-SiC Homoepitaxy 111
  K. Kojima, K. Masumoto, S. Ito, A. Nagata, H. Okumura

3C-SiC on Si Hetero-Epitaxial Growth for Electronic and Biomedical Applications 119
  M. Reyes, C. Frewin, P. J. Ward, S. E. Saddow

Chapter 4
GaN Power Devices 1

III-Nitride Materials and Devices for Power Electronics 129
  A. Dobrinsky, G. Simin, R. Gaska, M. Shur

Normally-Off GaN Transistors for Power Switching Applications 145
  O. Hilt, E. Bahat-Treidel, F. Brunner, A. Knauer, R. Zhytnytska, P. Kotara,
  J. Wuerfl

High Performance Normally-off GaN MOSFETs on Si Substrates   155
 H. Kambayashi, N. Ikeda, T. Nomura, H. Ueda, Y. Morozumi, K. Harada,
 K. Hasebe, A. Teramoto, S. Sugawa, T. Ohmi

Characterization and Performance of D-Mode GaN HEMT Transistor Used in a   167
Cascode Configuration
 T. MacElwee, J. Roberts, H. Lafontaine, I. Scott, G. Klowak, L. Yushyna

Voltage Switching Limits of Lateral GaN Power Devices   179
 K. Shenai

Chapter 5
Power Device Reliability 1

GaN-Based Power HEMTs: Parasitic, Reliability and High Field Issues   187
 G. Meneghesso, M. Meneghini, D. Bisi, R. Silvestri, A. Zanandrea, O. Hilt,
 E. Bahat-Treidel, F. Brunner, A. Knauer, J. Wuerfl, E. Zanoni

True Figure of Merit (FOM) of a Power Semiconductor Switch   199
 K. Shenai

Progress in SiC MOSFET Reliability   211
 D. R. Hughart, J. D. Flicker, S. DasGupta, S. Atcitty, R. J. Kaplar, M. J. Marinella

Reliability of GaN HEMTs: Electrical and Radiation-Induced Failure Mechanism 221
 T. J. Anderson, A. D. Koehler, M. J. Tadjer, K. D. Hobart, P. Specht, M. Porter,
 T. R. Weatherford, B. Weaver, J. K. Hite, F. J. Kub

Chapter 6
Power Semiconductor Curriculum

Systems-Driven Power Semiconductor Education   229
 K. Shenai

Power Semiconductor Device Education: Which Topics and What Depth   237
 W. P. Robbins, N. Mohan

Power Semiconductor Device, Course Contents Revisited   245
 I. M. Abdel-Motaleb

Power Electronic Module Packaging at UA   253
    S. S. Ang, H. A. Mantooth, J. C. Balda

Chapter 7
GaN Power Devices 2

III-N High-Power Bipolar Transistors   261
    R. D. Dupuis, J. Kim, Y. C. Lee, Z. Lochner, M. H. Ji, T. T. Kao, J. H. Ryou,
    T. Detchphrom, S. C. Shen

AlGaN/GaN MIS-HEMT Gate Structure Improvement Using Al_2O_3 Deposited by   269
PEALD and BCl_3 Gate Recess Etching
    R. Meunier, A. Torres, M. Charles, E. Morvan, M. Plissonier, F. Morancho

GaN Power Transistors with Integrated Thermal Management   279
    C. R. Eddy Jr., T. J. Anderson, A. D. Koehler, N. Nepal, D. J. Meyer, M. J. Tadjer,
    R. Baranyai, J. W. Pomeroy, M. Kuball, T. I. Feygelson, B. B. Pate, M. A. Mastro,
    J. K. Hite, M. G. Ancona, F. J. Kub, K. D. Hobart

Ammonothermal Bulk GaN Substrates for Power Electronics   287
    M. P. D'Evelyn, D. Ehrentraut, W. Jiang, D. S. Kamber, B. C. Downey,
    R. T. Pakalapati, H. D. Yoo

1000V Vertical JFET Using Bulk GaN   295
    Q. Diduck, H. Nie, B. Alvarez, A. Edwards, D. Bour, O. Aktas, D. Disney,
    I. C. Kizilyalli

Chapter 8
Manufacturing Challenges

Manufacturing Challenges in Wide Band Gap (WBG) Power Electronics   301
    K. Shenai

Production Readiness of AlGaN/GaN HEMT on 6"/8" Si   311
    D. S. Lee, J. Su, B. Krishnan, G. D. Papasouliotis, A. Paranjpe

Synchrotron X-ray Topography Studies of the Evolution of the Defect Microstructure   315
in Physical Vapor Transport Grown 4H-SiC Single Crystals
  M. Dudley, B. Raghothamachar, H. Wang, F. Wu, S. Byrappa, G. Chung,
  E. K. Sanchez, S. Mueller, D. Hansen, M. Loboda

Basal Plane Dislocation Mitigation Using High Temperature Annealing in 4H-SiC   325
Epitaxy
  N. A. Mahadik, A. Nath, E. A. Imhoff, R. E. Stahlbush, R. Nipoti

3D TCAD Simulations for More Efficient SiC Power Devices Design   331
  L. V. Phung, D. Planson, P. Brosselard, D. Tournier, C. Brylinski

Chapter 9
Power Device Reliability 2

Heat Dissipation in GaN Based Power Electronics   343
  Z. Su, J. A. Malen

High Voltage InAlN/GaN HFETs Achieved by Schottky-Contact Technology for   351
Power Applications
  Q. Zhou, W. Chen, S. Liu, B. Zhang, Z. Feng, S. Cai, K. J. Chen

Interaction of Defects with Quantum Well States: Electrostatic-Dependant Response   365
Time for Traps in AlGaN/GaN HEMTs
  M. J. Marinella, S. DasGupta, R. J. Kaplar, M. Sun, S. Atcitty, T. Palacios

Monolithic Integration of High Temperature Silicon Carbide Integrated Circuits   375
  M. Alexandru, V. Banu, J. Montserrat, P. Godignon, J. Millán

Chapter 10
Power Electronics Curriculum

Power Semiconductor Device Modeling and Simulation   391
  H. A. Mantooth, S. Ahmed, S. S. Ang

Application Engineering of Wide Bandgap Semiconductors   399
  B. Sarlioglu, D. Han, J. Noppakunkajorn, A. Ogale

xi

Studying the Performance of Series-Connected GaN FETs in Higher Voltage   413
Switching Applications
  A. Hasanzadeh, A. Khaligh

Chapter 11
Material Synthesis and Processing

Materials Issues for Vertical Gallium Nitride Power Devices   427
  A. D. Williams, T. D. Moustakas

Electrochemical Hydrogenation of Dimensional Carbon   439
  K. M. Daniels, S. Shetu, J. Staser, J. W. Weidner, C. Williams, T. Sudarshan,
  M. V. S. Chandrashekhar

Abrasive-Free Polishing of SiC Wafer Utilizing Catalyst Surface Reaction   447
  Y. Sano, K. Arima, K. Yamauchi

Growth of GaN by MOCVD on Rare Earth Oxide on Si(111)   455
  F. E. Arkun, R. Dargis, A. Clark, R. S. Smith, M. Lebby, J. M. Leathersich,
  F. Shahedipour-Sandvik

Author Index   463

xii

Facts about ECS

The Electrochemical Society (ECS) is an international, nonprofit, scientific, educational organization founded for the advancement of the theory and practice of electrochemistry, electronics, and allied subjects. The Society was founded in Philadelphia in 1902 and incorporated in 1930. There are currently over 7,000 scientists and engineers from more than 70 countries who hold individual membership; the Society is also supported by more than 100 corporations through Corporate Memberships.

The technical activities of the Society are carried on by Divisions. Sections of the Society have been organized in a number of cities and regions. Major international meetings of the Society are held in the spring and fall of each year. At these meetings, the Divisions and Groups hold general sessions and sponsor symposia on specialized subjects.

The Society has an active publication program that includes the following:

Journal of The Electrochemical Society — (JES) is the leader in the field of electrochemical science and technology. This peer-reviewed journal publishes an average of 550 pages of 85 articles each month. Articles are published online as soon as possible after undergoing the peer-review process. The online version is considered the final version and is fully citable with articles assigned specific page numbers within specific issues. The date of online publication is the official publication date of record.

Journal of Solid State Science and Technology — (JSS) is one of the newest peer-reviewed journals from ECS launched in 2012. JSS covers fundamental and applied areas of solid state science and technology including experimental and theoretical aspects of the chemistry and physics of materials and devices. Articles are published online as soon as possible after undergoing the peer-review process. The online version is considered the final version and is fully citable with articles assigned specific page numbers within specific issues. The date of online publication is the official publication date of record.

Electrochemistry Letters — (EEL) is one of the newest journals from ECS launched in 2012. It is dedicated to the rapid dissemination of peer-reviewed and concise research reports in fundamental and applied areas of electrochemical science and technology. Articles are published online as soon as possible after undergoing the peer-review process. The online version is considered the final version and is fully citable with articles assigned specific page numbers within specific issues. The date of online publication is the official publication date of record.

Solid State Letters — *(SSL)* is one of the newest journals from ECS launched in 2012. It is dedicated to the rapid dissemination of peer-reviewed and concise research reports in fundamental and applied areas of solid state science and technology. Articles are published online as soon as possible after undergoing the peer-review process. The online version is considered the final version and is fully citable with articles assigned specific page numbers within specific issues. The date of online publication is the official publication date of record.

Electrochemical and Solid-State Letters — (ESL) was the first rapid-publication electronic journal dedicated to covering the leading edge of research and development in the field of solid-state and electrochemical science and technology. ESL was a joint publication of ECS and IEEE Electron Devices Society. Volume 1 began July 1998 and contained six issues, thereafter new volumes began with the January issue and contained 12 issues. The final issue of ESL was Volume 16, Number 6, 2012. Preserved as an archive, ESL has since been replaced by SSL and EEL.

Interface— *Interface* is an authoritative yet accessible publication for those in the field of solid-state and electrochemical science and technology. Published quarterly, this four-color magazine contains technical articles about the latest developments in the field, and presents news and information about and for members of ECS.

ECS Meeting Abstracts— *ECS Meeting Abstracts* contain extended abstracts of the technical papers presented at the ECS biannual meetings and ECS-sponsored meetings. This publication offers a first look into the current research in the field. ECS Meeting Abstracts are freely available to all visitors to the ECS Digital Library.

ECS Transactions— (ECST) is the online database containing full-text content of proceedings from ECS meetings and ECS-sponsored meetings. ECST is a high-quality venue for authors and an excellent resource for researchers. The papers appearing in ECST are reviewed to ensure that submissions meet generally-accepted scientific standards. Each meeting is represented by a volume and each symposium by an issue.

Monograph Volumes — The Society sponsors the publication of hardbound monograph volumes, which provide authoritative accounts of specific topics in electrochemistry, solid-state science, and related disciplines.

For more information on these and other Society activities, visit the ECS website:

www.electrochem.org

CHAPTER 1

PLENARY SESSION

2

A Thermodynamic Interpretation of PVT Growth of Single Crystal SiC Material
and Challenges in Reducing Dislocations

T. Fujimoto, M. Katsuno, H. Tsuge, S. Sato, S. Ushio, K. Tani, H. Yashiro,
H. Hirano and T. Yano

Advanced Technology Research Laboratories,
Nippon Steel & Sumitomo Metal Corporation,
20-1 Shintomi, Futtsu, Chiba, 293-8511, JAPAN

The issues regarding the methods for the reduction of the densities
of elemental dislocations such as the basal plane dislocation (BPD)
and the threading screw dislocation (TSD) are discussed. We
show that the pressure-dependent Si-C binary phase diagram is
useful for the suppression of dislocation-generating foreign phases.
In addition, recent progress on dislocation reduction in viewpoint
of the process control of PVT growth is briefly reviewed. In
particular, a specific behavior of 1c TSDs observed in the PVT
growth of 4H-SiC is described in detail.

Introduction

Silicon Carbide (SiC) is a promising material specifically for power device applications.
Marked progress has been achieved in the research field of SiC power devices, giving
rise to remarkable demonstrations such as Metro trains in Tokyo which are equipped with
high energy-efficient systems utilizing SiC inverters (1). SiC single crystal wafers with
higher crystallinity are no doubt a key for realizing such power-efficient electronics, and
intense research has long been performed on the physical vapor transport (PVT) method,
which is so-called the modified Lely method. Although other growth processes such as
the top-seeded solution growth (TSSG) method (2) or the high-temperature chemical
vapor deposition (HTCVD) method (3) are also in rapid progress recently, the PVT is at
the moment a widely-adopted fabrication process for large diameter SiC single crystal
wafers, including recent successful demonstrations of 150mm-diameter 4H-SiC wafers.

Unlike other semiconductor crystals, SiC still includes dislocations. There have been
a number of literatures reporting that particular types of dislocations cause some effects
that cannot be ignored as for operational stability depending on the internal structures of
SiC devices. In particular, the development of JBSs and MOSFETs with improved
performance with the aid of PN junctions has stimulated material research for the
reduction of the basal plane dislocations (BPDs) (4). Further, it has been found that
threading dislocations such as threading screw dislocations (TSDs) cause deterioration of
device properties with regard to the leakage current in JBS as well as dislocation-induced
morphological disturbance of the wafer surface in MOS capacitance structures (5).

In order to reduce the BPD and TSD densities in PVT-grown 4H-SiC crystals, the
following two viewpoints are to be emphasized; (a) the establishment of the stable
growth conditions is necessary for the suppression of dislocations generated during the

growth as small in density as possible, and (b) the process for the reduction of existing dislocations is crucially desired, for example, elimination of the dislocations during the growth by means of annihilation or structural transformation. In this work, we first describe the pressure-dependent Si-C binary phase diagram (6), which helps much to avoid unwanted foreign phases. Progress in BPD and TSD reductions is then briefly described, and in particular, a specific extending behavior of TSDs observed in 4H-SiC single crystals is described in details. We emphasize that such phenomenon enables us to establish an in-process technology for further reduction of the dislocation densities.

Experimental Procedure

4H-SiC single crystals with various diameters up to 100 mm were grown by conventional PVT growth method. Details of the growth conditions employed in this study were described elsewhere (7). The growth temperature was set approximately in a range between 2300 and 2400°C. All the crystals grown were nitrogen-doped with concentrations up to around $1 \times 10^{19} cm^{-3}$. The crystals obtained were all sliced parallel to the growth direction, and then subjected to mirror-polishing using diamond-based abrasive slurry.

Observations of dislocations were carried out using BedeScan (Bede Scientific Instruments) X-ray topography with a resolution of 2.7μm. For investigations with higher resolution, observations using synchrotron monochromatic X-ray topography was also performed at BL15 of SAGA-LS in Japan, using (0004) and (11-20) diffractions of 4H-SiC, recorded onto nuclear plates.

Results and discussion

Macroscopic phase control for SiC single crystal growth

Figure 1(a) shows schematic of the PVT for SiC growth; a SiC seed crystal and source material such as SiC powders are both placed in a graphite crucible. At higher temperatures exceeding around 2000°C with an appropriate temperature gradient, the source starts decomposition into sublimed vapor ($v_1(g)$) and solid carbon (C(s)). The vapor is transported to the seed crystal in an inert Ar-N_2 gas ambient with reduced pressures. The SiC vapor $v_1(g)$ then becomes super-saturated at around the seed, followed by recrystallization of solid SiC (SiC(s)) onto the seed surface. It should be noted that the vapor $v_1(g)$ is Si-rich in composition, and therefore the residual vapor $v_2(g)$ shifts its composition further Si-richer after the solidification of SiC(s).

The pressure-dependent phase diagram of Si-C binary system we proposed (6) can elucidate the quasi-equilibrium aspect of the PVT. Figure 1(b) shows a part of the pressure-dependent Si-C phase diagram at the temperature of the grown crystal surface. The diagram varies depending on the temperature, and earlier thermodynamic calculations indicated that peritectic-like point B in Figure 1(b) changes its position the fine broken line toward E as the temperature increases (8). The vapor $v_1(g)$ is generated from the source material at higher temperatures, and hence it has to correspond to, for example, E in Figure 1(b). In such condition, $v_1(g)$ becomes unstable, and therefore SiC(s) precipitates with its composition shifting toward $v_2(g)$. Note that the effect of reactions with the graphite crucible is assumed to be smaller although it was shown experimentally that it definitely exchanges carbon with sublimed gaseous species (9).

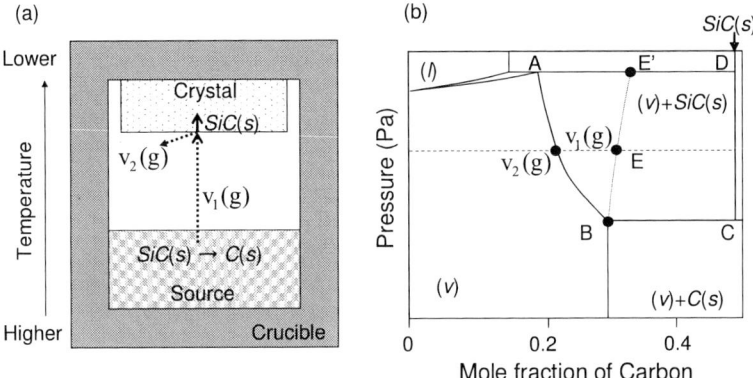

Figure 1. (a) Schematic of typical growth configuration of PVT method for SiC single crystals showing elemental processes at the source (SiC(s) → v_1(g) + C(s)) and the grown crystal (v_1(g) → SiC(s) + v_2(g)). (b) Pressure-dependent Si-C phase diagram for single crystal grown inside the crucible. The fine broken line corresponds to the trace of the peritectic-like point B with increasing the temperature (6).

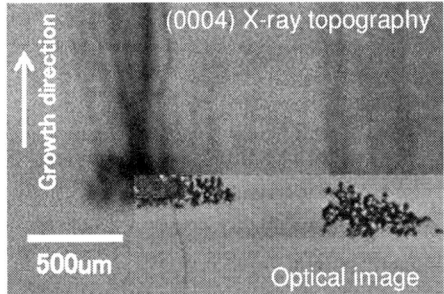

Figure 2. Cross-section view of Si-droplets formed in 4H-SiC single crystals observed using (0004) X-ray transmission topography. A part of the image was replaced with an image taken using optical microscopy in order to visualize the actual figure of the Si-droplets.

Macro-defects in SiC growth are in general formed mainly due to insufficiently optimized crystal growth conditions, leading to dislocation generations in the following growth stages. Figure 2 exemplifies the formation of Si-droplets in 4H-SiC crystals observed using optical microscopy and (0004) X-ray transmission topography. As is clearly seen in Figure 2, the Si-droplets generate a number of threading dislocations with screw components. The mechanism of Si-droplet formation is described in Figure 1(b). When the seed crystal is overcooled or the vapor pressure increases too much, the vapor v_1(g) reaches the line AD, indicated as E'. In such situation, Si-liquid phase starts being formed from v_1(g), giving rise to the Si-droplet formation. Thus, it is concluded that, in

5

order to avoid dislocation generation due to foreign phases, the growth parameters have to be controlled to have $v_l(g)$ be at least inside the region ABCD in Figure 1(b).

Reduction of dislocation densities: BPDs and TSDs

BPDs have drawn much attention with regard to long-term operation reliability of high-power devices utilizing PN junctions because it leads to the formation of Shockley-type stacking faults during forward biasing operation of the devices, resulting in severe deterioration of device properties (10). The shear stress component σ_{rz} inside the crystal is considered to be responsible for the BPD generation. Recent progress in process optimization has led a substantial reduction of the BPD density in 3 inch 4H-SiC crystals, demonstrating a remarkably small value of the BPD density of 1.05×10^2 cm^{-3} confirmed using both KOH etching and Synchrotron X-ray topography (11). It is well known that around 99% of the BPDs in PVT-grown crystals can be transformed into TEDs when epitaxial growth is performed onto 4° off oriented 4H-SiC substrates (12), and therefore, such substantial reduction of the BPD in PVT-grown crystals surely contributes to the realization of epitaxial 4H-SiC wafers with the larger areas of the BPD free-region in the epitaxial layer. For further reduction of the BPD density, numerical simulations based on the assumption that σ_{rz} is the dominant component for the BPD generation suggest that the optimization of growth process parameters to have the temperature gradients toward the growth direction be smaller realizes lower shear stress inside the crystal, likely to lead to larger diameter crystal growth with lower BPD densities (13).

It is known that the structural interaction with closed core TSDs induces the Frank-Read multiplication of BPDs (14). The TSD itself is also known to degrade PN junction reliability in viewpoints of both leakage current of JBSs and oxide layer reliability in MOSFETs (5). Therefore, although there is still debate for correlation between them, the suppression of the TSD density is important for avoiding both the BPD multiplication and the deterioration of device properties. Lower TSD densities of around 75/cm^2 have been reported for 3-inch 4H-SiC wafers grown by the conventional PVT method (11).

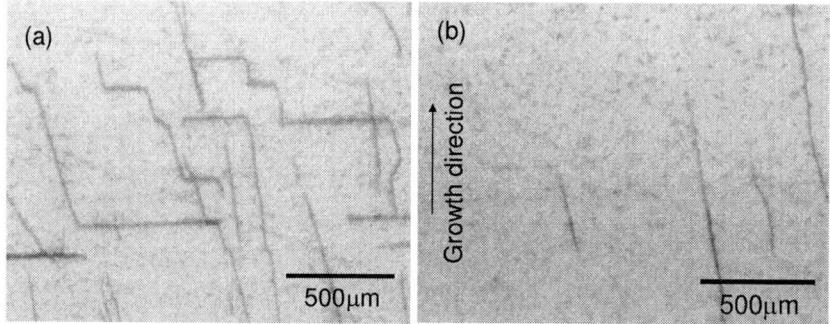

Figure 3. Cross section views of 3-inch 4H-SiC crystal using (0004) X-ray transmission topography obtained for the crystal regions (a) 5 mm above the seed/crystal interface, and (b) near the growth front.

The generation of the TSD is thought to be caused mainly by (i) foreign polytypes or other macroscopic inclusions such as Si-droplets, or (ii) crystal imperfections existing in seed crystals, for example, surface damage layers induced by mechanical scratches (15). In addition, we have observed that the TSDs show an intriguing behavior during the PVT growth. Figure 3(a) shows a (0004) X-ray topographic image obtained for 3-inch 4H-SiC crystals, indicating that the TSDs change their extending directions from the growth direction to the (0001) in-plane, or vice versa. As shown in Figure 3(b), the TSD density is markedly reduced in the following growth stage, suggesting that such reorientation of the TSD is likely to be responsible for the reduction of the TSD density. Some of the TSDs in Figure 3(a) seem to contact directly with each other, but closer views using synchrotron X-ray topography, as shown in Figure 4, clarify that almost all the TSDs do not exhibit the direct mutual interaction.

Figure 4. Cross section view of 3-inch 4H-SiC crystal observed using synchrotron (0004) X-ray transmission topography.

Similar behavior has been observed for TSSG-grown crystals and SiC films gown by metastable epitaxy, both known to be the structural conversion of TSD into Frank-type stacking faults (SFs). Such conversion mechanism exists in CVD-grown epitaxial layers and even in PVT-grown SiC crystals (16), suggesting that in both cases the overgrowth of vicinal surface macrosteps causes the reorientation of TSDs into Frank-type in-plane SFs. As is seen in Figure 3(a), all the TSDs bend into the in-plane almost parallel to the outer direction toward the crystal periphery (the direction to the left in Figure 3(a) and (b)), i.e., the macrostep flow direction. However, the overgrowth mechanism requires reverse-directing flows of the macrosteps in order to explain the reorientation of the TSDs back to the growth direction. In addition, the contraction of Frank-type in-plane SFs bounded by Frank partials into TSD structure has to occur simultaneously at the reorientation. The exact mechanism is still under investigation at the moment, and further investigation is necessary. By establishing the process control for the TSD reorientation, an in-process reduction of the TSD density would be possible anyway as is stated in Ref.(17), and further reduction of the BPD density is also expected upon the TSD reduction.

Summary

It is demonstrated that the pressure-dependent Si-C binary phase diagram we proposed is very useful for the suppression of dislocation-generating foreign phases. For the BPDs, the optimization of the temperature distribution inside the crystal is important in order to minimize the shear stress component of the internal stress. As for the TSDs, a specific behavior was found; the reorientation of the extending direction of the TSD occurs during the PVT growth, suggesting that in-process reduction of the TSD density could be possible in the PVT method.

Acknowledgments

This work is partly supported by New Energy and Industrial Technology Development Organization (NEDO) under the project of "Novel Semiconductor Power Electronics Project Realizing Low Carbon Emission Society".

References

1. Mitsubishi Electric Corporation, Press release No. 2749 on 26 March 2013.
2. K. Kusunoki, K. Kamei, N. Okada, N. Yashiro, A. Yauchi, T. Ujihara, and K. Nakajima, Mater. Sci. Forum, **527-529**, 119 (2006).
3. A. Ellison, B. Magnusson, B. Sundqvist, G. Pozina, J. P. Bergman, E. Janzén, and Vehanen, Mater. Sci. Forum, **457-460**, 9 (2004).
4. A. Agarwal, H. Fatima, S. Haney, and S. Ryu, IEEE Electron Device Lett., 28, 587 (2007).
5. H. Fujiwara, H. Naruoka, M. Konishi, K. Hamada, T. Katsuno, T. Ishikawa, Y. Watanabe, and T. Endo, Appl. Phys. Lett., **100**, 242102 (2012).
6. T. Fujimoto, N. Ohtani, H. Tsuge, M. Katsuno, S. Sato, M. Nakabayashi, and T. Yano, ECS J. Solid State Sci. Technol., **2**, N3018 (2013).
7. M. Nakabayashi, T. Fujimoto, M. Katsuno, N. Ohtani, H. Tsuge, H. Yashiro, T. Aigo, T. Hoshino, H. Hirano, and K. Tatsumi, Mater. Sci. Forum, **600-603**, 3 (2007).
8. J. Drowart, G. de Maria and M. G. Inghram, J. Chemical Phys., **29**, 1015 (1958).
9. Z. G. Herro, P. J. Wellmann, R. Püsche, M. Hundhausen, L. Ley, M. Maier, P. Masri, and A. Winnacker, J. Cryst. Growth, **258**, 261 (2003).
10. M. Skowronski and S. Ha, J.Appl.Phys., **99**, 011101 (2006).
11. H. Tsuge, S. Ushio, S. Sato, M. Katsuno, T. Fujimoto, and T. Yano, Mater. Sci. Forum, **740-742**, 7 (2013).
12. H. Wang, F. Wu, S. Byrappa, S. Sun, B. Raghothamachar, M. Dudley, E. K. Sanchez, D. Hansen, R. Drachev, S. G. Mueller, and H. J. Loboda, Appl. Phys. Lett., **100**, 172105 (2012).
13. R. Ma, H. Zhang, V. Prasad, and M. Dudley, Cryst. Growth & Design, **2**, 213 (2002).
14. N. Ohtani, M. Katsuno, H. Tsuge, T. Fujimoto, M. Nakabayashi, H. Yashiro, M. Sawamura, T. Aigo, and T. Hoshino, Jpn. J. Appl. Phys., **45**,1738 (2006).
15. M. Dudley, and X. Hauang, Mater. Sci. Forum, **338-342**, 431 (2000).
16. G. Dhanaraj, Yi Chen, H. Chen, W. M. Vetter, H. Zhang, and M. Dudley, Mater. Res. Soc. Symp. Proc., **911**, B05-27 (2006).
17. Yi Chen, Ph.D. Thesis, Stoney Brook University, 2008.

ECS Transactions, 58 (4) 9-15 (2013)
10.1149/05804.0009ecst ©The Electrochemical Society

Unexpected Sources of Basal Plane Dislocations in 4H-SiC Epitaxy

R. E. Stahlbush[a], and N. A. Mahadik[b]

[a] Power Electronics Branch, Naval Research Laboratory, Washington, DC 20375, USA
[b] Solid State Devices Branch, Naval Research Laboratory, Washington, DC 20375, USA

The suppression of degradation inducing basal plane dislocations (BPDs) in the critical drift layer of SiC power devices has occurred mainly by preventing BPDs in the substrate from propagating into the drift layer. As optimized epitaxial growth has produced drift layers free of BPDs over a large fraction of the wafer, other sources of BPDs have become important. Two alternate sources are discussed. The first is epitaxial inclusions, which mainly consist of grossly misoriented 4H-SiC. The local stress field around the inclusion introduces a cluster of BPDs. In low-BPD epitaxy, the outermost BPDs can glide centimeters forming half-loop arrays that have BPD segments along their whole length. The second source of BPDs not normally considered is BPDs that are converted to threading edge dislocation before reaching the drift layer. Experiments suggest that for high current levels and/or over time, device degradation is possible.

Introduction

The growth of 4H-SiC epitaxy that is free or nearly free of basal plane dislocations (BPDs) is necessary for reliable and stable operation of SiC power devices. Instability due to BPDs was first observed in PiN diodes (1). During the forward-biased operation of bipolar devices such as PiN diodes, BPDs in the drift layer cause an upward drift in the forward voltage. In forward bias, the drift layer is conductivity modulated and the resulting electron-hole recombination causes the BPDs to fault by the recombination enhanced dislocation glide mechanism (2). The expanding stacking faults degrade the drift conductivity by a combination of suppressing carrier lifetime and introducing conduction barriers (3-5). The stacking faults also increase the reverse-bias leakage and decrease the breakdown voltage (6,7). The same degradation is observed in unipolar devices, which during operation can have conductivity modulation such as MOSFETs in which the body diode is forward biased (6).

Over the last decade significant progress has been made in reducing the BPD density in the epitaxial drift layer and in the best wafers a majority of the wafer area is free of BPDs. In early work, it was recognized most of the BPDs in the epitaxy originated in the substrate. Even though about 90% of substrate BPDs are converted into threading edge dislocations (TEDs) at the substrate/epi interface (8), the remaining BPD density in the epitaxy was in the range of hundreds per cm^2, which is three to four orders of magnitude too high to use in power devices. Partially progress has come from lower BPD density in wafers (9). In 8° offcut wafers, various etching schemes either before or during epitaxial growth decreased BPDs by increasing the BPD-to-TED conversion (10,11). A big improvement occurred when the wafer offcut angle was changed from 8° to 4°. While the change was made to reduce boule waste when sawing it into wafers, it was soon reocgnized that epitaxy on 4° wafers appeared to have a much lower BPD density. Examination of the BPD density through the epitaxial thickness showed very different behavior in epitaxy grown on 8° and 4° wafers (12). In 8° wafers almost all of the BPDs, which were not converted to TEDs at the start of the growth, continued propagating on the basal plane through the whole epitaxial growth. In contrast, in 4° wafers, most of the BPDs present after the start of epitaxial growth were converted to TEDs within microns of growth. Epitaxial growths for power devices normally start with a buffer layer that is doped nearly as high as the substrate. BPDs that are converted to TEDs within the buffer layer do not enter the drift layer of the device and it had been assumed that they do not affect device operation. Other growth parameters that effect BPD conversion include the C/Si ratio, SiC etching and the details of the growth initiation (10-14). As these parameters have been optimized, it is possible to grow SiC epitaxy on 4° offcut wafers up to 100 μm and thicker in which BPDs from the substrate are sufficiently suppressed that a majority of the drift layer across the wafer area is free of BPDs.

As this level of perfection is achieved, BPDs and their detrimental effects arise from other unexpected sources. Two sources are discussed in this work. The first is BPDs that originate at inclusions in the epitaxy. Due to stress in the epitaxy some of these BPDs can form half-loop arrays (HLAs) that can glide centimeters away from the inclusions and degrade any devices fabricated along their path. The second unexpected source of degradation is due to BPDs confined to the epitaxial buffer, i.e. they are converted into TEDs within the buffer layer before reaching the drift layer. They can fault and the stacking fault can expand through the drift layer.

BPDs from Inclusions and Half-Loop Arrays

While dislocations are 1D and stacking faults are 2D, inclusions are 3D. A study of their structure has shown that they are primarily a volume of grossly misoriented 4H polytype interspersed with a small amount of 3C polytype (15). This conclusion is based on examining inclusions with x-ray diffraction and micro-Raman spectroscopy. As is the case for the arrow defect, it is possible that they originate from silicon droplets incorporated into the epitaxy (16). The inclusions disrupt the surface morphology and the stress field around them introduces a local cluster of BPDs.

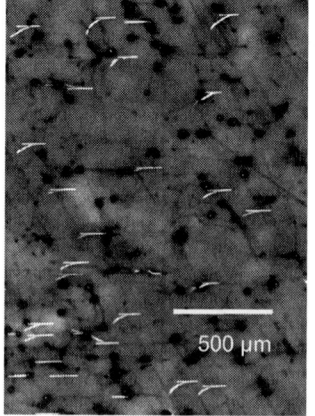

BPDs entering the lightly doped drift layer of the epitaxy from the substrate typically glide a distance comparable to the drift layer thickness as the epitaxy is grown. Typical curved BPDs due to the glide are shown in Fig. 1 (11), which shows a ultraviolet photoluminescence (UVPL) image of BPDs within a epitaxial layer doped in the $10^{14}/cm^3$ range. Characteristics of the UVPL technique are

Figure 1. UVPL image showing typical glide of BPDs that occurs during epitaxial growth.

that BPDs in the substrate are not visible and the entire path of BPDs through the lightly doped epitaxy is visible (17). The BPDs glide is in response to the slight lattice mismatch between the epitaxy and substrate and the resulting compression of the epitaxy. The lattice constant of the substrate is smaller due to its higher doping with nitrogen, which decreases the local bind lengths. The collective glide of all of the BPDs helps to release the strain.

As BPD density is decreased, individual BPDs tend to exhibit more glide, which compensates for the lower number of BPDs. This leads to a defect that complicates the effort to increase device yield. It is alternately named half-loop array (HLA) or pair array (18). It is a complicated defect that originates from a single BPD that glides in the basal plane during the epitaxial growth (19-21). While observed in epitaxy of all thicknesses, as the thickness increases, the BPD usually glides more and the size of the HLA increases. The extent of the BPD glide and the size of the HLA can be several centimeters in thick epitaxy, typically 100 µm or more. Any device through which the HLA passes is subject to threshold voltage drift and reversed-bias leakage (7).

A UVPL image of two HLAs is shown in Fig. 2. This example is from a commercially grown 100 µm thick epitaxy grown on an 8° offcut wafer. Figure 3 is a schematic of a single HLA. Each of the HLAs in Fig. 3 were formed from a BPD that entered the epitaxy from the substrate as shown in Fig. 3. As the BPD glides within the growing epitaxial layer the HLA is formed. Each HLA has three parts. The first is a sequence of half loops that appear as a nearly vertical series of dots. The second is an interfacial dislocation that forms a misfit dislocation between the substrate and epitaxy. Only one to the two interfacial dislocations is visible in Fig.2. It appears as a wavy vertical line and it is a misfit dislocation that releases the epitaxial strain. Note that the interfacial dislocation is not always at the substrate/epi interface. In many cases the BPD glide that forms the HLA starts after the beginning of the epitaxial growth. For example, the interfacial dislocation in Fig. 2 is 40 µm from the substrate. The third part of the HLA is a BPD that glides

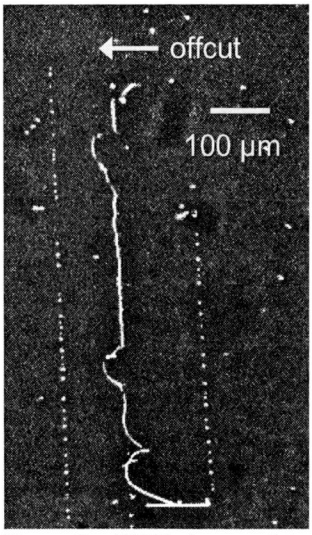

Figure 2. UVPL image of two HLAs. The one on the right shows its 3 parts: the series of half loops(dots), the interfacial dislocation roughly parallel to the dots and the gliding part near the bottom.

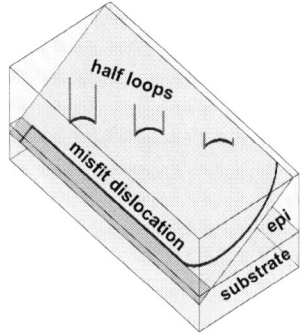

Figure 3. Schematic of HLA showing three half loops, the interfacial (misfit) dislocation and the gliding section. As it continues gliding during epi growth, additional half loops are formed.

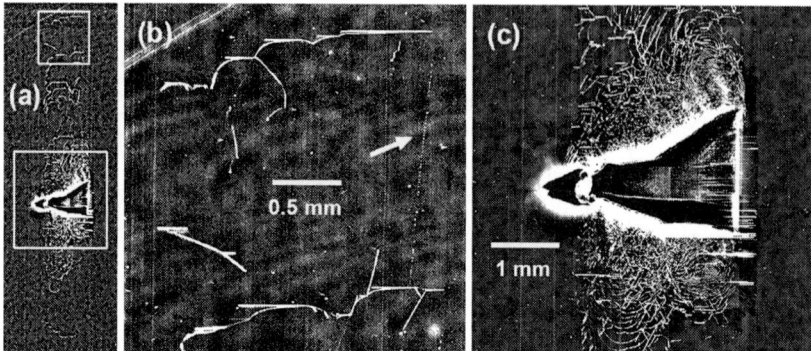

Figure 3. UVPL image of an inclusion. Its stress field induces a cluster of BPDs surrounding it and the outermost of the BPDs drift to form HLAs: (b) and (c) are magnified views of the boxed areas in (a). The arrow in (b) points to an string of dots where each dot is one of the half loops in the HLA that extends from the inclusion to its end shown in (b).

perpendicular to the offcut direction during the epitaxial growth. In Fig. 2 the glide direction is from the top to bottom. As it glides, it leaves a sequence of short – ~1 μm – segments whose ends are tied to the growth surface by two TEDs. Each of the dots in Fig. 1 is a half loop formed by two TEDs with a BPD segment at the bottom (8,21). The only portions of the HLA that form expanding faults are the BPD segments at the bottoms of the half loops. While initially small, the BPD segments at the bottom of the half loops are all on the same basal plane and during device operation they collectively form an expanding stacking fault that eventually spans the drift layer thickness along the length of the array of half loops.

When an inclusion forms in a lightly doped epitaxial layer that has very few BPDs, the outermost BPDs surrounding the inclusion experience the same force as an individual BPD and the result is the same. The outermost BPDs around the inclusion form HLAs that can glide centimeters. An example of HLAs originating at an inclusion is shown in Fig. 3. The full span of HLAs originating from this inclusion is two centimeters.

Stacking Faults from BPDs in the Buffer

Typical device epitaxy starts with a heavily doped buffer that is grown before the lightly doped drift layer. Recombination within the buffer is significantly lower than in the drift layer, and it has been widely assumed that BPDs converted to TEDs within the buffer would not introduce SFs into the drift layer. However, high power UV excitation has shown that BPDs confined to the buffer can form stacking faults that first expand slowly within the buffer, and once a stacking fault reaches the drift layer, it rapidly expands across the whole drift layer thickness (22).

The UVPL images in Fig. 4(a) shows BPDs confined to the buffer layer, as well as one BPDs that enters into the drift layer. The darker green arrow points to the BPD within the drift layer that appears as a bright line. In contrast to BPDs in the drift layer,

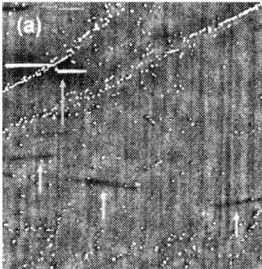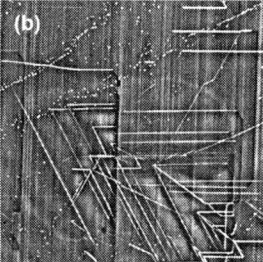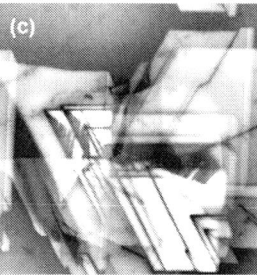

Figure 4. UVPL images of the epitaxial area before, (a), and after, (b) and (c), high intensity UV exposure. The darker green arrow points to a BPD with the drift layer and the three yellow arrows point to BPDs that did not reach the drift layer before being converted to TEDs.

BPDs in the buffer produce wider dark horizontal lines. Yellow arrows point to three of them. The dark contrast for the BPDs in the buffer is due to a reduction of the local carrier density surrounding it, which reduces the background luminescence. All of the bright-line BPDs have a dark broad line to their left – not shown – as they pass through the buffer before entering the drift layer. The dots in Figs. 4(a) and 4(b) are due to threading dislocations.

Figures 4(b) and 4(c) show the same area after high intensity UV exposure. Fig. 2(b) was taken in the 600-1000 nm range where the PL emission from the BPD partials bounding the stacking faults is brighter, whereas a 425 - 465 nm bandpass filter was used for Fig. 2(c), which shows luminescence from the stacking faults. After the UV exposure, more than 10 stacking faults are observed in this region that originated from BPDs that had converted to TEDs before reaching the epitaxial drift layer.

Making comparisons between recombination induced by UV excitation and that induced by typical current levels in devices is complicated (22). Calculations indicate that currents about an order of magnitude higher than typically used are required to initiate stacking fault formation from the BPDs in the buffer. However, the UV intensity dependence of BPD faulting within the buffer layer is also complicated. There appears to be an intensity threshold to initiate BPD faulting. The threshold increases as the depth of the BPD-to-TED conversion below the buffer/drift boundary increases. However, once the faulting starts, the stacking fault expansion rate is roughly linear with UV power at the stacking fault. This suggests that BPD confined to the buffer may degrade long term device reliability.

Conclusion

As SiC epitaxy is improved and it becomes possible to almost eliminate BPDs originating from the substrate from entering the device drift layer, device degradation from BPDs originating from other sources must be considered. Examples of two other sources have been presented. The first is from BPDs introduced by the local strain field around inclusions, areas of grossly misoriented epitaxy. With the absence of other BPDs, which collectively release the slight lattice mismatch between the substrate and drift layer epitaxy, the outermost BPDs in the cluster can glide centimeters. Each gliding BPD often forms a half-loop array that degrades any device fabricated along its path. The second source that is not normally considered is from BPDs that propagate into the epitaxial buffer layer below the drift layer, but are converted to TEDs before reaching the drift layer. UV exposure experiments suggest that these BPDs may be a source of long term device reliability.

Acknowledgments

This work was supported by the Office of Naval Research.

References

1. H. Lendenmann, F. Dahlquist, N. Johansson, R. Soderholm, P. A. Nilsson, J. P Bergman and P. Skytt, *Mater. Sci. Forum*, **353-356**, 727 (2001).
2. A. Galeckas, J. Linnros and P. Pirouz, *Appl. Phys. Lett*, **81**, 883 (2002).
3. J. P. Bergmann, H. Lendenmann, P. A. Nilsson, U. Lindefelt and P. Skytt, *Mater. Sci. Forum* **353-356**, 299 (2001).
4. R. E. Stahlbush, M. Fatemi, J. B. Fedison, S. D. Arthur, L. B. Rowland and S. Wang, *J. Elec. Mater*. **31**, 370 (2002).
5. A. Kuhr, J. Q. Liu, H. J. Chung and M. Skowronski, *J. Appl. Phys*. **92**, 5863 (2002).
6. A. Agarwal, H. Fatima, S. Haney and S. H. Ryu, *IEEE Elec. Dev. Lett*. **28**, 587 (2007).
7. R. E. Stahlbush, Q. Zhang, A. Agarwal, N. A. Mahadik, , *Mater. Sci. Forum* **717-720**, 387 (2012).
8. S. Ha, P. Mieszkowski, M. Skowronski, L. B. Rowland, *J. Cryst. Growth*, **244**, 257 (2002).
9. M. Dudley, N. Zhang, Y. Zhang, B. Roghothamachar, S. Byrappa, G. Choi, E. K. Sanchez, D. Hansen, R. Ddrachev and M. J. Loboda, *Mater. Sci. Forum*, **645-648**, 291 (2010).
10. H. Tsuchida, T. Miyanagi, I. Kamata, T. Nakamura, K. Izumi, K. Nakayama, R. Ishii, K. Asano and Y. Sugawara, *Mater. Sci. Fourm*, **483-485**, 97 (2005).
11. R. E. Stahlbush, B. L. VanMil, R. L. Myers-Ward, K. K. Lew, D. K. Gaskill and C. R. Eddy, Jr., *Appl. Phys. Lett.*, **94**, 041916 (2009).
12. R. L. Myers-Ward, B. L. VanMil, R. E. Stahlbush, S. L. Katz, J. M. McCrate, S. A. Kitt, C. R. Eddy, Jr. and D. K. Gaskill, *Mater. Sci. Forum*, **615-617**, 105 (2009).
13. W. Chen and M. A. Capano, *J. Appl. Phys.*, **98**, 114907 (2005).
14. J. J. Sumakeris, J. P. Bergman, M. K. Das, C. Hallin, B. A. Hull, B. Janzen, H. Lendenmann, M. J. O'Laughlin, M. J. Paisley, S. Ha, M. Skowronski, J. W. Palmour and C. H. Carter, Jr., *Mater. Sci. Forum*, **527-529**, 141 (2006).
15. N. A. Mahadik, R. E. Stahlbush, S. B. Qadri, O. J. Glembocki, D. A. Alexson, K. D. Hobart, J. D. Caldwell, R. L. Myers-Ward, J. L. Tedesco, D. K. Gaskill and C. R. Eddy, Jr., *J. Elec. Mater.*, **40**, 413 (2011).
16. X. Zhang, L. Li, M. Skowronski, J. J. Sumakeris, M. J. Paisley, and M. J. O'Loughlin, *J. Appl. Phys.*, **105**, 123529 (2009).

17. R. E. Stahlbush, K. X. Liu, Q. Zhang and J. J. Sumakeris, *Mater. Sci. Forum*, **556-557**, 295 (2007).
18. S. Ha, H. J. Chung, N. T. Nuhfer and M. Skowronski, *J. Cryst. Growth*, **262**, 130 (2004).
19. X. Zhang, M. Skowronski, K. X. Liu, R. E. Stahlbush, J. J. Sumakeris, M. J. Paisley and M. J. O'Loughlin, *J. Appl. Phys.*, **102**, 093520 (2007).
20. R. E. Stahlbush, B. L. VanMil, K. X. Liu, K. K. Lew, R. L. Myers-Ward, D. K. Gaskill, C. R. Eddy, Jr., X. Zhang and M. Skowronski, *Mater. Sci. Forum*, **600-603**, 317 (2009).
21. N. Zhang, Y. Chen, Y. Zhang, M. Dudley and R. E. Stahlbush, *Appl. Phys. Lett.*, **94**, 122108 (2009).
22. N. A. Mahadik, R. E. Stahlbush, M. G. Ancona, E. A. Imhoff, K. D. Hobart, R. L. Myers-Ward, C. R. Eddy, Jr., D. Kurt Gaskill and F. J. Kub, *Appl. Phys. Lett.*, **100**, 042102 (2012).

16

Correlation between Defects and
Electrical Characteristics/Reliability
Analyzed by Integrated Evaluation Platform for SiC

M. Kitabatake

R&D Partnership for Future Power Electronics Technology (FUPET)
1-1-1, Umezono, Tsukuba, Ibaragi 305-8568, Japan.

The Integrated Evaluation Platform for SiC wafers and epitaxial films is established. It provides information about the correlation between step-bunching defects and SiO_2/SiC electrical characteristics/reliability. The Observation-Recognition System discovers that the step bunching on the epitaxial wafer partly spreads along the scratches. The Q_{bd} measured in the Electrical-Characteristics Analyses has three types of distribution denoted as D1 (ideal SiO_2/SiC without defects), D2 (average Q_{bd} value half of D1 with step bunching), and D3 (small Q_{bd} value with large defects). The Defect-Structure Analyses elucidate that the oxide thickness fluctuates on the step bunching line. The local electric field concentration at thinner spots causes degradation of SiO_2/SiC reliability.

Introduction

We have every reason to believe that wide-bandgap semiconductor SiC can provide sophisticated semiconductor power devices for the next-generation ecological world. SiC power devices, when used as low-loss and high-speed semiconductor switches and rectifiers, realize energy-saving and compact power-electronics converters and inverters. Large-current (>100A), high-voltage (>10kV) and fairly large-size (>several mm square) devices are successively fabricated on a single-crystal 4H-SiC wafer.

Growth of single-crystal 4H-SiC had been difficult to implement because SiC crystalline bulk can only be obtained via very high-temperature (about 2300C) vapor-phase growth and other poly-type (such as 6H and 3C) inclusions easily occur. Moreover, there had been difficulties in cutting and polishing the brittle crystalline bulk to fabricate SiC wafers. Steady and continuous efforts have been made to make the device-ready SiC wafers with wafer diameter attaining 6 inches these days. Traditionally-observed large defects, which are wafer-penetrating holes called micro pipes, are almost eliminated while the commercially available SiC wafers still include thousands of defects. The threading screw dislocations (TSD), threading edge dislocations (TED), basal plane dislocations (BPD), and stacking faults (SF) have been reported as lattice defects in the SiC wafer.[1] The surface defects such as particles and scratches have been also reported.

On SiC wafers, epitaxial film must be grown, for example with thickness of 10μm, for the power device to achieve a breakdown voltage of 1kV. Typical epitaxial film is grown using the step-flow technique in CVD growth on the off-cut 4H-SiC(0001) inclined 4 degrees towards the [11-20] direction. Large progress in reduction of the defects in the SiC epitaxial films has also been made. The SFs, represented by other poly-type inclusion, and the BPDs are minimized under the optimized step-flow epitaxial growth conditions. The TSDs and the TEDs are usually transferred from the wafer to the epitaxial film. Epitaxial films may include additional defects which are reported as the extended defects (so-called carrot-shape, triangle-shape, bar-shape, and trapezoid-shape).[1] [2] The off-cut SiC wafer sometimes results in the formation of rough surface with step-bunching. Fortunately, the epitaxial film with fairly flat surface (almost bunching-free) is available today.

It is hard to say that SiC wafers are delivered to device makers with enough information to manage the SiC-device mass production. We know that no device can be fabricated without any defects. It is very important to evaluate correlation between the defects and the electrical characteristics/ reliability to manage the healthy mass production of the SiC power devices..

Integrated Evaluation Platform (IEP)

We have put steady effort to establish the Integrated Evaluation Platform (IEP) for the SiC wafers and epitaxial films.[3] Our IEP consists of the observation-recognition system (ORS), the electrical-characteristics analyses (ECA) and the defect-structure analyses (DSA) as shown in Fig. 1. The systematic evaluation and accumulation of the numerous data through IEP = ORS + DSA + ECA will give us deep detailed physics of SiC power devices and the necessary insight in improving device reliability.

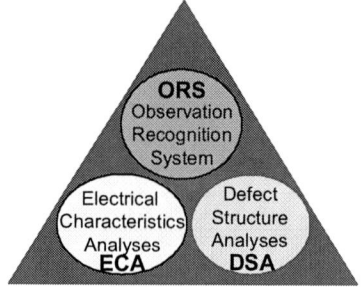

Figure 1. Integrated Evaluation Platform.

Observation/Recognition System (ORS) and the Example Data

Optical observation of SiC wafer is insufficient in recognizing defects on the CMP-polished flat SiC wafer. On the other hand, the defects in the epitaxial film usually exhibit surface irregularities which may assist the observation. The conventional observation systems are useful in counting large-sized defects such as micro pipes, but smaller-sized defects cannot be spotted as effectively. Our ORS is based on the confocal microscope with differential interference contrast on SICA (Lasertec) which enables high-resolution (depth resolution of <1nm and horizontal resolution of <1μm) and high-throughput (a few minutes for 3 inch wafer) observations. Entire high-resolution surface images visualizing detailed shapes of the defects are photographed and saved. Then the image processing sequence recognizes/classifies/locates the defects to output the corresponding data set of each defect. During this stage, classification of the defects is based on the surface image under the interference contrast. Defects are labeled as

Horizontal Line (HLine), Diagonal Line (DLine), Large, Latent Scratch (Latent), Black/White(B/W), White/Black(W/B) and so on. Example images of the defects are shown in Fig. 2. The classification of defects made by our ORS is considered to be equivalent to the reported defects as shown in Table I.

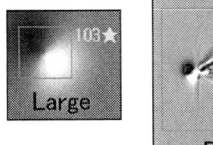 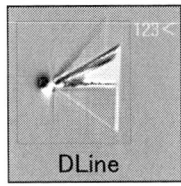 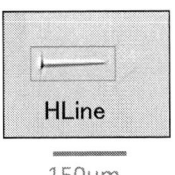  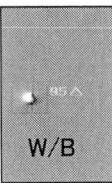

Figure 2. Example images of the recognized defects using the ORS.

Table I. Defect classifications of the ORS and equivalent reported defects

Classification in our ORS	reported defects
Horizontal line (HLine)	Carrot(-shape), Triangle(-shape)
Diagonal Line (DLine)	Triangle(-shape)
Large	DownFall
Latent Scratch (Latent)	(Latent) Scratch
Black/White(B/W)	Pit
White/Black(W/B)	Bump
Black	
White	

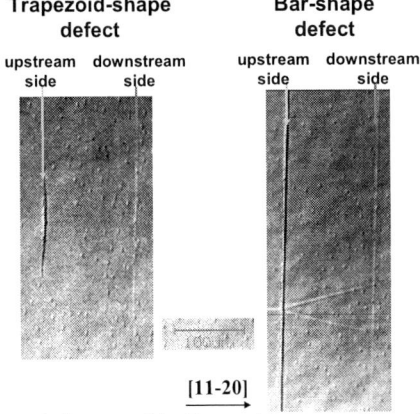

Figure 3. Images of the Trapezoid-shape and Bar-shape defects derived from the ORS.

Beyond these classifications, we discovered for the first time the trapezoid-shape defects and observed the bar-shape and triangle-shape defects on the fairly flat epitaxial-film surfaces. The sides of the triangle-shape, trapezoid-shape, and bar-shape defects are almost perpendicular to the step-flow direction of the 4H-SiC[11-20]. Two types of sides exist as the upstream side and the downstream side along the step-flow directions as shown in Fig. 3.

The ORS generates two kinds of defect maps which illustrate the locations of the recognized defects and the degree of small area step bunching. The step-bunching-induced surface-roughness map of typical fairly-flat SiC epitaxial film is shown in Fig. 4. It is discovered that the rough step-bunching area is not uniform and partly spread along the scratches. The rough area is composed of large numbers of the step-bunching lines, which can be expressed as the sides of the trapezoid-shape defects, stretching in a low along the scratches. The step-bunching lines grow almost perpendicular to the step-flow direction and >mm long depending on the epitaxial conditions.

The data about exact locations of the defects enable easy sample preparations for the DSA and two-image-comparison investigations as follows,

Inv. 1: trace of the defect shape before and after the epitaxial growth

Inv. 2: judge of existence of the defects under the test-element-group (TEG) electrodes

These TEG electrodes are used for the ECA and provide information about the correlation between defects and electrical characteristics/reliability as discussed in following section.

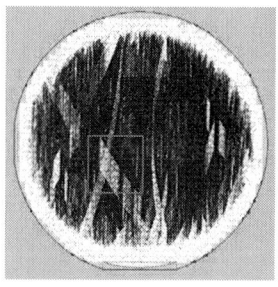

Figure 4. Step-bunching (surface-roughness) map derived from the ORS.

Electrical-Characteristics Analyses (ECA) and the Example Data

The numerous ECA data collected using a large number of small TEG electrodes, each of which contains a small number of defects (either zero or one as known from the ORS), provides simple information about the relationship between defects and the electrical characteristics / reliability. We started the ECA from the evaluation of SiO_2/SiC (MOS) interface. The epitaxial films grown on the n-type 4° off-axis 4H-SiC(0001) wafers were used as substrates of the MOS-capacitor TEG. The oxide films with thickness of 42 nm was formed by dry oxidation at 1200°C followed by post oxidation anneal in N_2 atmosphere at 1200°C. The time-dependent dielectric breakdown (TDDB) under constant stress current of $0.15mA/cm^2$ was measured to obtain Q_{bd} using thousands of the SiO_2/SiC MOS capacitors with small $\phi 150\mu m$ electrodes.

Figure 5 shows the Weibull distribution plot of the Q_{bd}, derived from the ECA, on the commercially-available fairly-flat 4H-SiC(0001) epitaxial film. [4] Few step bunching is observed on this epitaxial film. Three types of distributions of the Q_{bd} are observed as indicated by distribution1 (D1), distribution2 (D2), and distribution3 (D3) in Fig. 5.

The D1 exhibits the narrowest distribution at highest value of Q_{bd} (with the shape parameter m>10 and the average value of ~2.2 C/cm^2). Only MOS capacitors fabricated on a defect-free flat surface comprise the majority of D1. This D1 is thus found in the Q_{bd} test only when the small electrodes are used with high-possibility of making electrodes without defects. This highest Q_{bd} value is considered to be the ideal SiO$_2$/SiC value. The D1 also includes data with defects including BW:Pit (TED and TSD plotted as triangles in Fig.5). In this fairly-flat 4-degree-off epitaxial film, the TEDs and TSDs do not affect the Q_{bd}. Upstream sides of the trapezoid-shape and bar-shape defects can also be ignored in terms of Q_{bd}.

The D2 exhibits wider distribution with lower Q_{bd} value (with m~9 and the average value of ~1.3 C/cm^2) than the D1. Q_{bd} measured at the electrodes on the step bunching lines including the downstream sides of the trapezoid-shape and bar-shape defects are among D2. Although the D2 are affected by different types of the epitaxial defects, the D2 still shows a narrow distribution with average value half of that of the D1. The reliability of the large-size power device or TEG is determined by its weakest point. Density of the step-bunching lines have shrunk but has never been 0/mm^2 at this stage. Reliability of the SiO$_2$/SiC interface in typical SiC power devices so far should be discussed based on this D2 characteristics and reliability.

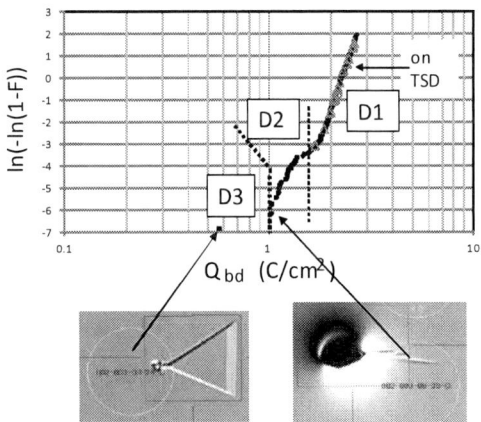

Figure 5. Weibull distribution plot of the Q_{bd}, derived from the ECA, on the example epitaxial film of the commercially-available fairly-flat 4° off-axis 4H-SiC(0001). D1:The highest Qbd on smooth surface, D2:Degraded a little by step bunching, D3:Only two data severely degraded by the large defects shown in lower images.

The much lower and wider distribution of Q_{bd} is observed in D3. The Q_{bd} as low as 1.02 C/cm^2 is measured with the correspondent electrode (indicated by the circle) which is located on the HLine: Carrot-shape defects next to the Large: DownFall as shown in the lower right image in Fig. 5. The smallest Q_{bd} of 0.57 C/cm² is also observed in the D3 on DLine: Triangle-shape defect whose image from the ORS is superimposed at the lower left in Fig.5. The electrode, which includes the pit apex of the triangle-shape defect, exhibit the smallest value of Q_{bd} and easy to fail under TDDB measurement.

Defects-Structure Analyses (DSA) and the Example Data

The DSA (TEM, SEM, AFM, PL, X-ray topography, etc.) are performed to investigate the defect structures. The step-bunching lines, which affect the reliability of the SiO₂/SiC interface as discussed in the ECA section, are investigated. The ORS elucidated that the local step-bunching lines are composed of the sides of the trapezoid-shape defects. There is difference between the effects by the upstream side and the downstream side which only degrades the Q_{bd}. [2] Figure 6 shows the AFM-profile change between as-grown and oxidized surface of the upstream side and the downstream side of the trapezoid-shape defect. Only on the downstream sides, surface profiles drastically change and the step bunching lines appear with 3-times larger surface roughness after the oxidation. There is little change on the upstream side.

The oxide thickness on the trapezoid-shape defect was determined using cross-sectional TEM as shown in Figure 7. The oxide thickness of SiO₂/SiC is almost uniform on the upstream side, while that on the downstream side is fluctuated. The step bunching of the 4H-SiC(0001) surface on the downstream side is clearly observed with the rough SiO₂ surface. The oxidation speed of the 4H-SiC(0001) terrace is much smaller than that of the step. Fast oxidation may result in the thicker oxide formation. The local electric field concentration at the thinner points than the average oxide thickness during the TDDB measurements causes degradation of the Q_{bd} in D2 distribution. The characteristics of the oxide on the bar-shape defect are almost the same as the trapezoid-shape defect discussed above.

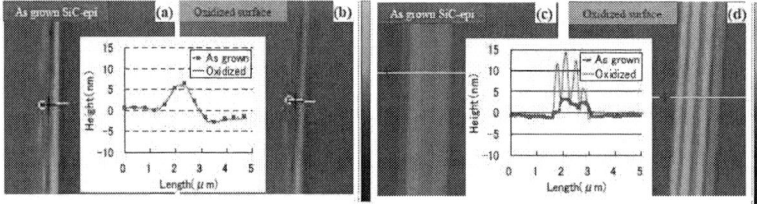

Figure 6. AFM surface-profile change of the trapezoid-shape defect.
(a)as-epitaxial (b)after-oxidized of the upstream side
(c)as-epitaxial (d)after-oxidized of the downstream side

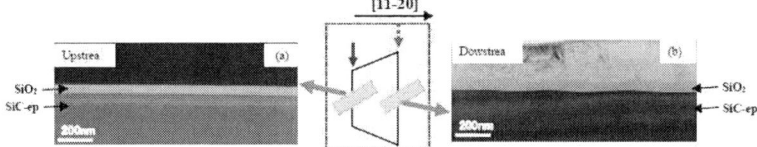

Figure 7. Cross-sectional TEM images of the the SiO₂/SiC interface
on the trapezoid-shape defect. (a) upstream side (b) downstream side.

Conclusions

IEP for SiC wafers and epitaxial films is established and provides information about the correlation between defects and electrical characteristics/reliability. As an example of the results, it is derived that SiC step-bunching defects on the epitaxial wafer partly spread along the scratches and degrade the SiC-MOS reliability. The Q_{bd} has three types of distribution such as D1 (ideal SiO_2/SiC without defects), D2 (half Q_{bd} average value of D1 with step bunching), and D3 (small Q_{bd} value with large defects such as HLine: Carrot-shape, DLine: Triangle-shape, and Large: DownFall). There are little effect to electrical characteristics / reliability by the small defects such as B/W: Pit:(TED and TSD) in compassion with the step bunching. The downstream sides of the bar-shape and trapezoid-shape defects result in formation of the step-bunching lines which degrade the Q_{bd} into D2. The oxide thickness is fluctuated on the step bunching line and the local electric field concentration at the thinner points causes degradation of SiO_2/SiC reliability.

Acknowledgments

This work is supported by Novel Semiconductor Power Electronics Project Realizing Low Carbon Emission Society under METI and NEDO in Japan.

References

1. H. Tsuchida et al, *Phys. Status Solidi B*, **246**, 1553 (2009).
2. J.Sameshima et al., *Materials Science Forum*, **Vols. 740-742**,745 (2013).
3. M. Kitabatake et al., *Materials Science Forum*, **Vols. 740-742**, 451(2013).

24

Growth of High-Quality GaN Template From Nanometer-Size Lattice Channels By Hydride Vapor Phase Epitaxy

A.Usui[1], H. Goto[1], T. Nakagawa[1], H. Sunakawa[1], T. Matsueda[1],
A. Okada[2], J. Mizuno[2], A. A. Yamaguchi[3], H. Shinohara[4] and H. Goto[4]

[1]Nitride Semiconductor Department, Furukawa Co., Ltd., Oyama, Tochigi, Japan
[2]Institute for Nanoscience and Nanotechnology,
Waseda University, Shinjuku, Tokyo, Japan
[3]Optoelectronic Device System R&D Center, Kanazawa Institute of Technology,
Nonoichi, Ishikawa, Japan
[4]Nano Processing System Division,
Toshiba Machine Co., Ltd., Numazu, Shizuoka, Japan

GaN is an attractive material in power devices for energy-saving measures in consumer products, automobiles, and industrial machines. To realize such GaN devices, however, high-quality GaN substrates are indispensable. In this paper, we describe the growth of high-crystalline-quality GaN template by HVPE with the nano-FIELO technique, where GaN growth starts by forming facet structures on nanometer-size channels which are opened on SiO_2 layer deposited on GaN/sapphire substrate. The lattice pattern consisting of 500 nm $\times$ 500 nm square SiO_2 masks surrounded by 80-nm-wide channels is used in this study. A nanoimprinting technique is applied followed by dry etching to fabricate lattice channels. From cross-sectional TEM observation, it is shown that the dislocation density is significantly reduced by the lattice pattern. Uniform GaN growth over 2-inch-diameter wafer is realized.

Introduction

Wide bandgap semiconductors such as SiC and GaN in power devices are attractive for energy-saving measures in consumer products, automobiles, and industrial machines [1]. To realize such GaN devices, however, high-quality GaN substrates are indispensable [2]. In order to obtain high-quality GaN crystal, we have proposed facet-initiated epitaxial lateral overgrowth (FIELO) [3]. The basic concept of FIELO is the growth of a high-quality layer by overgrowing crystals generated from small channels that are opened on a dielectric mask such as SiO_2. In particular, structures with facet planes appeared in the open channels were found to reduce the threading dislocation density by enhancing the bending of the dislocations [4]. However, when micrometer-size channels are used, it is inevitable that many dislocations will propagate from the GaN template through the channels. Furthermore, due to the formation of micrometer-size facets, a relatively thick layer growth (~30 μm) was required to recover a flat (0001) surface. A uniform dislocation distribution, regardless of the mask pattern, was not also achieved until the layer was about 100 μm thick. To overcome these problems, we proposed the nano-FIELO technique, which is based on nanometer-size channels [5]. A nanoimprinting technique was applied followed by dry etching to form these channels [6]. It was confirmed that GaN growth has started by forming nanometer-size facet structures on stripe channels of 50nm-width. As a result, uniform GaN layer has grown with the dislocation density as low as 5×10^7 cm^{-2} at the thickness of about 20 μm. However, the crystalline quality of the samples was anisotropic in nature between the channel direction and a perpendicular direction to the channel. Therefore, we developed in this study a novel nano-FIELO with lattice channel patterns consisting of 500 nm × 500 nm square masks surrounded by 80-nm-wide channels. The lattice channel nano-FIELO is shown to be very effective to eliminate the anisotropic nature of crystal quality [7]. We also show the dislocation behavior by the nanometer-size mask pattern by cross-sectional transmission electron microscopy (TEM) observation.

Experimental

Hydride vapor phase epitaxy (HVPE) method was used to deposit GaN film in this experiment. The Ga metal was loaded upstream in a horizontal-type quartz reactor, and the substrate was positioned downstream. GaCl was generated by the reaction between Ga metal and the HCl gas flowing over the Ga metal. The reaction efficiency to produce GaCl was kept at over 90 % by optimizing the flow conditions and the Ga metal temperature (~850 °C). GaCl was transported with a hydrogen carrier gas and mixed with NH$_3$ in the substrate region. The standard growth temperature in this study was 1040 °C. The nanometer-size pattern was fabricated as follows: (1) an ultraviolet (UV)-curable liquid resin was spin-coated onto the SiO$_2$ mask deposited onto 2-μm-thick MOCVD GaN template; (2) a patterned mold and the substrate were then pressed together by a nanoimprinting machine, and the resin was cured in UV light to transfer the pattern to the resist film; and (3) the SiO$_2$ film was dry-etched to form channels through the UV-cured resin using C$_3$F$_8$ and O$_2$ gases [8]. We used a patterned mold with a lattice pattern consisting of 500 nm × 500 nm square masks surrounded by 80-nm-wide channels. The lattice channels were aligned in the <11−20> and <1−100> directions. Figure 1(a) shows a bird's-eye scanning electron microscopy (SEM) image of the lattice channel pattern formed on the resist film. In Fig. 1(b), a cross-sectional SEM image of the pattern was shown. The channel width at the bottom was about 80 nm and the pitch was 580 nm. This means the mold pattern was imprinted precisely to the resist mask. After transferring the lattice channel pattern on the SiO$_2$ film by the dry-etching, GaN films were grown on these patterned substrates by HVPE to a thickness of about 20 μm. The crystalline quality of the grown films was characterized by X-ray rocking curve (XRC) analysis, and an evaluation of the dislocation density was carried out using cathodoluminescence (CL) and atomic force microscopy (AFM). Cross-sectional TEM images were taken to investigate the dislocation behavior near the region of the mask pattern.

Results and Discussion

Figure 2 shows a top-view SEM image of the initial growth stage, and facet structures can be seen clearly to have developed from every opening. However, the SiO$_2$ mask was still not covered in such short growth period and holes were remained in the center portion of the mask. A cross-sectional SEM image of this sample grown to a thickness of about 0.9 μm is shown in Fig. 3. The facet structures were completely buried at this thickness, and the development of flat (0001) surface was confirmed.

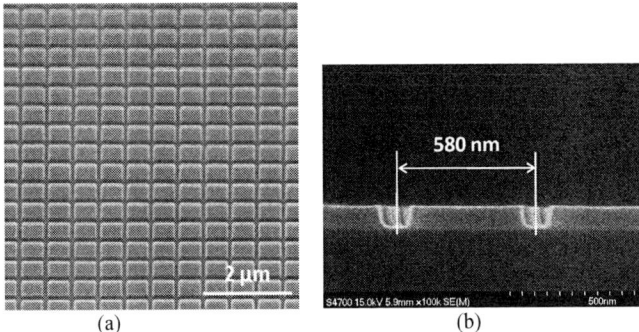

(a)   (b)

Figure 1. (a) Bird's-eye SEM image and (b) cross-sectional SEM image of the lattice channel pattern formed on the resist film.

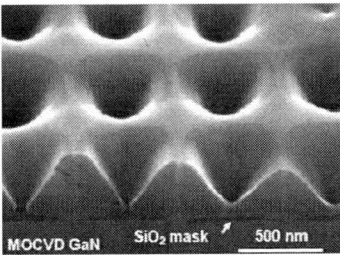

Figure 2. Top-view SEM image of the initial growth stage of a sample grown with the lattice channel pattern.

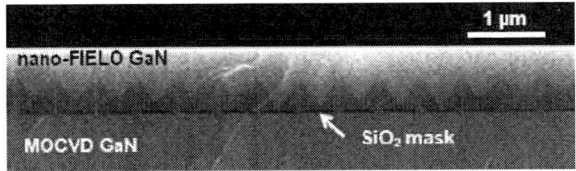

Figure 3. Cross-sectional SEM image of a sample with a thickness of about 0.9 μm grown with the lattice channel pattern.

The dislocation behavior for a sample grown with the lattice channel pattern was investigated by cross-sectional TEM observations. A bright-field image is shown in Fig. 4(a). Although a large dislocation density (mid. of 10^9 cm^{-2}) was found in the MOCVD GaN template, the density in the HVPE GaN above the mask pattern decreased significantly. Both the mask covering the propagation end of the threading dislocations and the narrow channels works effectively in the nano-FIELO. In order to distinguish the dislocation type, dark-field images were taken with g = 1−100 and g = 000−2, as shown in Fig. 4(b) and Fig. 4(c), respectively. From these results, we found that there were very few screw dislocations in both the GaN template and HVPE GaN layers. Pure-edge dislocations were found to be inclined and annihilated effectively above the mask region. A kind of dislocation propagating in the growth direction from the center of the SiO$_2$ mask was mainly mixed-type dislocation. These results are very similar to those observed previously by Sakai *et al.* in the FIELO sample grown with micrometer-size channels [9]. As described previously [5], it was found that the height of the facet structure affects on the dislocation density. The dislocation density had a tendency to decrease by increasing the height. Therefore, we tried to enhance the facet height in the present method. Figure 5 shows a cross-sectional SEM image of a sample grown with low III/V ratio. Although the detail about the role of III/V ratio on the facet formation will be discussed elsewhere, the facet structure height was successfully increased up to several microns. The dislocation density of the sample was evaluated by CL measurements. Figure 6(a) shows CL image for the sample shown in Fig. 5. The dark spot density (DD) corresponding to the dislocation density was found to be as low as 6.3 x 10^7 cm^{-2}. For comparison, a 20-µm-thick GaN layer on the MOCVD GaN template without the lattice channel pattern was deposited and the CL measurement was carried out. As shown in Fig. 6(b), DD for this sample (mid. of 10^8 cm^{-2}) was significantly higher than that of the nano-FIELO sample shown in Fig. 6(a). An advantage of the nano-FIELO over the conventional thick growth was clearly shown.

We reported previously that the dislocation density was also reduced to the similar value by using a stripe channel nano-FIELO method [5]. However, the crystalline quality of the samples grown using the stripe channel pattern was anisotropic in nature between the channel direction and a perpendicular direction to the channel. The full width at half maximum (FWHM) of (0002) XRC measured with a <11-20> incident beam was as narrow as 250 arcsec, but was broaden to ~500 arcsec when measured with a <1-100> incident beam. Sakai *et al.* showed previously that small angel tilt boundaries (SATB) were formed and these SATBs were aligned with the stripe mask edge in conventional FIELO with <11-20> stripe masks [9]. This brought the c-axis inclination along <1-100> direction. They confirmed that the formation of SATBs was related to the dislocation bending caused by the facet structures appeared in the stripe channels. This would be also a reason for the anisotropic crystalline of the nano-FIELO having the stripe channel. By the effect of inclined c-axis along <1-100> direction, the FWHM of (0002) XRC measured with <1-100> incident beam becomes broad, as compared with that of (0002) XRC measured with <11-20> incident beam. This is a reason why the lattice channel pattern was introduced in this study. FWHMs of the XRC for a 20-µm-thick nano-FIELO sample with the lattice channels were evaluated. These values were 214 arcsec and 226 arcsec for <11−20> and <10−10> incident conditions, respectively. This result indicates that the anisotropy was diminished by using the lattice channel pattern.

Using the lattice channel nano-FIELO, 2-inch-diamter template was successfully fabricated. The pattern was uniformly imprinted and processed on 2-inch wafer. HVPE growth resulted in specular surface. We think further enlargement of wafer size would be possible in future.

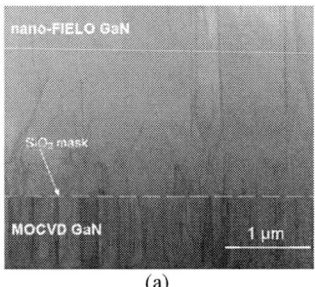

(a)

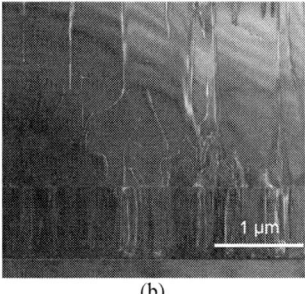

(b)

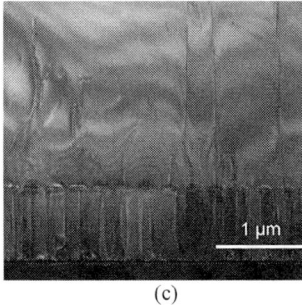

(c)

Figure 4. (a) Cross-sectional TEM bright-field image, (b) dark-field image taken with **g** = 1−100 and (c) dark-field image taken with **g** = 000−2 of a sample grown with the lattice channel pattern.

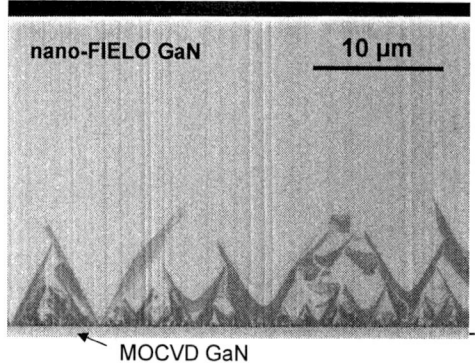

Figure 5. Cross-sectional SEM image of a sample grown with low III/V ratio.

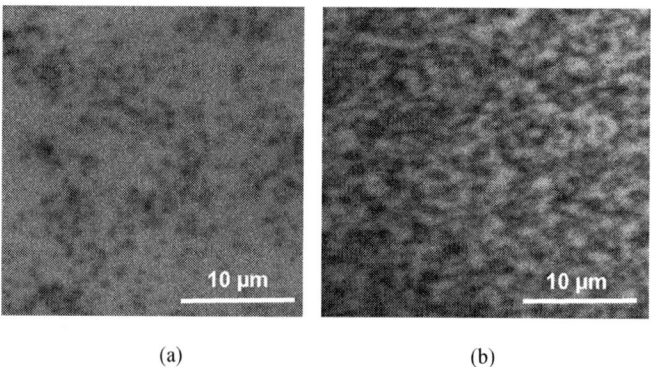

(a)   (b)

Figure 6. CL images of (a) a sample grown with the lattice channel pattern (DD: 6.3 x 10^7 cm^{-2}) and (b) a sample of 20-μm-thick GaN layer grown without the lattice channel pattern (DD: mid. of 10^8 cm^{-2}).

Summary

The growth of high-crystalline-quality GaN template by HVPE with the nano-FIELO technique was studied, where GaN growth has started by forming facet structures on nanometer-size lattice channels which were opened on SiO$_2$ layer deposited on GaN/sapphire substrate. The lattice pattern consisting of 500 nm × 500 nm square masks surrounded by 80-nm-wide channels was used. A nanoimprinting technique was applied followed by dry etching to form these lattice channels. From cross-sectional TEM observation, it was shown that the dislocation density was significantly reduced by nanometer-size lattice channels. The dislocation density evaluated by CL measurements was found to be as low as 6.3×10^7 cm^{-2}. This value was about one order of magnitude lower than that of GaN layer grown without the nano-FIELO method. Uniform GaN growth on 2-inch-diameter wafer was realized with the lattice channel nano-FIELO method. Such high-quality GaN templates will make it easier to fabricate high-performance light-emitting diodes and electron devices.

Acknowledgments

This study was partially supported by the New Energy and Industrial Technology Development Organization (NEDO) of Japan.

References

1. H. Okumura, *Jpn. J. Appl. Phys.*, **45**, 7565 (2006).
2. H. Amano, *Jpn. J. Appl. Phys.*, **52**, 050001 (2013).
3. A. Usui, H. Sunakawa, A. Sakai and A. A. Yamaguchi, *Jpn. J. Appl. Phys.*, **36**, L899 (1997).
4. A. Sakai, H. Sunakawa, A. Kimura and A. Usui, *J. Electron Microscopy*, **49**, 323 (2000).
5. A. Usui, T. Matsueda, H. Goto, H. Sunakawa, Y. Fujiyama, Y. Ishihara, A. Okada, S. Shoji, A. A. Yamaguchi, H. Nishihara, H. Shinohara, H. Goto and J. Mizuno, *Jpn. J. Appl. Phys.*, **52**, 08JB02 (2013).
6. A. Okada, S. Shoji1, H. Shinohara, H. Nishihara, H. Goto, H. Sunakawa, T. Matsueda, A. Usui, A. A. Yamaguchi and J. Mizuno, *J. Photopolym. Sci. Technol.*, **26**, 69 (2013).
7. A.Usui, *ECS J. Solid State Sci. Technol.*, **2**(8), N3045 (2013).
8. M. Fukuhara, J. Mizuno, M. Saito, T. Homma and S. Shoji, *IEEJ Trans.*, **2**, 307 (2007).
9. A. Sakai, H. Sunakawa and A. Usui, *Appl. Phys. Lett.*, **73**, 481 (1998).

ECS Transactions, 58 (4) 33-47 (2013)
10.1149/05804.0033ecst ©The Electrochemical Society

Packaging Techniques for Compact SiC Power Modules
Operable in an Extended Tj Range

Satoshi Tanimoto[a,b], Kinuyo Watanabe[a], Hidekazu Tanisawa[a,c],
Kohei Matsui[a,d], and Shinji Sato[a,c]

[a]Research Center, FUPET, c/o AIST, Tsukuba, Ibaraki 305-8568, Japan
[b]EV System Laboratory, Nissan Motor Co., Ltd., Yokosuka, Kanagawa 237-8523, Japan
[c]Adv. Tech. Develop. Div, Sanken Electric Co., Ltd., Niiza, Saitama 352-8666, Japan
[d]Corporate R&D Headquarters, Fuji Electric Co., Ltd., Hino, Tokyo 191-8502, Japan

Packaging technology applicable to SiC power devices operated in
an extended junction temperature range (Tjmax > 200°C) must be
developed in order to create much more compact and cost-effective
SiC power modules. This paper describes some of the technical
challenges involved in improving the reliability of the critical
package components—die attachment system, Al wire bonds and
encapsulation—in direct contact with SiC devices inside the power
module. Two numerical targets, (I) 3000 hours for a storage test at
250°C and (II) 3000 cycles for thermal cycling between -40°C and
250°C, were achieved through optimization and various
improvements.

Introduction

Because of their significant ability to operate with very low conduction and switching
loss even at junction temperatures (Tj) of > 200°C, SiC unipolar power devices, such as
MOSFETs and VJFETs, offer tremendous benefits for 0.6 to 1.8 kV-blocking voltage-
class power applications, including smaller size, lighter weight, greater robustness and
superior cost effectiveness (1, 2). Electric and hybrid electric vehicles (EVs/HEVs) and
consumer power electronics are just three possible applications (3). However, power
module packages aimed at these potential applications are still in an early phase of
development. One of the most crucial concerns is to ensure long-term reliability under
severe thermal stress (4). High Tj and large thermal excursions (ΔTj) often cause package
components to deteriorate in a short period of time (5, 6). Conventional high-temperature
high-Pb solder die attachments broke down due to fatigue within 500 cycles when
subjected to a small thermal cycling test (TCT) between 0°C and 165°C (7).

This paper describes various technical challenges involved in improving the
reliability of the critical package components—die attachment, wire bonding and
encapsulation—in direct contact with SiC devices inside the power module. Two
preliminary numerical targets were set: (I) 3000 hours for a storage test at 250°C and (II)
3000 cycles for a thermal cycling test (TCT) between -40°C and 250°C. The test results
showed that all of the components achieved these targets.

Experimental

This section describes the materials, conditions and procedures commonly used in
fabricating and testing the critical package components noted above.

Test Sample Preparation

Four different test samples, shown in Fig. 1, were prepared in accordance with the intended use: (a) storage tests and thermal cycling tests for die attachments (8), (b) the same tests for wire bonding (9), (c) power cycling tests for die attachments and wire bonding (9) and (d) storage tests and TCTs for the encapsulation system (10). In these samples, SiC dies (2 mm (W) × 2 mm (D) × 0.20-0.25 mm (H)) were mounted using a Pb-free eutectic high-temperature solder, Au-Ge (m.p. = 362°C, 0.03-mm-thick preform, Tanaka Kikinzoku Kogyo Co. Ltd., Japan). The samples in Fig. 1a had eight dummy dies and those in Figs. 1b and 1c had eight Schottky barrier diodes (SBDs: #S6005TCS, rated 600 V, 12 A, ROHM Co., Ltd. Japan or #CPW4-1200-S008B, rated 1200 V, 7.5 A, Cree, Inc., USA). The sample in Fig. 1d had a single SBD, assembled in a TO-254 ceramic package having a layered Cu/Mo/Cu base plate (Kyocera Corp., Japan). The samples in Figs. 1a to 1c had a 0.31-mm-thick SiN ceramic substrate attached with 0.3-mm-thick conductor foils on both sides by active metal brazing. The conductor was either oxygen-free Cu or CIC clad material, but Cu unless otherwise specified. These substrates are denoted here as "Cu (CIC)-SiN." The surfaces of the conductor foil and the metal base plate were electroless-plated with thick Ni-P and then finished by Au flashing. The wire bonding material was 200-μm-diameter lightly Ni-doped Al (#TANW, Tanaka Kikinzoku Kogyo Co., Ltd., Japan). A schematic cross-section of the samples in Figs. 1a to 1c is depicted in Fig. 2. Note that the sample in Fig. 1a had no Al pad and no wire bonding at the top because dummy SiC dies were used. The sample in Fig. 1b was ultrasonically wedge-bonded with two wires per die, while the eight SBDs of the sample

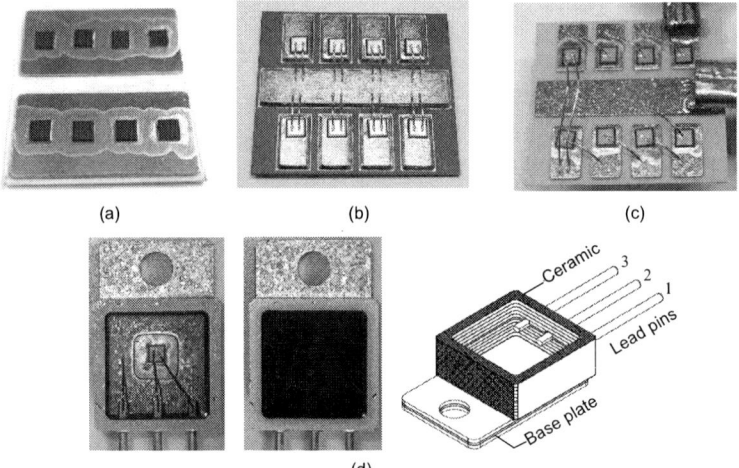

(a)　　　　　　　　　(b)　　　　　　　　　(c)

(d)

Fig. 1 Test samples used in: (a) storage tests and thermal cycling tests (TCTs) for die attachments, (b) the same tests for wire bonds, (c) power cycling tests (PCTs) for die attachments and wire bonds and (d) storage tests and TCTs for the encapsulation system.

in Fig. 1c were serially connected with Al wire. Die attachment and subsequent wire wedge bonding were carried out under optimal process conditions, respectively, using an SRO-707 vacuum reflow system (ATV Technology GmbH, Germany) and a 360C automatic wire bonder (Orthodyne Electronics Corp., USA). The first bonding of the Al wire was done on the SBD dies (Al pad) in Figs. 1b and 1c and on the lead terminals of the package case in Fig. 1d.

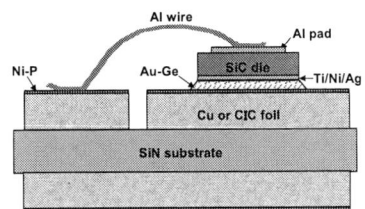

Fig. 2 Schematic cross-section common to samples in Fig. 1a to 1c.

Reliability Assessment

High-temperature storage tests were conducted for 3000 hours or more using STH-120 forced-air ovens (ESPEC Corp., Japan) where the test temperatures were 250°C and optionally 300°C. On the other hand, TCTs were performed in air at least until 3000 cycles between -40°C and 250°C and optionally between -40°C and 300°C. A TSE-11-AS system (28 min/cycle, ESPEC Corp., Japan) and an HT-TAKE-1 system (approximately 45 min/cycle, TAKEKEN Corp, Japan) were used respectively for the former and latter temperature range.

For traditional Si-based packages, the long-term reliability of wire bonds has been quantified and compared through power cycling tests (PCTs) (11-13). With this in mind, PCTs of Al wire bonds were optionally implemented using the samples (serially connected 8 SBD dies, with one bond per die) in Fig. 1c, enabling a comparison between present and previous results. Figure 3 outlines the experimental PCT set-up used. The first step was to set the primary test conditions, Tjmax, ΔTj and t_F (current forced interval time), at 200°C, 165°C and 2 s, respectively, and the second step was to record an I_F-V_F reference curve of the SiC-SBDs at an ambient temperature of 200°C, identical to Tjmax. Finally, a forced current during the forced interval, I_F, and a cooling time interval, t_C,

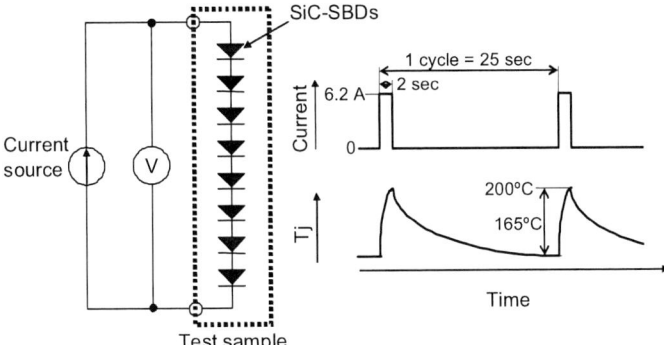

Fig. 3 Experimental setup for power cycle test: (left) circuit diagram and (right) applied current pulse and change in junction temperature.

were determined experimentally with an asymptotical approach so as to fully achieve the primary test conditions. The actual values obtained were $I_F = 6.2$ A and $t_C = 23$ s.

A standard die shear test and wire pull test were performed to quantify the mechanical strength of the die attachment and wire bonding system at various temperatures by using a Dage 4000 bond tester (Nordson Corp., USA). The electrical resistance of the Al wire bonds in Fig. 1b and the low-voltage I-V characteristic of the SiC-SBDs in Figs. 1b and 1d were measured with best accuracy with the 4-terminal Kelvin method using a 4156C semiconductor parameter analyzer (Agilent Technologies Inc., USA). The high current forward I-V characteristic (pulse mode) of the SiC-SBD in Fig. 1d was measured with a CS-3200 curve tracer (Iwatsu Test Instruments Corp., Japan) while its high voltage reverse characteristic was measured with a 2410 source meter (Keithley Instruments, Inc., USA). All the mechanical and electrical measurements were conducted at room temperature if not specified.

New Packaging Techniques for Extended Tj Range Application

Die attachment

Conventional die attachments using high-temperature Pb-based solders have problems with respect to solderability and reliability, not to mention their bio-hazardous property. McCluskey and coworkers recently indicated their poor fatigue resistance (7). A new die attachment system must be created that is harmless to biological bodies, compatible to semiconductor devices and packaging processes, and operable at elevated temperatures and even under larger repeated thermal excursions. The authors have so far developed SiC die attachment systems using various attach materials such as eutectic solders: AuGe (m.p. = 356°C) (8, 14), ZnAl (m.p. = 382°C) (15) and AuSn (m.p. = 280°C) (14), an off-eutectic solder, BiAgX (X = Sn, m.p. = 263°C) (16), and silver nano-materials, principally focusing on joint strength reliability. The work done on AuGe die attachments is described here because they have achieved the best results at this point in time. AuGe die attachments have been widely used in feasibility studies of SiC devices and circuits at elevated temperatures of < 300°C (17). Nevertheless, only fragmentary results have been reported concerning reliability and applicability (18).

For the purpose of obtaining acceptable joint strength, the cleaning procedures (SiC dies, AuGe preforms and SiN substrates), back metallization of SiC dies, and solder reflow conditions were optimized in the early stage of the present work. From the middle stage, the light-load application jig shown in Fig. 4 was routinely utilized during reflow soldering with the aim of reducing the solder layer thickness and the emergence of voids, leading to a significant reduction in solder layer thermal resistance (15). As a result, the AuGe layer became typically less than 20 μm in thickness.

As-soldered dies on the samples in Fig. 1a were sheared at various elevated temperatures. Figure 5 shows the average

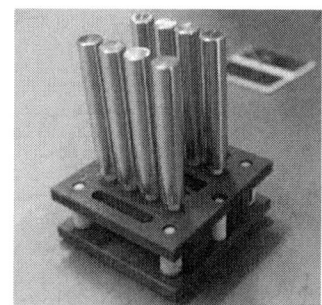

Fig. 4 Carbon jig used to apply a light load to the SiC dies.

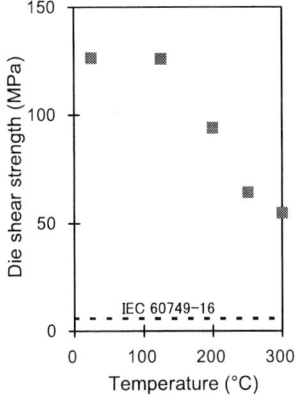 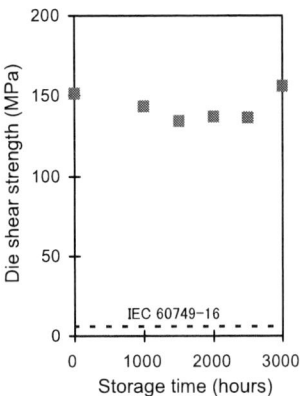

Fig. 5 Average shear strength of the AuGe die attachments (N = 7) in Fig. 1a as a function of shear test temperature.

Fig. 6 Average die shear strength of the AuGe die attachments (N = 7) in Fig. 1a as a function of storage time at 250°C.

shear strength as a function of shear test temperature (14). The results indicate that the optimized AuGe die attach system is usable at least for a short period of time in a temperature range up to 300°C, although shear strength decreased with increasing temperature. The die shear strength was 54.5 MPa even at 300°C, which is nine times higher than the shear strength criterion (6.2 MPa) for 2×2 mm^2 dies specified in IEC 60749-19.

There were a couple of previous reports about long-term storage of AuGe die attachments at 300°C (14, 19). They revealed that high-temperature storage caused the die attachments to degrade severely, resulting in a steep drop in joint strength. Figure 6 presents our test result for storage at 250°C of the samples in Fig. 1a, where the average die shear strength is plotted as a function of storage time. It is seen that there was no decline in shear strength during 3000 hours of storage within acceptable experimental error, signifying that adequately fabricated AuGe die attachments can withstand storage for more than 3000 hours in a temperature range up to 250°C without any problems. Cross-sectional scanning electron micrographs (X-SEM, not presented here) revealed that the AuGe solder layer still continued to tightly join the SiC die to the Cu-SiN substrate even after 3000 hours, although a rather thick NiGe intermetallic compound (IMC) developed at the AuGe solder/Cu foil interface. Recently, Leinenbach and his coworkers have pointed out on the basis of microhardness measurements that this AuGe reaction layer appears to be relatively soft (30).

Some TCTs between -40°C and 250°C were conducted on the AuGe die attachments in Figs. 1a and 1b prepared on Cu-SiN. Die shear strength and the electrical resistance between the Al anode pad on the SiC-SBDs and the Cu foil under the AuGe solder layer were measured as a function of the number of thermal cycles. As the results plotted in Fig. 7 indicate, the shear strength of the attachments decreased linearly and, after 3000 cycles, barely satisfied the IEC 67449-19 criterion (20). The X-SEM of the die attachment after 3000 cycles in Fig. 8b reveals a long crack in the solder layer. A comparison with the X-SEM of the as-soldered sample in Fig. 8a indicated that a steep decline in shear strength resulted from thermo-mechanical fatigue (cracking) that developed in the solder layer due to repetitive huge stress induced by greater thermal excursion and by the large coefficient of thermal expansion (CTE) mismatch between the SiC die (~5 ppm/°C) and the Cu foil (16.8 ppm/°C). Figure 7 also plots the change in electrical resistance. It is seen that the electrical resistance stayed virtually constant and did not properly reflect the profound progress of fatigue in the solder layer. This is because electrical resistance corresponded to the combination of very big on-resistance of the SiC-SBD and very small resistance of the die attachment.

In an effort to overcome the rapid degradation in shear strength, the Cu foil on the SiN substrate was replaced by a CIC (Cu/Invar/Cu clad) foil with a combined CTE of 5.1 ppm/°C, a C/I/C thickness ratio of 1/8/1 and a total thickness of 0.3 mm (20). Note that the extremely low CTE = 1.2 ppm/°C of Invar

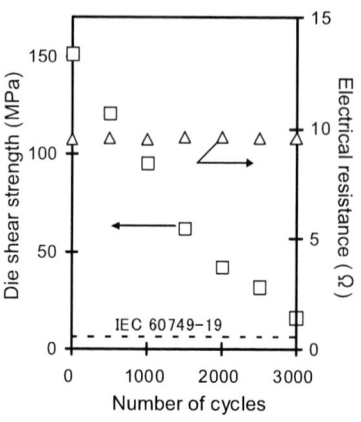

Fig. 7 Average die shear strength (Fig. 1a, N = 7) and electrical resistance (Fig. 1b, N = 24) of the AuGe die attachments on Cu-SiN substrate as a function of number of thermal cycles between –40°C and 250°C.

(Fe-32%Ni) plays a conclusive role in this optimal CTE matching between the conductor and the SiN substrate. Two sets of AuGe die attachments (Fig. 1a) on CIC-SiN were independently subjected to TCTs between -40°C and 250°C and between -40°C and 300°C. Figure 9 presents the test results along with the previous results for Cu-SiN, where the average shear strength (N = 7) is plotted as a function of the number of thermal cycles. The -40°C to 250°C TCT results indicate that the shear strength decreased at a rate much slower than that for the Cu-SiN substrates and still stayed at a level of 78 MPa after 3000 cycles, a value that is 15 times higher than the IEC 60749-19 criterion. This

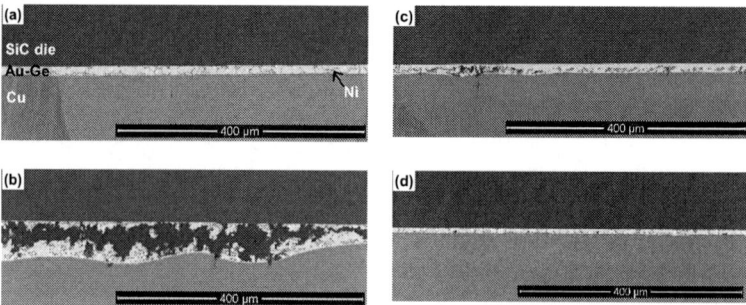

Fig. 8 Cross-sectional SEM images of the Au-Ge die attachments: (a) as-prepared, (b) after 3000 thermal cycles between –40°C and 250°C, (c) on CIC-SiN after 3000 thermal cycles between –40°C and 250°C and (d) after 37421 power cycles at $\Delta Tj = 165°C$ and $Tjmax = 200°C$.

means that our die attachment has exceptionally high reliability and a prolonged -40°C to 250°C thermal cycle life. A linear regression for the -40°C to 250°C TCT results revealed that the estimated cycle life expectancy was more than 8500 cycles. Figure 8c displays the cross-sectional SEM of a AuGe die attachment on CIC-SiN after a 3000-cycle TCT. This micrograph shows that the AuGe die attachment on the CIC-SiN substrate fully maintained a normal joint condition with minimal cracking. The -40°C to 300°C TCT results in Fig. 9 show that the rate of decline in shear strength was obviously faster than that for the -40°C to 250°C TCT but more gradual than that for the -40°C to 250°C TCT of the die attachments on Cu-SiN. The die attachments on CIC-SiN had sufficient shear strength of 52 MPa even after 3000 cycles.

The power cycle reliability of AuGe die attachments is described here probably for the first time. PCTs were carried out under the conditions of Tjmax = 200°C and ΔTj = 165°C (Tjmin = 35°C) using the samples in Fig. 1c on CIC-SiN. The tests were intended to assess the overall power cycle reliability of the die attachment and wire bonding. All faults observed were related to Al wire bonds rather than to die attachments (9). Hence, the maximum number of power cycles was limited by the wire bond life (maximum of approximately 35000 cycles). Figure 10 shows die shear strength as a function of the number of power cycles. The results reveal that, within acceptable experimental error, there was little decline in die shear strength until the test ended at 37421 cycles. A cross-sectional SEM of a AuGe die attachment after 37421 PCT cycles is shown in Fig. 8d. There are no signs of thermo-mechanical fatigue, such as micro-cracks and folding, in the solder layer. Previously, Si-IGBT die attachments on DCB (direct-Cu-bonded) alumina

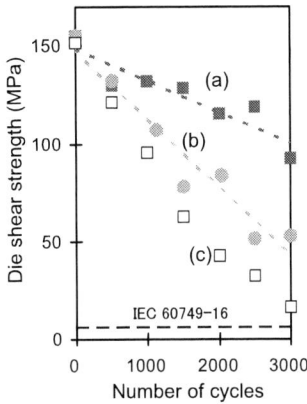

Fig. 9 Average die shear strength of the AuGe die attachments (N = 7) in Fig. 1a on: (a), (b) CIC-SiN and (c) Cu-SiN substrate, as a function of the number of thermal cycles. The tests were conducted between –40°C and 250°C for (a) and (c) and between –40°C and 250°C for (b).

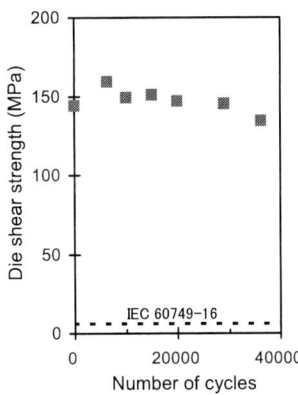

Fig. 10 Average die shear strength of the AuGe die attachments (N = 7) in Fig. 1c as a function of the number of power cycles where power cycling was carried out under conditions of ΔTj = 165°C and Tjmax = 200°C.

substrates were subjected to PCTs under relatively greater thermal excursion, ΔTj = 100°C, where a SnAg-based solder (Sn > 95 mass%) and a SnPbAg-based solder (Pb > 93 mass%) were used as the die attach materials. Morozumi and his coworkers reported that the lifetime was 25000 cycles for SnAg and 17000 cycles for SnPbAg (21). These test results indicate that the AuGe solder SiC die attachment has overwhelmingly better reliability against power cycle stress compared with these conventional solder Si die attachments.

Wire bonds

Although the long-term reliability of thick Al wire bonds (Al wire-Al bond pad system) has been investigated by many researchers for a long time (11-13), there are only a limited number of reports regarding reliability investigations in the over-200°C Tj range (22-24). In this work, the reliability of thick Al wire bonding was comprehensively assessed in an extended Tj range up to 300°C.

A commercially available, slightly Ni-doped Al wire (#TANW) of 200 μm in diameter was selected as the most adequate material at present. After soldering commercial SiC-SBD dies with AuGe onto Cu-SiN or CIC-SiN substrates, a pair of Al wires (Fig. 1b) or a single Al wire (Fig. 1c) was ultrasonically wedge-bonded under optimal conditions so as to connect the Al pad (first bond) on the SBD with the conductor foil (second bond) on the substrate.

Figure 11 shows the relationship between the average bond pull strength (gf: gram-force) and Tj (Al wire temperature) for pull tests performed on Al wire bonds in Fig. 1b (9). Fracture (breaking) occurred either on the beam (bulk) or at the heel for Tj < 50°C and only at the heel for Tj >100°C. The results indicate that the average bond pull strength decreased linearly with increasing Tj due to softening, but even at 300°C it still more than satisfied the IEC criterion (50 gf) for 200-μm-diameter Al wire specified in

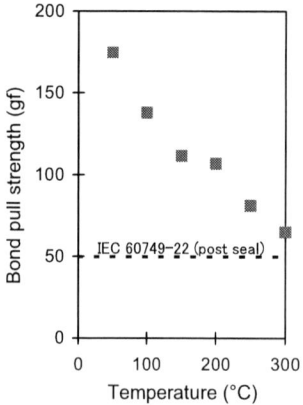
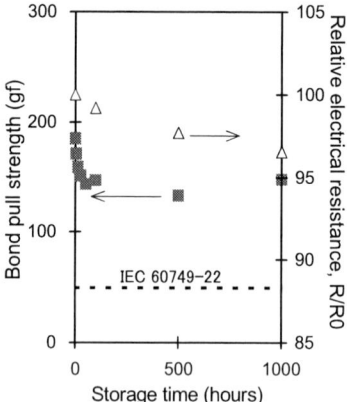

Fig. 11 Average pull strength of the Al wire bonds (N = 14) in Fig. 1b as a function of pull test temperature.

Fig. 12 Average pull strength (N = 14) and average electrical resistance (N = 7) of the Al wire bonds in Fig. 1b as a function of storage time at 250°C where R0 = 2.56 mΩ.

Document 60749-22.

A 250°C storage test for 1000 hours was conducted on the Al wire bonds in Fig. 1b (9). The test results are shown in Fig. 12 where the average bond pull strength and electrical resistance of the Al wire bonds are plotted as a function of storage time. The results show that the electrical resistance continued to decrease sublinearly with increasing time,

Table 1 Average pull strength (N = 14) of Al wire bonds in Fig. 1b subjected to storage tests for 3000 hours.

Storage time & temperature	0 hour	3000 hours	
		250°C	300°C
Pull strength (gF)	241	180	164
Relative strength (%)	100	75	68
Ratio to IEC 60749-22	3.4	2.6	2.3

while the bond pull strength dropped quickly in the early stage but afterward stayed at a level approximately 2.5 times higher than the IEC criterion. We found that Al wires broke mostly at the heel and rarely on the beam. Note that the storage test temperature, 250°C, was rather higher than the recrystallization temperatures of Al, 150°C to 200°C, depending on the purity and the degree of work hardening. Predictably, some cross-sectional SEM observations of test samples (not presented here) revealed that recrystallization of Al occurred early on the beam and at the heel and that the crystalline grains grew at a rapid rate as storage time passed (9). It is well known that grain growth can inherently reduce the electrical resistance of metals, including Al. In addition, Al is known to obey the Hall-Petch relationship (25), where yield strength falls in general as the average grain size becomes bigger. Thus, the electrical resistance and bond pull strength behavior in Fig. 12 can be qualitatively explained as the result of Al grain growth. Recently, Tanisawa and coworkers discussed the reason why electrical resistance does not correlate with bond pull strength (9).

Since favorable prospects were observed in Fig. 12, uninterrupted storage tests for 3000 hours were then carried out independently at 250°C and 300°C. The test samples

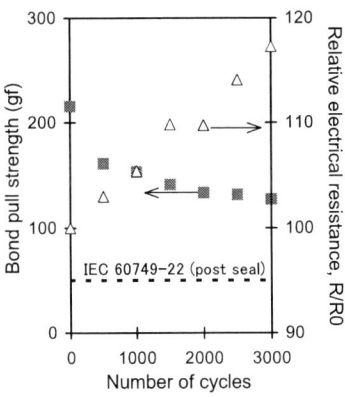

Fig. 13 Average pull strength (N = 14) and average electrical resistance (N = 7) of the Al wire bonds in Fig. 1b as a function of the number of thermal cycles between –40°C and 250°C where R0 = 2.59 mΩ.

were those in Fig. 1b. The test results are listed in Table 1. Compared with the IEC criterion, the results show that our Al wire bonding still maintained a sufficient level of bond pull strength after storage for 3000 hours not only at 250°C, but also surprisingly even at 300°C. It can be concluded that optimally assembled Al wire bonding can survive storage for 3000 hours at temperatures up to 300°C.

Next, we present the results regarding TCTs performed between -40 and 250°C. The test samples were the wire bonds in Fig. 1b. As shown in Fig. 13, electrical resistance continued to increase with increasing cycles but stayed under 17% until the test ended, which is acceptable, though not negligible. On the other hand, bond pull strength declined with increasing thermal cycles, albeit the rate of decrease lessened. However, after 3000 cycles, it was at a level of 127 gf, still satisfying the IEC criterion by a sufficient margin. There was a somewhat steep decline in bond pull before 500 cycles. This stemmed from the annealing effect that automatically occurred during the TCT heating period. The break mode of the wire bond samples subjected to the bond pull test was either beam break or heel break on the CIC foil or heel break on the SiC-SBD. Previously, Yamada et al. (24) and Ayakwa et al. (26) have independently investigated crack propagation along the bonding interface on power devices. However, interface breaks were not among the major break modes in the thermal cycling experiments conducted in this work.

Lastly, the PCT results are presented here. The sample in Fig. 1c was used in the tests. All the tests were implemented under the conditions of Tjmax =200°C and ΔTj = 165°C (Tjmin = 35°C).

First, 15 samples were subjected to a life test. The test results were plotted on a Weibull chart, as shown in Fig. 14, indicating the relationship between the cumulative failure probability, F, and lifetime (the number of power cycles to failure). The test data were well expressed in a straight line with a steep slope, suggesting that failure would be attributed to a single wear-out event. This hypothesis was consistent with the experimental observation that failure happened unexceptionally in the mode of Al wire bond bond lift-off on SiC-SBD dies. From the data, the lifetime (Nf) of the Al wire bond samples (i.e., the median number of cycles to failure) was determined to be 3.0×10^4 cycles. To the best of our knowledge, there has never been any report describing the results of an Al wire bond PCT conducted under a very large ΔTj of 165°C. Now, let us compare roughly our results with previous results in terms of lifetime, based on the assumption above. It is commonly known that the PCT lifetime of thick Al wire bonds on Si-IGBT devices decreases exponentially with increasing ΔTj in accordance with an empirical formula,

$$Nf = Ae^{-k\Delta Tj},$$

where A and k are fitting parameters and ΔTj is in the range of 40°C to 80°C. Concrete data supporting this relationship are presented in Refs. (11-13). For example, Cova and Fantini obtained a formula,

$$Nf = 1.01 \times 10^8 e^{-0.0966\Delta Tj},$$

by means of linear regression (12). Assuming this formula to be still applicable in a larger ΔTj range, Nf = 12 cycles can be estimated as the lifetime for conventional thick Al wire bonds on Si-IGBT when ΔTj = 165°C. Thus, we found that the proposed Al wire bonding technology on SiC devices apparently provides a PCT lifetime three orders of magnitude higher than conventional bonding on Si-IGBT.

Secondly, PCTs were conducted again with a view to finding the progression of wire deterioration. Test samples were picked out one by one

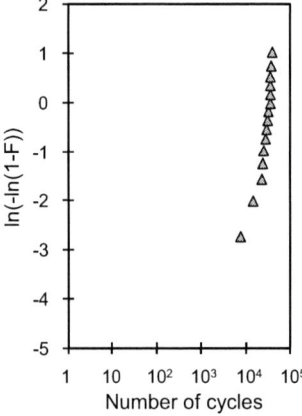

Fig. 14 Results of power cycling tests carried out for Al wire bonds in Fig. 1c under conditions of ΔTj = 165°C and Tjmax = 200°C where cumulative failure probability, F, is plotted as a function of the number of cycles on the Weibull chart. The samples were the same as those used in the PCT of the die attachments in Fig. 10.

according to a planned schedule and their wire bond pull strength was measured. It was found that breaks occurred in the beam or heel break mode until 10000 cycles, after which interface breaks appeared and became dominant. Figure 15 shows the average bond pull strength as a function of the number of power cycles. It is seen that the pull strength declined linearly and nearly reached the IEC criterion.

Encapsulation

The purpose of encapsulation is to protect the power chips, embedded passive components and interconnections from external environmental damage and to prevent arcing between the chip surface and adjacent metal features. Today, thermoset polymeric materials, mostly

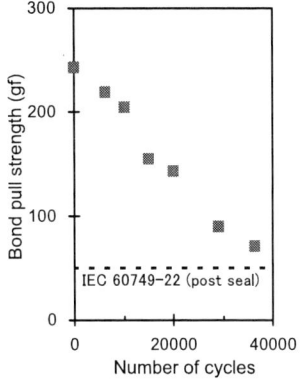

Fig. 15 Average pull strength of the Al wire bonds (N = 6 per 1 point) in Fig. 1c as a function of the number of power cycles under conditions of $\Delta Tj = 165°C$ and Tjmax = 200°C.

silicone gels, are widely used as encapsulants in power modules (4, 27). However, they are designed for use in a temperature range below 175°C. Recently, several research groups have investigated high-temperature oriented encapsulants, reporting that the heat resistance was still at level of less than 800 hours for storage at 250°C (6, 28, 29). More reliable encapsulants have to be developed for next-generation compact SiC power modules.

Cooperative studies for improving reliability under severe thermal stress were independently but concertedly conducted with four vendors (10). Silicone-based encapsulants (mostly prototype products) were improved step-by-step by each vendor and potted into the TO-254 packages in Fig. 1d to assemble one SiC-SBD die, after which they were comparatively tested in six runs. Four of the runs were storage tests at 250°C and the other two runs were thermal cycling tests from -40°C to 250°C. Three or five samples per encapsulant were subjected to storage and thermal cycle tests. A set of three samples with no encapsulant was also tested as a control. According to a planned schedule, the test was interrupted and the samples were checked for failure by microscopic inspection and electrical characterization. The latter included measurement of V_F (V) at +12 A (ROHM SBD) and +7.5 A (Cree SBD) and reverse leakage current I_L at –600 V (ROHM) and –1100 V (Cree). Lifetime is defined here as the median time to failure in the storage test and as the median number of cycles to failure in the TCT. Test materials are listed in Table 2. Only the #DCT-WR series were solidified materials and the rest were all gels. The #RX-157A and #DCT-WR series samples were white in color, the #DCT-BG series were black, and the others were clear and colorless. Almost all the encapsulants used were tested beyond the tentative specifications recommended by the vendors. Therefore, it should be noted that no mention is made here of one vendor's material being superior to the others within their range of specifications.

Storage tests were conducted four times in all. It was found that only structural failures occurred and there were no electrical failures at all. The photographs in Fig. 16

Table 2 Encapsulation materials and results of storage tests at 250°C and TCTs between –40°C and 250°C. The failure mode DC stands for "delamination at case sidewall," CR "cracking," AB "air bubbles" and WD "wire disconnection."

Vendor	ID of Material	250°C Storage		-40 to 250°C TCT	
		Life (Hr)	Mode	Life (cyc)	Mode
ADEKA	BYX-001G	100	DS, CR	200	CR
	BYX-002G	100	DS	500	DS, CR
	RX-157A	500	CR	500	WD
SEC	B1	24	DS		
	B2	24	DS, AB		
	X-32-3318-A/B	200	CR	1000	WD, CR
	X-32-3318-2A/B	500	CR	1500	WD, DS
WAS	920 LT A/B	24	DS		
DCT	D1	24	CR, AB		
	DCT-BG001-NT	> 3000		500	WD
	DCT-BG002-1-NT	> 3000		3000	DS
	DCT-BG002-2-NT	> 3000		2500	WD
	DCT-WR001-NT	> 3000		200	WD
	DCT-WR002-NT	> 3000		500	WD
Control	No encapsulation	> 3000		> 3000	

Note: ADEKA: ADEKA Corp., Japan; SEC: Shin-Etsu Chemical Co., Ltd., Japan; WAS: Wacker Asahikasei Silicone Co., Ltd., Japan; DCT: Dow Corning Toray Co., Ltd., Japan

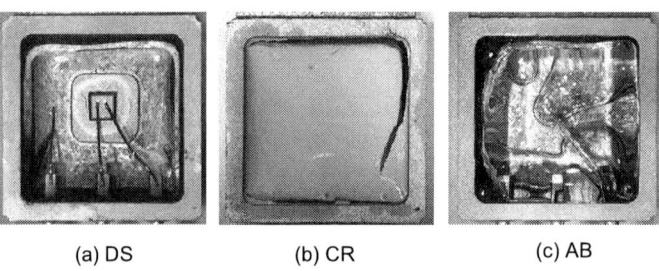

(a) DS   (b) CR   (c) AB

Fig. 16 Typical structural failures of encapsulated package samples: (a) delamination at case sidewall (DS), (b) cracking (CR) and (c) air bubbles (AB).

show the typical appearance of structural failures: (a) delamination at the case sidewall (denoted by DS), (b) cracking (denoted by CR) and (c) air bubbles (denoted by AB). DS and CR failures resulted from contraction of the encapsulant itself due to thermal depolymerization. In many cases, air bubbles were accompanied by cracking. Storage test results are shown in Table 2. The results indicate that both the #DCT-BG and #DCT-WR series withstood 250°C storage for 3000 hours. #RX-157A and #X-32-3318-2A/B achieved better results but failed in the DS mode after 500 hours.

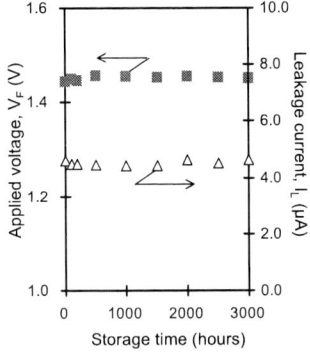 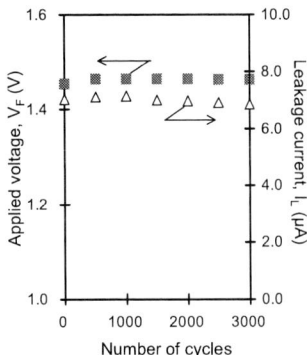

Fig. 17 Forward applied voltage at +7.5 A and reverse leakage current at -1.1 kV of an SiC-SBD sample in Fig. 1d encapsulated with #DCT-BG002-1-NT as a function of storage time at 250°C.

Fig. 18 Forward applied voltage at +7.5 A and reverse leakage current at -1.1 kV of an SiC-SBD sample in Fig. 1d encapsulated with #DCT-BG002-1-NT as a function of the number of thermal cycles between -40°C and 250°C.

On the other hand, the -40°C to 250°C TCTs yielded somewhat severe results (Table 2). Nine of the ten test sets experienced trouble before the test ended. It was found that all these samples, exclusive of the #BYX series, failed due to Al wire disconnection (denoted by WD, probably due to heel break or bond lift-off), making it impossible to measure the electrical characteristics of the SiC-SBD with the 4-terminal Kelvin method. Only the #DCT-BG002-1-NT sample reached 3000 cycles without any trouble.

In summary, the #DCT-BG002-1-NT encapsulant passed both the 3000-hour storage test at 250°C and the 3000-cycle TCT between -40°C and 250°C. Figure 17 plots the change in V_F at +7.5 V and I_L at -1100 V of this sample subjected to the storage test, while Fig. 18 shows the change in the sample subjected to the TCT. The devices used were SiC-SBD #CPW4-1200-S008B (Cree, Inc.). No anomalous change was observed in SBD electrical properties from the start to the end of the test. However, it should be noted that, after the storage test, the encapsulant gel was clearly solidified and appeared to increase in elasticity; a lot of fine wrinkles developed at the encapsulant surface as a result of the 3000-cycle TCT.

Conclusions

New packaging technology for SiC power modules operated in an extended junction temperature range (Tjmax > 200°C) should be developed in order to put into practical use a more compact and cost-effective SiC power conversion system. This paper described in detail some of the technical challenges involved in improving the reliability of three critical components in direct contact with SiC devices inside the power module: (1) die attachment system, (2) Al wire bonds and (3) encapsulation. A eutectic AuGe solder die attachment system on a SiN ceramic substrate with a Cu-Invar-Cu foil proved to be highly reliable to repeatedly large temperature excursions. Reliability tests revealed that lightly Ni-doped thick-Al wire bonds on SiC power devices withstood elevated temperatures and thermal cycle stress of more than 200°C for a long period of time without experiencing heel break or bond lift-off. A heat-resistant silicone-based encapsulant gel was newly developed in this work in cooperation with several venders. Two numerical targets, (I) 3000 hours for a storage test at 250°C and (II) 3000 cycles for thermal cycling between -40°C and 250°C, were completely achieved by using the new packaging techniques developed in this work.

Acknowledgments

This work was supported in part by the Green IT project directed by the New Energy and Industrial Technology Development Organization (NEDO) of Japan. The authors wish to thank Dr. Sumio Ashida (NEDO), Dr. Tomoyuki Yamada and Dr. Toshimi Wada (FUPET), Dr. Haruhito Mori and Dr. Masaki Nakano (Nissan Motor), and Dr. Hazime Shimizu and Dr. Hajime Okumura (AIST, Japan) for their encouragement, Ms. Ai Kawashima (FUPET) and Ms. Yukie Hirose (NRC) for their technical support, Dr. Yoshihiro Ishikawa (ADEKA, Japan), Dr. Makoto Ohara, Dr. Saneyuki Tanaka (Shin-Etsu Chemical, Japan), Dr. Shigeru Sasaki (Wacker Asahikasei Silicone, Japan), Dr. Yasuaki Futagami and Dr. Masayuki Onishi (Dow Corning Toray, Japan) for their provision of silicone gel prototypes, Mr. Takuya Kitamura and Mr. Kouji Hayakawa (Kyocera Corp., Japan) for their fabrication of the special SiN substrates and Dr. Kazuteru Takahashi (Nano Science Corp., Toshima, Japan) for his fruitful discussions concerning SEM-EDX.

References

1. J. Hornberger, A. B. Lostetter, K. J. Olejniczak, T. McNutt, S. Magan Lal and A. Mantooth, 2004 IEEE Aerospace Conference Proceedings (Big Sky, MT, USA), 2583 (2004).
2. P. Friedrichs, D. Stephani, *Microelectron. Eng.*, **83**, 181 (2006).
3. K. Hamada, *phys. stat. sol.* (b), **245**, 1223 (2008).
4. F. D. Barlow and A. Elshabini, in *High-Temperature High-Power Packaging Techniques for HEV Traction Applications* (ORNL/TM-2006/515), M. Olszewski, Program Manager, p. 1, (2006).
5. D. W. Palmer and R. Heckman, *IEEE Transact. Comp. Hybri. Manuf. Tech.*, **CHMT-1**, 333 (1978).
6. J. D. Scofield, J. N. Merrett, J. Richmond, A. Agarwal and S. Leslie, Proceedings of IMAPS HiTEC 2010 (Albuquerque, NM, May 2010), p. 289 (2010).
7. F. P. McCluskey, M. Dash, Z. Wang and D. Huff, *Microelectron. Reliab.*, **46**, 1910 (2006).
8. S. Tanimoto, K. Matsui, Y. Zushi, S. Sato, Y. Murakami and T. Yamada, Proceedings, The 18th Symposuim on Microjoining and Assembly Technology in Electronics (Yokohama, Japan, 2012), p. 107 (2012) (in Japanese and partially in English).
9. S. Tanisawa, S. Tanimoto, K. Watanabe, S. Sato, K. Matsui, Y. Murakami and K. Sasaki, Proceedings, The 19th Symposuim on Microjoining and Assembly Technology in Electronics (Yokohama, Japan), p. 105 (2013) (in Japanese and partially in English).
10. S. Tanimoto, K. Watanabe, H. Tanisawa, K. Matsui and S. Sato, Proceedings, The 27th Spring Meeting (Sendai, Japan), The Japan Institute of Electronics Packaging, p. 316 (2013) (in Japanese).

11. T. Schütze, H. Berg and M. Hierholzer, Proceedings, Industry Application Conference, the IAS Annual Meeting (St. Louis, MO, USA), **Vol. 2**, 1022 (1998).
12. P. Cova and F. Hantini, *Microelectron. Reliab.*, **38**, 1347 (1998).
13. K. Sommer, J. Göttert, G. Lefranc and R. Spanke, Proceedings, EPE 1997 (Trondheim, Norway) p. 1.112 (1997).
14. S. Tanimoto, K. Matsui, Y. Murakami, H. Yamaguchi and J. Okumura, Proceedings, IMAPS HiTEC 2010 (Albuquerque, NM), p. 32 (2010).
15. S. Tanimoto, K. Matsui, Y. Zushi, S. Sato, Y. Murakami, M. Takamori and T. Iseki, Proceedings, IMAPS HiTEC 2012 (Albuquerque, NM), p. 110 (2012).
16. S. Tanimoto, N. Hirama, I. Watanabe, H. Tanisawa, K. Matsui and S. Sato, Extended Abstracts, The 73rd JSAP Autumn Meeting, 2012 (Matsuyama, Ehime, Japan), p. 15-285 (2012) (in Japanese).
17. P. Alexandov, E. Wright, M. Pan, M. Weiner, L. Jiao and J. H. Zhao, *Solid-State Electron.*, **47**, 263 (2003).
18. D. R. Olsen and H. M. Berg, *IEEE Transact. Comp. Hybri. Manuf. Tech.*, **CHMT-2**, 257 (1979).
19. M. J. Palmer and R. W. Johnson, Proceedings, IMAPS HiTEC 2006 (Santa Fe, NM), p. 119 (2006).
20. S. Tanimoto, H. Tanisawa, K. Watanabe, K. Matsui and S. Sato, *Mater. Sci. Forum*, **740-742**, 2040 (2013).
21. A. Morozumi, K. Yamada, T. Miyasaka, S. Sumi and Y. Seki, *IEEE Transact. Industry Appl.*, **39**, 665 (2003).
22. P. Brusius, Proceedings, 1998 Fourth International High Temperature Electronics Conference (Albuquerque, NM), p. 151 (1998).
23. T. Matsunaga and Y. Uegai, Proceedings, 2006 Electronics System-integration Technology Conference (Dresden, Germany), 276 (2006).
24. Y. Yamada, Y. Takaku, Y. Yagi, I. Nakagawa, T. Atsumi, M. Shirai, I. Ohnuma and K. Ishida, *Microelectron. Reliab.*, **47**, 2147 (2007).
25. W. Wyrzykowski and N. W. Grabski, *Philosophical Magazine* A, **53**, 505 (1986).
26. P. A. Agyakwa, M. R. Corfield, L. Yang, J. F. Li, V. M. F. Marques and C. M. Johnson, Microelectron. Reliab., 51, 406 (2011).
27. F. P. McCluskey, R. Grzybowski and T. Podlesak (Eds.), in *High Temperature Elecronics* (CRC Press, New York), p. 149 (1996).
28. D. C. Katsis and Y. Zheng, Proceedings, 39th IEEE Power Electronics Specialists Conference, 2008 (Rhodes, Greece), p. 290 (2008).
29. T. Izumi, T. Hemmi, T. Hayashi and K. Asano, *Mater. Sci. Forum*, **740-742**, 669 (2013).
30. C. Leinenbach, F. Valenza, D. Giuranno, H. R. Elsener, S. Jin and R. Novakovic, J. Emectron. Mater., 40, 1533 (2011).

48

CHAPTER 2

SiC MOS POWER DEVICES

50

Channel Transport in 4H-SiC MOSFETs: A Brief Review

Sarit Dhar

Physics Department, Auburn University, Auburn, Alabama 36849, USA

This paper reviews the channel mobility problem in 4H-SiC MOSFETs. The importance of fundamental investigations on dielectric-SiC interfaces for next generation SiC power devices are highlighted and state-of-the-art interface passivation processes are discussed. Transport limiting mechanisms in current 4H-SiC MOSFETs are discussed and some key areas that need to be investigated for further improvement of this technology are identified. A necessary shift of focus from interface trap density reduction to other transport limiting mechanisms is motivated by discussing recent results.

INTRODUCTION

Low channel mobility in 4H-SiC power DMOSFETs is a critical weakness in SiC power electronics technology. The primary impact of this is higher conduction losses in the on-state [1]. The problem is mitigated by intelligent device design with short channels and tight cell layout which enables the superior performance of SiC DMOSFETs for blocking voltages ≥1200 V. Furthermore, the channel resistance decreases at higher temperatures and contributes significantly less to the device resistance compared to room temperature [1]. The remaining resistance comes from the bulk of the 4H-SiC, which is 10x lower than Si at any voltage rating owing to the wide band-gap of the material. Therefore, it might appear that further improvement of the channel mobility may not be a critical factor for next-generation devices. However, higher channel mobility is extremely desirable for the following reasons: (a) Further lowering of device on-resistance leading to smaller chip size and lower cost (b) More importantly, device operation at oxide fields lower than ~4 MV/cm which is typical for current generation devices. This leads to oxide instability and reliability concerns especially at higher temperature [2]. This can be

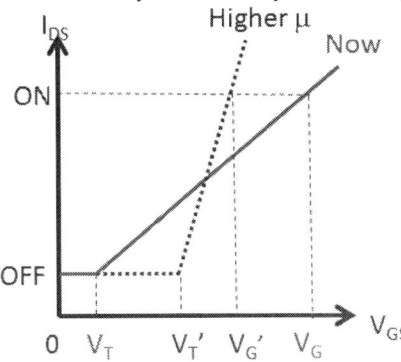

Figure 1. Schematic of transfer characteristics of MOSFETs with low and high mobility

understood by the simplistic schematic shown in Fig.1. (c) A threshold voltage (V_T) of about ~5 V is required to assure normally-off operation and adequate noise-immunity in high temperature and high noise environments. Current generation DMOSFETs suffer from a small V_T, about 2-3 V at 25°C, with further lowering at higher temperatures [1] making their reliability questionable. A higher V_T can be obtained by increasing the surface doping of the DMOS p-well. Unfortunately, this is not feasible as the channel mobility drops significantly at higher surface doping [2]. Therefore, in order to get an acceptable on-resistance with a higher V_T and lower operating oxide field in next-generation 4H-SiC DMOSFETs (as shown in Fig.1), a significant effort needs to be made towards improvement of channel mobility. This may also mitigate additional problems such as short-channel effects [3] and possibly relax cell designs which can also help in additional reliability and yield related problems. All these factors folded together could have a major impact on the device cost, which is currently one of the main obstacles for widespread adoption of this technology. The channel field-effect mobility is about 20 cm^2 V^{-1} s^{-1} at a gate oxide field of 4 MV cm^{-1} in commercial MOSFETs. Nitric oxide (NO) post-oxidation annealing [4] is currently the most established process to obtain this mobility along with adequate oxide robustness. While a few other processes have shown promise, none of them have the maturity or robustness to replace NO at this stage. Developments in this area have typically been using a trial and error approach, mainly due to the lack of detailed understanding of the electronic properties and chemical structure of dielectric/SiC interfaces at the nanoscale, as well as the mobility limiting mechanisms. In this paper, the critical importance of understanding these issues for further improvement of channel mobility will be highlighted. At the outset, the current understanding of the SiO$_2$/4H-SiC interface and the role of nitrogen will be reviewed. Next, new results that highlight the limitation of N in further improving the mobility will be discussed. Following this, promising results obtained recently using phosphosilicate glass as a gate dielectric will be detailed. Current understanding of channel transport will be reviewed, and the need for a breakthrough in understanding of the channel transport will be stressed.

THE SiO$_2$-SiC INTERFACE

The SiO$_2$-SiC interface lies at the heart of the channel mobility problem. Historically, enhancement mode 4H-SiC MOSFETs were difficult to turn on and suffered from single digit mobility in gate oxides grown thermally at high temperature (>1100°C). This behavior was attributed by numerous works [6] to a very large number of interface traps (>10^{13} cm^{-2}) at the SiO$_2$/SiC interface with energies in the SiC band gap. In 1999, Schorner et al. [7] compared channel mobility and D_{it} between various polytypes with identically processed gate oxides and demonstrated a clear correlation between D_{it} and mobility. They suggested that the 4H-polytype (largest band-gap ~3.3 eV) exhibits the highest channel resistance due to an extremely high D_{it} near the conduction band-edge (E_c). Due to smaller band-gaps, these states lie above the E_c of other polytypes and as a consequence do not affect the transport to the same degree. They

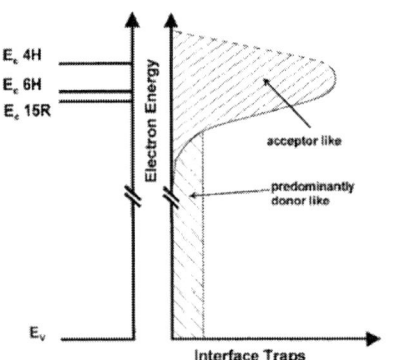

Figure 2. Interface states model postulated Ref. [7]

speculated a large D_{it} peak near the conduction band of 4H-SiC as shown in Fig. 2. The existence of such a peak has been suggested by low temperature capacitance voltage measurements [8] but the chemical nature of these interface traps have not been identified conclusively. The defect models that exist can be broadly classified into (i) Interfacial carbon related (ii) Near-interfacial oxide related (iii) SiC surface related. Most of the early work on defect identification focused on 'excess' carbon [9] at the interface. While oxidation of SiC results in a carbon free oxide, (with the bulk properties of the oxide similar to that grown on Si) the possibility of a small amount of remnant C at the interface cannot be precluded. Various C- dangling bond defects [10], as well as C clusters were theoretically predicted to be the root of the problem. Subsequent theoretical studies indicate that most of the C-related defects lead to traps in the lower half of the band-gap (hole traps) or in the large mid-gap region [11]. However, there are some unique dangling bond configurations at the SiC sub-surface or dimers in the oxide that could potentially lead to traps near E_c [10, 11]. Electron paramagnetic resonance (EPR) studies, which have been very effective in the case of Si for the detection of interfacial defects have not proven as effective for SiC. EPR studies on porous SiC indicate a small density of C dangling bonds [12], but it is difficult to correlate these results to devices due to the non-standard nature of the studied material. It should be noted that the trapping problem is not just confined to thermal oxides but is equally severe in deposited oxides. Shen and Pantelides [13] have recently postulated that C interstitials injected into the bulk of the SiC during oxidation to form di-interstitials near the surface could also be a transport limitation factor in the channel. While the role of C remains unclear, the limit of maximum 'excess C' has been very carefully quantified by medium energy ion scattering to only 1.8×10^{14} cm^{-2} [14].While this number is not small with respect to defects, it does question the role of C to some extent. Afanase'ev et al. [9] used internal photoemission studies to argue that that the large D_{it} near E_c is associated with an intrinsic oxide defect (most likely an O vacancy) that is present in oxides grown on silicon as well. The presence of such near-interfacial traps have been confirmed by detailed thermally stimulated dielectric relaxation current [15] and low temperature capacitance-voltage and deep level transient spectroscopy studies [16]. A Si-vacancy defect in the channel region has also been identified by magnetic resonance, but its role in limiting transport is not quite clear [17]. More recently, transmission electron microscopy studies have indicated the existence of a transition layer on both sides of the interface [18, 19]. While the chemical nature of this transition layer is yet to be understood in detail, it does point towards a non-abrupt interface that could suffer from local fluctuations of surface potential and surface roughness scattering. While 'as-oxidized' surfaces are not particularly useful for device development, it needs to be highlighted that they present a unique platform for the study of MOS physics and interface science. SiC is the only semiconducting material other than Si that thermally oxidizes to SiO_2, the most common insulator for electronics. The uniqueness of the SiC system lies in the fact that the various polytypes with different band-gaps provide a unique knob to study MOS interface physics – one where the semiconductor band gap can be tuned without changing the interfacial chemistry.

INTERFACIAL NITRIDATION

Incorporation of nitrogen atoms at the SiO_2/SiC interface, particularly using NO post-oxidation annealing is currently the most established method of obtaining acceptable channel mobility in SiC MOSFETs [5]. This method consists of annealing thermal or deposited oxides in NO at high temperatures (1175 °C - 1300 °C) following oxidation. The process leads to accumulation of about ~6 x 10^{14} cm^{-2} of N atoms with an extremely sharp profile (~1 nm width) at the interface [5] without any N incorporation in the bulk of oxide. Figure 3 shows that the N is predominantly bonded to Si as indicated by x-ray photoelectron spectroscopy studies (XPS) [20]. It is interesting to note that the N is resistant to HF etching- if the oxide is etched , ~90% of the N still remains on the surface [20], which suggests that some of the N may be buried in the very near surface region of 4H-SiC. Most importantly, the amount of N incorporated at the interface scales with the reduction of D_{it} near E_c as well as the maximum field-effect mobility obtained in MOSFETs [21]. To understand that improvement of channel conductance associated with

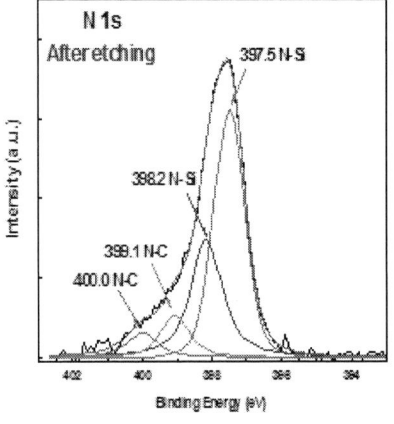

Figure 3. Typical N 1s spectrum from XPS[20].

Figure 4. Free electron concentration vs. gate voltage as a function of temperature in NO annealed devices [22].

NO annealing, the following equation for channel conductance σ_{ch} needs to be considered, where n_s and μ_{ch} are the channel free electron concentration and conductivity (or Hall) mobility, respectively.

$$\sigma_{ch} = qn_s\mu_{ch} \qquad\qquad \text{Eq. 1}$$

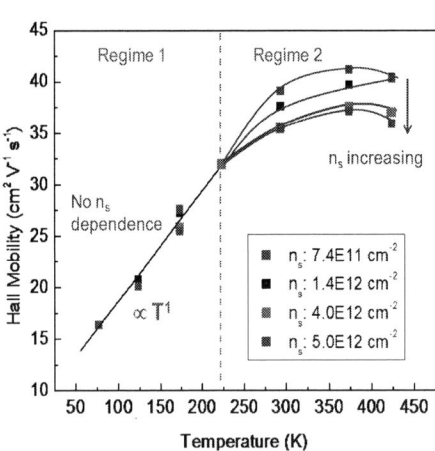

Figure 5. Temperature dependence of μ_H in NO annealed MOSFETs [22].

Typically, field-effect mobility μ_{fe}, which is an important device figure of merit (related to transconductance) is loosely referred to as channel mobility in the literature. However, to understand channel transport in more detail, it needs to be pointed out that μ_{fe} is a convoluted quantity which is a function of n_s and μ_{ch} as well as their dependence on gate voltage. Figure 4 shows independent measurements of n_s from Hall measurements on NO annealed MOS Hall bars with lightly doped p-wells (~$6x10^{15}$ cm^{-3}) as a function of gate voltage at various temperatures [22]. It can be seen from the figure that at any temperature, the slope of the n_s-V_G curve is equal to

C_{ox}/q above a certain V_G, indicating a clear strong inversion regime. At lower temperatures, higher interfacial trapped negative charge causes the n_s-V_G curves to shift to the right. Comparison with the ideal charge sheet model [23] reveal that at 293 K, for V_G= 15 V (E_{ox}~ 3 MV cm^{-1}), more than 80 % of the ideal free carrier concentration is available for conduction. On the other hand, at 77 K, only about 55% is available. This is in stark contrast to the results Hall results reported on devices without nitridation [5], where no strong inversion regime is observed due to severe electron trapping, even at high temperature. On the other hand, the maximum Hall mobility is about ~50 cm^2 V^{-1} s^{-1} which is only about 10% of the 4H-SiC bulk mobility indicating that while NO annealing drastically improves trapping, it does not improve μ_{ch} significantly. These results conclusively indicate that the improvement of conductivity via nitridation is primarily associated with the increase of the free carriers, but μ_{ch} also requires improvement. The Hall mobility μch as a function of n_s and temperature, as shown in Fig. 5., reveals two regimes. For 77 K < T < 223 K, μ_H shows a linear T^1 dependence, independent of n_s, in

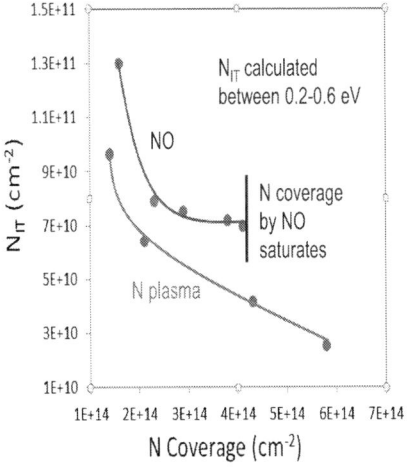
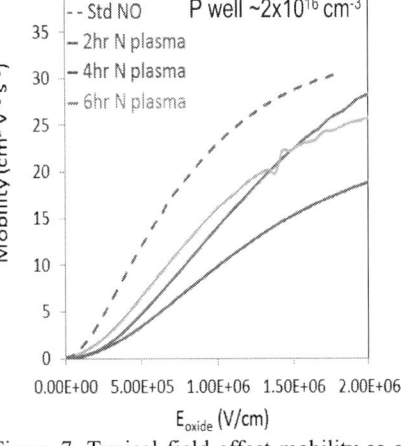

Figure 6. Total interface traps between 0.2 and 0.6 eV from Ec versus interfacial nitrogen coverage for NO annealing [21] and the N plasma process [28].

Figure 7. Typical field-effect mobility as a function of oxide field for NO annealing compared to various N plasma annealing times

stark contrast with Si MOSFETs, where mobility increases with decreasing temperature due to phonon scattering. Above 223 K, the dependence becomes sublinear up to ~ 373 K with an inverse dependence on n_s. The T^1 dependence is a signature of a Coulomb scattering [24] from interfaces charges and a large amount of positive fixed charge (~5x10^{11} cm^{-2}) tyically observed in nitrided interfaces. The behavuor at the lower temperatures can also be interpreted as a thermally activated transport due to hopping between localized states at the band-tail or thermal excitation to extended states [25], consistent with transport in through a disordred transition layer at the interface. At room temperature and above, the mobility decreases with increasing n_s indicating that surface roughness scattering [26] starts to play a key role, especially at higher temperatures and higher n_s (or oxide field).

LIMIT TO SCALING BETWEEN D_{it} AND μ USING NITROGEN

The monotonic decrease of D_{it} with higher interfacial nitrogen, accompanied by higher maximum μ_{fe} [21] raises an interesting question about the limitations of nitrogen. Interfacial N uptake via NO annealing at 1175°C (typical anneal temperature) saturates at about $\sim 5 \times 10^{14}$ cm^{-2} of N atoms [20]. Longer annealing times result in competition between N uptake and additional oxide growth at the interface which is believed to cause the saturation [27]. In order to investigate the effects of higher N coverage at the interface, an 'oxidation free' plasma nitridation process was investigated [28]. In this process, a thin ~150 A thermal oxide was subjected to annealing at 1160°C in a down-stream N$_2$ plasma system where N atoms are generated by a microwave source. Following this, a ~400 A LPCVD oxide was deposited to form a composite stack. Interestingly, this process can result in higher N coverage (measured by XPS) at the interface compared to NO annealing as shown in Fig.6. Plasma annealing for 6h results in ~50% higher N coverage compared to the N saturation amount by NO. These oxides also exhibit ~4x lower interface traps close to the conduction band-edge as measured on MOS capacitors. However, in spite of the improvement in D_{it}, μ_{fe} was not improved compared to NO annealing as shown in Fig. 7. These results indicate that there is a limit to improving mobility by solely decreasing the D_{it}, at least using the N plasma process. A possible mobility inhibitor could be higher Coulomb scattering associated with increased fixed charge formed in the oxide due to the plasma process. This mechanism will be discussed more later.

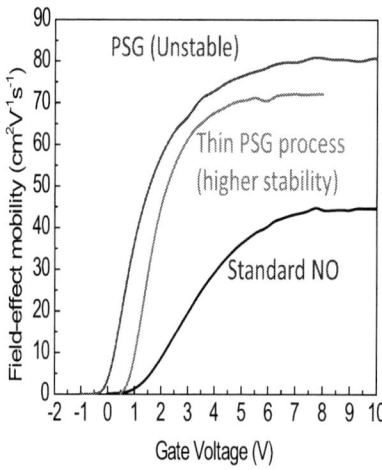

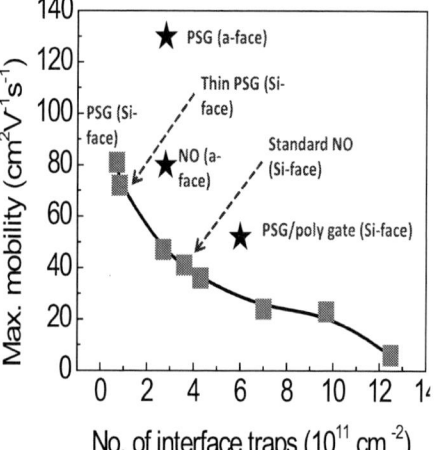

Figure 8. Field-effect mobility (MOSFETs with 6E15 cm-3 p-well doping) for different gate processes: NO, Thick PSG and Thin PSG. [38]

Figure 9. Maximum field-effect mobility versus Nit between 0.2-0.6 eV from Ec. The trendline is used to show the qualitative scaling as reported in [21]. The star symbols indicate recent results as discussed in the text that do not exhibit the scaling.

PHOSPHOSILICATE GLASS (PSG) PROCESS

Over the past decade, the main focus of the community has been on nitridation based processing methods, which seem to have reached their limit. Non-nitrogen processes such as sodium enhanced oxidation [29], post-metallization hydrogen annealing following NO [30], device fabrication on alternate crystal faces [30-32] and Al_2O_3 based gate stacks [33, 34] have demonstrated high channel mobility. Some of these methods are not attractive from a commercial point of view while others need more maturity. While the mobile ion contaminated sodium process obviously cannot be commercialized, it sets a minimum upper limit of μ_{fe} to greater than 200 cm^2 V^{-1} s^{-1}. In 2010, Okamoto et al. [35] reported that doping SiO_2 with 5-10% P results in a significant enhancement of the channel mobility to around ~100 cm^2 V^{-1}s^{-1} on lightly doped p-epitaxial layers. In this work, a thermal oxide was subjected to annealing at 1000°C in a flowing O_2/N_2 mixture bubbled through a liquid $POCl_3$ cell. Subsequently, similar results were obtained using SiP_2O_7 solid diffusion sources at 1000°C [36]. Unfortunately, while the μ_{fe} enhancement is substantial, converting the oxide to Phospho-Silicate-Glass (PSG) results in severe device instability. The instability is primarily associated with large amounts of polarization charge in the PSG [37]. Recently, a modified process that exclusively incorporates P near the oxide-SiC interface has been reported [38]. In this work, a composite stack consisting

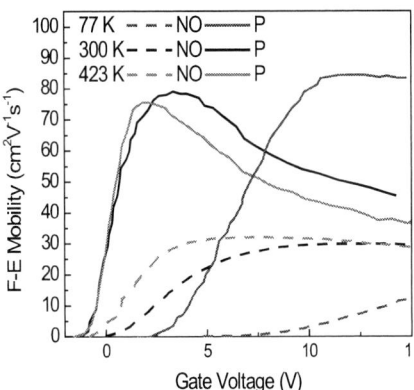

of a ~10 nm thin PSG layer and a ~40 nm LPCVD oxide film resulted in an impressive field-effect mobility of ~70 cm^2 V^{-1} s^{-1} (Fig. 8), in conjunction with significantly higher stability compared to thick PSG. According to [37], in such a composite stack, the instability associated with the polarization charge in PSG should scale linearly with the ratio of the thicknesses of the PSG and the undoped oxide. As evident from Fig. 8, the stability improvement comes at an expense of a 20% lower mobility compared to PSG, but the mobility is still 2x higher than state-of-the-art NO processes, which is significant. The maximum μ_{fe}, (which occurs near threshold, where trapping and Coulomb scattering effects are expected to dominate)

Figure 10. Temperature dependence of field-effect mobility for PSG and NO passivated 4H-SiC MOSFETs

from the PSG processes qualitatively scales with the total number of traps (N_{it}) near E_c. This is shown in Fig. 9, where $\mu_{fe,max}$ is plotted as a function of N_{it}. The other points on the trend line have been obtained from [21], where the N_{it} was varied using various NO annealing times/ interface N content. This is consistent with previous work where it was suggested that the primary effect is associated with a more effective trap passivation mechanism by P [39]. Recent results (star symbol labeled PSG a-face in Fig. 9) obtained using PSG in MOSFETs fabricated on the (11-20) a-face of 4H-SiC, strongly suggest a parallel counter-doping mechanism at play [40]. In this experiment, it was shown that for the a-face, NO annealing and PSG show the similar D_{it},

but the maximum μ_{fe} for the PSG devices (~130 cm^2 V^{-1} s^{-1}) is about 60% higher than the NO devices (~80 cm^{2}V^{-1}s^{-1}), consistent with 'non-scaling' observed in the N plasma results described above. In the case of the a-face, this effect was attributed to n-type doping (counter-doping) of the MOSFET p-well by a thin surface diffused layer (~1 nm) of P at the surface of 4H-SiC. The counter-doping converts the lightly doped p-type surface into n-type in a way that the surface n-type channel is fully depleted at zero gate bias and the device remains off. Once the gate voltage exceeds V_T, a surface accumulation channel is formed. A higher mobility is expected in such a channel due to the lowering of surface electric field (at the same electron concentration) leading to a reduction in surface roughness scattering. X-ray photo-electron spectroscopy; in conjunction with medium energy ion scattering confirm the presence of about 1.6-1.8 x 10^{14} atoms/cm^2 of P on the 4H-SiC surface after etching the PSG film (similar to N) [40]. Therefore, it is very likely that a small amount of P gets incorporated in a very thin layer of SiC (<2 nm, with the depth limit set by the XPS electron escape length) at the interface, of which a small percentage gets activated at the processing temperatures, converting the doping type from p to n. This is consistent with [41] where it was demonstrated that oxidation enhances P diffusion into SiC surfaces. The utilization of counter-doped or buried channels to improve channel transport 4H-SiC has been reported previously [42-44]. In these reports, the n-type doping was done via ion implantation to form buried channel depths of ≥ 1000 A and total dose $\leq 5 \times 10^{12}$ cm^{-3}. However, the threshold voltage obtained in such a process is intricately related to the oxidation method used, and the residual implantation damage and V_T instability is a concern if interface traps are not simultaneously passivated. The uniqueness of the PSG approach is that one can get an extremely thin surface counter-doped channel accompanied by passivation of near-interfacial traps. While counter-doping during NO processing cannot been ruled out, the higher effectiveness of PSG could be due to the higher dopant activation rate of P over N for n-type doping in the 1000-1200°C range [45]. A signature of D_{it} limited mobility is the positive mobility-temperature coefficient, i.e. mobility increasing with temperature as discussed earlier. Interestingly, the PSG MOSFETs exhibit a 'Si-like' negative temperature coefficient, as shown in Fig. 10, again indicating if D_{it} is below a certain value, it may no longer be the dominant limiting factor. This is also consistent with another point on Fig.9 (star symbol labeled PSG poly-gate), where a non-optimized polysilicon gate contact process was used in conjunction with PSG, resulting in higher D_{it} compared to standard NO, but with higher μ_{fe} as well. In summary, the available results strongly suggest that for PSG-based processing, both D_{it} reduction and surface doping play a role, but once D_{it} is reduced to a certain level, counter-doping dominates by mitigating surface roughness scattering.

CONCLUDING REMARKS

To date, the main focus of the SiC MOS community has been towards reduction of D_{it}. Recent results reviewed here strongly suggest that above room temperature, trapping is no longer be the dominating factor limiting transport for reasonably low D_{it} processes, but rather it is the electron mobility. Therefore, a change of focus is necessary within the community to identify other transport limiting mechanisms. Firstly, it should be noted that reduction of D_{it} does not necessarily guarantee a reduction of Coulomb scattering from of fixed charges. For wide band-gap materials like SiC, it is not an easy task to separate fixed charges from interface traps, especially if fixed charges of both polarities exist. These fixed charges not only affect mobility but also dictate V_T control. Results discussed above clearly indicate that surface roughness scattering is playing an increasingly important role in low D_{it} devices. This mechanism in the case of Si, is associated with the scattering of electrons at surface imperfections that results in fluctuations of the surface potential [24]. This leads to a decrease of mobility at higher transverse fields where the electrons are pushed towards the surface and dictates the universal mobility behavior in Si MOSFETs [46]. This is particularly relevant for SiC DMOSFETs with heavier doped p-wells, where the electric field in the semiconductor is high. Also, it is not clear from the literature, whether trapping increases as the p-well doping increases. Any process (such as PSG) that displays high mobility in MOSFETs for lightly p-well epilayers needs to be evaluated for higher doped p-wells, where surface roughness scattering is bound to play a larger role. Any correlations of counter-doping with surface roughness scattering and electron trapping (if any) needs to be resolved. While surface roughness scattering has been addressed by modeling [24], it is not clear if it related to the physical roughness of the surface. Rather it is could be the 'electrochemical' roughness of the surface that is more relevant. In this regard, properties of nanoscale transition layers, especially on the SiC side of the interface needs to be understood in greater detail. Finally, it is important to identify whether a Si-like 'universal mobility curve' exists for 4H-SiC MOSFETs.

Acknowledgments

The author would like to acknowledge Y. Sharma, A.C. Ahyi, A. Modic, T.-I. Smith, J. R. Williams (Auburn University), L. C. Feldman, Y. Xu and G. Liu (Rutgers University), A. Lelis (US Army Research Laboratory), A. K. Agarwal (US DOE), X. Shen and S. T. Pantelides (Vanderbilt University). Thanks to U.S. Army Research Laboratory (W911NF-07-2-0046), the U.S. National Science Foundation (MR-0907385) and the II-VI Foundation Block-Gift Program for providing support.

References

1. L. Cheng et al., *J. Elec. Mat.*, **41**, 910 (2012).
2. R. Green et al., *Mat. Sci. Forum*, **717-720**, 1085 (2012).
3. S.-H Ryu et al., *Mat. Sci. Forum*, **615-617**, 1085 (2009).
4. M. Noborio et al., *IEEE Trans. Elec. Dev.*,**52**, 1954 (2005)
5. S. Dhar et al., MRS Bull., **30**, 208 (2005) , references therein
6. V. V. Afanas'ev et al. *J. Phys.: Condens. Matter*, **16**, S1839 (2004), refs. therein
7. R. Schorner et al., *IEEE Elec. Dev. Lett.*, **20**, 241 (1999).
8. S. Dhar et al., *Appl. Phys. Lett.*,**92**, 102112 (2008).
9. V. V. Afanas'ev et al., *Phys. Status Solidi A*, **162**, 321 (1997), refs. therein
10. S. T. Pantelides et al., *Mat. Sci. Forum*, **527-529**, 935 (2005), refs. therein
11. P. Deak' et al., *J. Phys.D: Appl. Phys.*, **40**, 6242 (2007), refs. therein
12. H. J. Von Bardeleben, *Mat. Sci. Forum*, **457-460**, 1457 (2004).
13. X. Shen and S. T. Pantelides, *Appl. Phys. Lett.*,**98**, 053507 (2011).
14. X. Zhu et al., *Appl. Phys. Lett.*,**97**, 071908 (2010).
15. T. E. Rudenko et al., *Solid-State Elec.*,**49**, 545 (2005).
16. A. F. Basile et al., *J. Appl. Phys.*, **109**, 064514 (2011).
17. M. S. Dautrich et al., *Appl. Phys. Lett.*,**89**, 223502 (2006).
18. T. L. Biggerstaff et al., *Appl. Phys. Lett.*, **95,** 032108 (2009).
19. J. A. Taillon, *J. Appl. Phys.*, **113**, 044517 (2013).
20. Y. Xu , Rutgers University dissertation, under preparation (2013)
21. J. Rozen et al., *IEEE Elec. Dev. Lett.*, **58**, 3808 (2011).
22. S. Dhar et al., *Mat. Sci. Forum*, **717-720**, 712 (2012).
23. E. Arnold and D. Alok, *IEEE Trans. Elec Dev.*, **48**, 1870 (2001).
24. S. Potbhare et al., *IEEE Trans. Elec. Dev.*, **55**, 2029 (2008), refs therein
25. T. Ouisse, *Phys. Stat. Sol. (a)*, **162**, 339 (1997)
26. V. Tilak et al., *IEEE Trans. Elec. Dev.*, **54**, 2823 (2007).

27. S. Dhar et al., *J. Appl. Phys.,* **97**, 074902 (2011).
28. A. Modic et al., *J. Elec. Mat.,*submitted July 2013.
29. G. Gudjonsson et al., *Mat. Sci. Forum,* **483-485**, 837 (2005)
30. S. Dhar et al., *Mat. Sci. Forum,* **527-4529**, 949 (2006).
31. J. Senzaki et al, *IEEE Elec. Dev. Lett.,* 23, 13 (2002).
32. K. Fukuda et al, *Appl. Phys. Lett.,* **84**, 2088(2004).
33. T. Hatayama et al., *IEEE Elec. Dev. Lett,* 55, 2041 (2008).
34. D. Lichtenwalner et al., *Appl. Phys. Lett.,***95**, 152113 (2009).
35. D. Okamoto et al., *IEEE Elec. Dev. Lett.* **31**, 710 (2010)
36. Y.K. Sharma et al. *Solid-State Elec.* **68**, 103 (2012).
37. E.H. Snow and B.E. Deal, *J. of the Electrochem.Soc.* **113**, 263 (1966).
38. Y.K. Sharma et al., *IEEE Elec. Dev. Lett.,* **34**, 175 (2013)
39. D. Okamoto et al. *Appl. Phys. Lett.,* **96**, 203508 (3 pp.) (2010)
40. G. Liu, et al., *IEEE Elec. Dev. Lett.,* **34**, 181 (2013)
41. C. C. Tin et al., *Thin Solid Films,* **518**, 118 (2010).
42. S. Harada et al., *IEEE Elec. Dev. Lett.,* **22**, 272 (2001).
43. S. Harada et al., *IEEE Elec. Dev. Lett.,* **25**, 292 (2004).
44. S. Dhar et al., *IEEE Trans. Elec. Dev.* **57**, 1195 (2010).
45. M. A. Capano et al., *J. Elec. Mat.,* **29**, 210, (2000).
46. S. Takagi et al., *IEEE Trans. Elec. Dev.* **41**, 2357 (1994).

ECS Transactions, 58 (4) 61-70 (2013)
10.1149/05804.0061ecst ©The Electrochemical Society

Properties of Al-SiO$_2$-SiC(3C) Structures with Thermally Grown and PECVD Deposited SiO$_2$ Layers

H. M. Przewlocki[a], T. Gutt[a], K. Piskorski[a], P. Borowicz[a], M. Bakowski[b]

[a] Institute of Electron Technology, Al. Lotnikow 32/46, Warsaw, 02-668, Poland
[b] Acreo AB, Electrum 236, Kista SE-164 40, Sweden

The 3C-SiC silicon carbide is a promising substrate material for MOS transistors. Parameters of MOS transistors are strongly dependent on the processing of the semiconductor-dielectric system. We study and compare the crucial features of the SiC(3C)-SiO$_2$ systems with thermally grown and PECVD deposited SiO$_2$ layers. For both types of systems, using the Al-SiO$_2$-SiC(3C) capacitors as test structures, we determine and compare the following properties: The leakage currents and breakdown voltages of SiO$_2$ layers, densities and distributions in energy of interface traps, band diagrams of the SiC(3C)-SiO$_2$ systems, as well as structural properties and mechanical stresses in both types of SiO$_2$ layers. Characterization of the above mentioned features is done using electrical, photoelectric and optical methods, including micro-Raman spectroscopy. The characterization results lead us to the conclusion, that PECVD deposition of SiO$_2$ followed by wet oxygen annealing is more advantageous for MOSFET fabrication than the thermal growth of the SiO$_2$ layer.

1. Introduction

The expected application of 3C silicon carbide as a substrate material for high and medium power MOS transistors (1,2) requires determination of properties of the various combinations of materials constituting MOS structures with 3C-SiC substrates. In a recent paper (3) we analyzed and compared the properties of MOS capacitors on 3C-SiC substrates with different gate materials. In this report, using the results of measurements made on the same batch of MOS structures that was analyzed in (3), we study and compare the properties of Al-SiO$_2$-SiC(3C) structures with SiO$_2$ layers grown by thermal oxidation and deposited by plasma enhanced chemical vapor deposition (PECVD).
In particular, the following properties of both types of structures are analyzed and compared:
-   Leakage currents and breakdown voltages, as measured by electrical methods;
-   Distributions of interface traps, determined by electrical measurements;
-   Band diagrams of structures with both types of SiO$_2$ layers, determined by a combination of photoelectric and optical methods;
-   Structural properties and mechanical stresses in both types of SiO$_2$ layers, compared by micro-Raman spectroscopy.

2. Preparation of samples

MOS capacitors were produced by Acreo AB on two 3" wafers of n-type (001) 3C-SiC. A 12 μm thick n-type epitaxial layer with nitrogen doping of $7 \cdot 10^{15}$ cm^{-3} was grown at Acreo on top of the 200 μm thick substrates from Hoya, with the doping density of $5 \cdot 10^{18}$ cm^{-3}. One of the wafers was thermally oxidized in wet oxygen for one hour, at a temperature of T = 1150°C, while the second wafer had the SiO_2 layer deposited by PECVD at T = 300°C. This second wafer was further annealed for 3 hours in wet oxygen at a temperature of T = 950°C (1,2). The oxide thickness on both wafers was ca. 60 nm. The wafers were divided into quarters and different metals (Al, Au, Ni), were deposited on each quarter by ion beam sputtering, as well as polysilicon which was deposited by chemical vapor deposition (CVD). Circular gates of 0.7, 0.6, 0.5, 0.4, 0.3 and 0.2 mm diameters were formed by lift-off in case of metal gates and by etching in case of polysilicon. Top view of one segment of the so obtained MOS capacitor test structures is shown in Fig.1a with the diameters of the Al gates given in millimeters and the cross section of the resulting gate stack, connected to the measurement circuit is shown in Fig.1b.

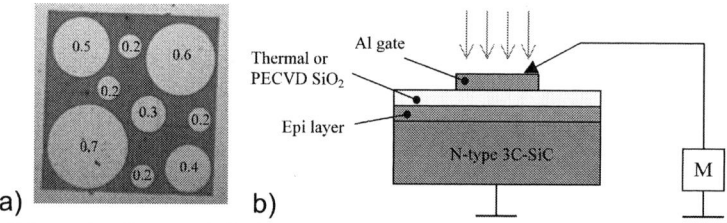

Fig.1. Schematic picture of a) Al gate contacts shown with their diameters given in millimeters in one module of the test structure and b) cross section of the gate stack connected to the measurement circuit.

In this article only the properties of Al-SiO_2-SiC(3C) structures, with Al gate thickness of t_{Al} = 25 nm and with SiO_2 layers deposited by thermal oxidation and by PECVD, are considered. Properties of structures with different metal gates were discussed in (3).

3. Oxide layer leakage currents and breakdown voltages

The leakage currents and breakdown voltages of dielectric layers are of crucial importance for the quality of advanced MOS devices. In this investigation a comparison was made between leakage current density, J, vs. gate voltage, V_G, characteristics of MOS capacitors with thermally grown SiO_2 layers on wafer no. 1 and capacitors with PECVD deposited SiO_2 layers on wafer no. 2. The J vs. V_G characteristics were taken for structures on both wafers, in the range of V_G = 0…42 V, making use of the measurement equipment comprising among others the Agilent B1500 semiconductor measurement and analysis system and the Cascade Summit 12000B-AP semiautomatic probe station. Characteristics taken for 14 capacitors of 0.2 mm diameter on each wafer are shown in Fig.2. It is shown in Fig.2 that capacitors with SiO_2 layers deposited by PECVD demonstrate better uniformity of the J vs. V_G characteristics than capacitors with oxides grown by thermal oxidation.

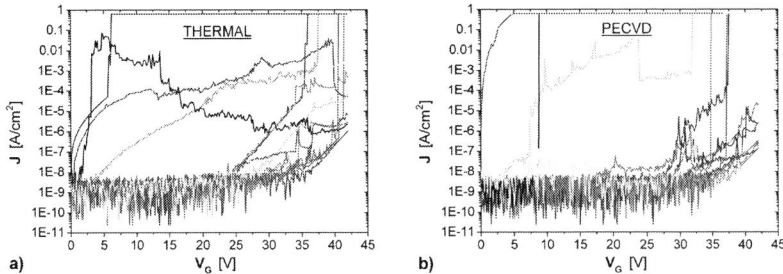

Fig.2. Current density J vs. gate voltage V_G characteristics of MOS capacitors with SiO_2 layers obtained by a) thermal oxidation, b) PECVD deposition.

To compare breakdown voltages of both kinds of structures it was arbitrarily assumed that the current density of $J = 10^{-3}$ A/cm^2 is the breakdown condition. The resulting distributions of breakdown voltages of both kinds of SiO_2 layers are shown by the histogram in Fig.3. This histogram clearly demonstrates that in case of SiO_2 layers deposited by PECVD the breakdown voltage is as good or even better than in case of thermal oxide layers.

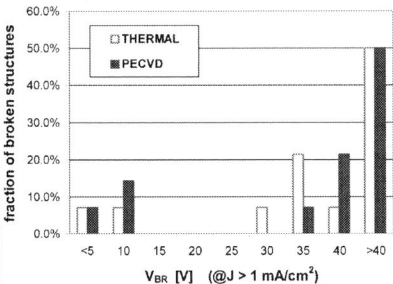

Fig.3. Histogram of the breakdown voltages of MOS capacitors with SiO_2 layers fabricated by thermal oxidation and by PECVD deposition.

4. Density of traps and their distributions in energy

The density and distribution in energy of the interface traps are strongly dependent on the technology of the semiconductor-dielectric system fabrication. Hence, it was important to compare the distributions of traps in MOS capacitors with thermally grown and PECVD deposited SiO_2 layers. The energy distributions of traps in both types of structures were measured by the conductance method. The C-V_G characteristics and the conductance spectra of MOS capacitors were measured at room temperature using Agilent 4294A precision impedance meter. The C-V_G data were used to determine doping level and ϕ_S vs. V_G relationship. The G_p/ω spectra were measured in the frequency interval from 100 Hz to 1 MHz and at the gate voltages from weak accumulation to the onset of inversion. The same spectra were used in the classic conductance method and in the MPAS (Multiple Parameter Admittance Spectroscopy) analysis.

The distribution of interface traps density D_{it} in the function of the distance in energy of the trap level E_t from the conduction band bottom E_C, $D_{it} = f(E_C–E_t)$, is shown in Fig.4.

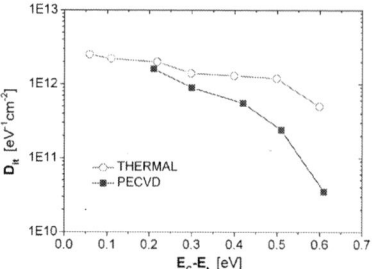

Fig.4. Example of density of traps D_{it} vs. energy depth E_C-E_t characteristic in Al-gate samples: circles – THERMAL oxide, squares – PECVD oxide.

As shown in Fig. 4, density of traps in the thermal oxide sample is higher than D_{it} in the PECVD sample. This difference is even more pronounced at deeper energy levels, where D_{it} of the PECVD sample falls below 10^{11} $eV^{-1}cm^{-2}$ and is lower by an order of magnitude than the D_{it} of the thermal oxide sample at the same energy depth.

The energy distribution of trap capture cross-sections can be inferred from the MPAS maps presented in Fig. 5. The maps of $log(G_p/\omega)$ in ϕ_S and $log(\omega^{-1})$ coordinates demonstrate two peaks of conductance linked together by a straight ridge (6). As it was demonstrated in (7) the slope of the ridge is equal to $-q\phi_S/k_BT$, where ϕ_S is band bending and k_BT/q is thermal potential, provided the capture cross section σ_n is constant. Drawing parallel lines having that slope for different constant σ_n one can easily estimate the value of the capture cross-section using the MPAS maps. However, the G_p/ω peaks in Fig. 5a for Al-gate sample lie on a line having a different slope than that of a constant σ_n, as can be noticed comparing it with the MPAS map determined for the Au-gate sample, shown in Fig. 5b. It means that σ_n is not constant in energy but it is rising with increasing E_t. Anyway its approximate value can be read as $10^{-14} \div 10^{-13}$ cm^2, which is slightly higher than for the P_b centres in Si:SiO$_2$ systems.

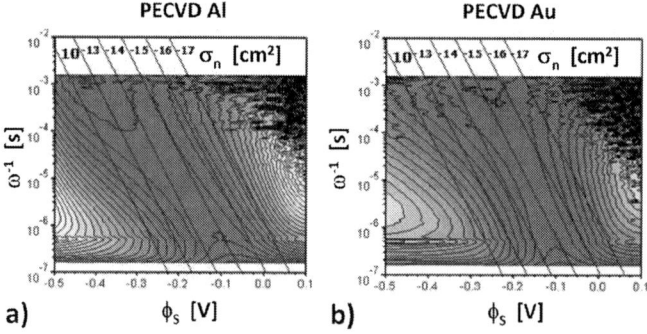

Fig.5. MOSC conductance distribution maps $G_p/\omega = f(\phi_S, \omega^{-1})$ of PECVD samples: a) Al-gate, b) Au-gate.

As shown in Fig. 5, the conductance ridge resulting from the interface traps switching activity can be used to evaluate trap capture cross-sections. The constant capture cross-section value of 10^{-15} cm^2 in PECVD Au-gate sample in Fig. 5b can be compared with the trap capture cross-section in the PECVD Al-gate sample in Fig. 5a, increasing from 10^{-15} cm^2 at $\phi_S = -0.05$ V to 10^{-14} cm^2 at $\phi_S = -0.25$ V, demonstrating the dependence of σ_n on energy.

5. Determination of band diagrams

The band diagrams of structures with both types of SiO$_2$ layers were determined by a combination of photoelectric and optical measurements. Photoelectric measurements were made in a measurement circuit schematically shown in Fig.1b. In this circuit photocurrents generated by the light beam of variable wavelength, λ, illuminating the device under test (DUT), from the gate side are measured. The origin of the photocurrent depends on the polarity of the gate bias supplied by the measurement system. In case of positive gate bias supplied by the system the photocurrent consists of electrons photoemitted from the substrate, over the potential barrier, E_{BS}, at the substrate-SiO$_2$ interface, drifting toward the gate, while in case of negative gate bias the photocurrent consists of electrons photoemitted from the gate, over the potential barrier, E_{BG}, at the gate-SiO$_2$ interface, drifting toward the substrate.

Due to the interference of light in the SiO$_2$ layer, the light power absorbed by the electron emitter (substrate or gate) changes with changing the wavelength of light illuminating the DUT. Hence, the dependence of light power absorbed by the emitter on the wavelength of light has to be found. This is done by the optical methods, using the ellipsometric determination of the gate and SiO$_2$ layer thicknesses, as well as, of the refractive, n, and extinction, k, coefficients of materials constituting the MOS structure, in function of the photon energy, hv. Once the thicknesses of the layers and the n(hv) and k(hv) characteristics of the materials of gate, dielectric and substrate are known, the dependence of the fractions of light power reflected from the structure, R(hv), transmitted through the gate, T(hv), and absorbed by the substrate, as well as the fraction absorbed by the gate, A(hv), can be calculated by the methods described e.g. in (8,9). The example of such characteristics, called RTA characteristics, determined for the Al-SiO$_2$-SiC(3C) structures is shown in Fig.6 (3).

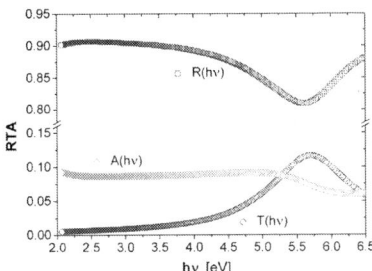

Fig.6. RTA characteristics determined for the Al-SiO$_2$-SiC(3C) structure with the Al gate thickness of $t_{Al} = 25$ nm and the SiO$_2$ layer thickness of $t_{OX} = 59$ nm.

The Fowler measurement method (11,12) is applied to determine the barrier heights at the gate-SiO$_2$ interface, E_{BG}, and at substrate-SiO$_2$ interface, E_{BS}. Using this method the photocurrent, I, vs. wavelength, λ, characteristics are taken for a range of gate voltage values and transformed into the photoemission yield, Y, vs. photon energy, hv, characteristics, using the relation:

$$Y = \frac{I \times h\nu}{P}$$
[1]

where I is the photocurrent in amperes, hv is the photon energy in electronvolts and P is the power in watts of the light absorbed by the emitter. The characteristic given by [1] is next converted into the $Y^{1/p}$ dependence on photon energy, in which p is a factor equal 2 for photoemission from metal and equal 3 for photoemission from semiconductor (12). An example of a $Y^{1/3}$ vs. hv characteristics is shown in Fig.7.

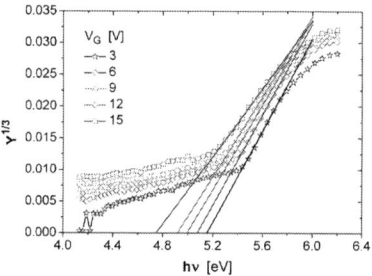

Fig.7. Characteristics of the cube root of photoemission yield, $Y^{1/3}$, vs. the photon energy, hv, determined for one of the Al-SiO$_2$-SiC(3C) structures.

As shown in Fig.7, each of the characteristics taken for different V_G values exhibit a range of measurement points lying on the straight lines. These straight lines extrapolated to the Y = 0 axis determine the values of hv at Y = 0, for each applied gate voltage. Next, these hv(Y=0) points are plotted in function of $V_{OX}^{1/2}$, where V_{OX} is the voltage drop in the dielectric of the MOS structure. The so obtained points in the, hv(Y=0), vs. $V_{OX}^{1/2}$ coordinates determine another straight line, which extrapolated to the V_{OX} = 0 axis determines the barrier height value (E_{BS} or E_{BG}), as illustrated in Fig.8.

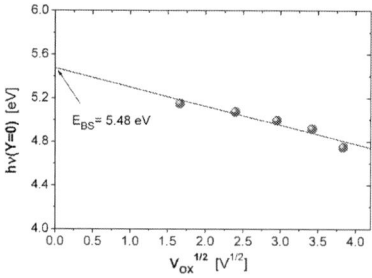

Fig.8. The dependence of, hv(Y=0), on the square root of the voltage drop on the SiO$_2$ layer, $V_{OX}^{1/2}$, determined for one of the Al-SiO$_2$-SiC(3C) structures.

The V_{OX} values corresponding to each of the applied V_G values, can be found by standard methods, as described in (3).

Complete characterization of the band diagram requires determination of potential values corresponding to two characteristic situations of the DUT, the flat-band situation in the semiconductor, when $\phi_S = 0$, $V_G = V_{FB}$ and the flat-band situation in the dielectric, when $V_{OX} = 0$, $V_G = V_{G0}$. The V_{G0} value can be found by measurements of the photocurrent I vs. gate voltage V_G characteristics at different wavelengths λ of the light beam in the vicinity of the I = 0 points and making use of the relations resulting from the theory of photoelectric phenomena in the MOS system at low electric fields in the dielectric, as described in (13). The remaining potentials which are necessary for the construction of the complete band diagrams are determined, mostly by standard methods as explained in (3).

The general representation of the so determined band diagrams is shown in Fig.9a for the state of flat-band in the dielectric ($V_G = V_{G0}$) and in Fig.9b for the state of flat band in the semiconductor ($V_G = V_{FB}$). Values of parameters indicated in Fig.9 are given for structures with both types of SiO$_2$ layers in Table 1.

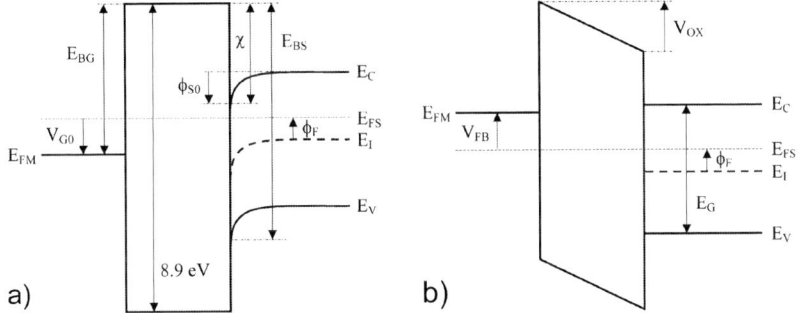

Fig.9. Band diagrams of the Al-SiO$_2$-SiC(3C) structure a) for $V_G = V_{G0}$, $V_{OX} = 0$ and b) for $V_G = V_{FB}$, $\phi_S = 0$. Values of the potentials marked in the figures are given in Table 1.

Table 1. Average values of the potentials indicated in Fig.9 for both types of SiO$_2$ layers in the MOS structures. Measurement results of 6 MOS structures were used to determine average values in the table.

E_{BG} [eV]	E_{BS} [eV]	χ [eV]	ϕ_F [V]	ϕ_{MS} [V]	Q_{eff} [C/cm^2]	N_{eff} [cm^2]	ϕ_{S0} [V]	V_{G0} [V]	V_{FB} [V]	V_{OX} [V]
					THERMAL					
3.466	5.897	3.536	-0.981	-0.038	5. 42·10^{-8}	3.38·10^{11}	0.082	0.044	-0.914	-0.877
					PECVD					
3.425	5.455	3.094	-0.961	0.028	7.14·10^{-8}	4.46·10^{11}	0.097	0.126	-1.196	-1.225

Values of parameters shown in Fig.9, were determined as described above, using also some values known from literature (E_G(SiO$_2$) = 8.9 eV (12), E_G(3C-SiC) = 2.38 eV (15)). The overall accuracy of barrier height values given in Table 1 is estimated at ± 0.150 eV. Values of the E_{BS} barrier heights at the semiconductor-dielectric interface, given in Table 1 are slightly lower than obtained by other authors, which results from the presence of a band of high density of interface states in the lower part of the band gap (in the vicinity of the valence band top) of 3C-SiC. The photoelectrons emitted from these states slightly disturb the photocurrents of electrons emitted from the valence band which results in apparent lowering of the barrier height.

6. Raman spectra

To determine if significant differences exist in the structure and/or in the mechanical stresses in thermally grown and PECVD deposited SiO_2 layers, Raman spectra were measured for both types of oxide layers. The spectra were measured making use of the MonoVista 2750i micro Raman spectrometer (Spectroscopy and Imaging GmbH) in the setup configuration as described in (16). The ultraviolet laser line of $\lambda = 266$ nm was used for excitation and the power of light on the sample did not exceed 400 μW. The spectrum measured in the range of Raman shift from 110 cm^{-1} to 1250 cm^{-1} is shown in Fig.10. The intensity of the strongest one phonon line of 3C-SiC (at ca. 975 cm^{-1}) was used as the normalization condition. The ordinate of this spectrum is expanded 100 times to emphasize the small signal in the range of 200 cm^{-1} to 800 cm^{-1} where the scattering of bulk SiO_2 is observed (17). The signal in the range of 100 cm^{-1} to 600 cm^{-1} is similar to data reported for the mixture of amorphous and nanocrystalline SiC (18) which is due to the penetration of the excitation light into the silicon carbide substrate. The domination of Raman scattering from mixture of amorphous and nanocrystalline 3C-SiC is in agreement with strong two phonon signal recorded for the substrate. Normalized two phonon spectra of 3C-SiC are shown in Fig.11.
As results from Fig.10 and Fig.11, no significant differences in the structure and/or in the mechanical stresses are observed in the Raman spectra of samples, with both types of SiO_2 layers.

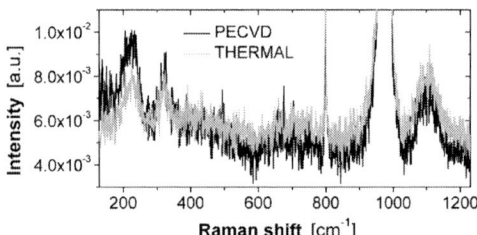

Fig.10. Normalized Raman spectra of thermally grown SiO_2 layer – black line and of PECVD deposited SiO_2 layer – gray line, both on 3C-SiC substrate. The ordinate is expanded 100 times in order to show the small signal below 800 cm^{-1}.

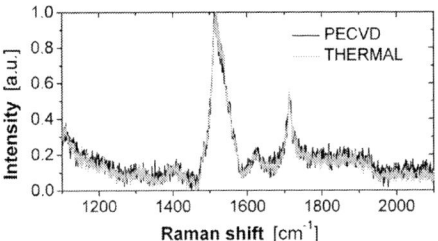

Fig.11. Normalized two-phonon Raman spectra of 3C-SiC substrate: black line – PECVD deposited SiO_2 layer, gray line – thermally grown SiO_2 film.

7. Conclusions

A comparative study was made of the properties of $Al-SiO_2-SiC(3C)$ structures with SiO_2 layers fabricated by thermal growth or by PECVD deposition followed by annealing in wet oxygen. It was found that:

- The leakage currents are, on the average, slightly lower and the breakdown voltages are slightly higher in case of structures with PECVD deposited SiO_2 layers, as shown in Fig. 2 and Fig. 3.
- Measurements of interface trap density and distribution in energy have shown that the density of traps is considerably lower in case of structures with SiO_2 layers deposited by PECVD. The difference is particularly noticeable for traps lying deeper in the band gap of silicon carbide, as shown in Fig. 4.
- Minor differences were found by photoelectric and optical methods in the band diagrams of both types of MOS structures, as shown in Fig. 9 and Table 1. These differences should not significantly influence electrical parameters of MOSFET structures.
- Micro-Raman spectroscopy was used to compare the structural properties and/or mechanical stresses in the samples with both types of SiO_2 layers. As shown in Fig. 10 and Fig. 11, no significant differences were found by Raman spectra in the structure and the mechanical stresses in samples of both types.

Finally it has to be stressed that the PECVD deposition of SiO_2 layers allows to significantly reduce the thermal budget of MOSFET processing.

To summarize, the PECVD deposition of SiO_2 layers, followed by annealing in wet oxygen is more advantageous than the high temperature oxide growth for the technology of MOSFET fabrication.

Acknowledgments

The research was partially supported by the European Union within European Regional Development Fund, through grant Innovative Economy (POIG.01.03.01-00-159/08, "InTechFun")

References

1. M. Bakowski et al, *J. Telecom. Inf. Techn.* **2**, 49 (2007).
2. H. Nagasawa et al, *Mat. Sci. Forum* **600-603**, 89 (2009).
3. H.M. Przewlocki et al., *ECS Transactions* **50**(3), 231 (2012).
4. R. Esteve et al., *Material Science Forum* **645**, 829 (2010).
5. R. Esteve et al, *J. Appl. Phys.* **106**, 044513 (2009).
6. J. Piscator et al, *J. Appl. Phys.* **106**, 0544510 (2009).
7. T. Gutt et al, *Microelectronic Eng.* **109**, 94 (2013).
8. C.N. Berglund, R.J. Powell, *J. Appl. Phys.* **40**, 5093 (1969).
9. E.H. Nicollian, J.R. Brews, MOS Physicsa and Technology, Wiley, N.Y. (1982).
10. K. Piskorski et al, *Opto-Electronics Rev.* **20**, 67 (2012).
11. R.H. Fowler, *Phys. Rev.* **38**, 45 (1931).
12. V.V. Afanas'ev, Internal Photoemission Spectroscopy, Principles and Applications, Elsevier (2008).
13. H.M. Przewlocki, *Solid State Electron.* **45**, 1241 (2001).
14. S.M. Sze, K.K. Ng, *Physics of Semiconductor Devices*, Wiley, Hoboken (2007).
15. M. Shur, *SiC Parameters Handbook*, http://www.ioffe.ru/SVA/NSM/Semicond/SiC
16. P. Borowicz et al, ISRN *Nanomaterials* 2012, 852405 (2012).
17. B. Champagnon et al, *J. Non-Crystalline Solids* **354**, 569 (2008).
18. M. Makowska-Janusik et al, *J. Physics: Condensed Matter* **17**, 5101 (2005).

Influence of ion implantation in SiC on the channel mobility in lateral n-channel MOSFETs

C. Strenger[a, c], V. Uhnevionak[a, c], A. Burenkov[a, c], A. J. Bauer[a, c],
P. Pichler[a, b, c], T. Erlbacher[a, b], H. Ryssel[a, b], and L. Frey[a, b]

[a]Fraunhofer IISB, Schottkystrasse 10, Erlangen 91058, Germany
[b]Chair of Electron Devices, Cauerstrasse 6, Erlangen 91058, Germany
[c]The Wide Bandgap Semiconductor Alliance (WISEA)

In this work, the impact of ion implantation into the MOSFET channel region on the channel mobility of electrons is investigated. For this purpose, a systematic investigation of the interface trap density, the interface trapped charge and the field-effect mobility is carried out for MOSFETs with different doping profiles and concentrations in the channel region. It will be demonstrated that implantation into the SiC MOSFET channel does not increase the amount of bulk defects. It leads, however, to a change of the doping concentration, and consequently of the bulk potential (ϕ_B). The latter crucially determines the amount of charged interface states, and as a result, represents one determining factor for the channel electron mobility. The experimental data indicates that bulk traps as suggested by Agarwal and Haney (1), are insufficient to explain the low electron mobility in SiC MOSFET inversion channels.

Introduction

As a consequence of the poor SiC/SiO$_2$ interface quality, a distinct reduction of the electron mobility (μ_e) in the inversion or accumulation layers of planar 4H-SiC n-channel MOSFETs and thus in their electrical performance can be observed (2). Particularly the interfacial quality is determined by the high density of interface states (D_{it}) that dramatically limit μ_e. Agarwal and Haney (1), however, have disclosed a speculative concept predicting that bulk traps in SiC beneath the interface as well could affect the μ_e in n-channel MOSFETs much in the same way as the interface traps at the SiC/SiO$_2$ interface. Thereby it was assumed that one very probable cause for the formation of these bulk traps is related to ion implantation of p-wells. Although this concept was put forward with no direct experimental evidence, it attracted a great deal of attention within the scientific community (3, 4). If the aforementioned concept and particularly the formation of bulk traps as a result of ion implantation proves to be true, then devices with p-wells grown by epitaxy should be considered, to avoid ion implantation in the MOSFET channel region. In this work, a systematic analysis of the influence of ion implantation in SiC on μ_e in n-channel MOSFETs will be presented.

Experimental

Lateral n-channel MOSFETs with a channel length (L) of 5 and 10 μm and a channel width (W) of 100 μm were fabricated on n- and p-type 4° off-axis (0001) Si-face 4H-SiC epitaxial layers. Fig. 1 shows the schematic cross-sections of the manufactured MOSFETs.

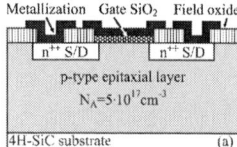 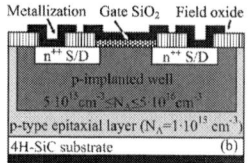 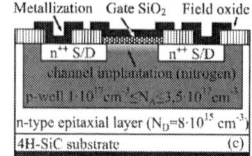

Figure 1. Schematic cross sections and doping levels of the lateral n-channel MOSFETs fabricated on (a) a p-type epitaxial layer, (b) an Al-implanted p-type epitaxial layer and (c) an Al-implanted n-type epitaxial layer. Note that the n-doping concentration (N_D) of the n-type epitaxial layer is small ($\approx 1,6\%$) compared to the p-doping concentration (N_A) of the implanted well.

Tab. 1 summarizes the relevant doping concentrations and doses for all investigated samples in this work. Only one MOSFET (#E1) has been fabricated on aluminum (Al) doped ($\approx 5e17$ cm^{-3}) p-type epitaxial layer without an additional implantation of Al in the channel region. For the fabrication of all other n-channel MOSFETs (#W1-#W5 and #CW1-#CW5), p- wells with different doping concentrations in the range of $5e15$ cm^{-3} to $3,5e17$ cm^{-3} were formed by Al-implantation. Furthermore, the MOSFETs #CW1-#CW5 received a shallow implant with nitrogen (N) in the channel region. Note that the implanted Al-concentration is identical for all MOSFETs with shallow N implant.

TABLE I. Doping concentrations and implantation doses for the investigated MOSFETs

Sample	Doping concentration epitaxial layer (cm^{-3})	Doping concentration p-type well (cm^{-3})	Nitrogen dose implanted in channel (cm^{-2})
#E1	N_A=5e17	x	x
#W1	N_A=1e15	N_A=5e15	x
#W2	N_A=1e15	N_A=1e16	x
#W3	N_A=1e15	N_A=5e16	x
#W4	N_D=8e15	N_A=1e17	x
#W5	N_D=8e15	N_A=3,5e17	x
#CW5	N_D=8e15	N_A=3,5e17	5e12
#CW4	N_D=8e15	N_A=3,5e17	1e13
#CW3	N_D=8e15	N_A=3,5e17	2e13
#CW2	N_D=8e15	N_A=3,5e17	3e13
#CW1	N_D=8e15	N_A=3,5e17	5e13

MOSFET fabrication process

Subsequent to the cleaning of the wafers, the adjustment marks were selectively dry etched into the SiC epitaxial layers, a scattering oxide was deposited and the source and drain regions were box-implanted with N. Thereafter, multiple Al-implantations were carried out to obtain suitable, box-shaped, p-type wells (Tab. I) for all MOSFETs to be fabricated except for #E1. Five samples (#CW1-#CW5) were then low energy implanted with different N ion doses (Tab. I) in the MOSFET channel region. Following the implantations and the removal of the scattering oxide, the wafers were annealed in argon (Ar) ambient at a temperature of 1700 °C for 30 min. For the annealing step, a capping layer was employed to reduce surface degradation. A 500 nm thick field oxide was deposited by CVD and patterned to open the gate and contact regions of the devices. Thereafter the gate oxide was grown by dry oxidation in N$_2$O atmosphere at 1280°C for 150 min and subsequently annealed at the same temperature for 30 min under N$_2$ ambient. The resulting interface between the SiO$_2$ and the SiC epitaxial layer was located in the vicinity of the maximum N concentration of the Gaussian like channel implantation profile. The oxide thicknesses of the MOSFETs vary between 25.5 nm and 27.5 nm. Due to a longer thermal oxidation (180 min) for sample #E1, its oxide thickness is 34 nm. Phosphorus-doped polycrystalline silicon was deposited by LPCVD and patterned to form the gate electrodes. For the fabrication of the source, drain, bulk, and p-well contacts, SiC was alloyed with nickel by rapid thermal annealing at 1100°C for 2 min and subsequently a metallization stack containing titan and platinum was deposited and patterned. Finally, a 500 nm thick SiO$_2$/SiN stack was deposited and patterned.

Results and Discussion

The results of the electrical characterization of the n-channel MOSFTEs with different p-type doping concentrations are presented in this paragraph. The transfer characteristics of the samples #W1, #W2, #W3, #W4, #W5 and #E1 are shown in Fig. 2a).

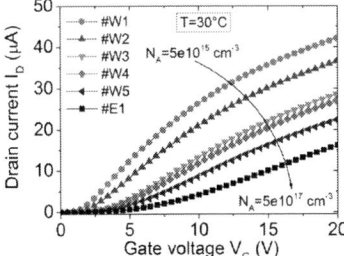
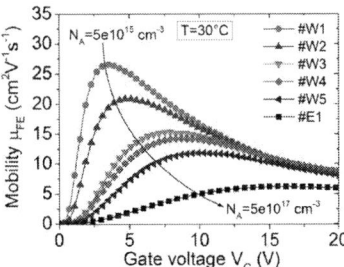

Figure 2 a). Transfer characteristics as a function of V$_G$ for MOSFETs with different p-type doping concentration.

Figure 2 b). Field-effect mobility as a function of V$_G$, determined from the I$_D$-V$_G$ characteristics shown in figure 2 a).

During the measurements the drain-source voltage (V$_{DS}$) was kept constant at 100mV to ensure that the MOSFETs are operated in the linear region. In this region, the channel acts as a resistor wherein the drain current (I$_D$) is proportional to V$_{DS}$ (5). For this case only, the field-effect mobility (μ_{FE}) for electrons in the MOSFET channel may be extracted from the collected I$_D$-V$_G$ characteristics by

$$\mu_{FE} = \frac{L}{W \cdot C_{ox} \cdot V_{DS}} \cdot \frac{\partial I_D}{\partial V_G}. \qquad [1]$$

Resulting field-effect mobilities are shown in Fig. 2 b). From these μ_{FE}-V$_G$ characteristics, we find that the peak field-effect mobility ($\mu_{FE,max}$) decreases with increasing N$_A$ and its absolute value amounts only to about 0,8 - 4,5% of the bulk mobility. The value of the latter is about 650cm^{2}V^{-1}s^{-1} for a p-doping concentration N$_A$ equal to 1e17 cm^{-3} at 30°C.

The low μ_{FE} values are a consequence of the different scattering mechanisms that can influence the mobility at SiC/SiO$_2$ interface and which are combined by

Matthiessen`s rule: bulk mobility, surface-roughness mobility, surface-phonon mobility, and Coulomb mobility (6). As the bulk mobility is up to several orders of magnitude higher than the determined μ_{FE} values, it cannot represent the dominant mobility limiting component. Additionally, H. Naik et al. (6) calculated that the surface-phonon mobility for MOSFETs on the Si-face of 4H-SiC is in the range of $450 - 650 cm^2 V^{-1} s^{-1}$, depending on the effective surface electric field (E_{eff}). Thus, in direct contrast to silicon MOSFETs, the surface-phonon mobility may not account for the low values of μ_{FE} presented in this work either. The second mobility component that significantly depends on E_{eff}, the surface-roughness mobility, is increasingly important for high E_{eff} values (starting from a gate voltage of approximately 7-12 V depending on the Coulomb mobility component and the gate oxide thickness). At low electric fields, its contribution to the reduction of $\mu_{FE,max}$ is small (7). We conclude, therefore, in agreement with other reports (8, 9, 10), that only the Coulomb mobility can be the predominant mobility component that reduces the field-effect mobility at low fields and in particular the peak-field effect mobilities of the investigated MOSFETs. This conclusion is further supported by the temperature dependence of μ_{FE} that is shown in Fig. 3: In all samples, μ_{FE} at low electric fields and especially $\mu_{FE,max}$, increases with temperature and gate bias and is thus consistent with a mobility model dominated by Coulomb scattering. Note that from the aforementioned mobility components only the Coulomb mobility possesses a positive temperature coefficient (10).

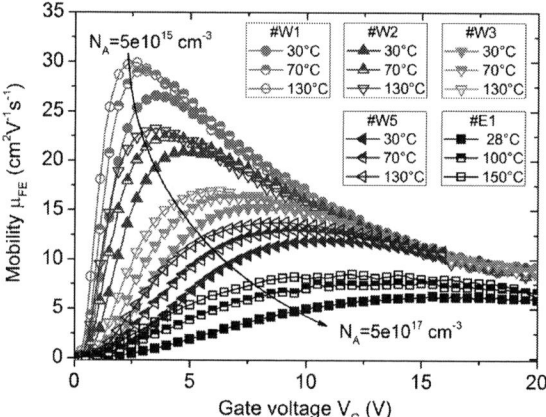

Figure 3. Field-effect mobilities as a function of V_G for the MOSFETs #E1 -#W5 for different temperatures in the range from 28°C − 150°C. Note that for the sake of clarity, the field-effect mobilities for #W4 are not shown.

A broad agreement exists in SiC related literature (9, 10, 11, 12) stating that Coulomb scattering in the MOSFET surface n-channel on the Si-face of 4H-SiC originates predominantly from trapped charges in acceptor-like interface states (in the following referred to as charged interface states) spatially located at the SiC/SiO₂ interface, or in its vicinity, in the oxide (13). The latter are frequently referred to as near interface traps (NITs). These charged interface states act as scattering centers for free electrons in the MOSFET inversion or accumulation channel, thus reducing their mobility (12). Therefore, as the number of charged interface states increases, channel electron mobility decreases.

Agarwal and Haney (1) assume that the amount of charged interface states is increased by the formation of additional states in the SiC epitaxial layer, close to the SiC/SiO$_2$ interface. It was speculated that one probable cause for the formation of these bulk traps might be related to the ion implantation of p-wells. Consequently, the reduction of the electron mobility with increasing Al-doping concentration (see Fig. 2 b)) could be explained by an increase of bulk states resulting from the increasing Al-implantation dose. According to Agarwal and Haney (1), these bulk traps would affect the mobility of electrons in n-channel MOSFETs in the same way as the interface traps at the SiC/SiO$_2$ interface. The μ_{FE} value in Fig. 2 b) for the Al-implanted sample #W5, however, is not lower than for epitaxial sample #E1 (#E1 not implanted with Al in the channel region) but higher by about a factor of 2. Note that for both samples, the Al-concentration in the channel region is virtually the same. Hence we believe that the reduction of μ_{FE} with increasing N_A cannot be fully explained by the formation of bulk traps as a consequence of ion implantation.

Our results are in agreement with the commonly accepted fact that the reduction of μ_{FE} is related to an increase in the amount of charged interface states. We believe that the amount of charged interface states is not increased by the formation of additional ion implantation induced bulk traps as suggested by Agarwal and Haney (1). Instead we support a model proposed by V. Tilak et al. (12). The model suggests that the total amount of charged interface traps predominately depends on the actual surface potential ϕ_S. The latter is significantly dependent on N_A as well as on V_G. The amount of charged interface states depends on the equilibrium density of filled interface states for a given voltage, applied to the gate contact of the MOSFET. In the following paragraphs, the methodology to determine the amount of the trapped charge at the SiC/SiO$_2$ interface of the investigated MOSFETs will be described briefly. The subsequently presented values for the interface trapped charge (Q_{it}) for the MOSFETs will be related to the variation of the channel mobility.

The gate voltage induced electric field in the oxide will influence, besides the semiconductor space charge (Q_{sc}), also the interface trapped charge (Q_{it}). As more and more acceptor like interface states become charged with increasing V_G, the relation between ϕ_S and V_G is altered (i.e. the ϕ_S - V_G characteristics are "stretched out"). The dependency of ϕ_S from V_G will consequently be expressed as

$$V_G = \phi_{MS} - \frac{Q_f}{C_{ox}} - \frac{1}{C_{ox}}\left(Q_{sc}(\phi_S) + Q_{it}(\phi_S)\right) + \psi_S. \qquad [2]$$

Therein ψ_S is the band-bending (note that $\psi_S = \phi_S - \phi_B$), ϕ_{MS} is the semiconductor-gate metal work function difference, and Q_f is the positive fixed charge at the SiO$_2$/SiC interface. From the measured sheet carrier concentration in the MOSFET inversion channel (not shown here), determined by Hall measurements on MOS-gated Hall bar structures, we computed the actual relation between ψ_S and V_G. The latter is compared to the ideal case for sample #W5 in Fig. 4.

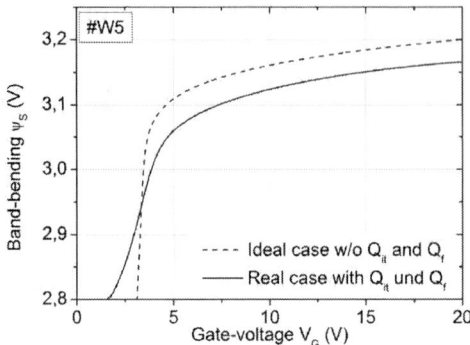

Figure 4. Comparison of the ideal and the real band-bending ψ_S as a function of V_G for MOSFET #W5.

These ψ_S - V_G characteristics allow for the extraction of the interface state density profile ($D_{it}(E)$) in the energy range from the conduction band (E_C) down to an energetic position 140 meV below E_C (Fig. 5 a)). Additionally, the $D_{it}(E)$ at energetic positions between E_c-200 meV and E_c-600 meV was determined on the basis of capacitance-voltage measurements on n-MOS structures by the "high-low" method first described by Castagné et al. (14). The extracted $D_{it}(E)$ in this energetic range is shown in Fig. 5 b).

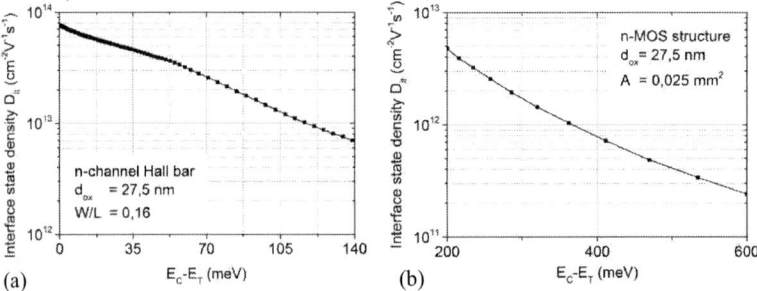

Figure 5. Interface trap density in the energetic range (a) from E_C down to E_C-140 meV as computed from the inversion sheet carrier concentration determined by Hall effect measurements and (b) from E_C-200 meV down to E_C-600 meV as extracted from capacitance-voltage measurements on n-MOS structures with the "high-low" method.

The knowledge of the interface state density profiles and the actual relation between the gate voltage and band-bending - and thus the position of the Fermi energy with respect to E_c - allows for the estimation of the interface trapped charge in the investigated MOSFETs. The interface trapped charge as a function of ϕ_S may be approximated by

$$Q_{it}(E_F) = q \cdot N_{it} = -\int_{E_n}^{E_F} D_{it}(E)dE . \qquad [3]$$

For the sake of better traceability, the $D_{it}(E)$ as determined for sample #W5 is assumed to be representative for all p-implanted samples. This assumption is justified as our preliminary calculations of the $D_{it}(E)$ for the other samples indicate that $D_{it}(E)$ doesn't change significantly, irrespective of implanted Al-dose. This is also in agreement with the findings of D. Okamoto et al. (15). Furthermore, only acceptor-like interface states will be considered whereby the neutrality level E_n is assumed to be at Ec-600 meV. Note that assuming the neutrality level to be equal to the intrinsic level (≈Ec-1610 meV) would not have a significant impact on the results. This is due to the fact that the D_{it} decreases exponentially with increasing energetic distance from E_C and is thus several orders of magnitude smaller at midgap than close to E_C. The bulk potentials ϕ_B in the p-wells with different N_A and in the epitaxial layer of #E1 can be determined by

$$\phi_B = \frac{kT}{q} \ln \left(\frac{N_A}{n_i} \right), \qquad [4]$$

where k is the Boltzmann constant and n_i the intrinsic carrier concentration of 4H-SiC. Following the results of Hattori et al. (16), incomplete activation of Al-doping atoms is taken into account for the Al-implanted samples. Furthermore, the electric field that penetrates into the MOSFET channel region allows for the assumption of a complete field induced ionization of all activated Al-doping atoms.

The computed amount of the trapped charge Q_{it} in acceptor-like interface states at the onset of inversion is presented in Fig. 6 as a function of the energetic distance between E_C and E_F.

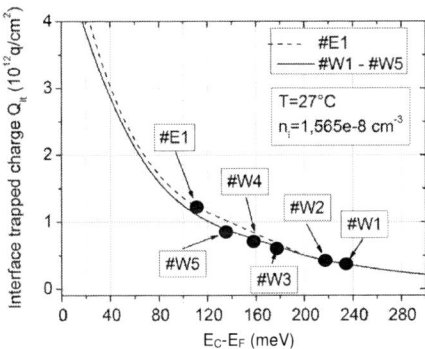

Figure 6. Interface trapped charge as a function of the energetic distance between E_C and E_F. Full circles represent the actual value of the interface rapped charge at the point of inversion for the MOSFETs #E1 - #W5.

As disclosed in Fig. 7, the MOSFET with the highest Q_{it} value of 1,2e12 q/cm^2 (#E1) possesses the lowest peak-field effect mobility. By contrast the MOSFET with the lowest Q_{it} value of 0,36e12 q/cm^2 (#W1) possesses the highest peak-field effect mobility. This close correlation between $Q_{it}(-\phi_B)$ and $\mu_{FE,max}$ is equally true for all other MOSFETs as presented in Fig 7.

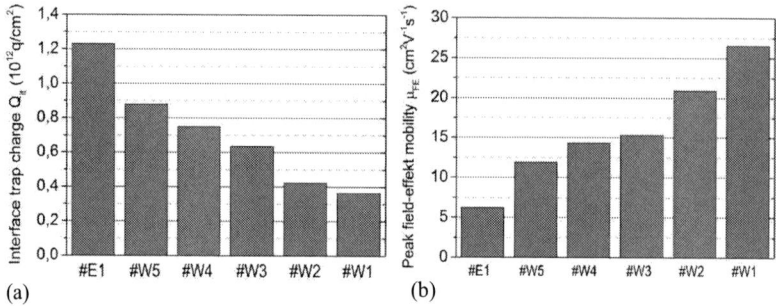

(a)   (b)

Figure 7. Bar chart for MOSFETs #E1 - #W5 displaying (a) the interface trapped charge when inversion of channel region takes place and (b) the peak-field effect mobilities of MOSFETs.

Noteworthy, this correlation was established solely by taking into account the surface potential dependent change in the filling of interface states. The change in the surface potential when inversion of the MOSFET channel occurs was explained by the change of the bulk potential induced by different values for N_A of the p-type wells. This relationship is illustrated in Fig. 8. In Fig. 8 a) the SiC band-diagram of a n-channel MOSFET for flat-band condition is shown. Only a small amount of acceptor-like interface states are charged in this case. With increasing band-bending more and more interface states become charged and can then act as scattering center. As the bulk potential increases, the band-bending needed to invert the MOSFET channel region equally increases and hence the amount of scattering centers at the SiC/SiO₂ interface increases (compare Fig. 8 b) and 8 c)).

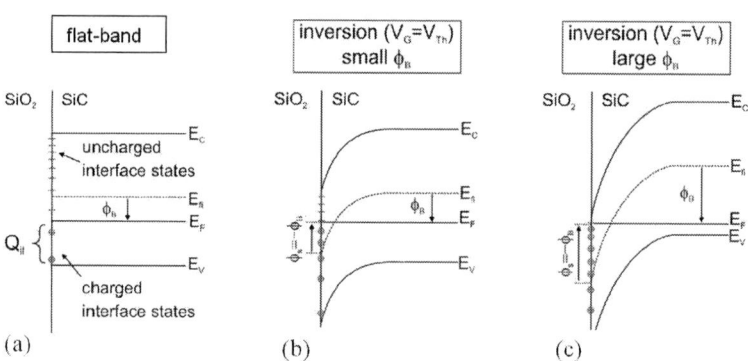

(a)   (b)   (c)

Figure 8. Schematic band diagrams and occupation of acceptor-like interface states of a n-channel MOSFET for the (a) flat-band case, (b) inversion case with small ϕ_B and (c) inversion case with large ϕ_B.

The extracted values of the interface trapped charge indicate that the Al-doping induced increase in the bulk potential from 1,39 V for sample #W1 to 1,52 V for sample #E1, leads to an increase of Q_{it} and thus to an increase in the number of scattering centers, of approximately 350 %. Hence we conclude that the main effect of ion implantation into SiC is related to the change of the doping concentration and consequently to the change of the bulk potential. As demonstrated, the latter crucially determines the amount of charged interface states and thereby the channel mobility as well. Our results show that the formation of ion implantation induced bulk defects, if any at all, as predicted by Agarwal and Haney (1) is not detrimental for the electron mobility in the MOSFET inversion channel.

To additionally substantiate this finding, the channel regions of the MOSFETs #CW1 - #CW5 (p-well doping concentration equal to that of #W5 – 3,5e17 cm-3) have been selectively counter doped with different doses of nitrogen (see Tab. 1). As expected, both the drain current (Fig. 9 a)) and the field-effect mobilities (Fig. 9 b)) increased with increasing N implantation dose. The peak-field effect mobilities are virtually the same as for the MOSFETs #W1 - #W5. The finding that the interfacial density of N, as determined by SIMS measurements, remained constant ($\approx$4e20 cm^{-3}) irrespective of the implanted N-dose, agrees with the finding that D_{it} close to the conduction band is almost identical for the samples #CW1- #CW5 (17). Should the formation of bulk traps due to ion implantation play a crucial role, then this result could not have been achieved.

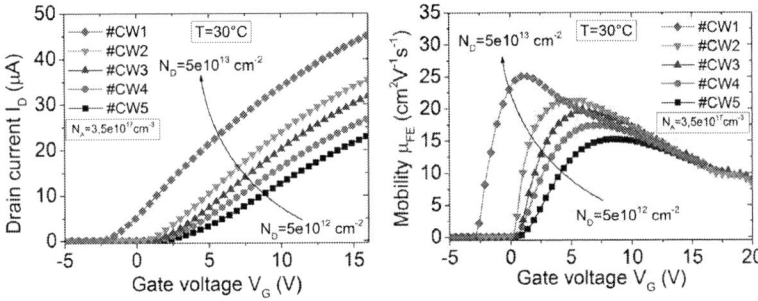

Figure 9 a). Transfer characteristics as a function of V_G for MOSFETs with identical p-type doping concentration and varying N dose implanted in the MOSFET channel region.

Figure 9 b). Field-effect mobilities as a function of V_G, as determined from the I_D-V_G characteristics in figure 2a).

Summary

In this work, we have investigated the impact of ion implantation into the MOSFET channel region on the channel mobility of electrons. Special attention was given to the investigation of the speculative concept offered by Agarwal and Haney (1), who suggested that bulk traps in the depletion region are formed as a consequence of ion implantation, can be dominant defects that limit the inversion layer mobility of SiC transistors. For this purpose, a systematic investigation of the interface trap density, the interface trapped charge, and the field-effect mobility was carried out for MOSFETs with different doping profiles and concentrations. Based on the demonstrated correlation of the aforementioned values, we concluded that the main effect of ion implantation into SiC is related to the change of the doping concentration and consequently to the bulk potential. It was demonstrated that the latter crucially determines the amount of charged interface states and hence the channel mobility as well. Thus, experimental arguments indicate that bulk traps, as speculated by Agarwal and Haney, are insufficient to explain the poor electron field-effect mobility in SiC MOSFET inversion layers.

Acknowledgments

This work has been performed by the French-German Consortium MobiSiC and supported by the Program Inter Carnot Fraunhofer (PICF) from BMBF (Grant 01SF0804) and ANR. The support from Robert Bosch GmbH as well as the financial support of the Hans L. Merkle-Stiftung is highly appreciated. Likewise, the authors are thankful for the support of M. Krieger, T. Sledziewski, G. Ortiz, E. Bedel-Pereira and V. Mortet for the execution of Hall measurements.

References

1. S. Dimitrijev, Understanding Semiconductor Devices, Oxford University Press, New York (2000)
2. A. Agarwal and S. Haney, Journal of Electronic Material, **37** (5), 646 (2008)
3. V. Tilak and K. Matocha, Journal of Electronic Material, **38** (4), 618 (2009)
4. A. Agarwal and S. Haney, Journal of Electronic Material, **38** (4), 621 (2009)
5. S. Sze and K. Ng, Physics of Semiconductor Devices, John Wiley & Sons, New York (2007)
6. H. Naik and T.P. Chow, Materials Science Forum,**679-680**, 595 (2011)
7. C. Strenger, V. Uhnevionak, A. Burenkov, A. J. Bauer, V. Mortet, E. Bedel-Pereira, F. Cristiano, M. Krieger and H. Ryssel, Materials Science Forum, **740-742** , 537 (2013)
8. X. Shen and S. Pantelides, Applied Physics Letters, **98**, 053507 (2011)
9. A. Poggi, F. Moscatelli, S. Solmi, A. Armigliato, L. Belsito and R. Nipoti, Journal of Applied Physics, **107**, 044506 (2010)
10. S. Potbhare, N. Goldsman and G. Pennington, Journal of Applied Physics, **100**, 044515 (2006)
11. V. Tilak and K. Matocha, G. Dunne, Material Science Forum, **645-648**, 1005 (2010)
12. K.Matocha and V. Tilak, Material Science Forum, **679-680**, 318 (2011)
13. P. Deák, J. Knaup, T. Hornos, C. Thill, A. Gali and T. Frauenheim, Journal of Physics D: Applied Physics, **40**, 6242 (2007).
14. R. Castagné, A. Vapaille, Surface Science, **28**, 157 (1971).
15. D. Okamoto, H. Yano, T. Hatayama, and T. Fuyuki, Materials Science Forum, **645-648** , 495 (2010)
16. R. Hattori, T. Watanabe, T. Mitani, H. Sumitani and T. Oomori, Material Science Forum, **600-603**, 585 (2009)
17. J. Rozen, S. Dhar, M.E. Zvanut, J.R. Williams, L.C. Feldman, Journal of Applied Physics, **105**, 124506 (2009)

On the Temperature Dependence of the Hall Factor in n-channel 4H-SiC MOSFETs

V. Uhnevionak[a,b], A. Burenkov[a], C. Strenger[a,b], A. J. Bauer[a], P. Pichler[a,b]

[a]Fraunhofer IISB, Schottkystrasse 10, Erlangen 91058, Germany
[b]Chair of Electron Devices, Cauerstrasse 6, Erlangen 91058, Germany

To interpret Hall-effect measurements in a range of temperatures, the Hall factor for the electron transport in the channel of a SiC MOSFETs was evaluated. The method of the Hall factor calculation is based on the interdependence with mobility components via the respective scattering relaxation times. For the first time, the temperature dependence of the Hall factor in n-channel 4H-SiC MOSFETs was calculated. The results of the calculation reveal a strong reduction of the Hall factor with increasing temperature. Depending on gate voltage and temperature, the values of the Hall factor vary between 1.2 and 1.5. In addition, the sheet carrier densities and drift mobilities derived from the Hall-effect measurements using our new temperature-dependent Hall factor show very good agreement with independent simulation results.

Introduction

Since SiC MOSFETs are well suited for operation at high temperatures due to the SiC material properties, it is important to understand the temperature dependence of the transport properties such as the drift mobility μ and the sheet carrier density n_{inv} in the inversion channel. The most straight-forward method for their characterization is the application of Hall-effect measurements. However, to obtain the sheet carrier density and the drift mobility from the Hall-effect measurements, the accurate value of the Hall factor is a prerequisite. The Hall factor can be defined as the ratio of the measured Hall-effect mobility μ_H and the drift mobility μ. If the mass anisotropic effect is neglected, the values of the Hall factor may vary between 1 and 1.93, depending on the scattering mechanisms involved (1). The highest value of 1.93 indicates the limiting case of Coulomb scattering. There are several works in which the Hall factor in bulk SiC was determined experimentally and investigated theoretically (2-4). For the significantly more complicated situation of electron transport in the channel of SiC MOSFETs, it is common practice to assume a Hall factor equal to unity. From the theory of Hall-effect measurements it is known that r_H approaches unity in strong magnetic fields B for which the condition $\mu B \gg 1$ is satisfied. As an example, for a magnetic field of $B = 1$ T, the mobility μ should be significantly higher than 10.000 cm^2/Vs. In SiC MOSFETs, the drift mobility of electrons in the inversion channel is typically on the order of 10 to 100 cm^2/Vs (5-7). As a consequence, the assumption of $r_H = 1$ leads to an overestimation of the drift mobility and to an underestimation of the sheet carrier density extracted from the Hall-effect measurements. To obtain r_H in the inversion layer of SiC MOSFETs, we recently introduced a new method for the calculation of the Hall factor (8). This method is based on the strong interdependence with mobility components via the respective scattering relaxation times. In this work, we use our method to investigate, for the first time, the Hall factor in n-channel MOSFETs as a function of temperature. We further apply it to improve the accuracy of the drift mobility and sheet carrier density values extracted from the Hall-effect measurements in the temperature range from 300 K till 400 K and show the consistency with TCAD simulations.

Experimental

In this study, lateral n-channel 4H-SiC MOSFETs have been fabricated on a p-type 4°-off 4H-SiC (0001) Si-face epitaxial layer with an aluminum concentration of $5 \cdot 10^{17}$ cm^{-3}. To characterize them electrically over a broad range of temperatures, current-voltage and Hall-effect measurements as well as numerical simulations were performed for a range of temperatures from 300 K till 400 K. The transfer characteristics were measured at a drain-source voltage of 0.1 V with gate voltages swept from 0 V to 20 V. The Hall-effect measurements were carried out using a permanent magnet with a magnetic field of 0.33 T. The drain-source voltage in the Hall-effect measurements was also set to 0.1 V and the gate voltages were varied from 0.25 V to 20 V. From the results of the Hall-effect measurements the drift-mobility μ and the sheet carrier density of electrons n_{inv} in the inversion layer were determined as a function of the gate voltage from the measurable quantities considering the Hall factor r_H by the formulas:

$$n_{inv} = r_H \frac{IB}{eV_H} \tag{1}$$

$$\mu = \frac{1}{r_H} \frac{V_H}{\rho IB} \tag{2}$$

Therein, I denotes the current, B the magnetic field, V_H the Hall-effect voltage, ρ the resistivity and e the elementary charge. The values of the Hall factor were obtained from [3], using our new method (8):

$$r_H = \frac{<\tau^2>}{<\tau>^2} \tag{3}$$

where $<\tau^2>$ and $<\tau>$ are scattering relaxation times specially averaged over the carrier energy.

The numerical simulations of the current-voltage and charge-voltage characteristics were done using Sentaurus Device of Synopsys by solving the drift-diffusion equations.

Results and discussions

The temperature dependence of the Hall factor in the channel of MOSFETs may be estimated by considering the effect of temperature on the mechanisms by which carriers are scattered in the inversion channel. Considering that the values of the Hall factor and the mobility depend on the same scattering mechanisms in the active area of the devices, the effect of temperature on the electron scattering can be easily obtained by evaluating the temperature dependence of the mobility components associated with the respective scattering mechanisms. On the basis of the temperature dependence of the mobility components, the respective scattering relaxation times can be obtained and used for the calculation of the Hall factor by the method reported in our previous work (8).

It is widely accepted that the device performance of lateral SiC MOSFETs is limited by Coulomb scattering at ionized impurities in the bulk and at interface charges as well as by surface-roughness scattering and phonon scattering (5, 7, 9). Each kind of scattering mechanisms has its specific temperature dependence. To calculate the mobility components and, from them, the scattering relaxation times as a function of temperature, current-voltage characteristics were simulated in the temperature range from 300 K till 400 K. In these simulations, near-interface trap and mobility degradation models were included, the full description of which can be found elsewhere (10). These models, as shown in Fig. 1, excellently reproduce the temperature dependence of the measured transfer characteristics. Using the results of the simulations and the method of small variations (11), the mobility components were determined as a function of gate voltage in the temperature range from 300 K till 400 K. In Fig. 2, the results of the extraction of the mobility components associated with Coulomb scattering at interface charges μ_C, surface-roughness μ_{SR} and phonon scattering μ_{PH} are shown for the temperatures of 300 K and 400 K.

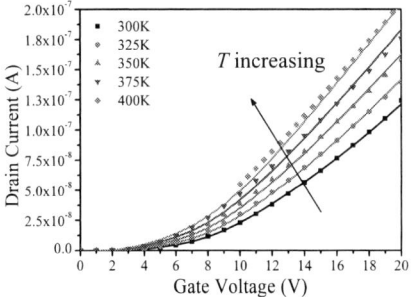 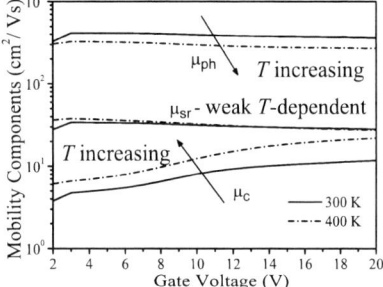

Figure 1. Temperature dependence of the simulated (lines) and measured (symbols) transfer characteristics.

Figure 2. Temperature dependence of the mobility components in n-channel SiC MOSFETs.

It should be noted that splitting up the total mobility into components at room temperature allows us to assess the relative contributions of the scattering mechanisms to the total mobility in the inversion channel. Additional calculation of the temperature dependence of the mobility components gives the opportunity to determine the effect of the temperature on each of the mechanisms by which electrons are scattered. From our results it was found that at room temperature Coulomb scattering at the interface charges and surface-roughness scattering are the dominant scattering mechanisms in the SiC MOSFETs investigated in this work. The mobility components associated with phonon scattering and Coulomb scattering at ionized impurities in the bulk (the latter is not shown here) are several orders of magnitude higher than the measured mobility. It means that these components do not contribute significantly to the total mobility of the inversion channel. With an increase in temperature from 300 K till 400 K the mobility components associated with phonon scattering and Coulomb scattering at ionized impurities in the bulk decrease approximately by a factor of 2. However, even at the higher temperature their contributions to the total mobility are negligible. This allows us to conclude that the impact of phonon scattering and Coulomb scattering at ionized impurities in the bulk is insignificant for the temperature performance of our SiC MOSFETs. Taking into account that surface-roughness scattering depends only weakly on temperature, as it can be seen in Fig. 2, for the temperature dependence of the SiC MOSFET characteristics only Coulomb scattering at the interface charges plays a fundamental role. This was also confirmed by others (9, 12). In accordance with our calculation, the mobility component associated with Coulomb scattering at the interface charges increases with an increase in temperature. This fact indicates a reduction of Coulomb scattering at the SiO_2/SiC interface at higher temperatures, which is reflected in the increase of drain currents shown in Fig. 1. The reduction of Coulomb scattering at the interface charges can be mainly explained by two coupled effects. The first one is the decrease in the amount of trapped charges due to the change of the Fermi level with temperature. The second one is the increase in screening of scattering centers due to the increase of the number of free electrons in the inversion channel.

Based on the fact that the value of the Hall factor is directly affected by the scattering mechanisms, the temperature dependence of r_H can be easily predicted. Considering that for the limiting case of Coulomb scattering, the Hall factor has its maximum and is equal to 1.93, the reduction of Coulomb scattering at the interface charges with increasing temperature will result, in our case, in the reduction of the Hall factor. This is confirmed by our calculation of the Hall factor as a function of temperature, where the newly introduced method of the calculation of r_H was used (8, 11). The results of the calculation of the Hall factor in the inversion channel as a function of the gate voltage for different temperatures are presented in Fig. 3.

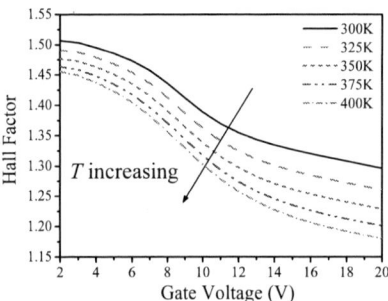

Figure 3. Temperature dependence of the Hall factor in n-channel SiC MOSFETs.

Taking the temperature dependence of the Hall factor, shown in Fig. 3, into account and using Eq. [1, 2], the sheet carrier density and the drift mobility were determined from the Hall-effect measurements. In Fig. 4 and Fig. 5 the respective values are compared with independent numerical simulation in the temperature range from 300 K till 400 K. The values of the sheet carrier density and the drift mobility, denoted in Fig. 4 and Fig. 5 as simulated, were derived from the results of the simulation of the transfer characteristics shown in Fig. 1.

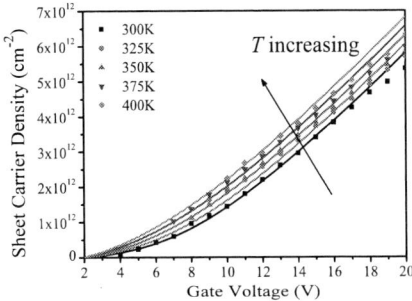
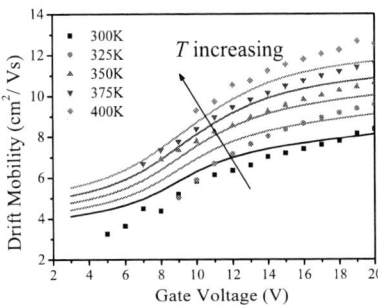

Figure 4. Temperature dependence of the simulated (lines) and measured (symbols) sheet carrier density in n-channel SiC MOSFETs.

Figure 5. Temperature dependence of the simulated (lines) and measured (symbols) drift mobility in n-channel SiC MOSFETs.

The results in Fig. 4 and Fig. 5 reveal that the sheet carrier densities and the drift mobilities in the inversion channel of 4H-SiC MOSFETs determined from the Hall-effect measurements using our new temperature-dependent Hall factor agree very well with independent simulation results. In contrast, if the assumption of a temperature-independent Hall factor $r_H = 1$ was used, an overestimation of the drift mobilities and underestimation of the sheet carrier densities between 20% and 50%, depending on gate voltage and temperature, were found.

Conclusion

In this work, our recently introduced method for the Hall factor calculation was validated by applying it to the interpretation of Hall-effect measurements at elevated temperatures. For the first time, the Hall factor in n-channel SiC MOSFETs was calculated as a function of temperature and gate voltage. Our results show that with an increase in temperature Coulomb scattering at the charges located at the SiO_2/SiC interface strongly reduces. This effect is reflected in the reduction of the Hall factor in the n-channel SiC MOSFETs with the temperature increase. In addition, it was shown that the sheet carrier densities and drift mobilities in the inversion channel of SiC MOSFETs derived from the Hall-effect measurements using our new temperature-dependent Hall factor agree very well with the simulation results performed with Sentaurus Device of Synopsys.

Acknowledgments

This work has been carried out in the framework of the project MobiSiC (Mobility engineering for SiC devices) and supported by the Program Inter Carnot Fraunhofer (PICF 2010) by the BMBF (Grant 01SF0804) and the ANR. The authors also would like to thank V. Mortet, E. Bedel-Pereira and F. Cristiano of LAAS-CNRS, Toulouse, France, for providing the electrical measurements.

References

1. S. M. Sze, *Physics of Semiconductor Devices*, Wiley, New York, 1981.
2. G. Rutsch, R. P. Devaty, W. J. Choyke, D. W. Langer, L. B. Rowland, E. Niemann and F. Wischmeyer, Mater. Sci. Forum 338, 733 (2000).
3. G. Ng, D. Vasileska, and D.K. Schroder, J. Appl. Phys. 106, 053719 (2009).
4. H. Iwata, K. M. Itoh, J. Appl. Phys. 89(11), 6228 (2001).
5. S. Dhar, S. Haney, L. Cheng, A.K. Agarwal, and K.P. Cheung, J. Appl. Phys. 108, 054509 (2010).
6. H. Linewih, S. Dimitrijev, K. Y. Cheong, Microelectron. Reliab. 43, 405 (2003).
7. V. Tilak, K. Matocha, and G. Dunne, IEEE Trans. Electron Devices 54, 2823 (2007).
8. V. Uhnevionak, A. Burenkov, P. Pichler, presented at the 9th Conference on Ph.D. Research in Microelectronics and Electronics, PRIME 2013, Villach, Austria, June 24th-27th, 2013
9. A. Perez-Tomas, P. Brosselard, P. Godignon, and J. Millan, J. Appl. Phys. 100, 114508 (2006).
10. V. Uhnevionak, C. Strenger, A. Burenkov, V. Mortet, E. Bedel-Pereira, J. Lorenz, and P. Pichler, submitted to the ESSDERC conference, Romania, 16-20 September, 2013.
11. V. Uhnevionak, A. Burenkov, C. Strenger, V. Mortet, E. Bedel-Pereira, F. Cristiano, A. J. Bauer, P. Pichler, submitted to the ICSCRM conference, Japan, September 29-October 4, 2013.
12. S. Dhar, A. C. Ahyi, J. R. Williams, S. Ryu and A. K. Agarwal, Mater. Sci. Forum, 717, 713 (2012).

Key Reliability Issues for SiC Power MOSFETs

A. Lelis, D. Habersat, R. Green, and E. Mooro[a]

[a] U.S. Army Research Laboratory, 2800 Powder Mill Road, Adelphi, MD 20783, USA

A brief review of the key results and issues regarding the threshold-voltage instability effect in SiC MOSFETs is presented. These include the basic effect, the strong dependence on measurement conditions, the effect of high-temperature bias stressing, and the implications for reliability testing.

Introduction

Even with the successful introduction of SiC power MOSFETs into the commercial market place, several key reliability issues have not been fully resolved. The main two issues are the stability of the device threshold voltage, V_T, and the reliability of the gate oxide. This work focuses on the V_T stability issue, which has been investigated by a number of different research groups in recent years [1-12].

Threshold-Voltage Instability

An excess negative shift of V_T under high-temperature reverse-bias (HTRB) conditions can lead to a critical increase in *OFF*-state leakage current and potential device failure [13-17]. This negative shift is illustrated in Fig. 1(a) [13, 17], which shows the instability in the I_D-V_{GS} characteristics, especially under bias-temperature stress conditions. Figs. 1(b) and 1(c) show the stress and measurement conditions for both the room-temperature and high-temperature back-and-forth bias-stress testing, respectively. Although a number of different types of interfacial charge are present either in the insulating gate oxide or at its interface with the SiC conduction channel—including interface traps and mobile ions—the primary defects are near-interfacial oxide traps [1, 15, 17].

Both research and commercial-grade devices exhibit a gate-bias induced V_T instability effect, wherein a positive-bias stress shifts V_T positively, and a negative-bias stress shifts V_T negatively. This effect is repeatable [18], and caused by the direct tunneling of electrons either into or out of these near-interfacial oxide traps, depending on the applied gate bias [1, 19]. These effects are exaggerated for longer bias-stress times or greater oxide electric fields [1-4, 8, 18]. When bias stressing at room temperature, the response is generally linear with log(t), where t is the stress time [1, 15, 17]. This is true whether a back-and-forth bias stress is applied [1], as indicated in Fig. 1(b), or whether a one-way bias stress is applied [3]. The results of a one-way stress are shown in Fig. 2. Fig. 2(a) shows the shift in the I_D-V_{GS} characteristics when a long-term negative-bias stress (as indicated in Fig. 3(a)) is followed by a series of increasingly longer positive-bias stresses (as indicated in Fig. 3(b)). The resulting shift in V_T versus cumulative stress time is shown in Fig. 2(b). Similar responses have been observed in MOS capacitor structures as well [20].

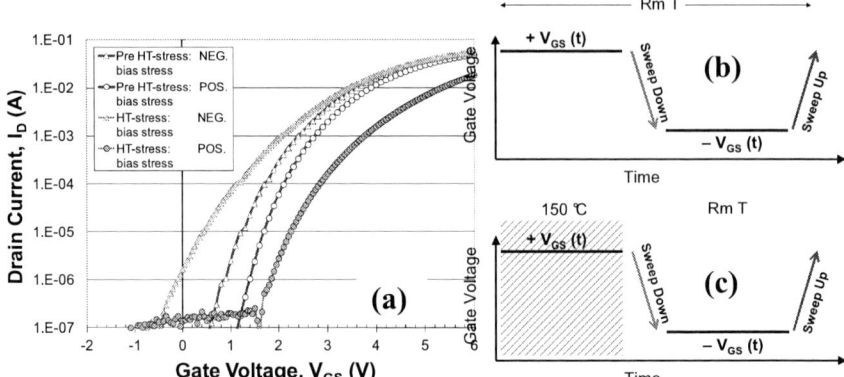

Figure 1: (a) Example of the instability in the I_D-V_{GS} characteristics of a power MOSFET as the result of a high-temperature bias stress test. Pre- high-temperature-stress measurements are shown as well [13, 17]. (b) Back-and-forth bias-stress sequence, with direction of gate-voltage ramp indicated. (c) Similar back-and-forth bias-stress sequence, except the positive-bias stress was at elevated temperature (150 °C).

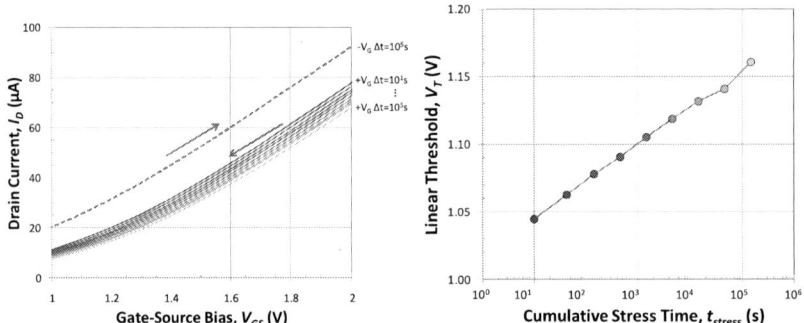

Figure 2: (a) I_D-V_{GS} characteristics for the one-way V_T instability stress and measure sequence indicated in Fig. 3. (b) V_T instability measured as a function of cumulative stress time. The response of the one-way stress is also linear with log(t).

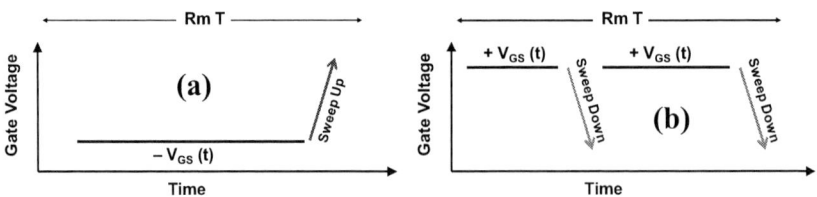

Figure 3: Stress and measurement conditions leading to the results in Fig. 2. (a) Long-term negative bias stress applied, followed by a positive gate-bias sweep to measure the effect of the stress. (b) Positive-bias stressing for increasing stress times, interrupted by a negative gate-bias sweep.

Not surprisingly, the magnitude of this V_T instability is affected by the device processing, in particular whether a nitrogen-based post-oxidation anneal was performed [1, 19, 21]. Even better results have been recently reported when using a phosphorus-based anneal [22].

Measurement Conditions

The measurement conditions following the gate-bias stress have a strong effect on the degree of V_T instability observed. For example, a much smaller V_T shift is seen when V_{GS} is swept positively in the conventional manner (from accumulation to inversion for n-channel devices) following a positive-bias stress than if V_{GS} is swept negatively. This result is illustrated in Fig. 4, applying the stress and measurement conditions indicated in Fig. 5. Likewise, much smaller V_T instabilities are observed if V_{GS} is swept slowly than for faster measurements (a 100-μs gate sweep typically shows a four times larger effect than a more conventional 1 to 10-s measurement) [1, 2]. This result is illustrated in Fig. 6. The reason for this is the acute sensitivity of the charge state of near-interfacial oxide traps to the applied gate bias during the measurement. The more different the measurement bias, and the longer it is applied, the more it counteracts the effect of the previously applied stress bias. A two-way tunneling model has been successfully developed that accounts for the dependence of the V_T instability on both the stress and measurement times by allowing for the simultaneous tunneling in and out of electrons, wherein a steady-state balance occurs in the wake of the tunneling front [23].

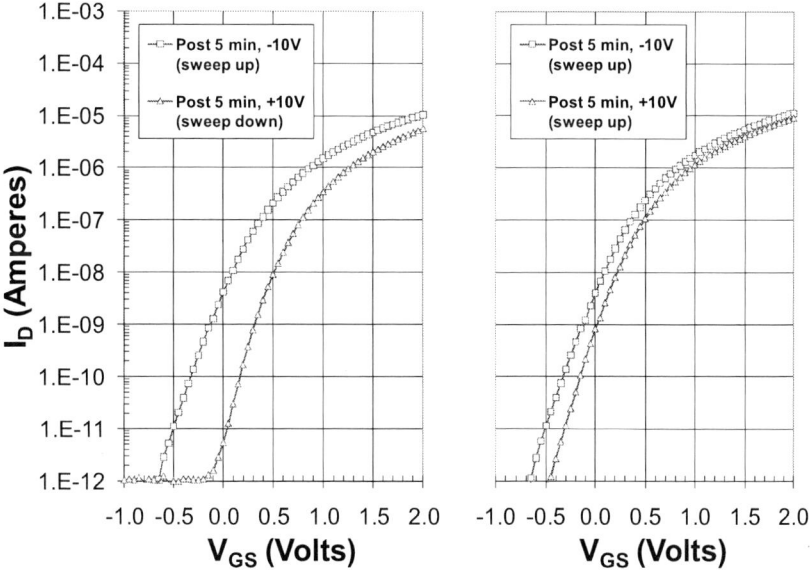

Figure 4: Variation in the back-and-forth V_T instability of a typical SiC MOSFET, depending on whether the gate bias is swept negatively (down) or positively (up) following a positive-bias stress.

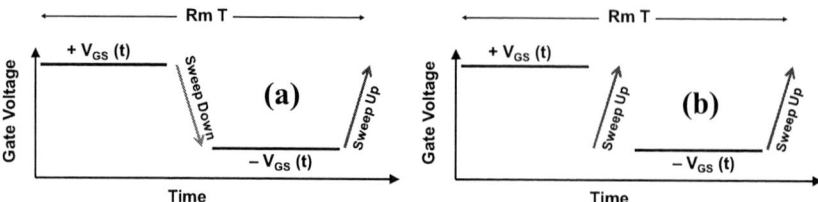

Figure 5: Stress-and-measure sequences used to obtain the results in Fig. 4. (a) Gate bias is swept negatively (down) following a positive-bias stress. (b) Gate bias is swept positively (up) following a positive-bias stress.

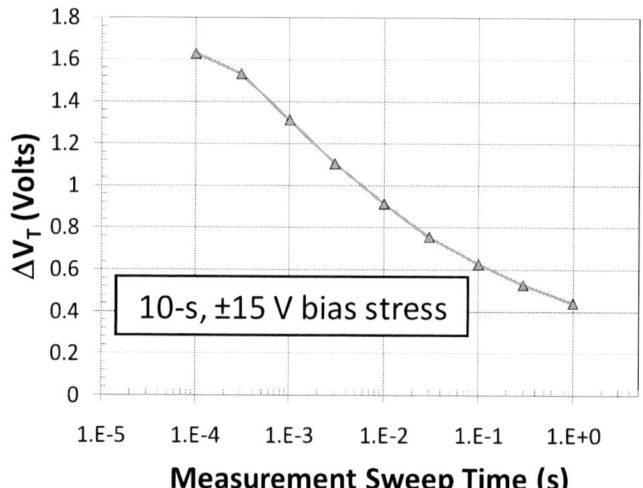

Figure 6. Magnitude of the back-and-forth V_T instability measured as a function of measurement speed, using a stress-and-measure sequence as in Fig. 5(a).

High-Temperature Stressing

Similar V_T instabilities are observed in both lateral test structures and fully-processed vertical power devices [15]. Similar instabilities are also observed in both 4H and 6H poly-types, in both MOSFETs and MOS capacitors, and in devices with either a thermal or deposited gate oxide [1, 18, 20]. Effects due to self-heating when performing ON-state stress and allowing the rated current to flow through the device are very similar to those when externally heating the device during a gate-bias stress [13, 15]. Much larger V_T instabilities are observed in either case, compared with room-temperature gate-bias stressing [13-17, 24]. This is very likely due to the activation of additional oxide traps, which can then participate in the oxide trap charging process [13, 15, 17]. It is likely that interface traps are being generated as well [25], but this is not the cause of the increased V_T instability since the same interface trap charge state should exist when measuring the threshold voltage. Additional difficulties in sorting out the bias-temperature response may be due to the presence of mobile ions in some sample sets [3, 26, 27].

The gate oxides of SiC MOSFETs are similarly sensitive to ionizing radiation as are Si MOSFETs for oxides of a similar thickness [28-30]. The radiation response may provide valuable insight into the bias-temperature response, since both effects are likely due to the activation of so-called E′ centers in the oxide [13, 15, 17, 24], which are related to an oxygen vacancy [1, 31].

Reliability Testing

Existing reliability test standards (based on Si technology) have been demonstrated to be inadequate for SiC given that the same device may be deemed to have both passed and failed, simply as a consequence of the delay in the measurement that is allowed under present standards [14, 16]. We have also shown that three different conclusions may be drawn as to the effect of bias temperature stress, depending on whether immediate high-temperature measurements, immediate room-temperature measurements (enabled by rapid cooling), or later room-temperature measurements are performed [24]. This is in addition to the effects of measurement speed and direction mentioned above.

Summary and Conclusion

Although significant improvements in V_T stability have been demonstrated in state-of-the-art devices, the issue has not yet been fully resolved. High-temperature bias-stressing likely increases the number of active oxide traps that can tunnel into or out of the oxide depending on the applied gate bias and oxide field. For a given number of active oxide traps, the measurement conditions will determine the magnitude of the V_T instability observed. Sweeping the gate bias positively (from negative to positive gate bias), instead of negatively, following a positive-bias stress can lead to a significant underestimation of the total number of active switching oxide traps present in the gate oxide of a SiC MOSFET. Likewise, sweeping the gate bias under standard measurement conditions (e.g., a 1-s measurement using a parameter analyzer) also results in a significant underestimation of the total number of active switching oxide traps present. Faster measurements on the order of 100 μs lead to instabilities about four times as great. The measurement temperature and delay time are critical as well. Therefore, it is of great importance that proper—and consistent—measurement conditions be applied when evaluating the reliability of SiC power MOSFETs.

References

1. A.J. Lelis, D. Habersat, R. Green, A. Ogunniyi, M. Gurfinkel, J. Suehle, and N. Goldsman: *IEEE Trans. Elec. Dev.*, **55(8)**, p. 1835 (2008).
2. M. Gurfinkel, H.D. Xiong, K.P. Cheung, J.S. Suehle, J.B. Bernstein, Y. Shapira, A.J. Lelis, D. Habersat, and N. Goldsman, "Characterization of Transient Gate Oxide Trapping in SiC MOSFETs Using Fast *I–V* Techniques," *IEEE Trans. Elec. Dev.*, **55(8)**, p. 2004 (2008).
3. T. Okayama, S.D. Arthur, J.L. Garrett, and M.V. Rao, "Bias-stress induced threshold voltage and drain current instability in 4H-SiC DMOSFETs," *Solid-State Electronics*, **52**, p. 164 (2008).
4. M. J. Tadjer, K. D. Hobart, E. Imhoff, and F. J. Kub, "Temperature and Time Dependent Threshold Voltage Instability in 4H-SiC Power DMOSFET Devices," *Mater. Sci. Forum*, **600-603**, p. 1147 (2009).

5. M. Grieb, M. Noborio, D. Peters, A. J. Bauer, P. Friedrichs, T. Kimoto, and H. Ryssel, "Comparison of the threshold-voltage stability of SiC MOSFETs with thermally grown and deposited gate oxides," *Mater. Sci. Forum*, **645-648**, p. 681 (2010).
6. H. Yano, Y. Oshiro, D. Okamoto, T. Hatayama, and T. Fuyuki, "Instability of 4H-SiC MOSFET Characteristics due to Interface Traps with Long Time Constants," *Mater. Sci. Forum*, **679-680**, p. 603 (2011).
7. S. DasGupta, R. Brock, R. Kaplar, M. Marinella, M. Smith, and S. Atcitty, "Extraction of trapped charge in 4H-SiC metal oxide semiconductor field effect transistors from subthreshold characteristics," *Appl. Phys. Lett.*, **99**, 023503 (2011).
8. M. Okamoto, Y. Makifuchi, M. Iijima, Y. Sakai, N. Iwamuro, H. Kimura, K. Fukuda, and H. Okumura, "Coexistence of Small Threshold Voltage Instability and High Channel Mobility in 4H-SiC(000_1) Metal–Oxide–Semiconductor Field-Effect Transistors," *Applied Physics Express*, **5**, 041302 (2012).
9. S. Noll, D. Scholten, M. Grieb, A. Bauer, and L. Frey, "Electrical Impact of the Aluminum p-implant Annealing on Lateral MOSFET Transistors on 4H-SiC n-epi," *Mater. Sci. Forum*, **740-742**, p. 521 (2013).
10. T. Aichinger, P. Lenahan, and D. Peters, "Interface defects and negative bias temperature instabilities in 4H-SiC PMOSFETs – a combined DCIV/SDR study," *Mater. Sci. Forum*, **740-742**, p. 529 (2013).
11. H. Watanabe, D. Ikeguchi, T. Kirino, S. Mitani, Y. Nakano, T. Nakamura, T. Hosoi, and T. Shimura, "Novel Approach for Improving Interface Quality of 4H-SiC MOS Devices with UV Irradiation and Subsequent Thermal Annealing," *Mater. Sci. Forum*, **740-742**, p. 741 (2013).
12. G. Pobegen and T. Grasser, "Efficient Characterization of Threshold Voltage Instabilities in SiC nMOSFETs Using the Concept of Capture-Emission-Time Maps," *Mater. Sci. Forum*, **740-742**, p. 757 (2013).
13. A.J. Lelis, R. Green, and D. Habersat, "High-Temperature Reliability of SiC Power MOSFETs," *Mater. Sci. Forum*, **679-680**, p. 599 (2011).
14. R. Green, A. Lelis, and D. Habersat, "Application of Reliability Test Standards to SiC Power MOSFETs," *2011 Intl. Rel. Physics Symposium Proc.* **11**, p. 756 (2011).
15. A. Lelis, R. Green, and D. Habersat, "Effect of Threshold-Voltage Instability on SiC Power MOSFET High-Temperature Reliability," *ECS Transactions*, **41(8)**, p. 203 (2011).
16. R. Green, A. Lelis, and D. Habersat, "Charge Trapping in SiC Power MOSFETs and its Consequences for Robust Reliability Testing," *Mater. Sci. Forum*, **717-720**, p. 1085 (2012).
17. A. Lelis, R. Green, M. El, and D. Habersat, "Effect of Stress and Measurement Conditions in Determining the Reliability of SiC Power MOSFETs," *ECS Transactions*, **50(3)**, p. 251 (2012).
18. A.J. Lelis, D. Habersat, F. Olaniran, B. Simons, J.M. McGarrity, F.B. McLean, and N. Goldsman, "Time-Dependent Bias Stress-Induced Instability of SiC MOS Devices" *Mater. Res. Soc. Symp. Proc.*, **911**, article 0911-B13-05 (2006).
19. V.V. Afanas'ev, A. Stesmans, F. Ciobanu, G. Pensl, K.Y. Cheong, and S. Dimitrijev, "Mechanisms responsible for improvement of 4H-SiC/SiO2 interface properties by nitridation," *Appl. Phys. Lett.*, **82(4)**, p. 568 (2003).

20. D. Habersat and A. Lelis, "Improved Observation of SiC/SiO2 Oxide Charge Trapping Using MOS C-V," *Mater. Sci. Forum*, **679-680**, p. 366 (2011).

21. D.B. Habersat, A.J. Lelis, J.M. McGarrity, F.B. McLean and S. Potbhare, "The Effect of Nitridation on SiC MOS Oxides as Evaluated by Charge Pumping," *Mater. Sci. Forum*, **600-603**, p. 743 (2009).

22. H. Yano, T. Araok, T. Hatayama, and T. Fuyuki, "Improved Stability of 4H-SiC MOS Device Properties by Combination of NO and POCl$_3$ Annealing," *Mater. Sci. Forum*, **740-742**, p. 727 (2013).

23. A.J. Lelis, D. Habersat, R. Green, and N. Goldsman, "Two-way Tunneling Model of Oxide Trap Charging and Discharging in SiC MOSFETs," *Mater. Sci. Forum*, **717-720**, p. 465 (2012).

24. Ronald Green, A. Lelis, M. El, and D. Habersat, "A Study of High Temperature DC and AC Gate Stressing on the Performance and Reliability of Power SiC MOSFETs," *Mater. Sci. Forum*, **740-742**, p. 549 (2013).

25. D.B. Habersat, A. J. Lelis, R. Green, and M. El, "Evaluation of PBTS and NBTS in SiC MOS Using In Situ Charge Pumping Measurements," *Mater. Sci. Forum*, **740-742**, p. 545 (2013).

26. A.J. Lelis, D. Habersat, R. Green, and N. Goldsman, "Temperature-Dependence of SiC MOSFET Threshold-Voltage Instability," *Mater. Sci. Forum*, **600-603**, p. 807 (2009).

27. D.B. Habersat, A. Lelis, and R. Green, "Detection of Mobile Ions in the Presence of Charge Trapping in SiC MOS Devices," *Mater. Sci. Forum*, **717-720**, p. 461 (2012).

28. S. Dixit, S. Dhar, J. Rozen, S. Wang, R. Schrimpf, D. Fleetwood, S. Pantelides,J. Williams, and L. Feldman, "Total Dose Radiation Response of Nitrided and Non-nitrided SiO$_2$/4H-SiC MOS Capacitors," *IEEE Trans. Nucl. Sci.*, **53(6)**, p. 3687 (2006).

29. C.X. Zhang, E. Zhang, D. Fleetwood, R. Schrimpf, S. Dhar, S.-H. Ryu, X. Shen, and S. Pantelides, "Effects of Bias on the Irradiation and Annealing Responses of 4H-SiC MOS Devices," *IEEE Trans. Nucl. Sci.*, **58(6)**, p. 2925 (2011).

30. A. Akturk, J. McGarrity, S. Potbhare, and N. Goldsman, "Radiation Effects in Commercial 1200 V 24 A Silicon Carbide Power MOSFETs," *IEEE Trans. Nucl. Sci.*, **59(6)**, p. 3258 (2012).

31. C.J. Cochrane, P.M. Lenahan, and A.J. Lelis, "An Electrically Detected Magnetic Resonance Study of Performance Limiting Defects in SiC Metal Oxide Semiconductor Field Effect Transistors," *J. Appl. Phys.*, **109**, 014506 (2011).

94

CHAPTER 3

SiC EPITAXY

96

Comparison of SiC Epitaxial Growth from Dichlorosilane and Tetrafluorosilane Precursors

Haizheng Song, Tawhid Rana, M.V.S. Chandrashekhar, Sabih U. Omar, Tangali S. Sudarshan

Department of Electrical Engineering, University of South Carolina, Columbia, South Carolina 29208, USA

As a novel Si-precursor in chemical vapor deposition epitaxial growth of 4H-SiC, tetrafluorosilane (TFS) is studied for both its advantages and disadvantages. Due to its high Si-F binding energy, TFS presents exceptional ability in the suppression of parasitic deposition and Si-droplet formation during SiC epitaxial growth. As compared with epigrowth using a chlorinated precursor, dichlorosilane (DCS), growth using TFS shows a lower growth rate most likely due to the TFS etching of SiC during growth and the lower reactivity of TFS, but the quality of the TFS grown epilayers is always superior. Epigrowth using TFS also shows better control of the uniformities of epilayer surface morphology and doping concentration, which offers excellent uniformity of SiC device performance. The influence of growth conditions including temperature, gas flow rate, Si/H_2 ratio, and C/Si ratio on growth rate and epilayer properties is investigated in growth using TFS and compared with growth using DCS.

1. Introduction

Silicon Carbide (SiC) epitaxial growth by chemical vapor deposition (CVD) is a well-established key step for the fabrication of SiC high power devices. Achieving high quality, low cost, thick SiC epilayers is essential for the commercialization of SiC devices. In the conventional SiC-CVD process, generally, silane (SiH_4) is employed as the Si-precursor. The major drawback associated with the silane chemistry in achieving high growth rates is the relatively weak bond strength of the Si-H (318 kJ/mol) bond in SiH_4 causing it to dissociate easily into elemental Si in the SiC-CVD process (1); the CVD process requires deposition temperature as high as 1500°C or above. The dissociated elemental Si with free dangling bonds can easily form the Si-Si bond during their collisions and initiate liquid Si-droplets and particulates on the growth surface and also on the walls of the reactor including any gas delivery tube located in a section of high temperature (1,2). Particulates in the grown epilayers arising from gas phase nucleation lead to severe morphological defects that make the epilayers unusable for power device fabrication (2). This is currently the major issue slowing the large scale commercialization of SiC power devices.

In recent years, chlorine-based CVD has attracted more attention in SiC growth with either HCl addition in the silane system or direct use of chlorinated Si or C precursors

(1). Very high growth rate and reduced Si-droplet formation have been reported in CVD growth of 4H-SiC (3-7). This is due to higher Si-Cl bonding energy (381 kJ/mol) compared to Si-H or Si-Si (8).

In the authors' lab, it was found that in the process of CVD growth using either silane or dichlorosilane (DCS, SiH_2Cl_2) precursors, there is always significant amount of depositions generated on the walls of the gas injector tube after each growth run (2). The deposition on CVD reactor walls is called parasitic deposition. As shown in Fig. 1, the yellowish deposition observed on the inner wall of the gas injector tube, after epigrowth using silane and DCS, is mainly due to Si related species (9,10). These depositions can be partially removed by scrubbing the surface.

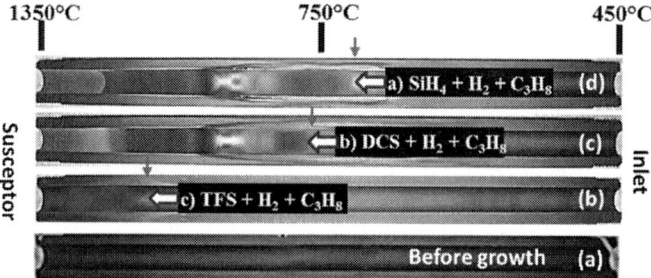

Figure 1. Images of gas injector tube (a) before epigrowth and after epigrowth using (b) tetrafluorosilane (TFS), (c) dichlorosilane (DCS), and (d) silane (SiH_4) precursors. The starting points of the parasitic deposition are shown by vertical arrows.(2)

It is believed that parasitic deposition is one of the major reasons that limit growth rate and epilayer quality because it will consume large amounts of the precursor gases and the loosely bound depositions can be carried to the growth surface causing defect inclusions during growth. The parasitic deposition occurs because of too early decomposition of the precursor gases in the transportation system, essentially due to the weak bond energy of the precursor molecules. Although the parasitic deposition often is not given sufficient attention in laboratory scale experiments, it cannot be ignored in commercial scale epitaxial production considering cost and product quality (11).

Recently, in the authors' group, a fluorinated silicon gas, tetrafluorosilane (TFS, SiF_4), was utilized as the Si precursor in the SiC-CVD process (2,10). Benefiting from high strength of the Si-F bond (565 kJ/mol), and thus high decomposition temperature of the Si precursor gas, significant reduction of parasitic deposition and related particles in the CVD process has been observed using TFS, as shown in Fig. 1b. High quality thick (~120 μm) 8° 4H-SiC epilayers have been demonstrated at a growth rate of ~30 μm/hr in a particle suppressed growth condition (2). A more important benefit from inhibiting parasitic deposition is that the reactor parts have a long lifetime. They can be used for more growth runs and yet achieve repeatable results; thus maintenance of the CVD reactor becomes easier.

As a novel precursor in 4H SiC-CVD, TFS needs to be extensively examined for both its advantages and disadvantages and whether this high Si-F bond strength precursor gas is, in practice, suitable for 4H SiC growth. In this paper, epigrowth on 4° SiC substrates using TFS precursor is addressed. The influence of growth conditions including temperature, gas flow rate, Si/H_2 ratio, and C/Si ratio on growth rate and epilayer properties is investigated and these results are compared with growth using a chlorinated precursor, DCS. Finally, Schottky diodes are fabricated on epilayers grown using TFS and DCS precursors and device properties are compared.

2. Experimental

The epitaxial growth was carried out in a home-built vertical hot wall CVD reactor. The gas injector tube in the reactor is axially split into two half parts (2). This split tube can be assembled to form a complete tube before growth and can be observed after the growth by separating them. The parasitic deposition generated during the growth run can be directly observed on the inner wall of the injector tube (as shown in Fig. 1).

The substrates used in this study were 8 mm × 8 mm pieces diced from commercial 4H-SiC wafers (Si face CMP polished, 4° and 8° off cut towards [11$\bar{2}$0] direction, n-type doing concentration of ~10^{18} cm^{-3}). Propane was used as the C-precursor; DCS (7% in Ar) or TFS (10% in H_2) was used as the Si-precursor; and H_2 was the carrier gas purified in a palladium hydrogen purifier. H_2 flow rate could be changed from 2~10 slm. Typical growth conditions were 1530~1650°C and 80~300 Torr.

The thicknesses of the epilayers were measured using the Fourier transform infrared reflectance (FTIR). The surface morphology of the epilayers was analyzed using Nomarski optical microscope (NOM) and atomic force microscopy (AFM). The mercury-probe C-V analysis was used for measuring the net doping concentration of the epilayers. X-ray diffraction (XRD) rocking curves and Raman-spectroscopy were used to determine the crystalline quality and the polytype inclusions, respectively.

Two epilayers were used to fabricate Schottky diodes to test I-V performance. One is 12 μm thick, grown using DCS on 8° offcut substrate, another is 25 μm thick, grown using TFS on 4° offcut substrate. Both epilayers have a net n-type doping concentration of ~10^{15} cm^{-3}. The surfaces of the epilayers were cleaned by a standard RCA procedure and were dry-oxidized at 1150°C for 4 hours. The (sacrificial) oxide layer was removed by wet etching in 1:10 HF acid solution. An ohmic contact was formed on the backside of the highly doped substrate by depositing Nickel and rapid annealing at 1000°C in Ar ambient. The epilayer surface was patterned by standard lithography to form circular 'dots' of three different diameters, 100, 150 and 250 μm. The Schottky contact was formed by depositing Nickel by e-beam evaporation at room temperature, followed by a post-lift-off rapid thermal annealing at 650°C in Ar gas. The Schottky annealing step was shown to produce highly reproducible, ideal Nickel Silicide/SiC Schottky contact (12).

3. Results and discussion

3.1. Effect of Si/H$_2$ ratio

First, the effects of DCS/H$_2$ and TFS/H$_2$ ratios were studied in the condition without propane flow. This study is similar to the case for in-situ etching or pretreatment prior to the initiation of epigrowth, but here, we will focus on the analysis of the different behaviors of DCS and TFS. Fig. 2 shows the deposition or etching rates at various H$_2$ flow rates (2~10 slm) with constant flow of a gas precursor (propane, DCS, and TFS). The experiments were performed on 8° 4H-SiC substrates at 1600°C, 300 Torr. The deposition and etching rates were determined by mass measurement before and after the experiment.

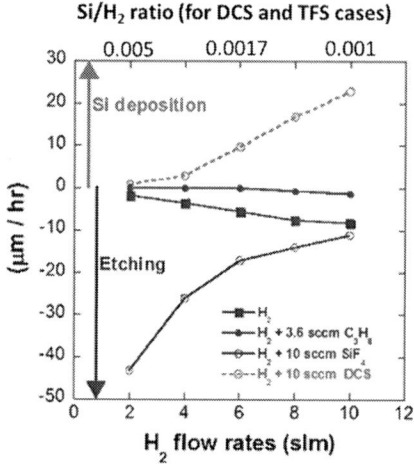

Figure 2. Etch/Si-deposition rates using various precursor gases as a function of H$_2$ gas flow rate at 1600°C 300 Torr. (8)

As expected, pure H$_2$ leads to etching of the SiC substrate; the etching rate increases with increasing of the H$_2$ flow rate, and the etching can be suppressed by adding a small amount of propane (C$_3$H$_8$). In the gas system of H$_2$ + 10 sccm DCS, the Si deposition rate increases with increase in H$_2$ flow rate and the deposition is due to Si-droplets formed on the SiC substrate. Thus DCS was found to be ineffective in terms of etching of SiC due to the domination of Si-droplet deposition. This is due to the decomposition of DCS at ~1000°C. Fig. 2 also indicates an increased DCS decomposition rate at higher H$_2$ flow rates (8). In the case of epitaxial growth using DCS, with propane addition, the epilayer surface starts to degrade at a DCS flow rate of 7 sccm (will be discussed in the next section). The optimal growth condition using DCS was found to be 4.5 sccm DCS + 6 slm H$_2$.

As seen in Fig. 2, TFS, however, shows strong etching of the SiC substrate in H_2 environment. The etching rate is as high as 43 μm/hr when H_2 flow rate is low (2 slm). An earlier report from the authors' group indicates that TFS reacts with Si on the SiC surface, enhancing H_2 etching by faster Si removal from the surface in the form of SiF_2 gas (8). The etching reaction is:

$$SiC + SiF_4 + H_2 \rightarrow 2SiF_2 + CH_2 \qquad [1]$$

The etching rate is reduced by diluting TFS with H_2 gas, i.e., reducing the SiF_4 partial pressure. SiC-CVD growth represents a competition between growth and etching (13). The results in Fig. 2 imply that in epigrowth using TFS, high H_2 flow rate is required to avoid too aggressive an etching. The TFS etch rate is independent of the off-cut angle (4° and 8°) and is expected to increase at higher temperatures (8). It is noted that a suitable TFS etch rate of SiC substrate improves the surface morphology of the substrate. The optimal growth condition using TFS was found to be 9~10 sccm TFS + 9~10 slm H_2.

3.2. Effect of temperature and precursor flow rate on growth rate and surface morphology

The epigrowth using TFS and DCS precursors was performed in the temperature range of 1530~1630°C. As shown in Fig. 3, in all of the precursor systems, the growth rate of 4° 4H-SiC decreases with increasing growth temperature. This behavior indicates that epigrowth in this temperature range is desorption limited. For high TFS flow rate (9.6 sccm), a large decrease of growth rate at higher temperatures may be attributed to the increase of TFS etching rate, as expected from the results of Fig. 2.

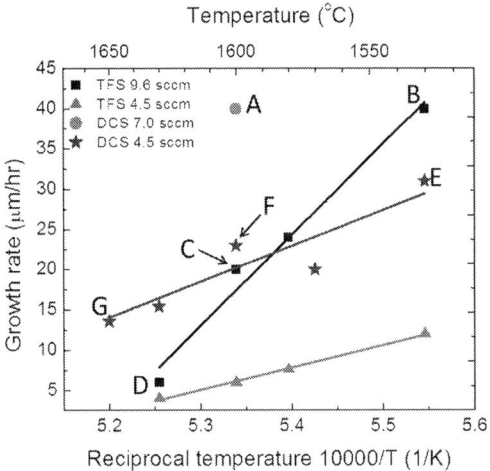

Figure 3. Growth rates of 4° 4H-SiC at various temperatures. C/Si ratios are fixed at 1.5 for TFS growth and 1.3 for DCS growth. Lines are drawn to guide the eye.

It is noted that at the same Si precursor flow rate (e.g., 4.5 sccm), growth using TFS has lower growth rates compared to the growth using DCS. This may be explained from two possible reasons. First, as discussed in the last section and reference (8), TFS is a strong etchant for SiC, but DCS is an ineffective SiC etchant. The SiC epitaxial growth is a competition between growth and etching. Stronger etching in the TFS system reduces the net growth rate. Second, due to the strong Si-F bound, reactivity of TFS on the SiC surface is lower than that of DCS which also limits the growth rate.

The change of growth rate with increasing precursor flow rates is investigated in both DCS and TFS precursor systems at 1600°C for a fixed C/Si ratio. The results in Fig. 4 also show lower growth rates in the TFS growth system. The slopes obtained by linear fitting indicate that TFS growth has a lower surface reaction rate constant. The intercepts on the y axis (zero flow rate of the precursors) gives the H_2 etching rate prior to the growth. The values are 5.6 and 6.6 μm/hr at 6 and 9 slm H_2 flow rates respectively, which are close to the earlier reported values by the authors' group shown in Fig. 2. The different intercepts on the x axis (zero growth rate) are attributed to the etching effect of the two precursor gases in the growth condition. Further, there is only H_2 etching in the DCS growth system; net growth will start at a DCS flow rate of 0.9 sccm. In TFS growth, both H_2 and TFS etch the SiC surface so the net growth has to be initiated at a higher flow rate (2.3 sccm) of the precursor.

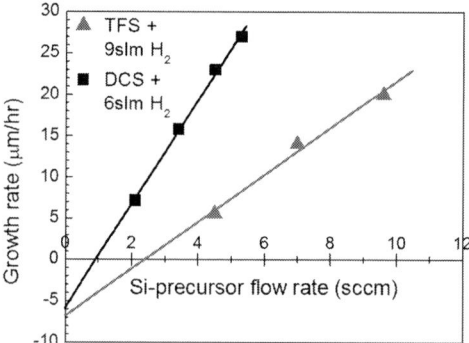

Figure 4. Growth rates of 4° 4H-SiC at various Si-precursor flow rates at fixed growth temperature 1600°C, and fixed C/Si ratios of 1.5 for TFS growth and 1.3 for DCS growth.

The surface morphology of epilayers grown at different temperatures in DCS and TFS growth is analyzed by NOM and AFM. The selected samples are marked A~G in Fig. 3, and the NOM and AFM images are shown in Figures 5 and 6. For a high DCS flow rate of 7.0 sccm (sample A in Fig. 3), the epilayer surface significantly degrades; comet-like features and increased roughness are observed on the surface (Fig. 5a). However, all of the epilayers (samples B~D in Fig. 3) grown using TFS at an even higher flow rate, 9.6 sccm, present specular surfaces (Fig. 5b~d). Only slight increase of step-bunching is found with increasing of the growth temperature. The effect of further increase in TFS flow rate is under investigation.

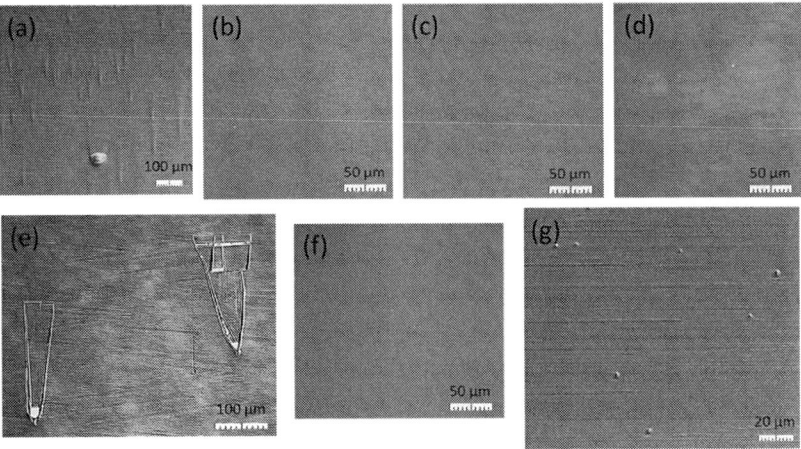

Figure 5. NOM images of epilayers grown at different temperatures and precursors. (a) – (g) correspond to samples A – G in Fig. 3.

Generally, increasing growth temperature will reduce the incorporations of morphological defects, 3C inclusions and IGSFs (14,15). For epigrowth using DCS (4.5 sccm) at the temperature of 1530°C (sample E in Fig. 3), high densities of triangular and carrot defects ($\sim$200 cm^{-2}) and zigzag surface steps are observed on the epilayer (Fig. 5e). In TFS growth at the same temperature (sample B), specular surface with minor step-bunching is achieved (RMS=1.2 nm in 20 μm × 20 μm area) (Fig. 5b). This epilayer is free of triangular defects and only has carrot defects with density 3$\sim$8 cm^{-2}. However, in all of the epilayers grown at this low temperature, 8H-IGSFs are found to be in the densities of 100$\sim$400 cm^{-2}. Elevating temperature ($>$=1600°C) will significantly reduce the IGSF density on epilayers in the DCS growth system ($<$10 cm^{-2}). However, on TFS grown epilayers, the IGSF density is still high (30$\sim$90 cm^{-2}). Understanding of the reasons that contribute to high IGSF density will be a subject of future studies.

At an elevated growth temperature of 1650°C, the epilayer grown using DCS (sample G) shows a wavy surface with strong step-bunching and growth pits (Fig. 5g). The epilayers grown using TFS still preserve a specular surface with few shallow growth pits ($\sim$10 cm^{-2}), but the growth rates are low.

The optimal growth is found to be around 1600°C in both TFS and DCS growth systems: sample C and sample F respectively. Both samples have similar growth rates, $\sim$20 μm/hr, and very smooth surfaces as shown in Fig. 5c and f. The surface roughness and step structure of these two epilayers ($\sim$10 μm thickness) was measured by AFM. Fig. 6 shows that the epilayer grown using TFS has a smaller RMS value (1.3 nm) of surface roughness. The step-bunching occurs on both samples. The epilayer from TFS growth has a macro step height of ~2 nm (~8 bilayers), but in DCS growth, the epilayer has more severe step-bunching; the RMS value is 2.2 nm and the macro step height is ~8 nm (~32 bilayers). It is very likely that the smoother surface and weaker step-bunching is attributed to the positive etching effect of TFS. As discussed in reference (8), moderate TFS etching of SiC can improve the surface morphology of the substrate; thus a smoother epilayer surface is achieved.

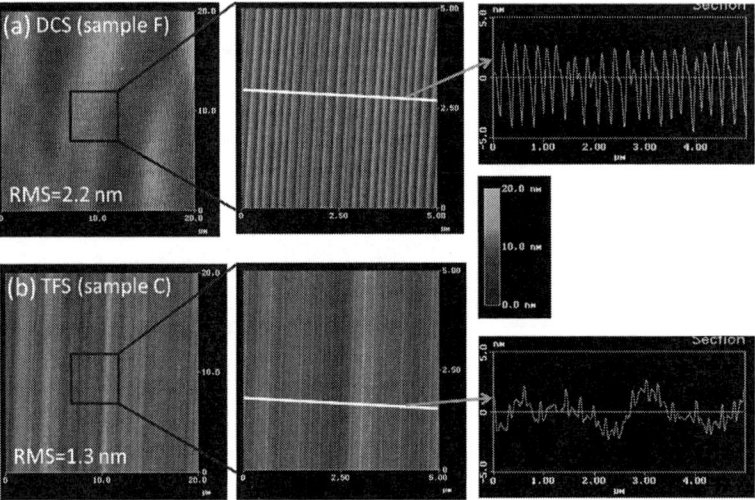

Figure 6. AFM images of (a) sample F and (b) sample C in Fig. 3, grown using DCS and TFS, respectively. (Both epilayers are about 10 μm thick and 10^{15} cm^{-3} n type doping.)

3.3. Effect of C/Si ratio on growth rate and doping concentration

The influence of C/Si ratio (at fixed Si precursor flow rate) on growth rate in both DCS and TFS growths are shown in Fig. 7a and b, respectively. The DCS growth is C supply limited when C/Si < 1 and becomes Si supply limited when C/Si > 1. This trend is similar to the growth using silane precursor (16). However, in TFS growth, the growth rate steadily increases with increasing C/Si ratio until the surface degrades at above C/Si ratio of 1.6. This suggests a C supply dependent growth mechanism.

The unintentional doping concentration at various C/Si ratios is shown in Fig. 8. Transition from n to p type doping is due to the site-competition mechanism (17). The doping behavior in epigrowth using DCS is very sensitive to C/Si ratio (2×10^{16} cm^{-3} n-type to 5×10^{15} cm^{-3} p-type for C/Si ratio ranging 0.8~1.4). However in epigrowth using TFS, over a wider C/Si ratio, 0.3~1.6, the epilayers show doping concentrations about one order of magnitude less (1.6×10^{15} cm^{-3} n-type to 1.2×10^{15} cm^{-3} p-type). It is proposed that the above behavior is related to the different gas decomposition mechanisms of DCS and TFS leading to different actual (effective) C/Si ratios at the growth surface (18,19). It is noteworthy that in epigrowth using TFS, there is a larger window of C/Si ratio for achieving high quality low doped SiC epilayers desired for high power SiC devices.

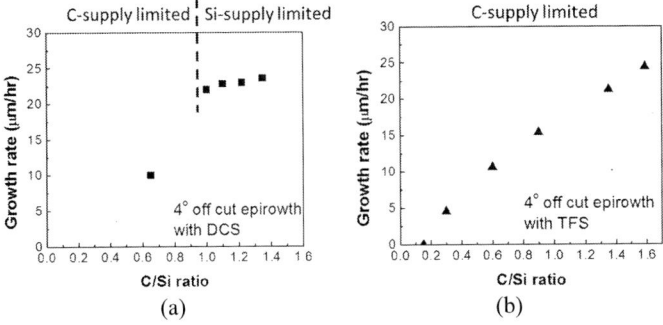

Figure 7. Growth rate as a function of C/Si ratio (at fixed Si precursor flow rate) in epigrowth using (a) DCS and (b) TFS.

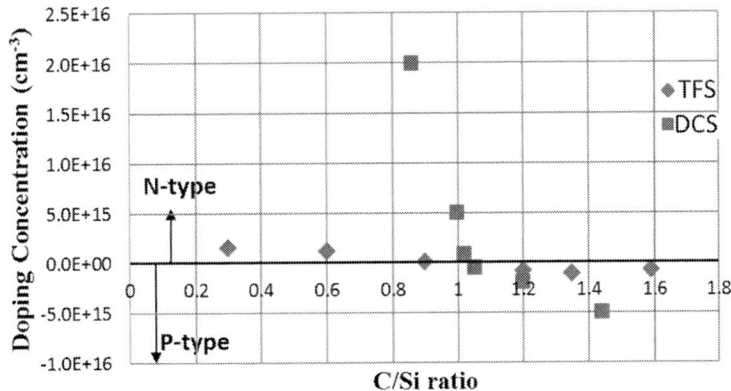

Figure 8. Doping concentration versus C/Si ratio for epilayers grown using TFS and DCS precursors.

3.4. Comparison of the quality of epilayers grown using DCS and TFS

The crystallinity of epilayers grown using DCS and TFS was measured by X-ray rocking curve of the (0008) plane diffraction. As shown in Fig. 9, the full-width half maximum (FWHM) of (0008) peaks are 12.7 and 7.5 arcsecond for epilayers from DCS and TFS growth respectively, indicating a better crystalline quality of the epilayer from TFS growth.

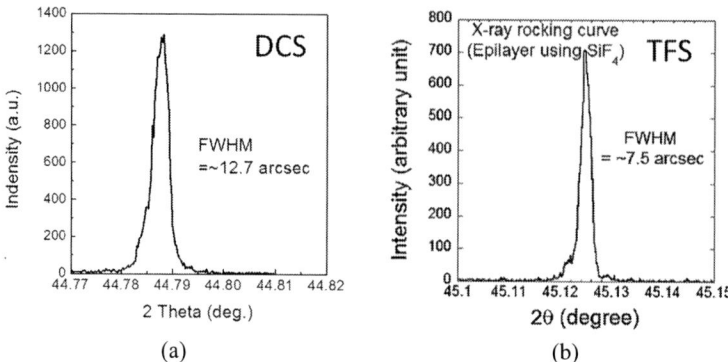

(a)

(b)

Figure 9. Comparison of X-ray diffraction rocking curves of epilayers grown using (a) DCS and (b) TFS precursors.

Raman spectroscopy, in the back scattering geometry, is utilized to analyze 3C polytype inclusion in the epilayers. The possibility of 3C inclusion in 4H polytype is analyzed by the intensity ratio of 4H peak (E_2, transverse optic (TO) mode at 776 cm^{-1}) to 3C peak (E_1, TO mode at 796 cm^{-1}) (20). Fig. 10 shows the Raman spectra of two 4° 4H-SiC epilayers grown using DCS and TFS respectively. The epilayer from TFS growth shows a 4H/3C ratio of 90, much higher than that in epilayer from DCS growth (4H/3C=40), indicating a lower possibility of 3C generation during the growth using TFS precursor.

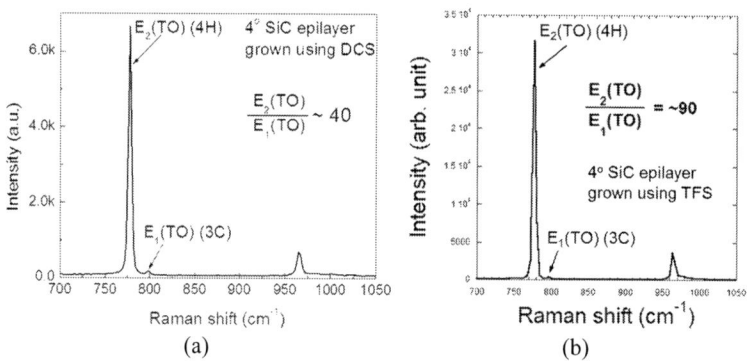

(a)

(b)

Figure 10. Comparison of Raman spectra of epilayers grown using (a) DCS and (b) TFS precursors.

3.5. Comparison of device performance for epilayers grown using DCS and TFS

Schottky diodes were fabricated on two epilayers, a 12 μm thick 8° offcut epilayer grown by DCS and a 25 μm thick 4° offcut epilayer grown by TFS. Both epilayers have a net n-type doping concentration $\sim 10^{15}$ cm^{-3}. Fig. 11 shows comparison between the I-V characteristics obtained from epilayers grown by DCS (a) and TFS (b). The average values for barrier heights (Φ_b) and ideality factor values (n) were similar for these two samples; however, the standard deviations from the average values were an order of magnitude larger for the DCS-grown sample than those for the TFS grown sample. The 'tighter' distribution of the surface barrier heights for the TFS-grown epilayer infers a more uniform surface morphology and doping levels over the entire epi surface. It is to be noted that the TFS data were obtained for a 4° offcut epilayer, which is expected to show a greater amount of non-uniformity in surface morphology because of the presence of surface step bunching as compared to an 8° offcut epilayer.

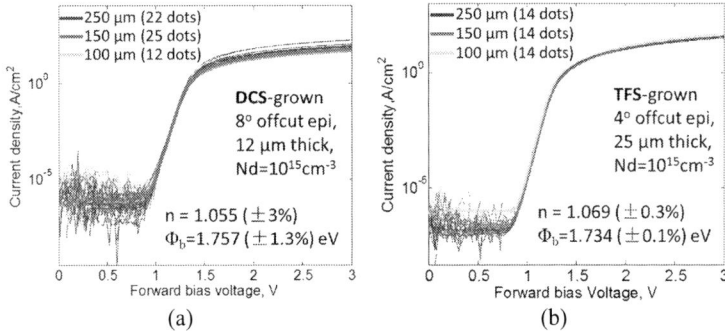

Figure 11. Forward I-V characteristics for Schottky diodes fabricated on (a) an epilayer grown using DCS on 8° offcut substrate and (b) an epilayer grown using TFS on a 4° offcut substrate. The latter shows a tighter distribution of the Schottky barrier parameters among the Schottky dots. The DCS data (a) were obtained from 22 Schottky 'dots' of 250 μm diameter, 25 of 150 μm and 12 of 100 μm. The TFS data (b) were obtained from 14 dots of each size.

4. Conclusions

As a novel Si precursor having high Si-F bond energy, TFS presents exceptional ability in the suppression of parasitic deposition that cannot be avoided in CVD growth using silane and chlorinated Si precursors. In SiC epigrowth using TFS, Si-droplet formation in the gas phase (i.e., gas-phase nucleation) is completely eliminated even at high TFS flow rates, which is one of the main challenges in epigrowth using silane and chlorinated Si precursors. TFS provides a clean environment for low cost and repeatable epigrowth. Excellent control of the uniformities of morphology and doping concentration will make TFS especially suitable for large area (e.g., 6 inches) and multi-wafer (e.g., in planetary CVD reactor) CVD growth to achieve SiC devices with excellent performance uniformity. Although in epigrowth using TFS, the growth rate is lower than that in DCS growth, due to the TFS etching of SiC and the lower reactivity of TFS, this issue can be compensated by increase of the TFS flow rate during the growth (due to absence of Si-droplet formation). In terms of epilayer quality including morphology, roughness, crystallinity, polytype inclusion etc., epigrowth using TFS is found to be superior to growth using DCS or silane precursors.

Acknowledgments

This work was supported by the Office of Naval Research, grant no. N000141010530. The authors thank the program manager Dr. H. Scott Coombe for his support of this research. This research is partially being performed using funding received from the DOE Office of Nuclear Energy's Nuclear Energy University Programs.

References

1. H. Pedersen, S. Leone, O. Kordina, A. Henry, S. Nishizawa, Y. Koshka, and E. Janzen, *Chemical Reviews*, **112**, 2434 (2012).
2. T. Rana, M. V. S. Chandrashekhar, and T. S. Sudarshan, *Physica Status Solidi A*, **209**, 2455 (2012).
3. F. La Via, G. Izzo, M. Mauceri, G. Pistone, G. Condorelli, L. Perdicaro, G. Abbondanza, L. Calcagno, G. Foti, and D. Crippa, *Journal of Crystal Growth*, **311**, 107 (2008).
4. S. Kotamraju, B. Krishnan, and Y. Koshka, *Physica Status Solidi RRL*, **3**, 157 (2009).
5. S. Leone, F. C. Beyer, H. Pedersen, O. Kordina, A. Henry, and E. Janzen, *Crystal Growth and Design*, **10**, 5334 (2010).
6. I. Chowdhury, M. V. S. Chandrasekhar, P. B. Klein, J. D. Caldwell, and T. S. Sudarshan, *Journal of Crystal Growth*, **316**, 60 (2011).
7. A. Henry, S. Leone, F. C. Beyer, H. Pedersen, O. Kordina, S. Andersson, and E. Janzen, *Physica B*, **407**, 1467 (2012).
8. T. Rana, M.V.S. Chandrashekhar, and Tangali S. Sudarshan, *Journal of Crystal Growth*, **380**, 61 (2013).
9. T. Rana, H. Song, M. V. S. Chandrashekhar, and T. S. Sudarshan, *Materials Science Forum*, **717-720**, 153 (2012).
10. T. S. Sudarshan, T. Rana, H. Song, and M.V.S. Chandrashekhar, *ECS Journal of Solid State Science and Technology*, **2** (8), N3079 (2013).
11. M. J. O'Loughlin, M. J. Paisley, and J. J. Sumakeris, US7118781 B1 (2006).
12. F. Roccaforte, F. La Via, V. Raineri, P. Musumeci, L. Calcagno, and G.G. Condorelli, *Applied Physics A*, **77** (6), 827 (2003).
13. B. L. VanMil, K.-K. Lew, R. L. Myers-Ward, R. T. Holm, D. K. Gaskill, C. R. Eddy Jr., L. Wang, and P. Zhao, *Journal of Crystal Growth*, **311** (2), 238 (2009).
14. C. Hallin, A. O. Konstantinov, B. Pecz, O. Kordina, and E. Janzen, *Diamond and Related Materials*, **6**, 1297 (1997).

15. S. Izumi, H. Tsuchida, I. Kamata, and T. Tawara, *Applied Physics Letters*, **86**, 202108 (2005).
16. W. Chen and M. A. Capano, *Journal of Applied Physics*, **98**, 114907 (2005).
17. D. J. Larkin, P. G. Neudeck, J. A. Powell, and L. G. Matus, *Applied Physics Letters*, **65**, 1659 (1994).
18. A. Fiorucci, D. Moscatelli, and M. Masi, *Journal of Crystal Growth*, **303**, 345 (2007).
19. S. Nishizawa and M. Pons, *Chemical Vapor Deposition*, **12**, 516 (2006).
20. H. Harima, S. Nakashima, and T. Uemura, *Journal of applied physics*, **78**, 1996 (1995).

110

Reducing the wafer off angle for 4H-SiC homoepitaxy

K. Kojima[a,b], K. Masumoto[a,b], S. Ito[b], A. Nagata[b] and H. Okumura[a,b]

[a] R &D partnership for Future Power Electronics Technology 2-9-5 Toranomon, Minato-ku, Tokyo 105-0001, Japan
[b] Advanced Power Electronics Research Center, National Institute of Advanced Industrial Science and Technology
Central 2 1-1-1 Umezono, Tsukuba, Ibaraki 305-8568, Japan

We have investigated key factors for controlling the polytype and surface morphology of 4H-SiC homoepitaxial growth on less than 4° off-axis substrates. In addition, we characterized the crystal quality and surface quality of the epitaxial layer in whole 3-inch vicinal off angled substrate. The results suggested that the control of surface energy, control of the vicinal off angle itself, and high temperature growth, is highly important in controlling the surface morphology and polytype stability of the epitaxial layer grown on a vicinal off angled substrate. We also obtained a high-quality epitaxial layer grown on a 3-inch vicinal off angle substrate, which was comparable to those on 4° off-axis substrates. These results suggest that the wafer off angle for epitaxial growth can be reduced to less than 4 degrees.

Introduction

Silicon carbide (SiC) is a potentially useful material for high-power, high-frequency, high-temperature, and low-power-loss electronic devices. Bulk crystal growth of SiC ingots is usually performed using sublimation growth method. This method suffers, however, from a lack of control over doping concentrations, and such control is necessary for the fabrication of electronic devices on semiconductor wafers. It is therefore necessary to employ the epitaxial growth method, which can provide effective reproducible control of doping concentration on SiC wafers grown by the sublimation method.

In early studies of SiC epitaxial growth, it was difficult to control the polytype. SiC has many polytypes, such as 3C, 6H, 4H, and 15R. These polytypes differ in stacking sequences of three Si-C layers along the c-axis. However, the crystal structure on the {0001} plane of α-SiC (as in 6H and 4H polytypes) is identical to that on the (111) plane of the 3C polytype. For this reason, 3C-SiC can easily be made to include α-SiC epitaxial layers grown on (0001) substrates by chemical vapor deposition. However, step-controlled epitaxy, in which the growth is performed on off-axis substrates, tilted several degrees toward the [11-20] direction,[1,2] could solve this problem and markedly increases in the stability of the polytype. In addition, a specular surface morphology can also be obtained under various growth conditions. As a result, high-quality SiC epitaxial wafers can be reliably produced. At present, 4° off-axis substrates are in common use. However, existence of such off angle makes some problems. In the case of SiC wafers, wafer size has reached 150mm, and the significant waste resulting from the off cutting of an ingot is required to be reduced. In addition, lowering the off angle can reduce the density of basal plane dislocations propagated from the substrate into the epitaxial layer.[3] In the case of a trench MOSFET, the wafer off-angle influences the anisotropy of the MOS channel characteristics.[4,5] Reducing the wafer off angle would be of great benefit in addressing such challenges.

In this study, we have investigated key factors for controlling the polytype and surface morphology of 4H-SiC homoepitaxial growth on less than 4° off-axis substrates, and have characterized the respective crystal and surface quality.

Experiments

Epitaxial growth was performed using two different CVD reactors. One was a conventional horizontal hot wall type chemical vapor deposition with $H_2 - C_3H_8 - SiH_4$ gas system and 1 x 2-inch wafer capacity. The H_2 flow rate was 40 slm and the typical growth temperature, growth pressure, SiH_4 flow rate, and C/Si ratio, were 1600 °C, 250 mbar, 6.67 sccm, and 0.6 - 0.9, respectively.

The other was similar horizontal hot wall CVD reactor, with similar gas system, but with 3 x 6 inch wafer capacity. The growth conditions were optimized based on those of the smaller system. Here, the flow rate of H_2 and SiH_4 was 100 slm and 50sccm, respectively, the typical growth temperature was 1700°C, and the C/Si ratio was varied from 0.7 to 1.0 by changing the C_3H_8 flow rate.

Vicinal off angled 2-inch and 3-inch wafers, with off-angles of less than 1°, were used. The wafer off angle was confirmed by XRD measurements. The surface morphology was investigated using a Nomarski microscope and atomic force microscope (AFM). Polytype inclusions were investigated using a photoluminescence image system with a UV lamp as the excitation source and a color CCD camera as the detector. Ni/SiC Schottky barrier diodes (SBDs) were formed on the epitaxial layers, to investigate the surface quality based on the I-V characteristics of the SBDs. The diameter of the Ni Schottky metal was 1 mm. Al was used as the backside contact. There were 65 measurement points over the entire 3-inch wafer.

Results and discussions

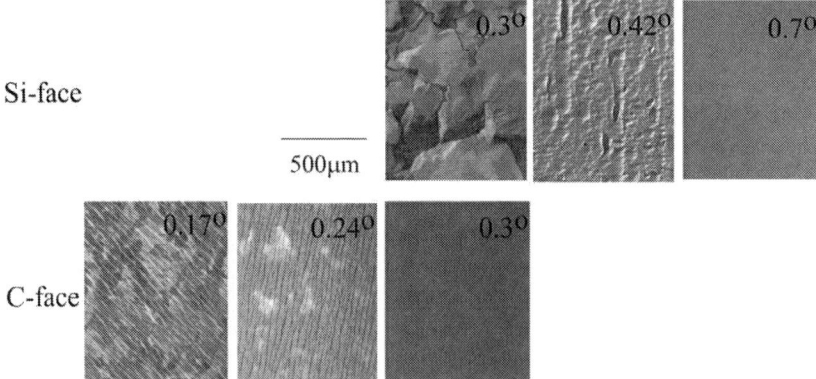

Figure 1 Surface morphology of grown epitaxial layers with varying vicinal off angle and pace polarity.

Figure 1 shows the surface morphologies of grown epitaxial layers with varying vicinal off-angle and pace polarity. The growth temperature, growth pressure, SiH$_4$ flow rate, and C/Si ratio, were 1600 °C, 250 mbar, 6.67 sccm and 0.5, respectively. The thickness of these epilayers was about 10 μm. As is well known, the surface morphology was degraded as the wafer off angle from 8 ° to 4 ° due to the generation of macro step bunching.[6] It was clear that surface morphology depended strongly on the vicinal off angle, and controlling the latter led to good surface morphology. The critical off angle for obtaining good surface morphology differed with the face polarity. For the C-face, more than 0.3° was required in order to obtain good surface morphology. However, the critical off angle on the Si-face was 2.5 times larger than that on the C-face. As is well known, the surface energy of the C-face is less than that of the Si-face,[7] and step bunching is effectively suppressed at very low off-angles as shown in Figure 1. In the case of the Si-face, significant step bunching is easily generated, due to the greater surface energy in comparison to the C-face, and 2D nucleation is also generated on the wider terraces of the bunched step. This increases the critical off angle significantly as shown in this figure.

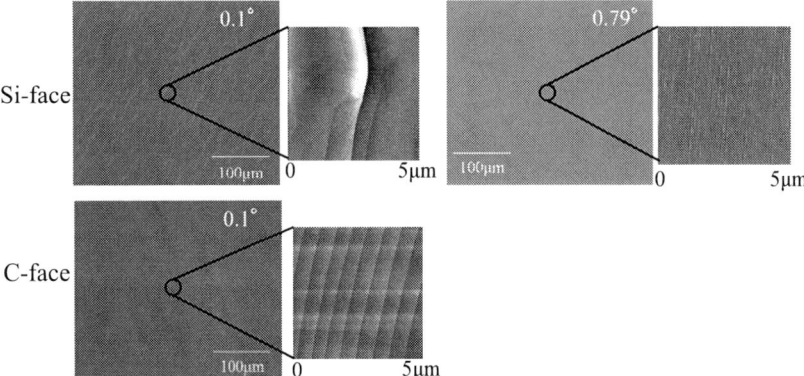

Figure 2 Surface morphology of substrates just after in-situ H$_2$ etching. AFM images are 5 μm square.

The substrate also showed the same surface morphology just after in-situ H$_2$ etching as shown in Figure 2. When the wafer off-angle was close to the on-axis (0.1°), macro step bunching which can be observed by optical microscope was generated on the substrate surface of the Si-face before epitaxial growth. In the case of the C-face, the generation of macro step bunching was effectively suppressed for on-axis substrates, due to the lower surface energy in comparison with the Si-face. However, increasing the vicinal off-angle to 0.79 ° effectively suppressed such step bunching. AFM observation confirmed a step terrace structure, with a step height of less than 1 nm, on the sample. This result indicates that macro step bunching is already generated on the substrate surface during the in-situ H$_2$ etching process. If this is the cause, we can easily assume that the epilayer surface exhibits the same step bunching structure or 2D nucleation owing to the existence of large terraces. This situation is consistent with the observed relationship between the vicinal off-angle and the surface morphology of grown epilayers as shown in Fig.1. Therefore, it is necessary to control the vicinal off-angle in order to obtain epitaxial layers with a specular surface on vicinal off angled substrates.

Si-rich conditions, with their low C/Si ratio, also suppress the generation of step bunching. Figure 3 shows the surface morphology with various C/Si ratio and face

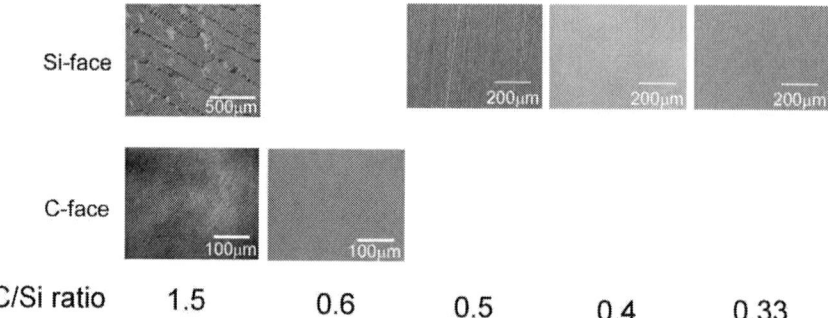

| C/Si ratio | 1.5 | 0.6 | 0.5 | 0.4 | 0.33 |

Figure 3 Surface morphology with varying C/Si ratio and face polarity.

polarity. When the C/Si ratio was 1.5, macro step bunching was generated on both faces. However, by decreasing the C/Si ratio, the generation of such bunched step structure could be effectively suppressed, and good surface morphology could be obtained at a C/Si ratio of 0.6 on the C-face. On the other hand, the Si-face required a lower C/Si ratio of 0.4 to obtain specular surface morphology without macro step bunching. Si-rich conditions are considered to be effective in reducing surface energy.[8] In general, the interfacial free energy between a liquid phase and a solid phase is less than that between a gas phase and a solid phase. We assume that the environment just above the substrate is similar to the interface between a liquid phase and a solid phase under these Si-rich conditions. Indeed, under too Si-rich growth conditions, Si droplets are generated, corresponding to a liquid phase. C-rich conditions, on the other hand, do not affect the surface free energy, because carbon does not exhibit a liquid phase. Then, macro step bunching was generated under C-rich conditions to reduce the surface energy as shown at the surface morphology with a C/Si ratio of 1.5. Therefore, control of the wafer off angle and surface energy is important key factors to obtain good surface morphology on vicinal off angled epitaxial wafers.

We subsequently applied the above results to a similar process involving a horizontal hot wall CVD reactor with 3 x 6 inch wafer capacity, and characterized the epitaxial layers grown on a 3-inch vicinal off-angled Si-face substrate. The results of the

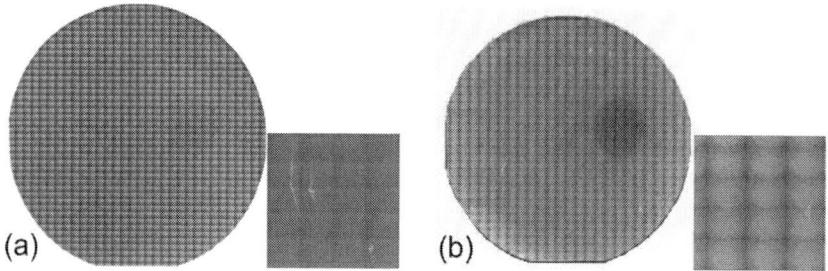

Figure 4 Surface morphology and polytype stability in whole 3-inch wafer.
(a) Nomarski image, (b) photoluminescence image.
Inset : High magnification image of 1cm square.

optimization suggested a typical growth temperature and C/Si ratio of 1700°C and 0.7, respectively. The uniformity (σ/mean) of thickness and carrier concentration of a typical epitaxial layer was investigated. We used 137 points to measure thickness, and 56 points to measure carrier concentration in whole 3 inch wafer. The uniformity of thickness and carrier concentration were 1.7% and 5.6%, respectively, when the mean thickness and doping concentration were 7.1 μm and 6.1 x 10^{15} cm^{-3}, respectively. Using AFM measurement, steep bunching was confirmed to be effectively suppressed, and step height to be effectively limited to below 1 nm for the area in the 10 μm square. These results were comparable to those of epitaxial layers with a 4° off-cut.[9]

Figure 4 shows (a) the Nomarski image and (b) the photoluminescence (PL) image in whole 3 inch epitaxial layer. Fig. 4(a) reveals some number of triangular defects. The PL image shows a greenish color, which indicates that the polytype is 4H. However, blue and yellow triangles area also observed as shown in Fig. 4 (b). This means that other polytypes such as 8H or 3C are included during epitaxial growth. Figure 5 shows the densities of triangular defects and polytype inclusions, with varying growth temperature and C/Si ratio, calculated for the whole 3-inch wafer. It is clear that the density of triangular defects decreases dramatically with an increase in growth temperature. However, the effect of a low C/Si ratio was found to be weak. This indicates that triangular defects are generated by 2D nucleation on the terraces. Increasing the temperature enables the surface atoms to reach steps, and 2D nucleation can thereby be suppressed. As the results, the density of triangular defects decreased at higher temperatures. On the other hand, polytype inclusions did not decrease with an increase in growth temperature, but decreased with a lower C/Si ratio. Si-rich conditions have been reported to enhance step flow growth.[10] In this situation, the stacking sequence can be informed from the step of the upstream side. As the result, the stacking sequence can be maintained and polytype inclusions decreased, as shown in this figure. As the results, combined density of triangular defects and polytype inclusions was obtained about 1.4 cm^{-2} with growth temperature of 1700°C and a C/Si ratio of 0.7. This means that we have successfully grown an epitaxial layer on vicinal off-angled 3-inch substrates, with good surface morphology and polytype stability.

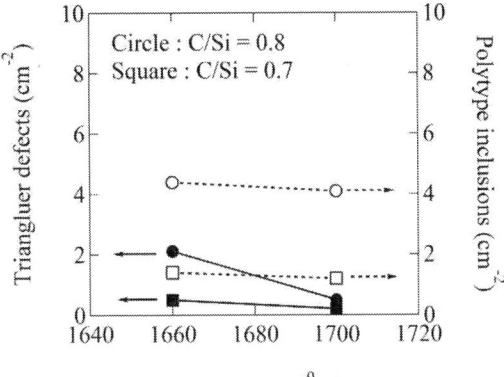

Figure 5 Density of epitaxial defects with varying growth temperature and C/Si ratio. Closed and open symbols indicate triangular defects and polytype inclusions, respectively. Circles and squares indicate a C/Si ratio of 0.8 and 0.7, respectively.

Finally, we investigated the I-V characteristics of Ni Schottky barrier diodes, in order to confirm surface qualities, such as step-bunching conditions and the degree of surface morphology degradation caused by triangular surface defects. A total of 65 devices, arranged in a radial pattern, were measured, with mean thickness and doping concentration as described above. With respect to the forward characteristics, the mean n value was 1.05, which indicates that the SBDs had ideal characteristics. A low leakage current of less than 1 x 10^{-6} A/cm^2 at 800 V was obtained, with a good yield of 78%. The mean blocking voltage of those 78 % of the devices was 960 V. The devices broke down due to an electric field concentration at the edge of the electrodes, because a portion of the edge of the electrodes was blackened after the I-V measurements. Therefore, the blocking voltage should be higher when a junction termination extension structure is fabricated.

These results suggest that the quality of the grown epitaxial layer was quite similar to that of a 4° off-axis epitaxial layer, and therefore, that the wafer off angle for epitaxial growth can be effectively reduced to less than 4 degrees.

Summary

In this paper, we have investigated key factors for controlling the polytype and surface morphology of 4H-SiC homoepitaxial growth on less than 4° off-axis substrate. The results suggest that the control of surface energy, control of the vicinal off angle itself, and high temperature growth, is highly important in controlling the surface morphology and polytype stability of epitaxial layers grown on a vicinal off angled substrate.

We also grew epitaxial layers on 3-inch vicinal off angle substrates based on the above investigation, and obtained the high quality 3-inch epitaxial layers grown on Si-face vicinal off angle substrates. These results suggest that the wafer off angle for epitaxial growth can be reduced to less than 4 degrees.

Acknowledgments

This work is partially supported by Novel Semiconductor Power Electronics Project Realizing Low Carbon Emission Society under New Energy and Industrial Technology Development Organization (NEDO).

References

1. K. Shibahara, N. Kuroda, S. Nishino and H. Matsunami, Jpn. J. Appl. Phys. **26**, L1815 (1987).
2. H. S. Kong, J. T. Glass and R. F. Davis, J. Appl. Phys. **64**, 2672 (1988).
3. K. Kosciewicz, W. Strupinski, D. Teklinska, K. Mazur, M. Tokarczyk, G. Kowalski and A. Olszyna, Mater. Sci. Forum **679-680**, 95 (2011).
4. S. Harada S. Ito, M. Kato, A. Takatsuka, K. Kojima, K. Fukuda and H. Okumura, Mater. Sic. Forum **645-648**, 999 (2010).
5. Y. Ueoka, K. Shingu, H. Yano, T. Hatayama, and T. Fuyuki, Jpn. J. Appl. Phys. **51**, 110201 (2012).
6. B. Thomas and C. Hecht, Mater. Sci. Forum, **483-485**, 141 (2005).
7. M. Syväjärvi, R. Yakimova and E. Janzén, Diamond Relat. Mater. **6**, 1266 (1997).

8. K. Kojima, S. Nishizawa, S. Kuroda, H. Okumura, and K. Arai, J. Crystal Growth **275**, e549 (2005).
9. C. Hecht, R. Stein, B. Thomas, L. Wehrhahn-Kilian, J. Rosberg, H. Kitahata, F. Wischmeyer, Mater. Sci. Forum, **645-648**, 89 (2010).
10. S. Nakamura, T. Kimoto and H. Matsunami, Jpn. J. Appl. Phys. **42**, L846 (2003).

3C-SiC on Si Hetero-epitaxial Growth for Electronic and Biomedical Applications

M. Reyes[1], C.L Frewin[1], P. J. Ward[2] and S. E. Saddow[1]

[1]University of South Florida, Dept. of Electrical Engineering, 4202 E. Fowler Ave., Tampa, FL 33620, USA
[2]Anvil Semiconductors Ltd, Birmingham Rd., Allesley, Coventry, CV5 9QE, UK

> The growth of cubic silicon carbide on silicon, namely 3C-SiC/Si, has been extensively studied at the University of South Florida over the past decade and numerous electronic and biomedical applications explored using this material system. The key step to 3C-SiC devices is the growth of high-quality epitaxial layers of 3C-SiC. In order to improve the manufacturability of future 3C-SiC devices, a simplified 3C-SiC growth process on 50 and 100 mm Si (100) substrates has been developed in a low-pressure horizontal hot-wall chemical vapor deposition (CVD) reactor. A simplified growth process consists of a single thermal ramp to the growth temperature followed by the 3C-SiC growth. The 3C-SiC epitaxial layers were characterized via optical microscopy, secondary electron microscopy (SEM), atomic force microscopy (AFM), X-ray diffraction (XRD), and secondary ion mass spectrometry (SIMS). Examples of biomedical devices realized with this material system are also introduced, including neural probes and in-vitro recording devices, as well as myoglobin and glucose biosensors.

Introduction

By virtue of its thermal, electronic, chemical, and physical properties, silicon carbide is an excellent semiconductor material for the fabrication of high power, high temperature, and high frequency devices. SiC also displays superior resistance to chemical attack, making it an attractive material for use in harsh environments, and may also be one reason it possesses excellent biological compatibility [1-3]. Consequently, all of these traits enable this material to be an excellent candidate for long-term, implantable biomedical applications [4-7].

In earlier work, the hetero-epitaxial growth of single crystal 3C-SiC on silicon (100) substrates (up to 50 mm wafer sample size) via CVD has been realized [8-10]. The growth processes developed during earlier work included the carbonization of the silicon surface followed by a slow growth thermal ramp including the addition of the silicon precursor (Figure 1.a). The latter process was then modified to include a controlled variation of the Si/C ratio during the second thermal ramp after carbonization (Figure 1.b) with the aim to improve the epitaxial film quality via reduced defect density [11]. However; although improvements were obtained, it was concluded that the epitaxial film quality was still comparable to films obtained in the original process (Figure 1.a). Therefore, the modifications implemented added unwanted complexity and cost to the deposition process. As a result, focus was shifted into developing a simplified 3C-SiC growth process with the goal to make it more reproducible, easier to control and more cost effective without compromising epitaxial film quality.

This paper reports on the progress made to date in the development of a simplified single crystal 3C-SiC growth process using the H_2-C_2H_4-SiH_4 precursor gas system on 50 mm and 100 mm (100) silicon substrates. It also reports on several 3C-SiC based biomedical device prototypes developed at the University of South Florida using the growth processes described in this work.

Experimental

Growth experiments were conducted in a low-pressure horizontal hot-wall CVD reactor described elsewhere [12]. The CVD reactor has been scaled up to accommodate 100 mm substrates and the CVD gas chemistry used in this work included the H_2-C_2H_4-SiH_4 precursor system. N-type silicon (100) 50mm and 100 mm wafers were used as substrates during the hetero-epitaxial growth. The process sequence of carbonization and growth has been simplified from those shown on Figure 1a-b. The current deposition process (Figure 1.c) uses a single thermal ramp to the growth temperature with a ramp rate of ~28 °C/min. The silicon substrate is heated under hydrogen carrier gas flow. At a temperature of 1100°C, C_2H_4 is introduced into the gas stream to facilitate carbonization, or the initial generation of SiC, and continues throughout the remaining ramp until the substrate reaches the growth temperature of 1370°C. At 1370°C, SiH_4 is introduced into the gas mixture and the C_2H_4 precursor flow is scaled for continued epitaxial growth. The growth stage consists of two stages. During Growth Stage 1, the precursors are scaled up to achieve a C/Si ratio of 1.2 and nitrogen doping is introduced into the gas chemistry. These conditions are held for 30 minutes. Nitrogen is added to the gas chemistry because the resulting films grown were used for electronic device fabrication. During Growth Stage 2, the C/Si ratio of 1.2 is held constant and nitrogen doping is removed. Growth follows for 60 minutes. The process pressure was 375 Torr during the entire growth schedule. The resulting epitaxial layers were analyzed via optical microscopy, SEM, AFM, XRD, and SIMS.

Results and Discussion

Growth on 50 mm wafers

The simplified 3C-SiC growth process was first developed in our CVD reactor using the 50 mm hot-wall configuration. The carbonization step was optimized via the compromise between a low carbon precursor flow to reduce surface roughness and the flow needed to avoid large line and square hydrogen etching surface defects in the silicon substrate. The results obtained suggested that there was a constant dynamic carbon adsorption and desorption process occurring throughout the temperature ramp. The epitaxial growth rate was determined to be ~ 3μm/h and the films grown displayed specular surface morphology. AFM characterization revealed that the surfaces grown using this process had roughness values of ~7 nm RMS. While performing optical and AMF characterization, it was noted that small microcracks were present in the 3C-SiC epitaxial film surface suggesting the presence of tensile stress in the epitaxial layer.

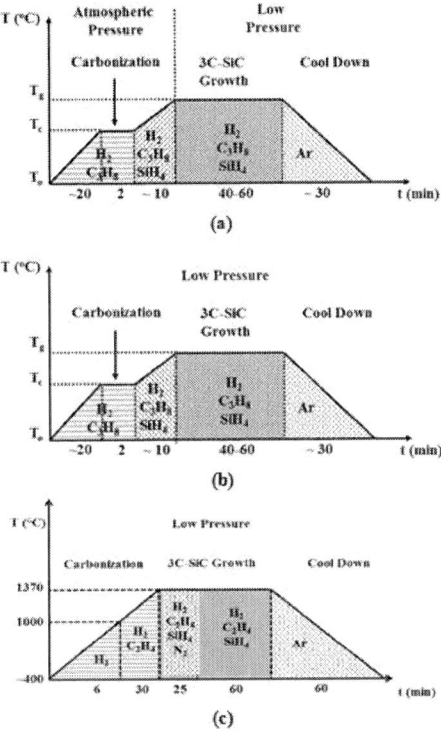

Figure 1: (a) Original USF 3C-SiC growth process, consisting of carbonization, slow grow temperature ramp and the growth step. (b) Modified 3C-SiC growth process, it differs from the original process in that the slow grow thermal ramp includes a variation of the Si/C ratio as a function of temperature. (c) Simplified 3C-SiC growth process, it eliminates the temperature plateau during the carbonization process. It includes a single thermal ramp to the growth temperature in the presence of the carbon precursor followed by the growth step.

Figure 2.a illustrates the XRD powder diffraction spectrum for a ~ 4 µm thick 3C-SiC representative epi film on a 50 mm substrate. The Theta-2Theta scan show the characteristic peak of the <002> 3C-SiC plane at 41.4 °. The XRD rocking curve scan of the <002> 3C-SiC diffraction peak revealed a FWHM of 0.13° (480 arcsec) as seen in Figure 2.b. This value is comparable to those obtained on films our past processes [8-11].

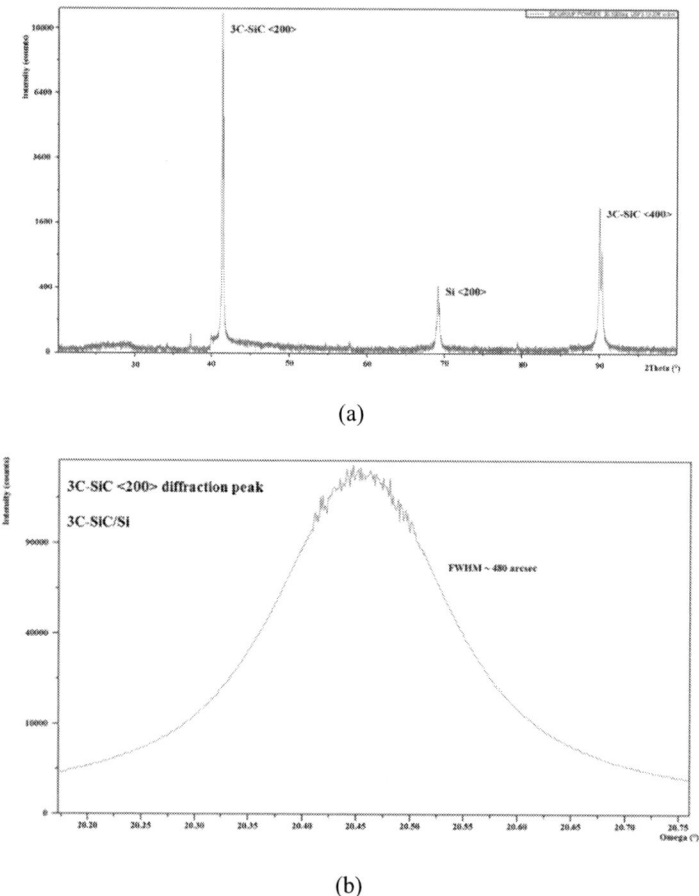

(a)

(b)

Figure 2: (a) Theta - Two Theta XRD scan showing the characteristic peak of the <002> 3C-SiC plane at 41.4 °. (b) Rocking curve scan of the <002> 3C-SiC peak, FWHM = 0.1333 ° (480 arcsec). Film thickness is ~ 4 μm grown on a 50 mm substrate at a rate of~ 3 μm/h.

The unintentional background doping concentration, determined by capacitance-voltage measurements on these 3C-SiC films was ~5.0 x 10^{15} cm^{-3}, similar to that seen in 4H-SiC homo-epitaxy. SIMS measurements were carried out to determine the depth profiles of nitrogen intentional doping during the growth of the SiC films. The profiles show the n^{+} nitrogen doped layer of ~1 micron followed by a shoulder of nitrogen out-diffusion which is probably following an extended dislocation network from the Si/SiC interface. The bulk of the film has a nitrogen concentration below the SIMS machine background.

Growth on 100 mm wafers

The simplified 3C-SiC growth process developed using the 50 mm CVD configuration was scaled up and applied to 100 mm substrates which employed an Anvil proprietary stress relief technique after the reactor hardware was modified to accommodate this size substrate. The epitaxial layers obtained in the scaled up process also displayed specular surface morphology. Optical microscopy characterization revealed the presence of two kinds of defects: voids at the 3C-SiC/Si interface (seen as small white dots) and APB's. Both these defects are typical on 3C-SiC films. Additional characterization via SEM showed that the void size was ~ 200 nm across. The small voids observed at the 3C-SiC/Si interface are formed during the carbonization stage over a small temperature range at ~1150°C. The literature suggests that at lower temperatures low carbon precursor mole fractions are established and there is not enough carbon to react with the silicon surface. As a result, the vertical diffusion of silicon is favored thus creating voids at the buffer layer/substrate interface. Similarly, low carbon precursor mole fractions may promote the formation of etch pits at the substrate due to a preferential etching rate process rather than the desired deposition.

AFM and XRD characterizations were also performed on films grown on 100 mm substrates. Surface roughness values of ~ 9 nm RMS were typically measured on the 3C-SiC films. The XRD rocking curve of the <200> 3C-SiC diffraction plane displayed a FWHM 0.45° (1625 arcsec) on this thinner layer ~ 3 µm grown at a rate of ~ 2 µm/h.

It is evident that thicker, higher quality epitaxial layers were obtained on 50 mm substrates using the same process input C/H_2, Si/H_2 and Si/C; however the Si/H_2 seems to be insufficient for the 100 mm process hence giving a lower growth rate. In this case, the ratio between the defective film thickness at the Si/SiC interface and low defect density bulk film results in a poorer XRD result.

3C-SiC biomedical device summary

To date several important device prototypes have been developed at the University of South Florida based on the 3C-SiC films described earlier (see Fig. 3). While the details are best left to separate reports to allow the reader to better understand the results obtained, it is hoped that a list of the devices under develop will be beneficial to report here. These devices fall into 2 general categories – neural engineering and biosensors. In the biosensor arena two devices have been fabricated and excellent results reported. The first is an in-vivo glucose sensor that is based on a shift in RF frequency resulting from a change in glucose concentration in the blood stream [13]. The preliminary results showed a -65 kHz shift in center ffrequency per 1 mg/dl change in glucose concentration that was added incrementally to pig blood [13, 14]. Measurements were made with a vector network analyser using the mechanism of a change in blood permittivity caused by increasing glucose concentration. The second SiC biosensor sensor was designed to provide an early warning for cardiac arrest via the detection of myoglobin, a protein prevalent in the blood stream during an acute myocardial infarction (AMI) [15]. The sensor utilizes self-assembled monolayers of amino-propyldiethoxymethylsilane (APDEMS) [16] that are functionalized with anti-myoglobin antibodies. When myoglobin interacts with the antibodies, a change in device impedance can be measured using an electrochemical cell [17].

The second class of devices involves both micro-electrode arrays (MEAs) and intracortical neural implants (INIs). In both cases, 3C-SiC serves as a biocompatible substrate on which electrodes are fashioned, allowing for bi-directional communication with neurons, either in-vitro (MEAs) or in-vivo (INIs) [18, 19]. A more complete discussion of this work can be found in the book chapter by Frewin *et al.* [20]. The work at the time of this publication focused on using SiC INIs to manufacture MRI-compatible DBS probes (deep brain stimulation) through a collaboration with the University of Sao Paulo, Sao Carlos, Brasil. The initial results show that 3C-SiC INIs are not adversely affected by the high power magnetic fields produced by MRI and therefore do not produce image artefacts during MRI stimulation [21].The true goal of the group is to use the semiconductor material to produce active electronics which will interact with the biological environment on a nano scale instead of the current micro-scale of the INI devices.

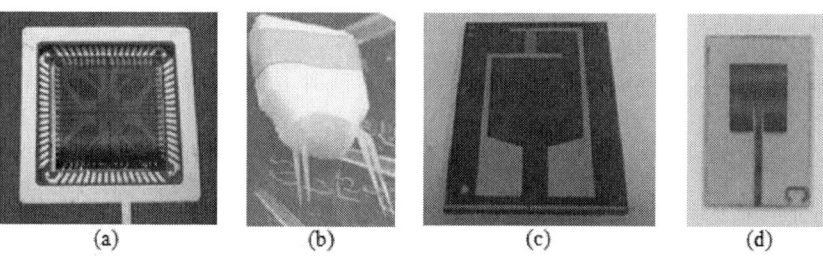

(a)   (b)   (c)   (d)

Figure 3: Examples of biomedical research devices made using the USF 3C-SiC epitaxial growth process. a) 64 channel microelectrode array (MEA) for in-vitro neural stimulation and recording. The pictured device includes the base material of 3C-SiC heteroepitaxially grown on Si with Ti/Au electrodes and amorphous SiC coated SiO_2 insulation. b) Passive planar neural implantable prosthetic probes micromachined from 3C-SiC on Si. The Si was removed to yield the free-standing 3C-SiC probes pictured above. c) A Myoglobin interdigitated electrode (IDE) biosensor constructed on 3C-SiC on Si with Ti/Au comb electrodes insulated using SU-8 photoresist. d) An in-vivo continuous monitoring RF antenna glucose sensor. The device was constructed from semi-insulating 4H-SiC with a metallic-like patch radiator of heavily-doped polycrystalline 3C-SiC.

Conclusions

A simplified 3C-SiC growth process consisting of a single ramp to the growth temperature has been developed on 50 and 100 mm Si(100) substrates. Single crystal 3C-SiC layers have been obtained in both size substrates. X-ray rocking curves taken on the <002> diffraction plane of layers grown on 50 and 100 mm substrates displayed FWHM values of 0.13° (480 arcsec) and 0.45° (1625 arcsec), respectively. Several examples of the use of 3C-SiC in biomedical devices have been introduced and references provided for the interested reader to learn more about this important technological application of 3C-SiC.

Acknowledgements

C. W. Locke of the USF SiC Group is acknowledged for his initial work on the 50 mm process.

References

[1]   C. Coletti, M. J. Jaroszeski, A. Pallaoro, A. M. Hoff, S. Iannotta, and S. E. Saddow, "Biocompatibility and wettability of crystalline SiC and Si surfaces," *Conf Proc IEEE Eng Med Biol Soc,* vol. 2007, pp. 5850-3, 2007.

[2]   C. L. Frewin, M. Jaroszeski, E. Weeber, K. E. Muffly, A. Kumar, M. Peters, A. Oliveros, and S. E. Saddow, "Atomic force microscopy analysis of central nervous system cell morphology on silicon carbide and diamond substrates," *J Mol Recognit,* vol. 22, pp. 380-8, Sep-Oct 2009.

[3]   S. E. Saddow, C. Coletti, C. L. Frewin, N. Schettini, A. Oliveros, and M. Jarosezeski, "Single-crystal Silicon Carbide: A Biocompatible and Hemocompatible Semiconductor for Advanced Biomedical Applications," *Materials Science Forum,* vol. 679-680, pp. 824-830, 2010.

[4]   S. E. Saddow, Ed., *Silicon Carbide Biotechnology: A Biocompatible Semiconductor for Advanced Biomedical Devices and Applications.* Amsterdam: Elsevier, 2011.

[5]   S. Santavirta, M. Takagi, L. Nordsletten, A. Anttila, R. Lappalainen, and Y. T. Konttinen, "Biocompatibility of silicon carbide in colony formation test in vitro. A promising new ceramic THR implant coating material," *Arch Orthop Trauma Surg,* vol. 118, pp. 89-91, 1998.

[6]   U. Kalnins, A. Erglis, I. Dinne, I. Kumsars, and S. Jegere, "Clinical outcomes of silicon carbide coated stents in patients with coronary artery disease," *Med Sci Monit,* vol. 8, pp. PI16-20, Feb 2002.

[7]   R. Yakimova, R. M. Petoral, G. R. Yazdi, C. Vahlberg, A. L. Spetz, and K. Uvdal, "Surface functionalization and biomedical applications based on SiC," *Journal of Physics D-Applied Physics,* vol. 40, pp. 6435-6442, Oct 21 2007.

[8]   S. Harvey, M. Reyes, Y. Shishkin, and S.E. Saddow, "High growth rate of single crystal 3C-SiC on 2 inch Si (001) wafers," in *Electronic Materials Conference (EMC)*, Penn State Universitty, Pennsylvania USA, 2006.

[9]   M. Reyes, Y. Shishkin, S. Harvey, and S. E. Saddow, "Development of a high-growth rate 3C-SiC on Si CVD process," *Mater. Res. Soc. Symp. Proc.,* vol. 911, pp. 79 – 84, 2006.

[10]   M. Reyes, Y. Shishkin, S. Harvey, and S. E. Saddow, "Increased Growth Rates of 3C-SiC on Si(100) Substrates via HCl Growth Additive," *Materials Science Forum,* vol. 556-557, pp. 191-194, 2006.

[11]   C. Locke, G. Kravchenko, P. Waters, J. D. Reddy, K. Du, A. A. Volinsky, C. L. Frewin, and S. E. Saddow, "3C-SiC Films on Si for MEMS Applications: Mechanical Properties," *Material Science Forum,* vol. 615-617, pp. 633-636, 2009.

[12]   R. L. Myers, Y. Shishkin, O. Kordina, and S. E. Saddow, "High growth rates (> 30 mu m/h) of 4H-SiC epitaxial layers using a horizontal hot-wall CVD reactor," *Journal of Crystal Growth,* vol. 285, pp. 486-490, Dec 15 2005.

[13]   S. Afroz, "A Biocompatible SiC RF Antenna for In-vivo Sensing Applications," Ph.D., Electrical Engineering, University of South Florida, 2013.

[14]   S. Afroz, S. W. Thomas, G. Mumcu, and S. E. Saddow, "Implantable SiC based RF antenna biosensor for continuous glucose monitoring," in *IEEE Sensors*, Baltamore, Maryland USA, 2013 (in press).

[15]   A. Oliveros, "Myoglobin detection on SiC: immunosensor development for myocardial infarction," Ph.D., Electrical Engineering, University of South Florida, 2013.

[16] S. J. Schoell, M. Sachsenhauser, A. Oliveros, J. Howgate, M. Stutzmann, M. S. Brandt, C. L. Frewin, S. E. Saddow, and I. D. Sharp, "Organic Functionalization of 3C-SiC Surfaces," *Acs Applied Materials & Interfaces,* vol. 5, pp. 1393-1399, Feb 27 2013.

[17] A. Oliveros, A. Guiseppi-Elie, and S. E. Saddow, "Silicon carbide: a versatile material for biosensor applications," *Biomed Microdevices,* vol. 15, pp. 353-68, Apr 2013.

[18] C. L. Frewin, A. Oliveros, C. Locke, I. Filonova, J. Rogers, E. Weeber, and S. E. Saddow, "The Development of Silicon Carbide Based Electrode Devices for Central Nervous System Biomedical Implants," *Mater. Res. Soc. Symp. Proc.,* vol. 1236E, pp. 1236-SS01-02, 2009.

[19] C. L. Frewin, C. Locke, S. E. Saddow, and E. J. Weeber, "Single-Crystal Cubic Silicon Carbide: An in vivo biocompatible semiconductor for brain machine interface devices," in *Engineering in Medicine and Biology Society,EMBC, 2011 Annual International Conference of the IEEE*, Boston, MA, 2011, pp. 2957 - 2960.

[20] C. L. Frewin, L. Abbati, E. J. Weeber, and S. E. Saddow, "SiC for Brain–Machine Interface (BMI)," in *Silicon Carbide Biotechnology: A Biocompatible Semiconductor for Advanced Biomedical Devices and Applications*, S. E. Saddow, Ed., ed Amsterdam: Elsevier, 2011.

[21] S. E. Saddow, "SiC biotechnology for advanced biomedical applications," in *Second Workshop on Advanced Cybernetics*, ed. University of Sao Paulo, Sao Carlos, Brasil, 2013.

CHAPTER 4

GaN POWER DEVICES 1

128

III-Nitride Materials and Devices for Power Electronics

A. Dobrinsky[a], G. Simin[b], R. Gaska[a], M. Shur[c],

[a] Sensor Electronic Technology, Inc. Columbia, SC 29209 USA
[b] Department of Electrical Engineering, University of
South Carolina, Columbia, SC 29208 USA
[c] ECSE and PAPA, Rensselaer Polytechnic Institute, NY 12180 USA

Group III-Nitride based devices are expected to compete and possibly outperform state of the art silicon carbide based devices for power electronic applications. GaN, AlGaN, AlInGaN and other III-Nitride compound materials offer a very high breakdown field. Devices based on III-Nitride compounds have very high electron density and mobility in the device channel, leading to record low device on-resistance, high temperature stability and low gate capacitance enabling high switching speed and low switching loss. In order to enable III-Nitride based devices for power applications continuous efforts are underway in the area of material growth, novel power device concepts, designs, modeling and simulations.

Introduction

Power devices such as RF and microwave amplifiers for telecommunication and other power applications require devices with ever increasing performance in terms of noise figure (NF), Johnson's Figure of Merit (FoM) or combined figure of merit (CFOM). These devices are required to operate at frequencies exceeding 100 GHz at high operational powers. To meet challenges of high power devices, group III-Nitride based devices become a viable new alternative to AlGaAs/InGaAs based pseudomorphic MODFETs. Group III-Nitride material's large bandgap, corresponding large dielectric breakdown field combined with good electron transport properties such as mobility and saturation velocity are necessary elements of high power devices. In addition to superior electronic properties, group III-Nitride devices grown on thermally conductive substrates (such as SiC) offer superior thermal performance. The requirements for the performance of power electronic devices are becoming more demanding. For example modern cell phones require wider bandwidth and improved efficiency. Continuous development of satellite communications and TV broadcasting requires amplifiers operating at higher frequencies and higher powers. These requirements result in significant investment in the development of new materials and new device designs for power electronics, with group III-Nitrides and SiC being the most promising candidate materials. The future and present applications of high-power group III-Nitride based devices such as Heterojunction Field-Effect Transistors (HFETs) include amplifiers operative at high power levels and high temperatures. Such applications include radars, missile and satellite electronics as well as low cost compact amplifiers for wireless base stations. Table 1 lists important parameters of key materials used for power electronics and the normalized Johnson's figure of merit (JM_n) (the normalization done by Johnson's figure of merit of silicon) [1,2]

TABLE I. Key Material Properties for Power Performance at High Frequencies, all the symbols are defined in Appendix

	Si	GaAs	4H-SiC	GaN	Diamond
E_g (eV)	1.1	1.42	3.26	3.42	5.45
ε_r	11.8	13.1	10	8.9	5.5
μ_n [cm²/(Vs)]	1350	8500	700	1200-2000	1900
v_{sat}[10^7 cm/s]	1.0	1.0	2	2.5	2.7
E_{br} [MV/cm]	0.3	0.4	3	3.3	5.6
K [W/(cm K)]	1.5	0.43	3.3	1.3	20
$JM_n = (E_{br}v_{sat}/(2\pi))^2$	1.0	1.8	400	756	50
CFOM = $K \varepsilon_r \mu_n v_{sat} E^2_{br}/$ $(K \varepsilon_r \mu_n v_{sat} E^2_{br})_{Si}$	1.0	3.6	193	220	8228

Table I shows that the leading contenders for high power electronic devices are 4H-SiC and GaN (not counting Diamond). The wide bandgap of these materials results in higher breakdown voltages due to high breakdown field necessary for band-to-band impact ionization. Both SiC and GaN have high electron saturation velocities, which allow high frequency operation. However, the unique benefit of GaN is the ability to form heterojunctions, which makes it superior to SiC. In addition, the polarization doping associated with AlGaN/GaN heterojunctions increases the carrier density in the 2D gas carrying channel, which is unique feature of group III-Nitride based devices. The advantages of the AlGaN/GaN based high electron mobility transistors (HEMTs) further includes higher electron mobility due to reduced ionized impurity scattering. The combination of high carrier concentration and high electron mobility results in a high current density and a low channel resistance, which are essential for high frequency operation, and power applications. We also give values for the Combine Figure of Merit (CFOM) given by CFOM =$k\varepsilon_r\mu v_s E^2_B/($ $k\varepsilon_r\mu v_s E^2_B)_{Si}$, k- being thermal conductivity, ε_r - dielectric constant, μ is low field mobility, v_s is saturation velocity and E_B is breakdown field.

In this article, we discuss the key components of group III-Nitride semiconductor based power electronic devices. First we review growth of high purity device layers by metal organic chemical vapor deposition (MOCVD) and molecular beam epitaxy (MBE). Next, we present transport properties of group III-Nitride materials as well as their thermal properties. Finally we conclude with discussion of state-of-the-art devices based on these materials.

Group III-Nitride Materials and Their Growth

Group III-Nitride semiconductor layers are typically grown on sapphire or SiC substrate by Metal Organic Chemical Vapor Deposition (MOCVD). MOCVD is the standard technique of growing nitride semiconductor films and heterostructures. Other growth techniques used include Molecular Beam Epitaxy (MBE) [3] and Reactive Molecular-Beam Epitaxy (RMBE) [4]. Recently Migration Enhanced Metalorganic Chemical Vapor Deposition (MEMOCVD®) technique [5] has improved the MOCVD process. In MEMOCVD® the durations and waveforms of precursor pulses are optimized and the pulses might overlap, providing a continuum of growth techniques ranging from

Pulsed Atomic Layer Epitaxy (PALE) to conventional MOCVD. In a MOCVD mode (at higher growth temperature), MEMOCVD® has a high growth rate. This growth mode is used for buffer layers, which are the layers grown over the substrate. Subsequent layers are grown at lower temperatures and have a much better quality due to a better mobility of precursor species on the surface, better atomic incorporation and improved surface coverage. MEMOCVD® grown layers exhibit about an order of magnitude reduction in the dislocation densities, and, as a consequence, have longer lifetimes and narrower photoluminescence (PL) lines.

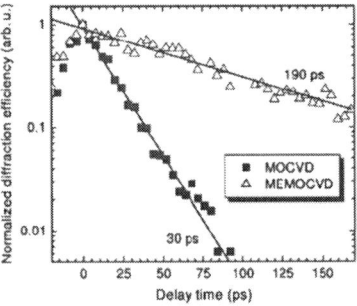

Figure 1. Light Induced Transient Grating (LITG) decay in MOCVD and MEMOCVD™-grown AlGaN epilayers for the grating period of 7.7 μm. Carrier lifetimes were estimated by fitting the decay transients with single exponents (lines) [6].

MEMOCVD systems are capable to operate in both conventional and atomic layer deposition regimes. When atomic layer deposition is required, metalorganic precursors are introduced in a cyclic or pulsed fashion. Epilayers are deposited on sapphire or SiC substrates placed on graphite susceptor, which is heated to the growth temperature by rf-induction.

Additional improvements in materials quality are achieved by combining MEMOCVD® with the Epitaxial Lateral Over Growth (ELOG) resulting in Migration Enhanced Lateral Over (MELEO) growth [7].

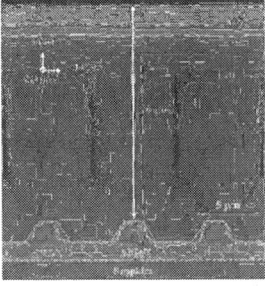

Figure 2. SEM micrographs of a fully coalesced 20 μm thick AlN sample grown by MELEO technique [7]

The lateral overgrowth technique used patterned sapphire as shown in Figure 2, and results in dislocation density reduction by orders of magnitude [7]. Additionally other techniques such as incorporating superlattices of group III-Nitride layers have been used to reduce stresses associated by growing on lattice mismatched substrates.

Transport Properties

We will focus on the transport properties of GaN, AlGaN, AlInN, and compare them with 4H-SiC, since these wide band gap materials are most important for high power applications. We also compare these transport properties with those of GaAs, which used to be a candidate material for high power applications.

Figure 3 compares the calculated temperature dependencies of the electron mobility in bulk GaN and GaAs with and without accounting for the impurity scattering [8]. The electron mobility in bulk GaN is less sensitive to ionized impurity scattering than that in GaAs.

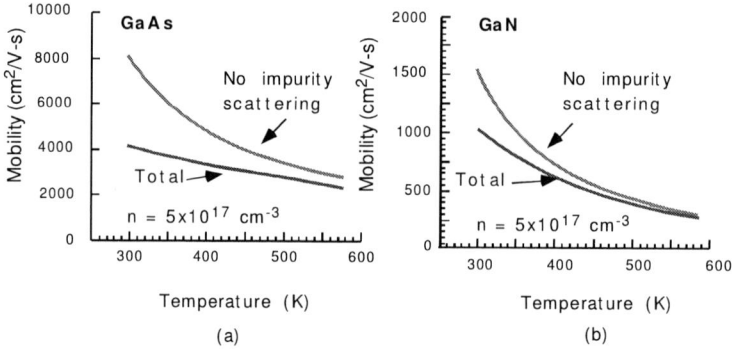

Figure 3. Comparison of mobility in GaAs and GaN materials [8]

Please note that the mobility of bulk semiconductor material may be very different from the mobility of a 2DEG formed at the heterointerface between AlGaN/GaN and AlGaAs/GaAs materials. Figure 4, for example, compares mobility as a function of temperature for bulk GaN and 2DEG formed at the AlGaN/GaN interface [10,11].

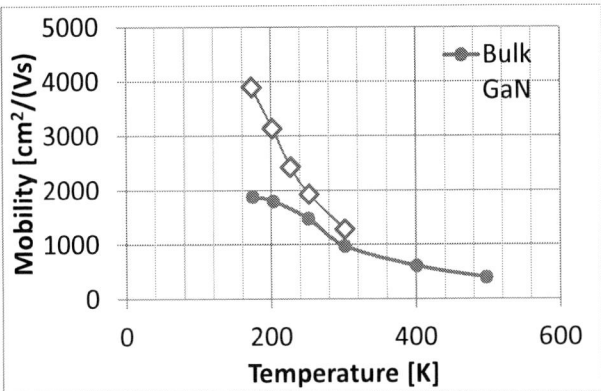

Figure 4. Drift mobility as a function temperature T after [10,11]

Great improvements in channel mobility of optimized AlGaN/GaN 2DEG channel-layer have yielded Hall mobility values of up to 2,000 cm²/(V s) as reported in [36] at room temperature. Gaska et al. [37] reported on the electron mobility in the 2D-gas electron gas at the GaN/AlGaN interface exceeding 10,000 cm²/(V s) at cryogenic temperatures and exceeding 2000 cm²/(V s) at room temperature. These values were observed in the samples with very high sheet carrier concentration (on the order of 10^{13} cm²). Variation in mobility of group III-Nitride materials is attributed to factors that include: phonon scattering, ionized impurity scattering, scattering on threading dislocations as well as alloy scattering.

Please note that while electron mobility in group III-Nitride materials can be high, the hole mobility in bulk GaN is much lower with typical values of about 100 cm²/(V s) or lower.

The typical electron mobilities in group III-Nitride semiconductors are greater than mobilities attained in SiC as can be seen from the figure 5(a) below obtained by fitting experimental data, plotted after [9] for donor concentration densities of $N_d = 5 \times 10^{14}$ [1/cm²]. For higher donor concentration densities the mobility is further reduced and is shown in figure 5(b).

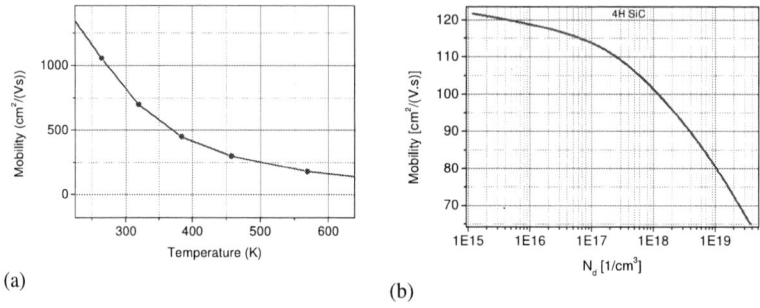

(a)   (b)

Figure 5. Hall Mobility of SiC (a) at $N_d = 5 \times 10^{14}$ [1/cm³], (b) at T=300K plotted after [9].

Figure 6 shows the computed velocity-field characteristics of GaN and GaAs at different temperatures using Monte Carlo simulations (see [12]). As can be seen from that figure, the electron velocity in GaN is less sensitive to temperature than that of GaAs. Hence, GaN devices should be more tolerant to self-heating and clearly more suitable for high temperature operation. Figure 6 demonstrates one aspect of superiority of group III-Nitride materials for high-temperature high power applications.

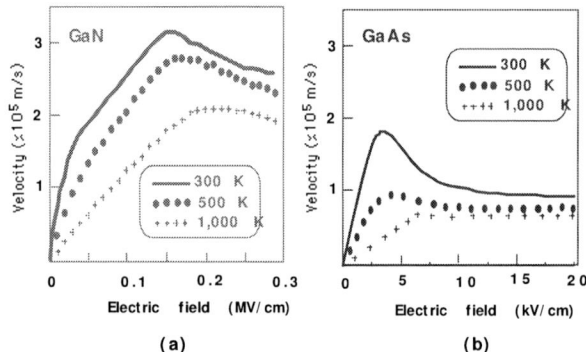

(a)   (b)

Figure 6. Drift Velocity at Room and Elevated Temperatures [12].

Figure 7 shows the bulk III–Nitride drift velocities, determined using Monte Carlo calculation [10]. The GaAs data determined using Monte Carlo is in agreement with experimentally measured velocity data calculated for T=300 K and the doping concentration of 10^{17} cm⁻³. Each of III–V semiconductors achieves a peak in its drift velocity. Note that InN achieves the highest steady-state peak drift velocity: 4.2×10^7 cm/s at an applied electric field of 65 kV/cm. The peaks for GaN correspond to, 2.9×10^7 cm/s at 140 kV/cm, and that of AlN, 1.7×10^7 cm/s at 450 kV/cm. For GaAs the peak drift velocity, 1.6×10^7 cm/s occurs at a much lower electric field values (4 kV/cm) than in the nitrides. The electric fields corresponding to drift velocity peaks represent the critical fields beyond which transient effects such as velocity overshoot become significant.

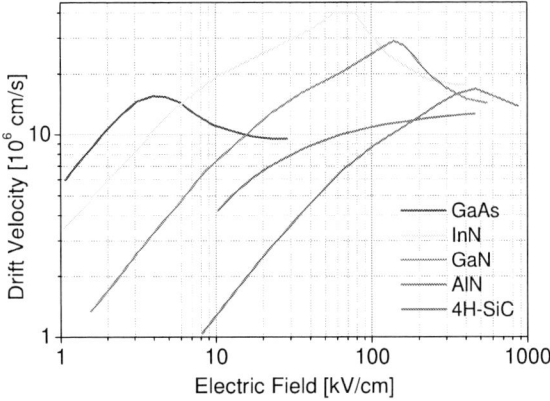

Figure 7. Comparison of Drift Velocity at 300 [K] plotted after [13,14]

Similar to doping effects on mobility characteristics of the group III-Nitride semiconductor layers, the increased doping of semiconductor layers decreased the drift electron velocity as shown in figure 8. Similar drift velocities as a function of doping concentration are demonstrated for other group III-Nitride materials in figures 9-11. Note that semiconductor alloys with high level of indium have a relatively high drift velocity, whereas semiconductor alloys with high aluminum content have a relatively low drift velocity.

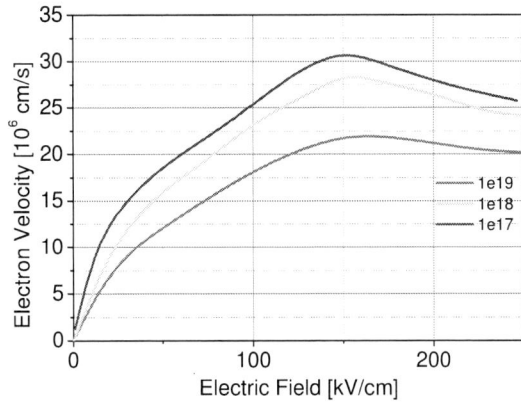

Figure 8. Drift Velocity for GaN at 300 [K] as a function of doping plotted after [15].

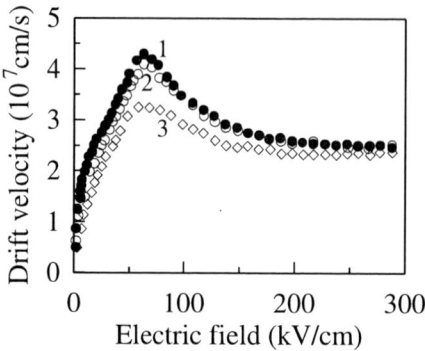

Figure 9. Drift Velocity for InN at 300 [K] as a function of doping 1 - 10^{17}; 2 - 10^{18}; 3 - 10^{18} [16].

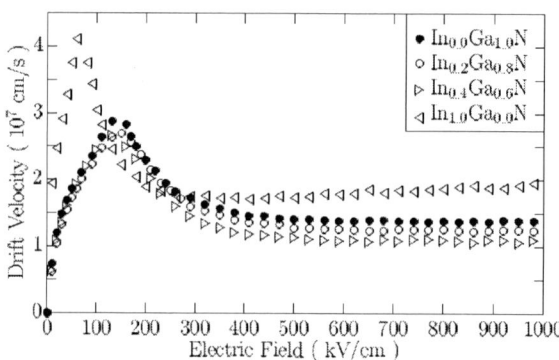

Figure 10. Drift Velocity for InGaN at 300 [K] as a function of In molar fraction [17]

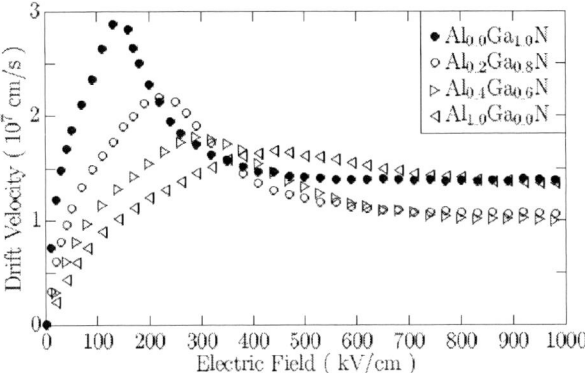

Figure 11. Drift Velocity for AlGaN at 300 [K] as a function of Al molar fraction [17]

Thermal Properties

We first consider the thermal conductivity of $Al_xGa_{1-x}N$ as a function of the temperature. The thermal conductivity was fitted based on the available experimental data [19,20]. Figure 12 shows experimental data for thermal conductivity of GaN as a function of temperature, while figure 13 shows the fit for experimental data for AlN. The thermal conductivity fit for GaN is approximated by the following expression:

$$\kappa_{GaN}(\tau) = 4300e^{-10.5\tau} + 180e^{-0.39\tau}, \ (0.3 < \tau < 1.3), \text{ and for AlN}$$

$$\kappa_{AlN}(\tau) = \frac{-7.2\tau^2 + 150\tau - 0.03}{\tau^2 - 0.5\tau + 0.3}, \ (0.1 < \tau < 3.5), \ \tau = \frac{T}{T_0}, \ T_0 = 300K,$$

with all the coefficients having units of W/(m·K) when needed. In order to fit thermal conductivity for $Al_xGa_{1-x}N$ we use the fit given by [34] page 23 and by [35],

$$\kappa_{Al_xGa_{1-x}N}(x,\tau) = \left[\frac{x}{\kappa_{AlN}(\tau)} + \frac{1-x}{\kappa_{GaN}(\tau)} + \frac{x(1-x)}{C_{AlGaN}}\right]^{-1}$$

Figure 14, shows the behavior of thermal conductivity as a function of aluminum mole fraction x.

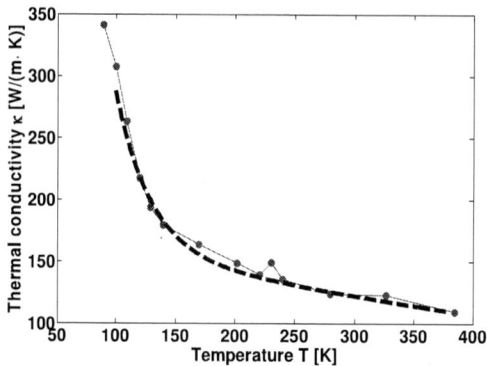

Figure 12. Thermal conductivity of GaN [18,19]

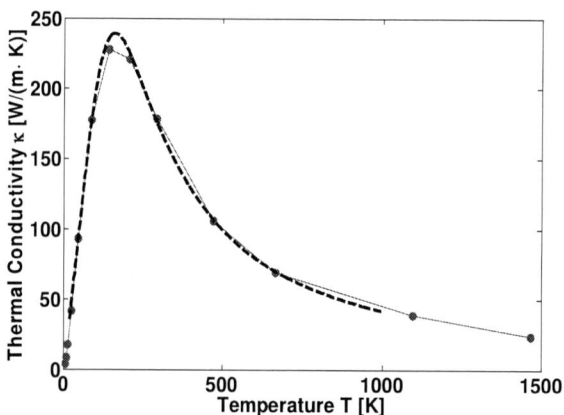

Figure 13. Thermal conductivity of GaN [18, 20]

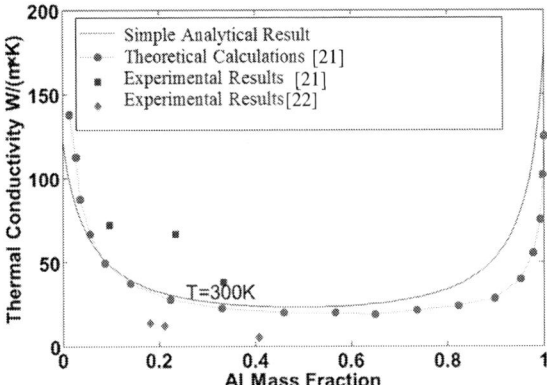

Figure 14. Thermal conductivity of AlGaN [18, 21, 22] as a function of aluminum molar fraction

Similar expressions are obtained for InGaN and AlInN composite alloy as shown in figures 15(a) and 15(b).

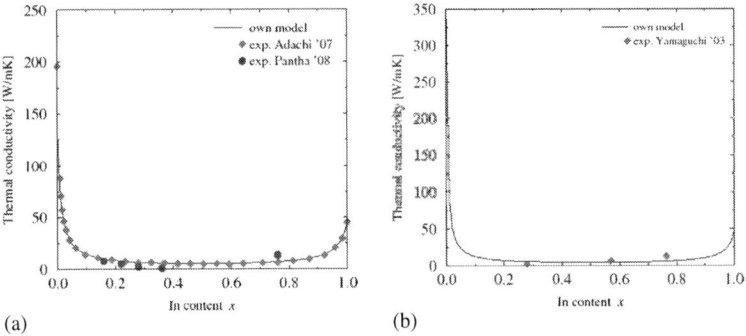

(a)   (b)

Figure 15. Thermal conductivity of InGaN and AlInN [23, 24, 25,26] as a function of aluminum molar fraction

Figure 16 compares the thermal conductivity of various group III-nitride alloys, as well as GaAs with thermal conductivity of 4H SiC. As can be seen, AlN has the highest thermal conductivity at lower temperatures, but similar thermal conductivity of SiC at room temperatures. On the other hand, the lowest thermal conductivity is that of GaAs, followed by GaN. While not shown, the thermal conductivity of sapphire is on the order of 0.3~0.4 W/(cm-K), which is significantly lower than the thermal conductivity of SiC even at room temperatures (~4 W/(cm-K)), as a result for temperature management, SiC is a preferred substrate for epitaxial layer growth.

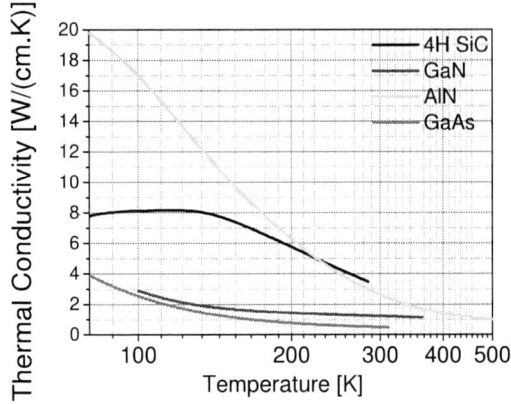

Figure 16. Thermal conductivity of group III-Nitride materials and SiC for various temperatures, plotted after: GaN [19], 4H SiC [27], AlN [28], GaAs [29].

Group III-Nitride Devices

III-Nitride power transistor switches have a potential to provide a leap-ahead performance improvement of power converters, power conditioners and other critical power electronics applications. The improvement can be achieved due to much higher breakdown field allowing for smaller device on-resistance, excellent temperature stability allowing operating temperature of 300C or higher, chemical and radiation hardness on III-Nitride materials. III-Nitride heterostructures offer additional benefits due to extremely low-resistive 2D electron channel formed at heterointerface. However, in spite of more than decade of intensive material and device technology development, significant gap exists between the achieved and projected device performance level. The proposed research aimed at developing robust and reliable technology to radically reducing if not eliminating completely the existing performance gap.

In power semiconductor switches, lowering the conduction loss requires minimal device on-resistance. However, as the operating voltage increases, the required length of the active region also increases, leading to higher on-resistance. Therefore, minimization of on-resistance and increasing the operating voltage are controversial requirements. Record low on-resistance is expected to be obtained in III-Nitride heterostructure field-effect transistors (HFETs or HEMTs) due to record high 2D-electron sheet density and high electron mobility at the heterointerface. However, III-Nitride HFETs have significant issues precluding their efficient application in power electronics. Being essentially lateral geometry devices, HFETs have the channel aligned along the semiconductor surface and located very close to the surface. The problem is illustrated by a typical state-of the high-voltage AlGaN/GaN HFET (Fig. 17).

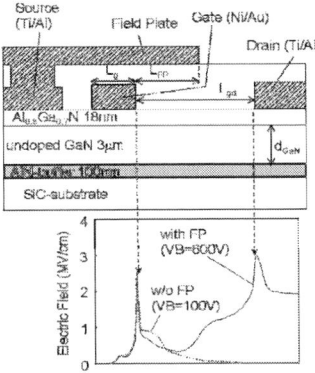

Figure 17. Electric field distribution along the gate-drain spacing in HFET with field plate (solid line) and without field plate (dash line). (After [30]).

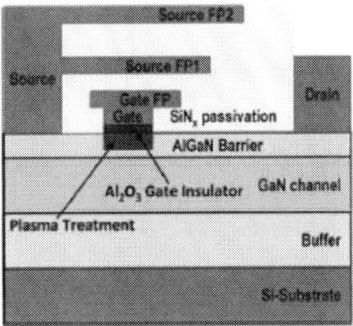

Figure 18. State-of-the-art multiple field-plate 1200 V HFET [31]. The gate – drain distance is 15 μm, yielding the average breakdown field of around 0.8 MV/cm as compared to the breakdown field in GaN of 2.6 MV/cm measured in [38].

First, in HFETs, the electric field in the gate - drain spacing is strongly non-uniform and as a result premature breakdown occurs limiting the device performance. Second, due to (i) high carrier concentration in the channel and (ii) close vicinity of the surface, efficient control over the space charge distribution in the gate – drain spacing is extremely challenging. This charge nonuniformity further increases electric field nonuniformity and hence reduces the operating voltage. Adding field-plate electrodes partially resolves the problem but does not fully eliminate it. Field-plated devices suffer from excessive capacitances and premature breakdown between field-plate and drain electrodes. As a result of the above limitations, even the best in class sophisticated state-of-the-art III-Nitride HFETs (Fig. 18) do not achieve the breakdown voltages predicted by fundamental material properties.

The surface effects can be radically reduced in devices that combine lateral source - gate region and vertical (perpendicular to the surface) gate - drain drift regions. Known solution of this type is illustrated in in Fig. 19 [32]. Devices of this type are typically made in Si, GaAs, SiC material systems and they all use buried p-doped layers isolate lateral and vertical current paths. This approach is hardly applicable to III-Nitride material system. The issues related to creating current blocking layers in III-Nitride materials are discussed in detail in literature, see, e.g. [33] While III-Nitride devices offer tremendous performance improvement due to a high breakdown field and other unique material properties, obtaining buried p-doped III-N layers leads to significant degradation of material properties and to considerable complications to the growth/fabrication technology preventing cost reduction. Lateral-vertical device with p-blocking layer shown in Fig. 19, was proposed back in 2007 and has not yet been commercialized. This example best illustrates challenges in obtaining high-performance low cost III-Nitride power switches using existing technologies.

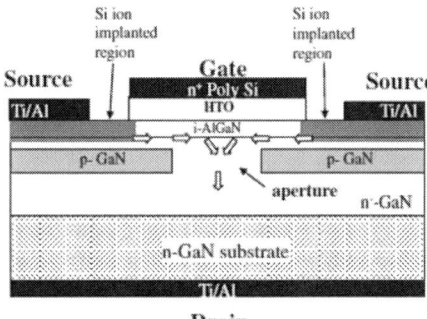

Figure 19. Lateral-vertical AlGaN/GaN power MOSHFET [32].

Conclusions

Superior transport properties and very large current carrying capability, and a very high breakdown field make group III-Nitride contenders for ultimate devices for power electronic applications. Advanced epitaxial techniques, such MEMOCVD® and MELEO have promise of lower defect and dislocation densities and better uniformity. Novel device designs are prerequisite for taking full advantages of nitride superior materials properties.

References

1. E. O. Johnson, RCA Rev., pp. 163–176, Jun. 1965
2. U. K. Mishra, et al., Proceedings of the IEEE, Vol. 96, No. 2, February 2008
3. M. E. Aumer, et al., Applied Physics Letters 75:3315 (1999)
4. H. Gotoh, et al., Japanese Journal of Applied Physics 20:L545–L548 (1981)
5. Fareed, et al., US Patent 7192849, (2007).
6. M. S. Shur, and R. Gaska, IEEE Transactions on Electron Devices 57:12–25 (2010)
7. R. Jain, et al., Applied Physics Letters 93:051113 (2008)
8. M. S. Shur, Solid State Electronics, 42, 2131 (1998)
9. T. T. Mnatsakanov, Semiconductors, Vol. 35, No. 4, 2001, pp. 394–397
10. U. Bhapkar, M. Shur, J. Appl. Phys. 82, 1649 (1997)
11. H.M. Ng et al, Appl. Phys. Lett., 73, 821, (1998)
12. M. S. Shur, Mat. Res. Soc. Symp. Proc., vol. 483, pp. 15-26 (1998)
13. B. E. Foutz et al., J. Appl. Phys. 85, 7727 (1999)
14. Khan, I.A, J.A. Cooper, Am. Sci. Forum 264-268 (1998), 509-512.
15. B. E. Foutz, et al., Appl. Phys. Lett., 70, No 21, pp. 2849-2851 (1997)
16. S. K. O'Leary et al., J. Appl. Phys. Vol. 83, p. 826-829 (1998)
17. B. E. Foutz, et al., Mat. Res. Soc. Proc. 572, (1999)
18. M. S. Shatalov, International Journal of High Speed Electronics and Systems Vol. 21, No. 1 (2012) 1250011
19. W. Liu et al., Physica status solidi (a), vol. 202, p. R135–R137, Sep. 2005.
20. K. Watari et al., Materials Research, vol. 17, no. 11, pp. 2940–2944, Jan. 2011.
21. W. Liu et al., Journal of Applied Physics, vol. 97, no. 7, p. 073710, 2005.
22. B. C. Daly et al., Journal of Applied Physics, vol. 92, no. 7, p. 3820, 2002.
23. S. Vitanov: Simulation of High Electron Mobility Transistors, PhD Thesis.
24. S. Yamaguchi, et al., Appl. Phys. Lett. vol. 83, no. 26, pp. 5398-5400, 2003.
25. S. Adachi et al., J. Appl. Phys., vol. 102, no. 6, p. 063502, 2007
26. B. Pantha et al., Appl. Phys. Lett. vol. 92, no. 4, p. 042112, 2008
27. Morelli et al., Silicon Carbide and Related Materials Eds. Spencer, M.G., et al., Institute of Physics Conference Series N137, 1993, 313-316.
28. Slack et al., J. Phys. Chem. Solids 48, 7 (1987), 641-647.
29. Carlson et al., J. Appl. Phys. 36, 2 (1965) 505.
30. W. Saito et al., IEEE Trans. El. Dev. V. 50, 2528-2531 (2003)
31. R. Chu et al., IEEE El. Dev. Lett., V. 32, 632-634 (2011)
32. M. Kanechika et al., Japanese Journal of Applied Physics, Vol. 46, No. 21, 2007, pp. L503–L505
33. S. Chowdhury et al., Semicond. Sci. Technol. 28 (2013) 074014 (8pp)
34. R. Quay, Gallium nitride electronics. Springer, 2008, ISBN 9783540718901
35. V. Palankovski and S. Selberherr, pp. 25–28, HITEN 99. Third European Conference on High Temperature Electronics. (IEEE Cat. No.99EX372) 7 July 1999
36. D. Storm et al., Electron. Lett. 40, 1226 (2004)
37. R. Gaska, et al., Appl. Phys. Lett. 74, 287 (1999)
38. N. V. Dyakonova et al., Electronics Letters, Vol. 34, No. 17, pp. 1699-1700, August 20, 1998

Normally-off GaN Transistors for Power Switching Applications

O. Hilt, E. Bahat-Treidel, F. Brunner, A. Knauer, R. Zhytnytska, P. Kotara and J. Wuerfl

Ferdinand-Braun-Institut, Leibniz-Institut fuer Hoechstfrequenztechnik, Gustav-Kirchhoff-Strasse 4, 12489 Berlin, Germany

Normally-off high voltage GaN-HFETs for switching applications are presented. Normally-off operation with threshold voltages of more than 1 V and 6 V gate swing has been obtained by using p-type GaN as gate. Different (Al)GaN-based buffer compositions using intentional doping and potential barriers have been used to obtain high blocking voltages. 1000 V blocking was obtained for devices with carbon-doped buffer structures, however, they suffer from a stronger increased dynamic on-state resistance as compared to devices based on an iron-doped GaN-buffer or an AlGaN-buffer. The best trade-off between low dispersion and high blocking strength was obtained for a modified carbon-doped GaN-buffer that showed only a 2.6x increase of the dynamic on-state resistance for 500 V switching.

Introduction

High voltage GaN-based power switching transistors enable efficient power converters with increased power density. High converter switching frequencies can be realized with lateral GaN-based HFETs due the low area-specific on-state resistance for a given blocking strength and the low gate charge required for switching. The GaN-HFET technology is also considered as cost-efficient since devices can be manufactured on 8 inch Si substrates using processing steps of a CMOS-Fab environment (1). Major challenges for market introduction are the realization of a robust normally-off device technology and the reduction of the increased dynamic on-state resistance (R_{ON}) when switching the device from high-voltage off-state bias to the on-state.

Lateral GaN-HEMTs use the 2-dimensional electron gas (2DEG) that is established at the hetero-junction between the GaN-buffer layer and the AlGaN-barrier layer as transistor channel. These devices have intrinsic normally-on properties since the transistor gate needs to be negatively biased to deplete the 2DEG. However, a normally-off characteristic is required for power-electronics application because of inherent power-system safety requirements and because of the preferred unipolar gate driver design.

High internal electric fields combined with high GaN material dislocation densities lead to device charging when applying high off-state drain voltages in the order of 600 V. The resulting increased dynamic on-state resistance (2) generates additional on-state losses for switching applications and may impede the principle advantages of GaN-based switches as compared to Si-based devices.

Normally-Off AlGaN/GaN HFETs

Recent attempts to convert AlGaN/GaN HEMTs into normally-off devices, using gate recess (3) or fluorine incorporation (4) showed limited applicability for power electronics due to their low threshold voltages $V_{th} < +1$ V, their low gate swing of ~ 2 V and their high on-state gate current. The introduction of a gate insulator suppresses the gate current and may extend the gate swing (1). The used extrinsic insulator layers are often ALD-based oxides like Al_2O_3 or HfO_2 and are deposited at temperatures < 300°C. The observed trap states in the insulator bulk or at the interface to the AlGaN barrier often deteriorate the device switching performance and limit the device reliability (5).

In the p-GaN gate transistors presented here, the gate consists of in-situ grown Mg-doped GaN with an ohmic contact for biasing (6). P-GaN gate GaN HFETs combine the high-mobility 2DEG transistor channel known from AlGaN/GaN HEMTs with robust normally-off operation, as required for applications in power electronics (6-8). Similar to enhancement-mode GaN MISFETs, a wide gate swing allows secure on-state and off-state condition even in applications with high EMI noise.

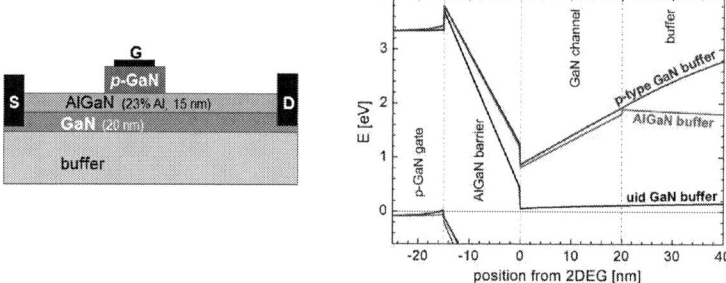

Figure 1. Simulated band diagram at the gate position of a p-GaN gate $Al_{0.23}Ga_{0.77}N$/GaN HFET for different buffer compositions: uid GaN buffer in black, $Al_{0.05}Ga_{0.95}N$ buffer in red and p-type GaN buffer in blue. The conduction band energy at the 2DEG position is significantly shifted above the Fermi level for the AlGaN buffer or p-type GaN buffer structure. The device structure is sketched on the left.

When unbiased, the p-type gate depletes the 2DEG of the transistor channel and a positive gate bias is needed to open the transistor. The simulated band diagram at the gate position (Fig. 1) confirms that the 2DEG channel at the heterojunction between GaN buffer and AlGaN barrier does not cross the Fermi level (black conduction band line in Fig. 1). Replacing the GaN-buffer by AlGaN and using uid GaN for the transistor channel generates a virtual p-type doping at the AlGaN back-barrier of the channel. The potential well of the 2DEG is then dragged above the Fermi level by almost 1 eV (red conduction band line in Fig. 1) and a more positive gate bias is required to again fill the 2DEG channel beneath the gate. Using an AlGaN-buffer structure is thus beneficial for a high threshold voltage. A similar high energy of the heterojunction potential well can be obtained by a GaN-buffer with p-type properties beneath the uid GaN-channel (blue conduction band line in Fig. 1). A high threshold voltage can in principle also be realized by reducing the electron concentration in the 2DEG (i.e., reduced Al-concentration in the AlGaN barrier). But this would give a higher on-state resistance as compared to using the proposed back-barrier (6).

Fabricated p-GaN gate transistors based on an $Al_{0.05}Ga_{0.95}N$-buffer, a carbon-doped GaN-buffer (GaN:C) and an iron-doped buffer (GaN:Fe) demonstrate that the predicted normally-off characteristics are valid for the different buffer compositions, as shown by the linear and logarithmic representation of the transfer characteristics in Fig. 2. Threshold voltages of up to 1.8 V have been realized. In on-state, the gate bias can be driven up to 6 V and even more but the gate current increases due to the pin-diode-like nature of the gate module. The gate bias for on-state condition should be chosen as $V_{GS} = 5$ V in order to limit the on-state gate current to approx. 10 µA/mm.

In off-state, at $V_{GS} = 0$ V, the drain leakage current is limited to < 1 µA/mm for the AlGaN-buffer and the GaN:C-buffer devices and to 30 µA/mm for the GaN:Fe-buffer devices. The higher off-state leakage of the GaN:Fe-buffer devices correlates with their low threshold voltage of $V_{th} = 0.7$ V. This is caused by the higher electron density in the 2DEG as can be concluded from the high on-state device current of the GaN:Fe-buffer devices as compared to both other buffer structures, see linear representation of the transfer characteristics in Fig. 2.

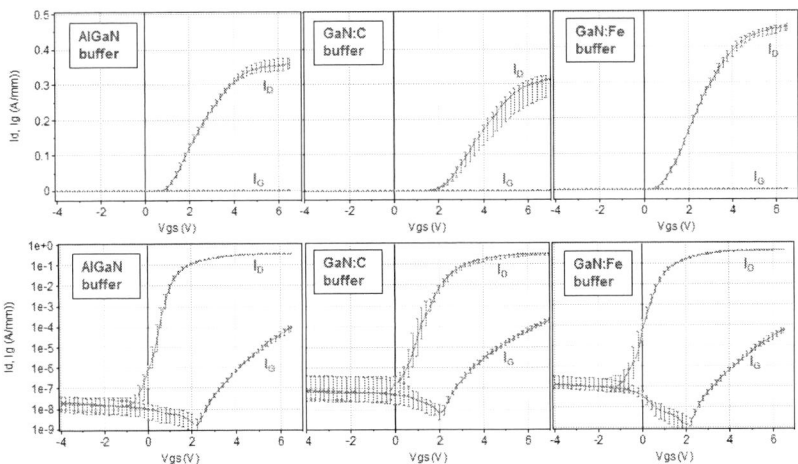

Figure 2. Transfer characteristics and gate current (linear representation on top, logarithmic representation on bottom) of p-GaN gate HFETs based on an AlGaN buffer (left), on an carbon-doped GaN buffer (middle) and on an iron-doped GaN buffer (right). The median currents with quantiles of 30 devices spread over a 3" wafer are shown. Device gate width is 2.1 mm, gate drain separation is 15 µm and $V_{DS} = 10$ V.

Buffer Concepts for High Voltage Operation

The breakdown voltage of GaN HFETs should ideally increase with the gate-drain separation d_{GD} since the electric field in off-state distributes between gate and drain. This breakdown voltage scaling is lost when electrons from the transistor channel bypass the gate control region via deep layers in the buffer. The resulting source-drain leakage current prevents higher blocking voltages, independently of d_{GD}. The major cause for this electron punch-through effect is the limited barrier height of the GaN buffer beneath the transistor channel with the two-dimensional electron gas (2DEG) (9). The Fermi level in uid-doped MOCVD-grown GaN-layers is usually shifted towards the conduction band (10) and the back-side barrier towards the buffer is in the order of 100 meV, see Fig. 1.

Punch-through effects and the associated source-drain leakage can get suppressed by a better electron confinement to the channel. An AlGaN-buffer beneath the GaN-channel creates an efficient barrier for the channel electrons (Fig. 1). The buffer conduction band is shifted upwards due to polarization charge differences and the band energy differences between the GaN channel and the AlGaN buffer (9). A breakdown voltage scaling of 42 V/μm gate-drain separation was measured for $Al_{0.05}Ga_{0.95}N$ back-barrier based p-GaN gate devices and 680 V breakdown voltage has been obtained (Fig. 3). GaN-buffer compensation doping with carbon (GaN:C) or iron (GaN:Fe) also shifts the buffer Fermi level towards the valence band and the resulting back-barrier (Fig. 1) prevents electrons in the channel to enter the buffer. More than 1000 V breakdown strength with 118 V/μm scaling has been obtained for GaN:C-based p-GaN gate transistors (Fig 3). Transistors based on GaN:Fe-buffer yield 50 V/μm blocking strength scaling (Fig. 3). Introducing a buffer structure that introduces a back-barrier to the device structure is thus beneficial for both, a high threshold voltage and a high blocking voltage.

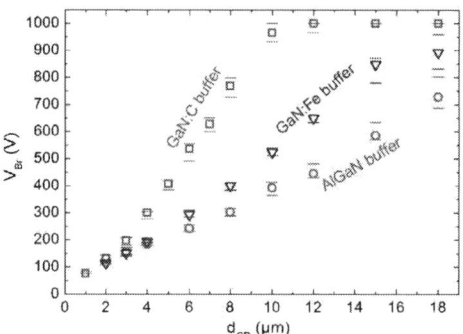

Figure 3. Transistor breakdown strength as a function of gate-drain separation for different buffer compositions. The n-SiC substrate was kept floating. Symbols are wafer medians and bars 25%/75% percentiles.

A sufficient vertical isolation of the GaN-based layers is required for a high transistor blocking voltage when conductive substrates are used. Compared to semi-insulating SiC, conductive substrate like n-type SiC or Si give a considerable cost advantage. Vertical leakage currents from a drain- or source-contact on top of the GaN-based layers towards the grounded n-SiC substrate are shown in Fig. 4 for the different buffer compositions.

There are particular high vertical leakage currents for the AlGaN-buffer and GaN:Fe-buffer devices that would limit the transistor blocking voltage to approx. 120 V if the substrate is on source potential. 600 V blocking voltage can be achieved for a configuration with the substrate on drain potential or for a floating (electrically isolated) substrate. Approximately two orders of magnitude lower vertical leakage currents were measured for the structures based on a GaN:C-buffer and 400 V blocking voltage can be obtained if the substrate is biased on source potential. Carbon-doped GaN buffer structures give significant advantage in terms of lateral and vertical blocking strength as compared to the two other buffer compositions.

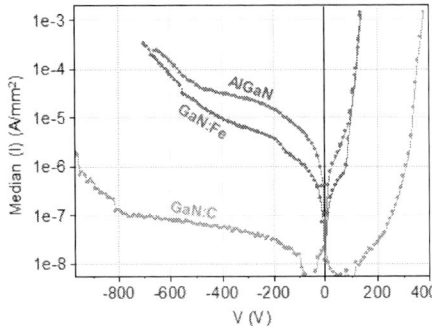

Figure 4. Vertical leakage current from a drain pad to the grounded conductive n-SiC substrate as a function of drain bias for three different buffer types. Wafer medians are shown.

Dynamic Characteristics and High Voltage Switching

Besides the transistor off-state properties, also the switching dynamics are strongly determined by the buffer composition. An increase of the dynamic R_{ON} immediately after switching from high-bias off-state condition is often observed in GaN-HFETs. Trap states in the (Al)GaN-based semiconductor stack or on the semiconductor surface are considered as root cause for this effect (11). Electron traps inside the buffer may get filled under high bias off-state condition and deplete the transistor channel after switching the transistor to on-state since electron detrapping time constants are longer than the switching-cycle time. Here, we compare the impact of non-intentionally doped (AlGaN-buffer) and doped (GaN:Fe, GaN:C) buffer structures on the dynamic R_{ON}.

To analyze the increase of dynamic R_{ON} pulsed on-state IV curves have been measured after the devices have been biased up to 65 V in off-state, see Fig. 5. The pulse time in on-state was 0.2 µs and the time in off-state in-between the pulses was 2500 µs. The measurements were performed on-wafer.

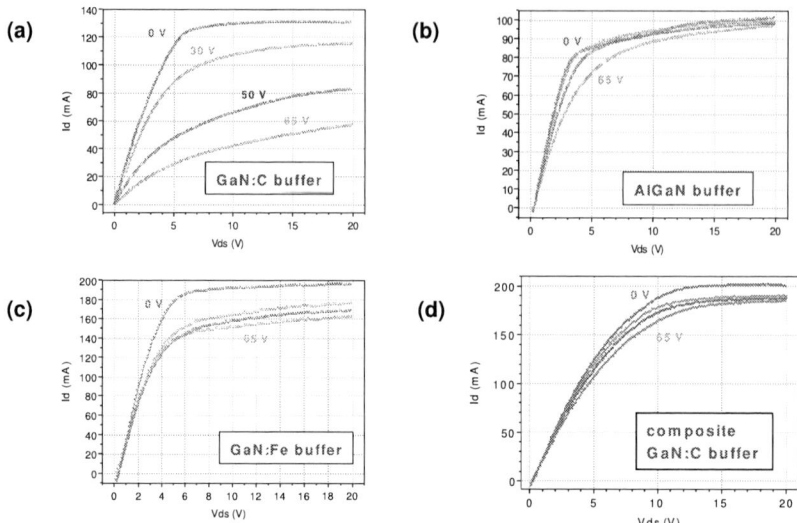

Figure 5. On-state IV curves pulsed from different off-state drain bias voltages between 0 V and 65 V. On-state pulses are 0.2 µs long. Test transistors with 0.25 mm gate width and different buffer compositions are compared.

For all buffer compositions, a reduced current has been obtained when switching from 30 V, 50 V or 65 V drain bias as compared to switching from the non-stressed condition of 0 V drain bias. The current degradation increases with off-state bias. Most severe reduction of the drain current is visible for the GaN:C buffer device (Fig. 5a) where the dynamic R_{ON} is increased 4-times for switching from 65 V off-state bias. Switching experiments of packaged device showed that the dynamic R_{ON} further increases with off-state bias and a more than 1000x increased dynamic R_{ON} has been measured for 800 V switching (12). The 500 V switching transient in Fig. 8 (left) corresponds to the wafer of Fig. 5a and has a 200x increased dynamic R_{ON}. The pulsed IV curves for the AlGaN buffer (Fig. 5b) and the GaN:Fe buffer (Fig. 5c) show significantly less dispersion as compared to the GaN:C buffer device but with a still 1.5x increased dynamic R_{ON} from 65 V off-state bias.

Obviously, the different buffer compositions have different strengths and weaknesses in terms of switching efficiency and blocking capability. Fig. 6 puts the increase in dynamic R_{ON}, as measured from 65 V off state bias in relation to the device blocking strength. Data have been collected for GaN:C-buffer devices with different carbon concentration [C] in the buffer. The increase in dynamic R_{ON} reduces from factor 8 to factor 2.8 when reducing [C] from 2e19 cm^{-3} to 1e18 cm^{-3} but the blocking strength also drops from 118 V/µm to 30 V/µm. For a given blocking strength of 40-50 V/µm, the AlGaN-buffer and GaN:Fe-buffer devices have less dispersion than GaN:C buffer devices with 1-2e18 cm^{-3} doping concentration. In particular, the GaN:Fe buffer shows the best trade-off with 50 V/µm blocking strength and an increase in dynamic R_{ON} of factor 1.4.

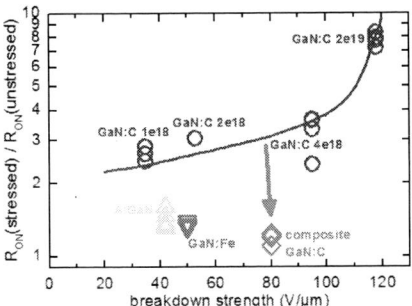

Figure 6. Increase in dynamic on-state resistance for 0.2 μs long pulses into on-state from 65 V off-state bias in relation to the device breakdown strength scaling. [C] in cm^{-3} is noted for the GaN:C devices. Test transistors with 0.25 mm gate width and different buffer compositions are compared.

Switching experiments with packaged GaN:Fe-buffer devices were performed for higher voltages (Fig. 7 left) and the increase of the dynamic R_{ON} was determined from the on-state voltage drop 5 μs after switching to on-state. Dynamic R_{ON} is increased 2.8x as compared to the DC value for 200 V off-state stress (Fig. 7 right).

In conclusion, a GaN:C buffer-based transistor as presented in Fig. 5a is despite its high blocking strength not suitable for high-voltage switching transistors. GaN:Fe buffer-based devices show an acceptable switching performance and may be used up to a few hundred Volts. The poor vertical isolation will exclude these devices from switching 600 V or more.

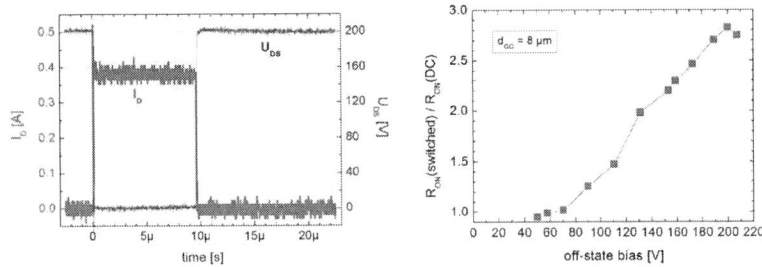

Figure 7. Switching characteristics of GaN:Fe buffer based GaN-HFETs. Left: 200 V / 0.4 A switching transient. Right: increase in dynamic R_{ON} as function of off-state bias. The dynamic R_{ON} was determined from the on-state voltage drop at 5 μs.

Potentials Inside the HFET Structure

The initial assumption that doping-related trap states are a major cause for increased dynamic R_{ON} (11) cannot explain the observed low increase in dynamic R_{ON} of the GaN:Fe-buffer devices. The complete semiconductor stack charge balance and the resulting potentials inside the device have to be taken into account for a deeper understanding of the observed differences in dispersion. Fe-doping is known to create acceptor states 0.5 eV below the conduction band in the upper half of the GaN band gap and the material can be considered as n-type rather than p-type (10). Carbon-doping creates acceptor-like deep states 0.9 eV above the valence band and GaN:C can be considered as p-type. Based on these trap energies, time-dependent physical-based simulations of the internal device potentials under switching have been performed in (10) that reproduced the observed dispersion differences between GaN:C and GaN:Fe based devices.

A pn-junction is present between the (n-type) 2DEG channel and the (p-type) GaN:C buffer beneath. For the GaN:C-buffer device a depletion layer is formed in the drift region between gate and drain for positive off-state bias. The buffer is electrically isolated from drain potential and the drain voltage drops between the transistor channel layer and the buffer which remains close to source potential. When switching to low drain bias in on-state, the GaN:C buffer remains isolated from the drain and the buffer potential shifts to negative potential which discharges on a very long time scale only. As consequence, the 2DEG gets depleted from the charged buffer and the on-state resistance is increased as observed.

On the other side, there is no pn-junction in the drift region for the GaN:Fe buffer devices and the buffer is in electrical contact with the drain in on-state and in off-state. The buffer potential can thus follow the drain potential during switching and the 2DEG density is not reduced when switching to on-state.

The sketched situation for the GaN:C-buffer devices is further supported by the observed high lateral blocking strength (Fig. 3) and the very low vertical leakage current (Fig. 4). In both cases, the depletion zone of the pn-junction between the transistor channel and the p-type buffer efficiently suppresses leakage currents. This pn-junction is absent for the AlGaN-buffer and the GaN:Fe-buffer structures explaining the observed vertical leakage currents. Additionally, lateral AlGaN/GaN Schottky-diodes based on a GaN:C-buffer according to Fig. 5a were manufactured. Fast switching has been observed (13) and no signature of a significantly increased dynamic R_{ON} was found for 300 V switching (14). In diode switching, the anode reverse voltage is negative. Then, the pn-junction between the GaN channel and the GaN:C buffer is closed and the buffer is electrically connected to the anode. In conclusion, the trap-state energy levels strongly determine the internal device potentials, that again control the dynamic R_{ON}.

GaN:C-Buffer Based HFET with Low Dynamic On-State Resistance

Increasing the distance between the transistor channel and the GaN:C-buffer related pn-junction may reduce the channel depletion due to the negatively charged GaN:C-buffer after switching from off-state. Recent switching experiments with a modified composite GaN:C-buffer device that additionally had an 800 nm thick AlGaN back-barrier and a GaN cap incorporated showed a successful combination of high blocking strength an low dispersion (15). The device with 80 V/μm blocking strength showed only a 1.2-times increase in dynamic R_{ON} for pulsing from 65 V (Fig. 5d and purple diamonds in Fig. 6). Drain voltage and drain current transients during 500 V / 0.5 A switching with a packaged device and source-connected substrate were taken (Fig. 8, right) and a 2.6x increase in dynamic R_{ON} after switching from 500 V was detected.

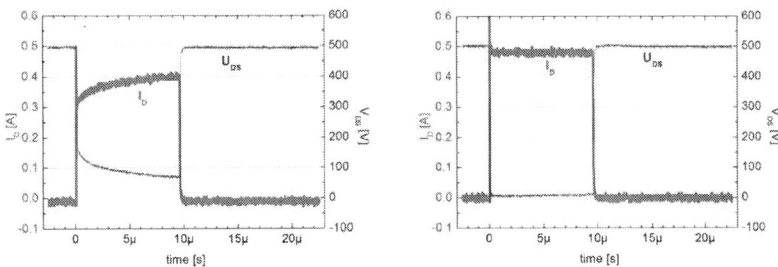

Figure 8. 500 V / 0.5 A switching transients for two different GaN HFET based on a GaN:C-buffer. The left device, corresponding to Fig. 5a, has a 100 nm thick uid GaN channel is on top of the GaN:C buffer and shows a more than 100x increased dynamic R_{ON}. The device on the right, corresponding to Fig. 5d has an 800 nm thick AlGaN layer inserted between the GaN:C-buffer and the uid GaN-channel. An only 2.6x increased dynamic R_{ON} was extracted from the on-state voltage drop.

Conclusion

Normally-off p-GaN gate transistors with 1 V threshold voltage and higher and 5-6 V gate swing have been demonstrated for different buffer compositions. Buffer structures that establish a back-barrier towards the transistor channel shift the threshold voltage towards higher values without impeding the transistor channel conductivity in the access regions. Further, buffer structures with a back-barrier effect are needed to preserve the V_{Br}-scaling with the gate-drain separation also for high voltages. GaN:Fe-buffer devices showed an acceptable increased dynamic R_{ON} but the high substrate leakage currents impedes operation above approximately 200 V. GaN:C-buffer devices demonstrated excellent lateral and vertical isolation. Strong dispersion must be avoided by careful engineering of the electronic properties of the (Al)GaN-based epitaxial layers. 500 V / 0.5 A switching with an only 2.6x increased dynamic on-state resistance has been demonstrated.

Acknowledgments

Support from the German BMBF (LES project PowerGaNPlus, contract no. 13N10908) and from EU FP7 project HipoSwitch (Grant no. 287602) is gratefully acknowledged.

References

1. B. De Jaeger, M. Van Hove, D. Wellekens, X. Kang, H. Liang, G. Mannaert, K. Geens and S. Decoutere, Proc. ISPSD 2012, Bruges, p. 49 (2012).
2. N. Ikeda, Y. Niiyama, H. Kambayashi, Y. Sato, T. Nomura, S. Kato and S. Yoshida, Proc. of the IEEE, **98**(7), 1151 (2010).
3. J. W. Saito, Y. Takada, M. Kuraguchi, K. Tsuda and I. Omura, IEEE Trans. on Electron Devices, **53**, 356 (2006).
4. Y. Cai, Y. Zhou, K. M. Lau, and K. J. Chen, IEEE Trans. on Electron Devices, **53**, 2207 (2006).
5. X. Sun, O. I. Saadat, K. S. Chang-Liao, T. Palacios, S. Cui, and T. P. Ma, Appl. Phys. Lett., **102**, 103504 (2013).
6. O. Hilt, A. Knauer, F. Brunner, E. Bahat-Treidel and J. Würfl, Proc. ISPSD 2010, Hiroshima, p. 347 (2010).
7. Y. Uemoto, M. Hikita, H. Ueno, H. Matsuo, H. Ishida, M. Yanagihara, T. Ueda, T. Tanaka and D. Ueda, IEEE Trans. on Electron Devices **54**(12), 3393 (2007).
8. I. Hwang, H. Choi, J. Lee, H. Choi, J. Kim, J. Ha, C. Um, S. Hwang, J. Oh, J. Kim, J. Shin, Y. Park, C. U-In, I. Yoo and K.Kim, Proc. ISPSD 2012, Bruges, p. 41 (2012).
9. E. Bahat-Treidel, O. Hilt, F. Brunner, J. Würfl, and G. Tränkle, IEEE Trans. on Electron Devices, **55**(12), 3354 (2008).
10. M. J. Uren, J. Möreke, and M. Kuball, IEEE Trans. on Electron Devices, **59**(12), 3327 (2012).
11. D. Jin and J.A. del Alamo, Microelectronics Reliability, **52**(12), 2875 (2012).
12. O. Hilt, E. Bahat-Treidel, E. Cho, S. Singwald and J. Würfl, Proc. ISPSD 2012, Bruges, p. 345(2012).
13. E. Bahat-Treidel, O. Hilt, R. Zhytnytska, A. Wentzel, C. Meliani, J. Würfl and G. Tränkle, IEEE Electron Device Lett., **33**(3), 357 (2012).
14. N. Badawi, E. Bahat-Treidel, S. Dieckerhoff, O. Hilt and J. Würfl., EPE '13-ECCE Europe, Lille, submitted for publication, (2013).
15. E. Bahat-Treidel, O. Hilt and J. Würfl, Proc. 37th Workshop on Compound Semiconductor Devices and Integrated Circuits (WOCSDICE), Warnemünde, p. 61 (2013).

High performance normally-off GaN MOSFETs on Si substrates

H. Kambayashi[a], N. Ikeda[a], T. Nomura[a], H. Ueda[b], Y. Morozumi[b], K. Harada[c],
K. Hasebe[c], A. Teramoto[d], S. Sugawa[d], and T. Ohmi[d]

[a]Furukawa Electric Co., Ltd. 2-4-3 Okano, Nishi-ku, Yokohama, Japan
[b]Tokyo Electron Ltd., Minato-ku, Tokyo, Japan
[c]Tokyo Electron Tohoku Ltd., Nirasaki, Yamanashi, Japan
[d]Tohoku Univ. 6-6-10 Aza-Aoba, Aramaki, Aoba-ku, Sendai, Japan

> The enhancement mode AlGaN/GaN hybrid MOS-HFETs on Si
> substrates have been demonstrated. The breakdown voltage of over
> 1.71 kV was achieved by investigating the epitaxial structure.
> Furthermore, a high integrity SiO_2/Al_2O_3 gate stack has been
> demonstrated for GaN MOSFETs. The SiO_2 film formed on GaN
> by the MW-PECVD exhibits good properties compared that by the
> LP-CVD. Then, by incorporating the advantages of both of SiO_2
> with a high insulating characteristics and Al_2O_3 with good
> interface characteristics, the SiO_2/Al_2O_3 gate stack structure has
> been employed in GaN MOS devices. It is shown that a low
> interface state density between gate insulator and GaN, a high
> breakdown field, and a large charge-to-breakdown by applying 3-
> nm Al_2O_3 in this structure. The SiO_2/Al_2O_3 gate stack has also
> been applied to AlGaN/GaN hybrid MOS-HFET and excellent
> properties with the threshold voltage of 4.2 V and the maximum
> field-effect mobility of 192 cm^2/Vs are shown in the transistor.

Introduction

GaN has excellent physical properties for power devices such as a high breakdown field, a high career mobility, and a high saturation velocity compared with Si and SiC [1]. For GaN-based transistors, AlGaN/GaN heterojunction field-effect transistors (HFETs) have been intensively reported. AlGaN/GaN HFETs can realize a low on-state resistance and a high breakdown voltage due to a high electron mobility of two dimensional electron gases (2DEG) generated at the interface of the AlGaN/GaN heterostructure. However, the operation of AlGaN/GaN HFETs is essentially normally-on mode. For power switching applications, normally-off operation with the threshold voltage above 3 V is strongly required from the fail-safe point of view [2]. A GaN MOSFET is one of the good candidates to achieve normally-off operation. We have previously demonstrated GaN MOSFETs on Si substrates using an ion implantation [3]-[5]. These transistors showed good normally-off operation, however, the on-state resistances were significantly high. This is because the activation of implanted ions was insufficient and then the resistance of implanted region was high [5]. By incorporating the advantages of both a MOS channel and high electron mobility of 2DEG in AlGaN/GaN drift region, normally-off AlGaN/GaN hybrid MOS (MIS)-HFETs have been proposed [6]-[8]. This structure can obtain a good normally-off operation by applying MOS channel and a low on-state resistance. Our previously fabricated AlGaN/GaN hybrid MOS-HFETs have shown a good normally-off operation with the threshold voltage of 2.7 V, the maximum drain current of over 100 A, specific on-state resistance of 9.1 $m\Omega cm^2$, and the breakdown voltage of over 600 V, respectively [8]. However, the breakdown voltage was lower and the on-state resistance was higher compared with normally-on AlGaN/GaN HFETs. To improve these properties, the investigation of epitaxial structure and application of a high quality gate insulator are essentially needed. In this work, high performance normally-off AlGaN/GaN hybrid MOS-HFETs on Si Si substrates with a high breakdown voltage of over 1.71 kV and a high maximum field-effect mobility of 192 cm^2/Vs have been demonstrated.

Investigation of epitaxial structure for high breakdown voltage of AlGaN/GaN hybrid MOS-HFET

Experimental

For increasing the breakdown voltage of the transistors, the thickness of epitaxial layers is one of the important parameters. In the case of AlGaN/GaN HFETs, thick epitaxial layers with a highly resistive buffer layer and a thin unintentional GaN (u-GaN) channel layer on a highly resistive carbon-doped GaN (C-GaN) layer have been reported as an effective method to increase the breakdown voltage [9]. By using this technique, improving the breakdown voltage of AlGaN/GaN hybrid MOS-HFETs has been examined.

AlGaN/GaN hetero-structures were grown on a 4-inch Si substrate by the metal organic vapor-phase epitaxy (MOVPE). The total thickness of epitaxial layer was over 7.3 μm. A heterostructure constructed with a GaN-based buffer layer, a highly resistive C-GaN layer, 50-nm or 400-nm u-GaN channel layer and a 20-nm-thick $Al_{0.25}Ga_{0.75}N$ were formed on a Si(111) substrates. Figure 1 shows the schematic cross-sectional view of the AlGaN/GaN hybrid MOS-HFET. After mesa etching, the channel region under the gate was defined by reactive-ion etching (RIE). The electrodes were deposited by using the sputtering. State-of-the-art Ti/AlSi/Mo-based ohmic electrodes were formed on the AlGaN layers as source and drain electrodes to reduce the contact resistance [10]. The SiO_2 films formed by Capacitive Coupled Plasma (CCP)-CVD were used as the gate insulators and field plate films. The thickness of SiO_2 gate insulator was 40 nm. Ti/Au was used as the gate electrode, including for the gate field plate structure. The DC characteristics were examined and the current collapse characteristics were investigated.

Results and Discussion

Figure 2 shows the transfer characteristics of a fabricated AlGaN/GaN hybrid MOS-HFET. This MOS-HFET has 50-nm u-GaN channel layer with the recess etching depth of

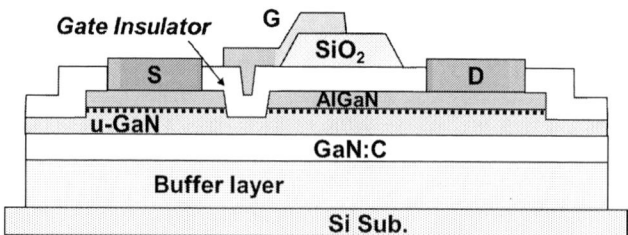

Figure 1. Schematic cross section of AlGaN/GaN hybrid MOS-HFET.

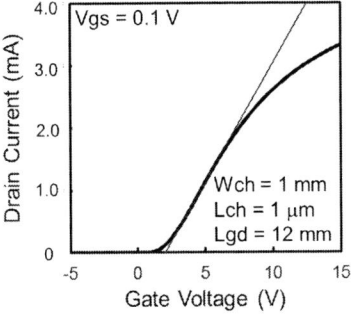

Figure 2. Transfer characteristics of AlGaN /GaN hybrid MOS-HFET.

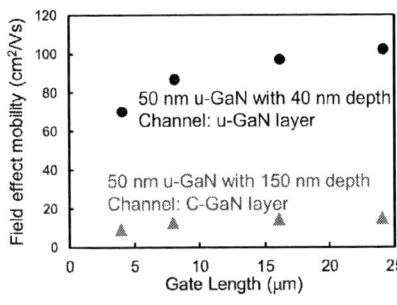

Figure 3. Field-effect mobility versus channel length of AlGaN/GaN hybrid MOS-HFETs with a recess etching depth of 40 nm and 150 nm.

40 nm. The threshold voltage (V_{th}) was extrapolated as 2.0 V, resulting in the achievement of a good normally-off operation. Figure 3 shows the maximum field effect mobility (μFE) dependence for the channel layer from gradual channel approximation. The μFEs with the recess etching depths of 40 and 150 nm are plotted on Fig. 3. In the case of the depth of 40 nm, μFE was estimated to be 102 cm^2/Vs. However, the μFE of MOS-HFET with the depth of 150 nm was 15 cm^2/Vs. It is considered that the channel was formed at C-GaN layer for the 150 nm recess case. This suggests that to control the etching depth under the gate is very important to achieve a lower on-state resistance. Figure 4 shows the output characteristics of AlGaN/GaN MOS-HFET with 50-nm u-GaN channel layer and the recess etching depth of 40 nm. The on-state resistance of this MOS-HFET was estimated to be 7.1 mΩcm^2.

Figure 5 shows the off-state characteristics of AlGaN/GaN MOS-HFETs with 50-nm u-GaN channel layer and the recess etching depth of 40 nm. Several transistors with different gate-to-drain distances (L_{gd}) were examined. As the L_{gd} increases, the breakdown voltage increases, resulting in obtaining 1.21 kV for an L_{gd} of 12 μm and over 1.71 kV for an L_{gd} of over 18 μm. Figure 6 shows a comparison of breakdown voltage and μFE of the AlGaN/GaN MOS-HFETs with (a) 400-nm u-GaN and recess etching

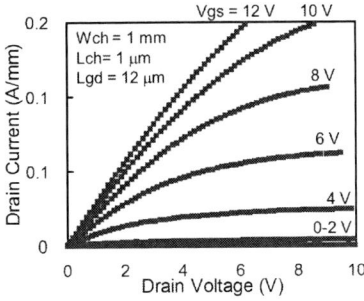

Figure 4. Output characteristics of AlGaN /GaN hybrid MOS-HFET.

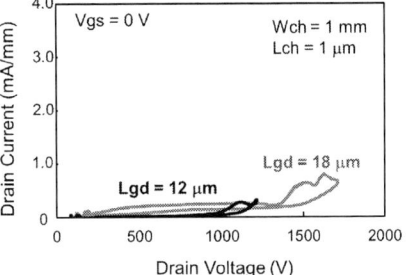

Figure 5. Off-state characteristics of AlGaN/GaN hybrid MOS-HFETs with the L_{gd} of 12 μm and 18 μm.

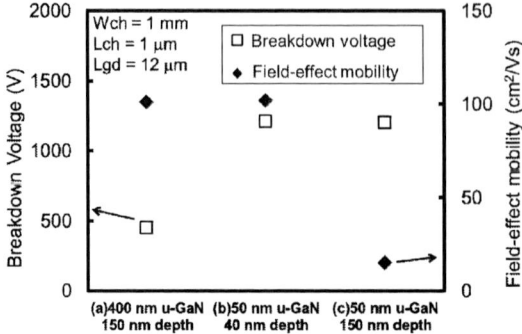

Figure 6. Comparison of breakdown voltage and field-effect mobility of the AlGaN/GaN hybrid MOS-HFET with different u-GaN thickness and recessed depth.

depth of 40 nm, (b) 50-nm u-GaN and recess etching depth of 40 nm, and (c) 50-nm u-GaN and recess etching depth of 150 nm. In the case of (a) and (b), the μFE are almost the same. The point in common between (a) and (b) is that the channel is fabricated on u-GaN layer. As the results, the μFE becomes higher compared with (c), the channel of which is formed on carbon doped layer. For breakdown voltage, (b) and (c) become higher. The point in common between (a) and (b) is that these MOS-HFETs have thin u-GaN channel layers. Therefore, breakdown voltage is indicated not to depend on the recess etching depth.

Next, a comparison of current collapse properties of AlGaN/GaN MOS-HFETs with and without gate-field plate structure is shown in Figure 7. The current collapse phenomena were observed using our own on-wafer measurement methods [11]. The current collapse was examined by three steps; (1) the on-state resistance (R_bf) was measured, next (2) the drain-to-source stress bias (Vds) was applied when the gate-to-source voltage was 0V, and then (3) the on-state resistance (R_af) was estimated again. An R_af / R_bf ratio was defined as a current collapse. The MOS-HFETs without a gate-field plate structure increased the current collapse ratio in a linear fashion as the off-stress

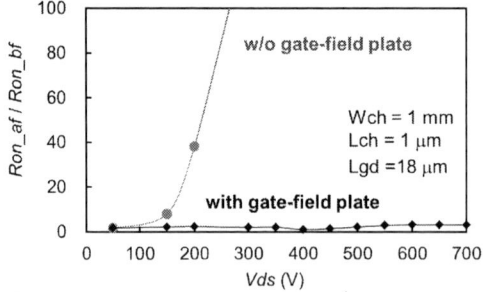

Figure 7. Comparison of current collapse ratios for AlGaN/GaN hybrid MOS-HFETs with and without gate-field plate. The Ron_bf is the on-state resistance before stress bias and the Ron_af is the on-state resistance after stress bias.

became higher. On the other hand, the ratio of the transistors with a gate-field plate is maintained at a constant and low value despite an increase in the off-stress bias.

SiO$_2$ gate insulator formation for GaN MOS devices

For GaN MOSFET, a high quality gate insulator is essentially required to be a low interface-state density (D_{it}) between gate insulator and GaN, a high breakdown voltage and a high reliability, respectively. Both SiO$_2$ and Al$_2$O$_3$ films are good candidates as the gate insulators of GaN MOS transistors since these insulators have large direct wide bandgaps, large conduction band and valence band offsets on GaN, respectively [12]. At first, we have first investigated the formation process of SiO$_2$ films by Microwave (2.45 GHz: MW) Plasma Enhanced Chemical Vapor Deposition (PECVD), LP (Low Pressure)-CVD, and Capacitive Coupled Plasma (CCP: 13.56 MHz) enhanced CVD for the gate insulator of GaN MOS devices. MW plasma is capable of exiting a low-electron temperature (<1 eV) and a high-electron density (>10^{12} cm^{-3}) at the substrate surface position [13]. LP (Low Pressure)-CVD is a formation process without using plasma and it has been reported as the gate insulator of GaN MOS devices [14]. In this work, we have first compared the formation process of SiO$_2$ films by MW-PECVD, CCP-CVD and LP-CVD as the gate insulator of GaN MOS devices.

Experimental

In order to investigate the interface properties between the gate insulator and GaN and the electrical characteristics of gate insulators, n-type GaN with Si-donor doping at a concentration of 2×10^{17} cm^{-3} on Si(111) substrates are applied for fabrication of GaN MOS capacitors. The n-type GaN layer was grown by the MOVPE. Ohmic electrode on the n-type GaN layer and the gate electrode on SiO$_2$ films were fabricated by a sputter deposition. SiO$_2$ films were deposited below 400 $^\circ$C by MW-PECVD and CCP-CVD at 800 $^\circ$C by the LP-CVD on the n-type GaN layer. Details of MW-PECVD equipment have been described in [15] and the CCP-CVD and the LP-CVD are conventional systems. The thickness of SiO$_2$ is around 50 nm. Ohmic electrode on the n-type GaN layer and the gate electrode on gate insulator were fabricated by sputter deposition equipment. We also compared the properties of GaN MOS capacitors with MW-PECVD SiO$_2$ between annealing and without annealing after deposition. The annealing process was performed at 800 $^\circ$C by a furnace in N$_2$ ambient. An annealing after SiO$_2$ deposition is effective for decreasing the D_{it} of SiO$_2$/GaN and improving the charge-to-breakdown Q_{bd} [3, 16].

Results and Discussion

Figure 8 shows the energy distribution of the D_{it} of SiO$_2$/GaN. The D_{it} was estimated by applying the Terman method [17] to the capacitance-voltage (C-V) characteristics measured at 150 $^\circ$C. The D_{it} of the GaN MOS capacitor with MW-PECVD SiO$_2$ is lower than that with CCP-CVD SiO$_2$ and LP-CVD SiO$_2$. Especially, a localized surface state was found at approximately Ec-0.6 eV for the GaN MOS capacitor with LP-CVD SiO$_2$. This result indicates that the nitrogen vacancy-related defects were formed at GaN surface because the surface of GaN was exposed at a high temperature when SiO$_2$ film was formed by the LP-CVD on GaN. Similar phenomena have been previously reported [18]. These results indicate that a low temperature fabrication of gate insulators is

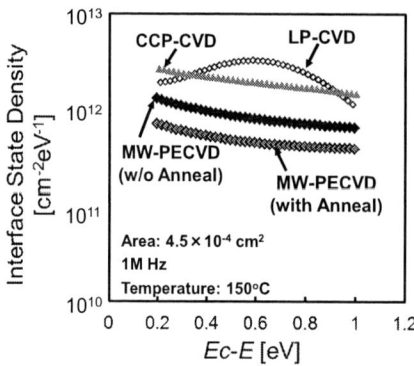

Figure 8. D_{it} of GaN MOS capacitors with MW-PECVD SiO$_2$, CCP-CVD SiO$_2$ and LP-CVD SiO$_2$ calculated from the C-V characteristics measured at 150 °C.

important for GaN MOS devices. The difference between MW-PECVD SiO$_2$ and CCP-CVD SiO$_2$ suggests to be caused by plasma damage such as ion bombardment and charge-up by CCP-CVD SiO$_2$ deposition on GaN since CCP excites electrons to very high temperature. Furthermore, by annealing after MW-PECVD SiO$_2$ deposition, the D_{it} was decreased to approximately half compared with the MW-PECVD SiO$_2$ without annealing. Figure 9 shows the typical current density-electric field (J-E) characteristics at 200 °C of these GaN MOS capacitors with MW-PECVD SiO$_2$, LP-CVD SiO$_2$ and LP-CVD SiO$_2$. Electric field E_{ox} is defined by

$$E_{OX} = \frac{V_g - V_{FB}}{EOT}, \qquad (1)$$

where V_g is gate voltage, V_{FB} is flatband voltage, and EOT is equivalent SiO$_2$ thickness. V_{FB} and EOT were evaluated from C-V characteristics. The MW-PECVD SiO$_2$ has a high breakdown electric field with over 11 MV/cm while the breakdown field of CCP-PECVD SiO$_2$ and LP-CVD SiO$_2$ are lower than 8 MV/cm. These results are considered to reflect the quality of SiO$_2$ films. That is, SiO$_2$ formed by MW-PECVD has higher quality compared with SiO$_2$ films formed by the other methods. The breakdown properties of GaN MOS capacitors with the MW-PECVD SiO$_2$, which were annealed and not annealed after deposition, were also evaluated and confirmed to be almost the same. Figure 10 shows the charge-to-breakdown Q_{bd} of these GaN MOS capacitors. Constant current stress 10 mA/cm^2 was applied to the MOS capacitors. The Q_{bd} with MW-PECVD SiO$_2$ is

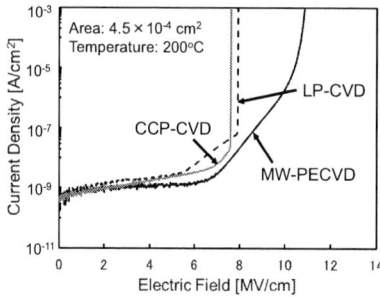

Figure 9. J-E characteristics of GaN MOS capacitors with MW-PECVD SiO$_2$, CCP-CVD SiO$_2$ and LP-CVD SiO$_2$ measured at 200 °C.

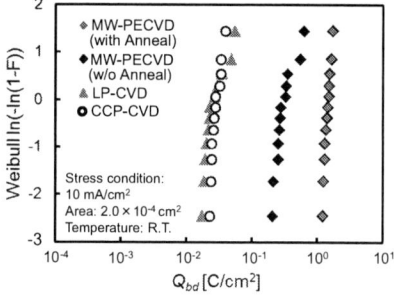

Figure 10. Charge-to-breakdown Q_{bd} of GaN MOS capacitors with MW-PECVD SiO$_2$, CCP-CVD SiO$_2$ and LP-CVD SiO$_2$ measured at room temperature.

over one order of magnitude higher than that with CCP-PECVD SiO_2 and LP-CVD SiO_2. Moreover, by annealing after MW-PECVD SiO_2 formation, The Q_{bd} was increased. These indicate that MW-PECVD SiO_2 is more reliable than CCP-PECVD SiO_2 and LP-CVD SiO_2 as the gate insulator for GaN MOS devices, and the annealing after SiO_2 deposition on GaN is effective for decreasing D_{it} of SiO_2/GaN and improving Q_{bd} of SiO_2.

SiO_2/Al_2O_3 gate stack for GaN MOS devices

We have shown a high quality SiO_2 gate insulator for GaN MOS devices. However, the D_{it} of SiO_2/GaN is still higher compared with Si devices and it should be reduced to realize further high efficiency power transistor. Al_2O_3 is also very attractive as the gate insulator for GaN MOS transistor as well as SiO_2 since it has wide bandgap (7-9 eV) [12, 19]. We have previously shown that bandgap of the Al_2O_3 formed by atomic layer deposition (ALD) is 7.1-7.8 eV by estimating using the x-ray photoelectron spectroscopy (XPS) [20]. Here, we report on the properties of the GaN MOS capacitors with Al_2O_3 gate insulator formed by ALD and show the advantage and disadvantage of Al_2O_3 for GaN MOS devices. We also show the results of the GaN MOS capacitors with SiO_2/Al_2O_3 gate stack. This structure has the advantages of both SiO_2 and Al_2O_3. Furthermore, SiO_2/Al_2O_3 gate stack has been applied to AlGaN/GaN hybrid MOS-HFETs.

Experimental

GaN MOS capacitors were fabricated on n-type GaN with Si-donor doping at a concentration of 2×10^{17} cm^{-3} on Si(111) substrates. Al_2O_3 films were formed by ALD using trimethylaluminum (TMA)/O_3/O_2 gases as Al and O sources at 300 °C. SiO_2 films were deposited by MW-PECVD at 400 °C. The thickness of SiO_2 is around 50 nm. Annealing process was performed by a furnace at several temperatures for 30 minutes in N_2 ambient. After that, AlGaN/GaN hybrid MOS-HFETs were fabricated. A heterostructure constructed with a GaN-based buffer layer, a highly resistive C-GaN layer, 50-nm u-GaN channel layer and a 20 nm-thick $Al_{0.25}Ga_{0.75}N$ was formed on a Si(111) substrate using MOVPE. After mesa etching was done, the channel region was defined by RIE. The etched depth was 35 nm. After gate insulators were formed, gate, source and drain electrodes were fabricated by a sputter deposition. The channel lengths are varied from 6 to 50 μm and the channel width is 0.84 mm.

Results and Discussion

The energy distributions of the D_{it} of GaN MOS capacitors with ALD Al_2O_3/GaN and MW-PECVD SiO_2/GaN have been estimated. ALD Al_2O_3 was annealed at 700 °C and MW-PECVD SiO_2 was annealed at 800 °C after deposition, respectively. As shown in Figure 11, The D_{it} of GaN MOS capacitor with ALD Al_2O_3 is approximately half compared with that with MW-PECVD SiO_2. Figure 12 (a) shows the depth profiles of Al, O, Ga in ALD Al_2O_3 and Figure 12 (b) shows the depth profiles of Si, O, Ga in MW-PECVD SiO_2 deposited on GaN , respectively. The profiles were measured by the secondary-ion mass spectrometry (SIMS). ALD Al_2O_3 film was annealed at 700 °C and MW-PECVD SiO_2 was not annealed respectively. As shown in Figure 12 (b), Ga is

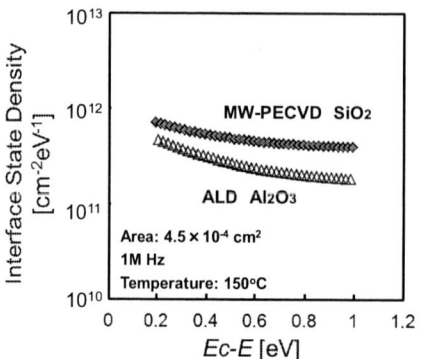

Figure 11. D_{it} of GaN MOS capacitors with ALD Al$_2$O$_3$ and MW-PECVD SiO$_2$ calculated from the C-V characteristics measured at 150 $^\circ$C. ALD Al$_2$O$_3$ film is annealed at 700 $^\circ$C and MW-PECVD SiO$_2$ is annealed at 800 $^\circ$C after deposition. The thicknesses of ALD Al$_2$O$_3$ and MW-PECVD SiO$_2$ are 58.4 and 52.5 nm, respectively.

diffused from GaN into SiO$_2$ film. Especially, by annealing after SiO$_2$ deposition, the diffused Ga concentration becomes high. On the other hand, the Ga diffusion is drastically suppressed in the case of Al$_2$O$_3$ film shown in Figure 12 (a) even if Al$_2$O$_3$ film was annealed after deposition. Though the direct relationship between diffusion of Ga and the interface property is not clear, the diffusion of Ga may generate the interface trap of MW-PECVD SiO$_2$/GaN becomes higher compared with ALD Al$_2$O$_3$/GaN. The interface properties of ALD Al$_2$O$_3$/GaN with 700 $^\circ$C annealing and without annealing a were almost the same. Figure 13 shows the current density-electric field (J-E) characteristics of these GaN MOS capacitors measured at room temperature. The breakdown electric fields with ALD Al$_2$O$_3$ without annealing and annealed at 700 $^\circ$C are below 7 MV/cm while the MW-PECVD SiO$_2$ has a high breakdown electric field with over 11 MV/cm. The leakage currents at 3 MV/cm of the GaN MOS capacitors with ALD Al$_2$O$_3$ without annealing and annealed at 700 $^\circ$C are three orders of magnitude larger compared with the MOS capacitor with MW-PECVD

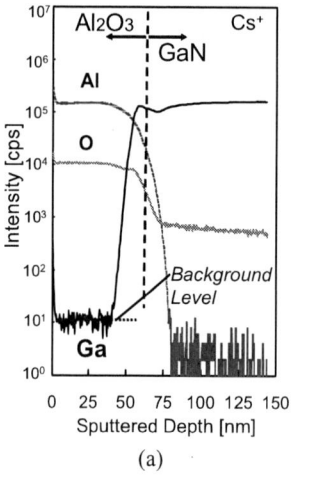

(a)

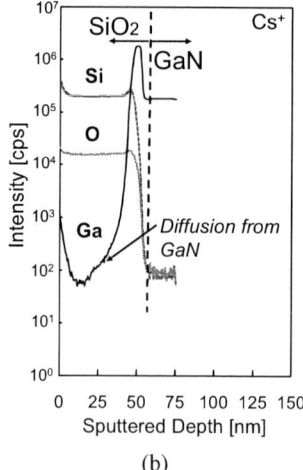

(b)

Figure 12. SIMS depth profile results of (a) Al, O, Ga in Al$_2$O$_3$/GaN and (b) Si, O, Ga in SiO$_2$/GaN, respectively. The Al$_2$O$_3$ film is annealed at 700 $^\circ$C and the SiO$_2$ is not annealed after deposition.

SiO_2. Furthermore, GaN MOS capacitor with ALD Al_2O_3 annealed at 800 °C has low

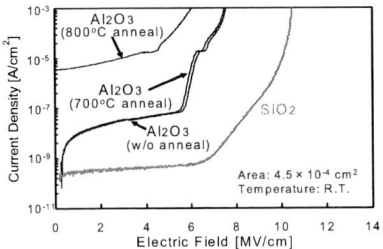

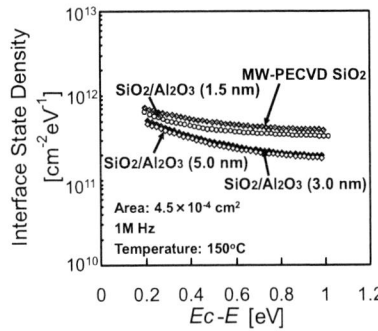

Figure 13. *J-E* characteristics of GaN MOS capacitors with ALD Al_2O_3 and MW-PECVD SiO_2 measured at room temperature. ALD Al_2O_3 films are not annealed, annealed at 700 °C, and 800 °C, respectively, and MW-PECVD SiO_2 is annealed at 800 °C after deposition. The thicknesses of ALD Al_2O_3 films without annealing, 700 °C annealing, and 800 °C annealing are 58.6, 58.4, and 52.1 nm, respectively. The thickness of MW-PECVD SiO_2 is 52.5 nm.

Figure 14. D_{it} of GaN MOS capacitors with SiO_2/Al_2O_3 and MW-PECVD SiO_2 calculated from the *C-V* characteristics measured at 150 °C.

breakdown electric field with below 6 MV/cm and very high leakage current. This result suggests that Al_2O_3 film might be crystallized by 800 °C annealing and then, high leakage paths are generated in Al_2O_3 film. Similar phenomena have been previously reported. From these results, we confirmed that the ALD Al_2O_3 film in this experiment could obtain lower D_{it}, however, the breakdown electric field was lower and the leakage current was higher compared with MW-PECVD SiO_2.

From the both advantages of SiO_2 and Al_2O_3, the SiO_2/Al_2O_3 stacked structure has been employed in GaN MOS devices. For this structure, the Al_2O_3 thickness is important to realize good interface property and high insulating characteristics. Figure 14 shows the energy distribution of the D_{it} of GaN MOS capacitors with SiO_2/Al_2O_3 and MW-PECVD SiO_2. SiO_2/Al_2O_3 stacked structures with 3.0 and 5.0 nm Al_2O_3 have low D_{it} and these properties are almost the same as Al_2O_3. On the contrary, the D_{it} of 1.5 nm Al_2O_3 is higher than the other structures. Figure 15 shows the depth profiles of Al, O, Ga in SiO_2/Al_2O_3 formed on GaN. In the case of 3.0 nm and 5.0 nm Al_2O_3, the diffusion of Ga of GaN is suppressed. However, in the case of the 1.5 nm Al_2O_3, Ga is diffused into insulator. This result suggests that 1.5 nm Al_2O_3 film doesn't have the effectiveness of suppressing the Ga diffusion and then, the D_{it} of 1.5 nm Al_2O_3 becomes higher than other structures. Figure 16 shows the typical *J-E* characteristics of these

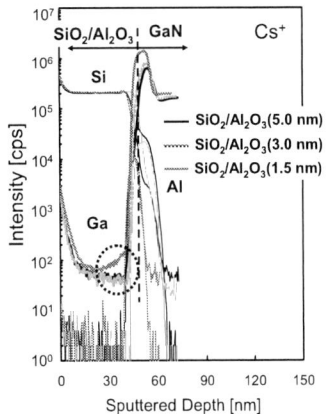

Figure 15. SIMS depth profile results of Al, Si, O, Ga in SiO_2/Al_2O_3 deposited on GaN.

GaN MOS capacitors with the SiO_2/Al_2O_3 stacked structures at 200 °C. In the case of 5

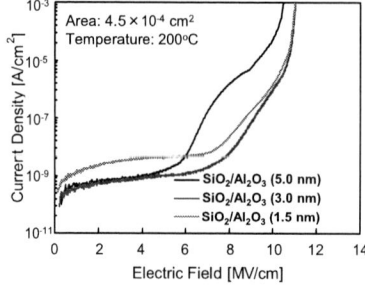

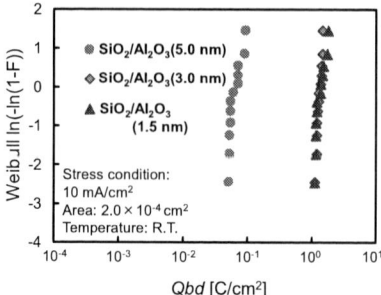

Figure 16. *J-E* characteristics of GaN MOS capacitors with SiO_2/Al_2O_3 measured at 200°C.

Figure 17. Charge-to-breakdown Q_{bd} of GaN MOS capacitors with GaN MOS capacitors with SiO_2/Al_2O_3 measured at room temperature.

nm Al_2O_3, the electric field at which the leakage current begins to increase is lower compared with other stack structures. The Al_2O_3 gate insulator has a low breakdown field and a high leakage current. Therefore, this phenomenon may be the influence of thicker Al_2O_3 film. In the case of 1.5-nm Al_2O_3, the leakage current is higher than the 3-nm Al_2O_3. Figure 17 shows the charge-to-breakdown Q_{bd} of these GaN MOS capacitors with SiO_2/Al_2O_3. The Q_{bd} with 3.0 nm and 5.0 nm Al_2O_3 are almost the same properties as that with MW-PECVD SiO_2 while the Q_{bd} with 1.5-nm Al_2O_3 is over one order of magnitude lower than that with other structures. These results lead us to the conclusion that 3-nm Al_2O_3 is effective for applying SiO_2/Al_2O_3 gate stack structure in this experiment.

Figure 18 shows the transfer characteristics of our fabricated AlGaN/GaN hybrid MOS-HFETs with SiO_2/Al_2O_3 (3.0 nm) and MW-PECVD SiO_2. Al_2O_3 was formed by ALD and around 50 nm SiO_2 was deposited MW-PECVD. SiO_2/Al_2O_3 was annealed at 700 °C and MW-PECVD SiO_2 was annealed at 800 °C after formation. Both MOS-HFETs show good normally-off operations with the threshold voltage of about 4.2 V. The on-state characteristic of MOS-HFET with SiO_2/Al_2O_3 is superior to that with MW-

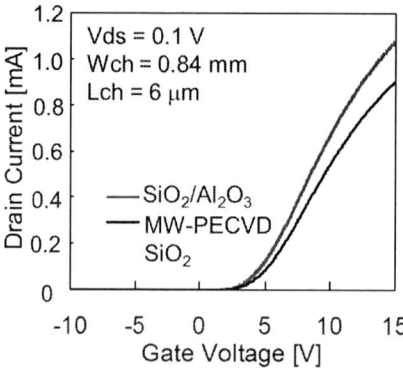

Figure 18. Transfer characteristics of AlGaN/GaN hybrid MOS-HFETs with SiO_2/Al_2O_3 (3.0 nm) and MW-PECVD SiO_2.

PECVD SiO_2. The field-effect mobility of MOS-HFET with MW-PECVD SiO_2 calculated from the transfer characteristics is 192 cm^2/Vs at the channel length of 50 μm, which is much superior to that with MW-PECVD SiO_2 of 161 cm^2/Vs.

Conclusion

In this work, high performance AlGaN GaN hybrid MOS-HFETs have been demonstrated. To achieve a high-breakdown voltage, the epitaxial structure of AlGaN/GaN hybrid MOS-HFETs has been investigated. We confirmed that the breakdown voltage of AlGaN/GaN hybrid MOS-HFET fabricated on 7.3 μm epilayer with a thin u-GaN channel layer on highly resistive C-GaN layer achieved over 1.71 kV with the L_{gd} of 18 μm. Furthermore, we confirmed that gate field structure is effective for AlGaN/GaN hybrid MOS-HFETs to suppress the current collapse. For increasing the on-state property, the gate insulator of AlGaN/GaN hybrid MOS-HFETs has been studied and it has confirmed that the SiO_2 gate insulator formed by the MW-PECVD on GaN has low D_{it} of SiO_2/GaN, high-breakdown electric field and high Q_{bd} compared with that by the CCP-CVD and LP-CVD. Furthermore, annealing after SiO_2 is effective to increase the properties of SiO_2 gate insulator. We also confirmed that ALD Al_2O_3 can realize a low D_{it} of Al_2O_3/GaN. However, the breakdown electric field of ALD Al_2O_3 is lower and the leakage current is higher compared with MW-PECVD SiO_2. By incorporating the advantages of both MW-PECVD SiO_2 and ALD Al_2O_3, SiO_2/Al_2O_3 gate stack structure has been employed. We confirmed that 3.0 nm Al_2O_3 could realize good properties as the gate insulator for GaN MOS devices in our experiments. We also have demonstrated a high performance AlGaN/GaN hybrid MOS-HFET by applying SiO_2/Al_2O_3 gate stack, which has a good normally-off operation with the threshold voltage of 4.2 V and a high maximum field-effect mobility of 192 cm^2 /Vs.

References

1. T.P. Chow, and R. Tyagi, *IEEE Trans. Electron Devices*, **41**, 1481 (1994).
2. M. Kodama, M. Sugimoto, E. Hayashi, N. Soejima, O. Ishiguro, M. Kanechika, K. Itoh, H. Ueda, T. Uesugi, and T. Kachi, *Appl. Phys. Express*, **1**, 021104 (2008).
3. H. Kambayashi, Y. Niiyama, S. Ootomo, T. Nomura, M. Iwami, Y. Satoh, S. Kato, and S. Yoshida, *IEEE Electron Device Lett.*, **28**, 1077 (2007).
4. T. Nomura, H. Kambayashi, Y. Niiyama, S. Otomo, and S. Yoshida, *Solid State Electron.*, **52**, 150 (2008).
5. Y. Niiyama. S. O otomo. J. Li, H. Kambayashi, T. Nomura, S. Yoshida, K. Sawano, and Y. Shiraki, *Jpn. J. Appl. Phys.*, **47**, 5409 (2008).
6. W. Huang, T. P. Chow, Y. Niiyama, T. Nomura, and S. Yoshida, *in Proc. 20th ISPSD*, 295 (2008).
7. T. Oka, and T. Nozawa, *IEEE Electron Device Lett.*, **29**, 668 (2008).
8. H. Kambayashi, Y. Satoh, S. Ootomo, T. Kokawa, T. Nomura, S. Kato, *Solid-State Electron.*, **54**, 660 (2010) .
9. N. Ikeda, S Kaya, J. Li, Y. Sato, S. Kato, and S.Yoshida, *in Proc.21th ISPSD*, 251 (2009).
10. N. Ikeda, K. Kato, K. Kondoh, H. Kambayashi, J. Li, and S. Yoshida, *Phy. Status Solidi A*, **204**, 2028 (2007).
11. N . Ikeda, Y. Niiyama, H. Kambayashi, Y. Sato, T. Nomura, S. Kato, and S. Yoshida, *Proc. of the IEEE*, **98**, 1151 (2010).

12. J. Robertsona and B. Falabretti, *J. Appl. Phys,*. **100**, 014111 (2006).
13. T. Ohmi, M. Hirayama and A. Teramoto, *J. Phys. D*, **39**, R1 (2006).
14. M. Kanechika, M. Sugimoto, N. Soejima, H. Ueda, O. Ishiguro, M. Kodama, E. Hayashi, K. Itoh, T. Uesugi, and T. Kachi, *Jpn. J. Appl. Phys.* **46**, 503 (2007).
15. H. Ueda, Y. Ohsawa, Y. Tanaka, and T. Nozawa, *J. Appl. Phys.* **48**, 126001 (2009).
16. H. Kambayashi, T. Nomura, S. Kato, H. Ueda, A. Teramoto, S. Sugawa, and T. Ohmi, *Jpn. J. Appl. Phys.*, **51**, 04DF03 (2012).
17. L. M. Terman, *Solid-State Electron.*, **5**, 285 (1962).
18. T. Hashizume, and H. Hasegawa, *Appl. Surf. Sci.* **234**, 387 (2004).
19. T. Hashizume, S. Ootomo, T. Inagaki, and H. Hasegawa, *J. Vac. Sci. Technol. B*, **21**, 1828 (2003).
20. H. Kambayashi, T. Nomura, H. Ueda, K. Harada, Y. Morozumi, K. Hasebe, A. Teramoto, S. Sugawa, and T. Ohmi, *Jpn. J. Appl. Phys.* **52**, 04CF09 (2013).

Characterization and Performance of D-mode GaN HEMT Transistor Used in a
Cascode Configuration

T. MacElwee, J. Roberts, H. Lafontaine, I. Scott, G. Klowak, and L. Yushyna
GaN Systems Inc., 300 March Road, #501 Ottawa, Ontario Canada K2K 2E2

Cascode configured D-mode GaN HEMT device performance is
reported in this paper. The basic parameters of the D-mode HEMT
will be covered as well as the integration of the cascode
configuration into the PQFN package. Finally 500 volt 3.3 amp
switching characteristics of the cascode will be discussed showing
the excellent switching performance with voltage slew rates as
large as 70 V/ns being measured.

Introduction

Gallium Nitride as a material system has been the subject of intensive research over the
last twenty years. The initial demonstration of this material system was focused on light
emitting structures as a potential source of compact blue/UV LED solid state emitters.
The first single crystal GaN transistor was demonstrated in 1993 by Khan et al (1). This
was quickly followed by the demonstration of an AlGaN/GaN HEMT device also by
Khan et al (2) in 1993. Over the years, the quality of the material from advancements in
growth deposition techniques and device processing has steadily improved. For the first
time now, GaN HEMT devices are starting to be considered for use in the high-voltage
power switching arena by systems designers.

GaN HEMT devices naturally form as depletion mode, normally-on devices. This
means that a negative gate bias must be supplied to turn the device off. From a safety
perspective, systems which use depletion mode power devices must pay special attention
to the power-on sequence to ensure that the gate bias is first established prior to turning
on the main power supply bus. If these criteria are not satisfied at the system level, the
depletion mode power transistor can present itself to the system as a dead short, resulting
in a safety as well as a system hazard.

Considerable effort has been devoted to the fabrication and demonstration of high
performance enhancement mode (E-mode) GaN HEMT devices. There are various
techniques used to try to extinguish the 2DEG under the gate at zero bias. This is done by
either placing sufficient negative charge directly under the gate close to the 2DEG using
low energy ion implantation techniques, or controlled and precise etching techniques, or a
combination of the above (3-5). The main issue that confronts these various approaches is
damage due to the implantation or etching steps and the resulting need to effectively
remove this damage through thermal annealing. These techniques present significant
challenges to the process and fabrication of repeatable, consistent and reliable E-mode
HEMT devices. In general the performance of E-mode HEMT devices is not as good as
measured for a D-mode HEMT device.

To avoid the deficiencies of the above approaches, a high voltage GaN D-mode HEMT device was combined with a low voltage E-mode NMOS device in the cascode configuration, delivering the high voltage, low on-resistance characteristics of the GaN D-HEMT device, simultaneously with the higher threshold voltage of a modern E-mode NMOS power device. Fabrication of the NMOS transistor in a CMOS process flow will allow the integration of additional functionality such as differential inputs, slew rate control for switching, high temperature and current monitoring. The resulting integrated device will exhibit the high voltage, high speed switching benefits desired, making its use and integration into systems much easier.

GaN Device Fabrication

The GaN D-mode HEMT device developed in this work was fabricated using a conventional RF GaN process flow on 3" 4H SI-SiC starting substrate. The SiC substrate should allow for excellent thermal performance and high voltage operation of the switching device due to the semi-insulating nature of the SiC substrate. The epitaxial layer stack includes an AlN nucleation layer followed by a SI GaN buffer layer and a UID channel layer. The total thickness of the GaN epitaxial layer is approximately 1.8 µm. Next, a 22 nm AlGaN barrier layer with ~22% aluminum mole fraction was grown, followed by an approximately 3 nm UID GaN capping layer. The surface passivation is accomplished using a thin, deposited SiN layer. Ohmic contacts to the source and drain regions are formed directly on top of the GaN cap. The Schottky gamma gate was formed with a gamma extension field plate towards the drain of the device. Two additional field plates are utilized in this design, one connected to the gate and the other connected to the source. The inter-metallic dielectrics used here are deposited SiN. A schematic cross section of the D-mode HEMT structure is shown in Figure 1.

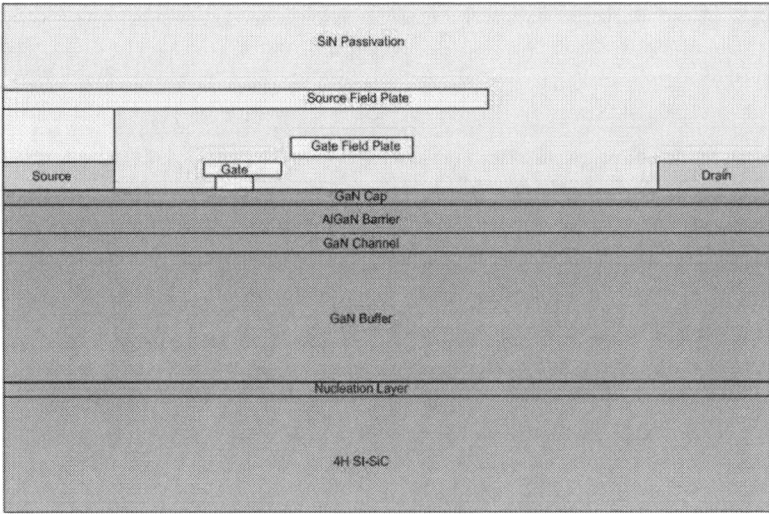

Figure 1. Schematic cross section of the AlGaN/GaN HEMT device with integrated field plates and epitaxial layer structure.

GaN HEMT Performance

Electrical characterization of the finished wafers showed that the source/drain contact sheet resistance was 0.24 Ω-mm with a channel sheet resistance and 2DEG mobility of 440 Ω/sq and 1672 cm^2/V·s respectively. From CV measurements, the 2DEG sheet charge was determined to be $\sim$ 9x10^{12} q/cm^2 yielding a threshold voltage of -3.2 volts, and transistors with 1 μm gate lengths demonstrated a maximum I_{dsat} of 566 mA/mm. The intrinsic breakdown capability of the technology as a function of the gate to drain spacing for devices with gate lengths of 1 μm and a total device width of 250 μm, are shown in Figure 2.

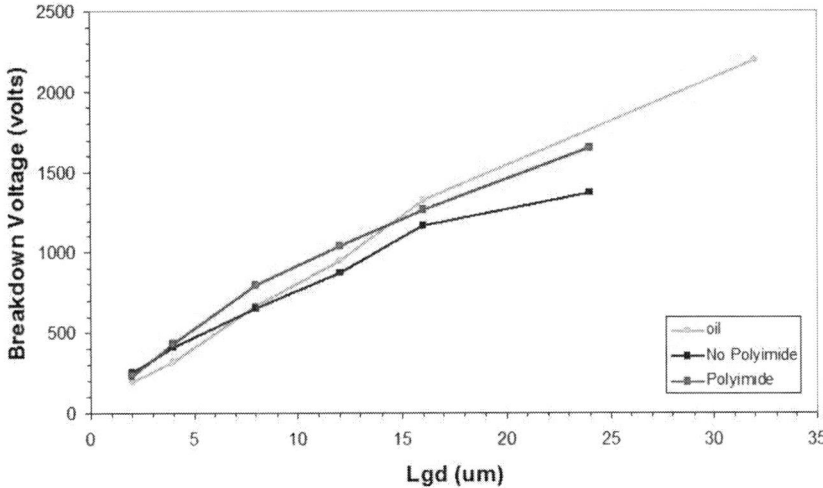

Figure 2. Intrinsic breakdown results for 250 μm wide devices with 1μm gate lengths. The breakdown voltage is seen to saturate with increasing L_{gd} probably due to surface related effects.

From the above measurements, the average breakdown electric field is over 100 V/μm for L_{gd}<8 μm. As L_{gd} is further increased however, the average breakdown electric field is seen to reduce and for L_{gd}>16 μm, it is found to have dropped to 78 V/μm. The maximum breakdown voltage that was measured for this material was 2200 volts for an L_{gd} spacing of 32 μm.

The I_d-V_d characteristics of a packaged GaN HEMT with an 80 mm gate width are shown in Figure 3. This device can source up to 33 amps and has an on-resistance of 150 mΩ and a measured breakdown voltage as high as 800 volts. In addition, the device makes use of backside via technology (BSV) taking the high voltage drain terminal connection through to the back side of the wafer. This removes the drain's high voltage from the front surface where the source and gate terminals are located. A 3 μm thin film of polyimide was also used to provide additional surface breakdown protection.

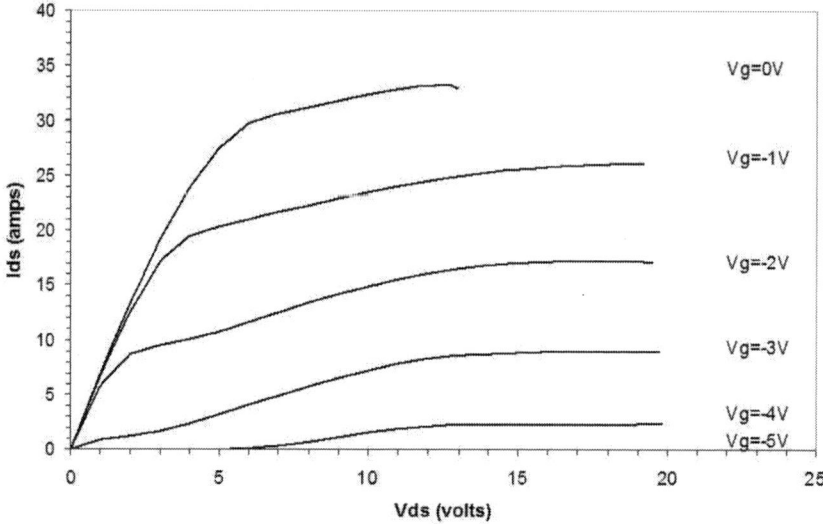

Figure 3. I_d-V_d characteristics of a 80mm wide GaN HEMT power switching device. This device makes use of BSV technology to move the high voltage drain terminal to the back side of the wafer and away from the source and gate terminals on the surface. This device has a DC R_{on} of 150 mΩ and a measured breakdown voltage as high as 800 volts.

Field plates have been used in this device to help control the lateral electric field in the channel region and at the interface between the barrier layer and the SiN passivation layer. These measures are primarily designed to help control dynamic on-resistance, which is an unexpected increase in the device on-resistance that can happen after a large voltage has been applied to the drain, when the device has been in the off-state. The origin of this increase in the on-resistance is thought to be charge trapping on the surface, at or near the interface between the barrier and the SiN passivation layer. Field plates however directly contribute to an increase in the device output capacitance, C_{oss}. Regardless of whether the field plate is connected to the gate or the source, the presence of the field plates can be measured as changes in C_{gd} and/or C_{ds} as a function of the drain bias. For the successful integration of the GaN HEMT into a cascode structure, this additional capacitance must be understood and modeled. Figure 4 shows the capacitance characteristics measured for the packaged device in Figure 3, measured in the off-state. The step terraces in the output capacitances are due to the presence of the field plates over the drain drift region of the device. The step reduction in capacitance is caused by the depletion of the underlying 2DEG as the drain bias is increased, starting first with the gamma portion of the gate followed by the gate strap and the top metal field plates.

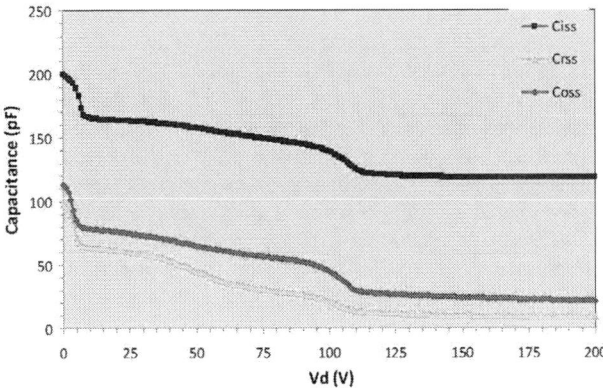

Figure 4. Measured device capacitances of the power switching device in Figure 3. The terraced nature of the C_{rss} capacitance is due to the presence of the field plates. This characteristic will directly influence the behavior of the cascode configuration during switching.

In order to maximize the current density efficacy, the die has been designed using a castellated island topology. This topology makes use of copper posts and through wafer vias to allow for a very compact and scalable layout strategy. The most significant feature of this design is the ability to direct the current off the die through the copper posts, resulting in a design that is typically more compact when compared to more traditional layouts. In addition, since there is no current being moved laterally, the parasitic inductance of the copper post connections can be very small. This should result in clean switching waveforms when the GaN die is used in the cascode configuration. The off-chip connections can be sized to adequately handle the large currents that are produced with this design. A typical example of the castellated design with copper posts and through wafer vias is shown in Figure 5. This device has a gate width of 80 mm and a die area of only 2 x 2 mm.

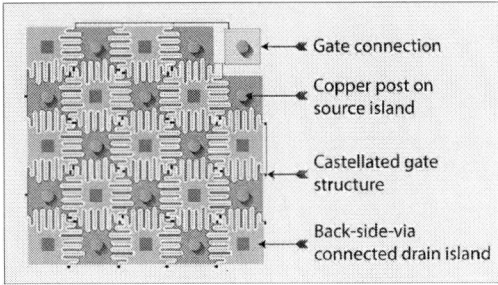

Figure 5. A schematic representation of the castellated design layout showing the location of the copper posts and the through wafer vias. The through wafer vias allow the drain connection to be made on the back side of the wafer, isolating the high voltage terminal from the lower voltage source and gate contacts.

The cascode configuration requires the integration of a low on-resistance E-mode MOS device. A custom CMOS die was designed to allow the GaN HEMT device with copper posts to be flipped onto, and integrated with, the CMOS die to form a 3D cascode assembly. The silicon technology used was 0.8 μm CMOS. The threshold voltage for this large NMOS device was 0.8 volts, on-resistance was 15 mΩ and drain to source breakdown voltage was 6 volts. The addition of the small on-resistance of the NMOS will increase the total cascode on-resistance by approximately 10 %. In addition, the CMOS technology allowed the integration on-chip of differential Schmitt trigger inputs for enhanced noise immunity, adjustable slew rate control to limit the switching dI/dt, and a maximum die temperature sensor for thermal shut down protection. The CMOS die including the NMOS power transistor measured 4.05 x 4.05 mm.

Packaging and Thermal

Careful consideration was given to the packaging requirements for the GaN cascode configuration. Standard QFN packages are not suited for high current and voltage applications. In addition, they are limited in their ability to adequately remove the heat that might be generated during operation. For these reasons it was necessary to customize the package leadframe reducing the die pad size along one edge to provide sufficient creep distance between the low voltage and high voltage connections. A clip bond has been used to connect the drain on the back side of the die stack to the package leadframe. This is a formed copper strip that replaces more traditional wire or ribbon bonds and essentially clips the die to the leadframe providing a low inductance, high current capable interconnect. The clip bond is assembled by simultaneously soldering to the back side of the GaN die and the leadframe using a reflow process. To aid in thermal management, the clip bond also incorporates some novel design features to enable it to extend as an exposed pad on the top of the package allowing for heat removal through the top of the package, as well as the bottom. A schematic cross section of the assembled die in the PQFN package is shown in Figure 6.

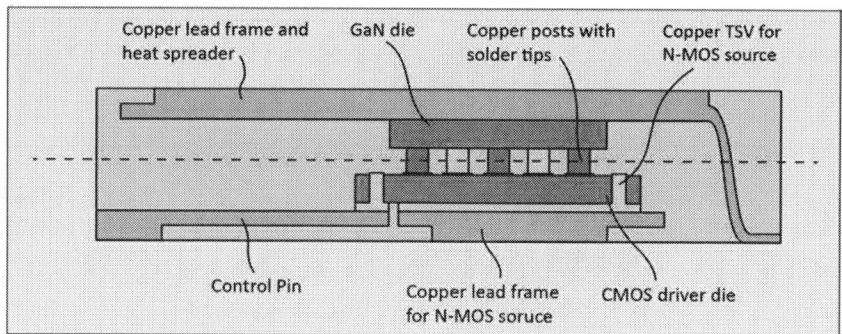

Figure 6. Cross section of PQFN package showing the silicon CMOS die on the bottom mounted to a heat slug and the GaN die flipped with the copper posts attached to the top of the CMOS.

To package these two die, a PQFN style has been used. The package is assembled with the silicon CMOS die mounted backside down while the GaN die has been flipped over with the copper posts soldered down onto the topside CMOS die metallization. The cross section shows the CMOS die consisting of the large NMOS power transistor and controller/driver circuitry mounted on a copper lead frame for bias and heat extraction. The GaN die with its copper posts have been flipped over and soldered on top of the NMOS device forming the cascode configuration. A second copper lead frame has been attached to the top of the GaN die to provide drain bias as well as help in the removal of any heat generated in the GaN die during operation.

To avoid excessive heat concentration, the channel must be distributed evenly over the entire die as is shown in Figure 5. To assess the thermal success of a particular layout strategy, thermal modeling using the finite element solver ElectroFlo® has been used. CAD design files have been used to provide an accurate representation of the spatial power density expected for a particular design. Detailed thermal simulation used not only the GaN die layout but also the details of the package, including, each layer of the assembly. The junction to case (top copper clip) thermal resistance is estimated to be 0.95 °C/W with the package held at 50 °C. To estimate the junction temperature as a function of current, electro-thermal simulations are performed where Joule heating due to an applied bias is solved self consistently with the heat equations in steady state. The results are shown in Figure 7. At low currents, the junction temperature is almost at the case temperature. As the current is increased beyond 10 amps, the GaN junction temperature rises and at 17.5 amps it reaches 150 °C. Under these conditions the CMOS die average junction temperature was less than 66 °C.

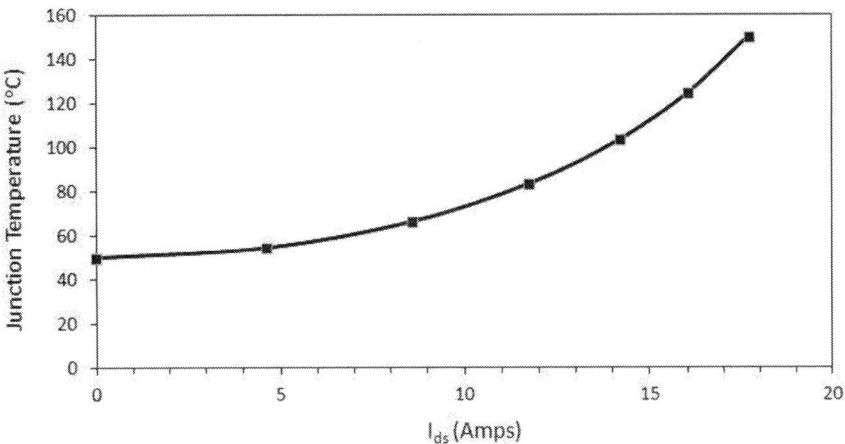

Figure 7. Thermal simulation including the Joule heating solved self consistently showing the GaN HEMT junction temperature increases to 150 °C at 17.5 amps.

Cascode Performance

Figure 8 shows a schematic of the cascode configuration used in this paper.

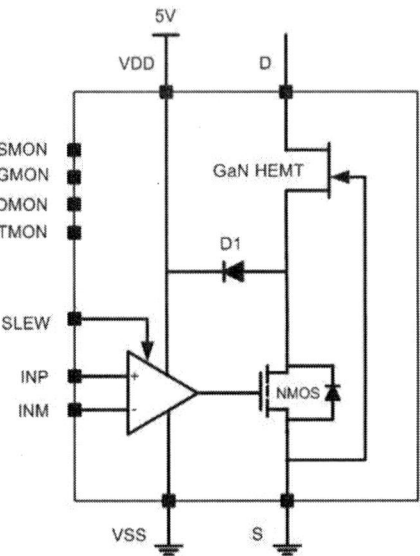

Figure 8. A schematic representation of the cascode circuit. Indicated are the NMOS, GaN HEMT, and the CMOS control circuitry.

The GaN HEMT and the NMOS device are indicated as well as the input control circuitry used to provide differential input, Schmitt triggering and slew rate control over the turn-on speed of the NMOS device. The cascode circuit uses the NMOS device to control the current through the GaN HEMT as needed, while the GaN HEMT is used to provide the high voltage capability required for the power switching operation. Since the gate of the HEMT is grounded, when the NMOS device is turned off, the source voltage of the HEMT will start to rise as the NMOS drain capacitance is charged. Once the HEMT source voltage rises above approximately 3 volts, the HEMT will shut off. At very high speeds, there can be significant charge transfer from the HEMT C_{ds} into the NMOS C_{ds} resulting in voltage spikes being imposed on the common node connecting the NMOS drain and the HEMT source terminals. While these voltage spikes are not a concern for the HEMT device, they can drive the NMOS past its breakdown voltage and result in damage to this device that could affect long term reliability. The slew rate control can be useful in controlling these voltage spikes as it can limit the dV/dt of the switching edges.

A SPICE model for the GaN HEMT has been developed for use in studying the operation of the cascode circuit and is shown in Figure 9.

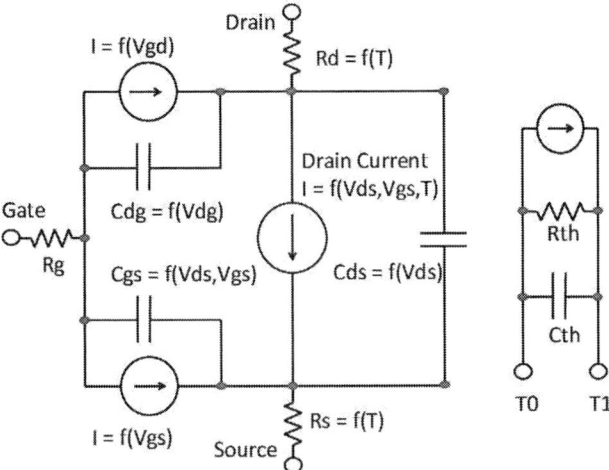

Figure 9. SPICE representation of the GaN HEMT device used in simulating the cascode configuration. The self consistent thermal model is shown on the right hand side of the figure. The capacitive effects of the field plates are also included in this model.

Special attention has been paid to capturing the HEMT drain capacitance as it is a complicated function of the drain to source voltage due to the presence of the field plates. The parasitic device resistances have also been included in this model. The SPICE model is complete with self heating capability. With self heating turned on, the effects of the operating conditions, duty cycle and instantaneous power dissipation can be included in the simulation. DC and transient simulations using this model and approach will show the dynamic effects including negative output conductance due to self heating.

To maximize the switching performance of the GaN device, every effort was made to allow the current flow to be vertical in the 2 transistor, cascode connected, die stack in the customized 10 x 10 mm PQFN package, and every effort was taken to allow for parallel connections wherever possible, thus minimizing resistance and inductance inside the package.

The custom CMOS die allowed the switching control signal to be standard 5V logic levels with less than 1 pF loading. This input stage included a Schmitt trigger function to safeguard against the significant electrical noise typical in high voltage, high current switching environments. Positive and negative going thresholds were set to 3.1 and 2.2 V respectively. The gate drive for the NMOS device was modified to allow for slew rate control by an external resistor.

The test PCB was designed with standard, double sided FR-4 of 1.6 mm thickness and the custom PQFN package was designed to be compatible with standard 72 pin PQFN pin locations – with some pins removed to meet creep distance requirements for the high voltages used in many of the testing activities. The test PCB was able to accept a socket with 0.5 mm pad centers and 0.2 mm spacing and this socket was used for the measurements referred to in this paper. Considerable care had to be taken with the physical location of the ground connections for the oscilloscope probes to ensure they were as close to the source of the NMOS device (which was ground for the high voltage supply). The load resistor was specially selected to be able to dissipate power but also having low inductance. The exceptionally clean waveforms shown in Figure 10 show no adverse impact to the ringing observed on the logic level input signal and the output voltage switching with over 70 V/ns performance.

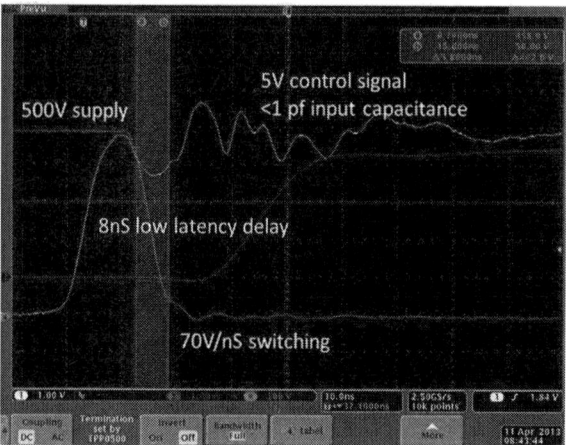

Figure 10. A scope capture of the GaN cascode switching 3.3 amps from a 500 volt supply. The propagation delay through the CMOS control circuitry is approximately 8 ns. Minimal ringing is observed on the switched current and voltage waveforms.

Total propagation delay through the CMOS control circuitry is 8 ns. Of particular note is the voltage dependency exhibited by the output capacitance of the GaN device which acts to slow down the output voltage once it has fallen below approximately 50 volts. This effect is due to the presence of the first metal and gamma field plates and their capacitive coupling to the underlying 2DEG in the GaN drift region. As the potential of the drift region between the gate and drain reduces, the 2DEG under the field plates accumulates and the capacitance abruptly increases. This phenomenon has the effect of softening the turn-on transition and results in less ringing due to the reduction in dI/dt and any parasitic inductance present in the circuit.

Conclusions and Summary

We have successfully demonstrated the integration of a high voltage and high current D-mode GaN HEMT device with a silicon NMOS transistor in a cascode configuration. The NMOS device has a threshold voltage of 0.8 volt, an on-resistance of approximately 15 mΩ, and a breakdown voltage of 8 volts. The D-mode GaN HEMT has an on-resistance of 150 mΩ, a threshold voltage of -3V and a breakdown voltage of approximately 800 volts. To maximize switching speed and minimize parasitic inductance which could result in spurious ringing, copper posts have been used to move the current vertically off the GaN die.

A custom designed CMOS integrated driver has been implemented to simplify the use and driving of this switching cascode device. Control is accomplished using common 5 volt logic input signals that drive input Schmitt triggers to suppress any noise that could be present when using this device in a noisy switching environment. In addition, there is the ability to control the switching slew rate of the NMOS device to suppress any unwanted spurious signals from developing during switching. A PQFN package was developed that allowed the silicon and GaN die to be stacked vertically, with the necessary creep distances required for the high voltage operation of the device. A custom designed clip was used to secure the stacked die to the leadframe used in the package which served to provide electrical contact to the drain of the GaN device as well as removal of heat out of the top of the package. The total size of the finished package is 10 x10 mm resulting in a very compact high voltage, high current switching device. Switching performance of 3.3 amps and up to 500 volts of applied bias with a voltage transition rate of up to 70 V/ns when driving a resistive load has been observed.

References

1. M.A. Khan, T.N. Kuznia, A.R. Bhattaraia, D.T. Olson, "Metal semiconductor field effect transistor based on single crystal GaN" Appl. Phys. Lett. 62, 1786, 1993.
2. M. Khan, A. Bhattarai, J. Kuznia, and D. Olson, "High Electron Mobility transistors Based on a GaN-AlxGa1-xN Hetrojunction" Appl. Phys. Lett. 63, 1214, 1993.
3. Yong Cai, Yugang Zhou, Kei May Lau and Kevin J. Chen, "Control of Threshold Voltage of AlGaN/GaN HEMTs by Fluoride-Based Plasma Treatment: From Depletion Mode to Enhancement Mode", IEEE Trans. Electron Devices, vol. 53, no. 9, pp 2207-2215, Sept. 2006.
4. Shuo Jia, Yong Cai, Deliang Wang, Baoshun Zhang, Kei May Lau and Kevin J. Chen, "Enhancement-mode AlGaN/GaN HEMTs on silicon substrate", Phys. Stat. Sol 3, no. 6, pp 2368-2372, 2006.
5. Masahito Kanamura, Toshihiro Ohki, Toshihide Kikkawa, Kenji Imanishi, Tadahiro Imada, Atsushi Yamada, and Naoki Hara, "Enhancement-Mode GaN MIS-HEMTs With n-GaN/i-AlN/n-GaN Triple Cap Layer and High-k Gate Dielectrics", IEEE Electron Device Lett., vol. 31, no. 3, pp 189-191, March 2010.

Voltage Switching Limits of Lateral GaN Power Devices

Krishna Shenai

Energy Systems Division, Argonne National Laboratory
9700 South Cass Avenue, Bldg. 362, Argonne, IL 60439-4815

Abstract

The dv/dt switching limitations of power semiconductor devices are evaluated in a boost (PFC) power converter using circuit simulations. State-of-the-art commercial silicon CoolMOS devices, commercial Silicon Carbide (SiC) Junction Barrier Schottky (JBS) diodes, and emerging Gallium Nitride (GaN) lateral power transistors are considered. It is shown that although SiC and GaN power devices have low stored charge and small capacitances, they experience high switching dv/dt stresses which may pose serious switching limitations especially in high-frequency power converters. This problem is likely to be further exacerbated by the presence of a high density of crystal defects in SiC and GaN materials which will manifest in the form of poor field-reliability. Specific guidelines for device selection are recommended in order to optimize both performance and field-reliability.

I.   Introduction

It has been well-known for more than two decades that power devices made from wide bandgap (WBG) semiconductors such as Silicon Carbide (SiC) and Gallium Nitride (GaN) offer much lower on-state resistance ($R_{DS(on)}$) than comparable silicon power devices because of their superior electrical and thermal conductivities, and breakdown field compared to silicon [1, 2]. Single-chip power Junction Barrier Schottky (JBS) diodes rated up to 1,700V/25A; power MOSFET's rated up to 1,200V/50A; and, power JFET's rated up to 1,700V/4A – all fabricated on 4H-SiC material are now commercially available. Although these devices are finding increasing applications in computer/telecom power supplies, motor control, and smart grid [3-5], serious concerns pertaining to long-term reliability of these devices in compact power converters under stressful field operating conditions remain [6]. For example, most SiC power devices are not dv/dt- and avalanche-rated, and the data sheets make no mention of the Safe-Operating Area (SOA), especially above room temperature. However, it has been known for some time [7] now that the dv/dt capability of SiC power devices is severely limited because of a high density of crystal defects that are present in the drift-region of the device.

Lateral GaN power transistors are fast-emerging as viable candidates in power converters designed for maximum voltage ratings below 1,000 volts. For example, GaN power transistors are being developed for point-of-load (POL) DC-DC power converters [8] and power inverters [9]. These devices offer the lowest $R_{DS(on)}$ for a given voltage rating, low capacitances, and do not store minority carrier charge. Consequently, GaN power transistors are ideally suited for high-frequency power conversion below 1,000 volts. However, GaN material inherently consists of orders-of-magnitude higher density of crystal defects and has much lower thermal conductivity than SiC; these limited material properties of GaN further raise concerns about the field-reliability when used in compact power converters. As of today, manufacturers do not specify dv/dt- and avalanche-ratings, and the SOA for GaN power transistors. Hence, detailed evaluation of the circuit-level electro-thermal stresses and field-reliability of both SiC and GaN power devices is of paramount importance.

In this paper, a detailed evaluation of dv/dt stresses on silicon, SiC, and GaN power devices are studied in DC-DC boost (PFC) converter topology using accurate yet simple physics-based circuit simulation models. It is shown that dv/dt stresses in excess of 100 V/ns are impressed across diodes and switches under repetitive field-switching conditions that can be deleterious for long-term converter reliability. Higher dv/dt stresses are primarily caused by reduced device capacitances.

II.   Boost Converter Design Results

The boost DC-DC converter circuit topology shown in Figure 1(a) is used over wide power range in computer/telecom power supplies, power factor correction (PFC), and renewable energy applications. In this circuit, when switch Q is turned on, the boost

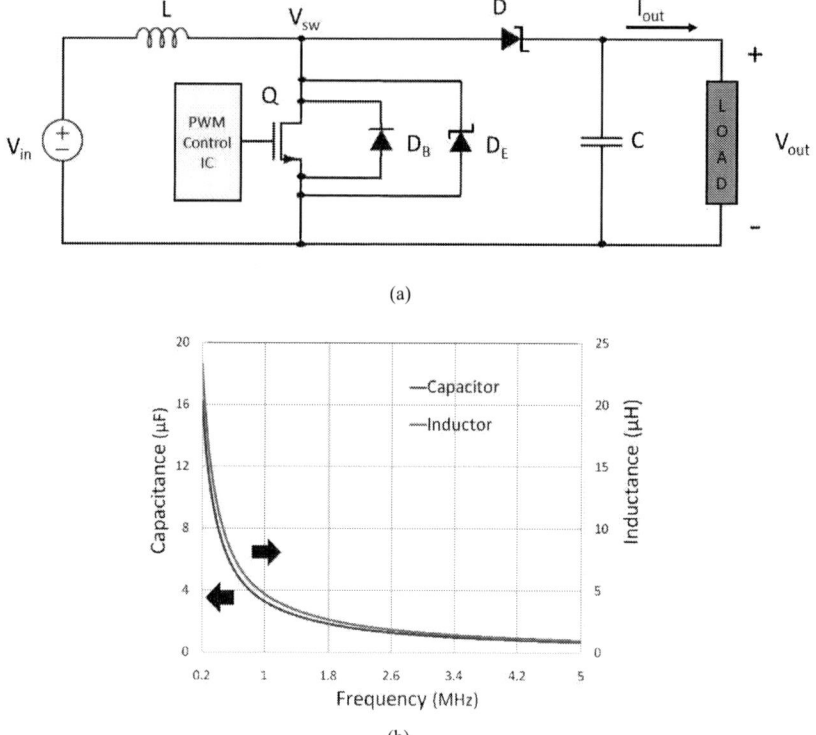

(a)

(b)

Figure 1: (a) Circuit schematic and (b) calculated inductor and capacitor values vs. frequency for a 200V/380V, 10kW DC-DC boost (PFC) converter.

inductor is charged by the input voltage source. During this time, diode D is reverse-biased. In order to avoid a voltage spike at the drain of Q, the reverse recovery of diode D is used in place of conventional P-i-N diodes. When the switch Q is turned off, the output capacitor C is charged from the energy stored in the inductor as well as by the electricity delivered from the source. The diode D_B represents the body PN junction diode of silicon CoolMOS devices; external diode D_E is used to provide the reverse conduction path for the lateral GaN power transistor which does not have an inherent body diode.

The minimum inductor value needed to maintain continuous current conduction as a function of switching frequency, and a corresponding capacitor value for the desired output ripple are calculated using standard closed form expressions [10, 11]; only power semiconductor devices and inductor are assumed to be the lossy elements in the circuit; other circuit and package parasitic elements are neglected. The sizes of magnetic components and capacitors in the circuit vary inversely with frequency; hence, higher switching frequency is desired for reducing the converter size and cost. The main advantage of using the GaN transistor for power switching is that switching frequency can be kept high (>1 MHz) while still maintaining lower conduction power loss (compared to a silicon power MOSFET with identical rating); higher switching frequency can reduce capacitor and inductor values exponentially as shown in Figure 1(b). By using wide bandgap (WBG) semiconductor power devices, DC-DC converters can be made more compact, thereby facilitating high density power converters with low cost and size.

For power switch Q, both silicon and GaN power transistors were studied; the best commercially available 600V/30A silicon CoolMOS device and a scaled lateral 600V/30A GaN power transistor were used. For diode D, the best commercially available silicon and SiC diodes were studied. Tables I(a) and I(b) list the key device parameters extracted from the data sheets. Various combinations of the switch and diode were evaluated in order to study their interactions in the boost (PFC) converter. Accurate physics -based models for silicon CoolMOS [12], silicon and SiC diodes [13], and GaN power transistor [14] were extracted from the manufacturer's data sheets and used in circuit simulations. The stimulated results clearly suggest 3% to 4% higher power conversion efficiency with GaN power transistors (in place of silicon CoolMOS); there is also a drop in efficiency by about 4% as the temperature is increased from -50°C to 150°C (see Figures 2(a) and 2(b)). Figure 3 summarizes the simulated values of dv/dt across the switch and diode for various converter configurations. It is clear that dv/dt stress increases with switching frequency and slightly decreases with increased temperature; however, high dv/dt stresses are likley to cause severe field-reliability concerns, and hence, must be investigated in detail.

TABLE I(A): COMPARISON OF IMPORTANT TRANSISTOR PARAMETERS USED IN THE DESIGN OF 200/380V, 10 KW DC-DC BOOST CONVERTER

Device	Reverse Breakdown Voltage V_{BR} (V)	Threshold Voltage V_{TH} (V)	$R_{DS(on)}$ (mΩ)	C_{ISS} (nF)	C_{OSS} (nF)	C_{RSS} (nF)	dv/dt rating (V/ns)
Best Commercial Silicon MOSFET	600	3	0.099	4.5	20	1.8	50
GaN E-Mode FET	600	1.4	0.025	1.6	2.1	0.3	-

TABLE I(B): COMPARISON OF IMPORTANT DIODE PARAMETERS USED IN THE DESIGN OF 200/380V, 10 KW BOOST DC-DC CONVERTER

Device	Reverse Breakdown Voltage V_{BR} (V)	Junction Capacitance at zero Bias C_{JO} (V)	Forward Voltage Drop (V_F)	Series Resistance R_S (mΩ)	Reverse Recovery Time T_{rr} (ns)
Best Commercial SiC Diode	600	480	1.8	40	0
Best Commercial Si Diode	600	61	2.6	60	30

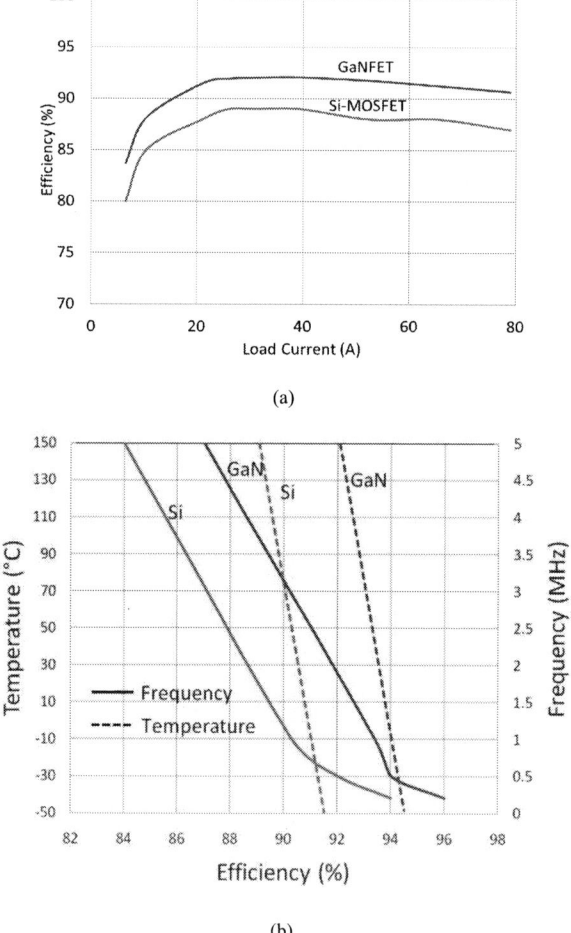

(a)

(b)

Figure 2: Simulated (a) load regulation at 25°C and 1 MHz and (b) efficiency vs. temperature and frequency for 200V/380V boost DC-DC converter with device parameters listed in Tables I(a) and I(b).

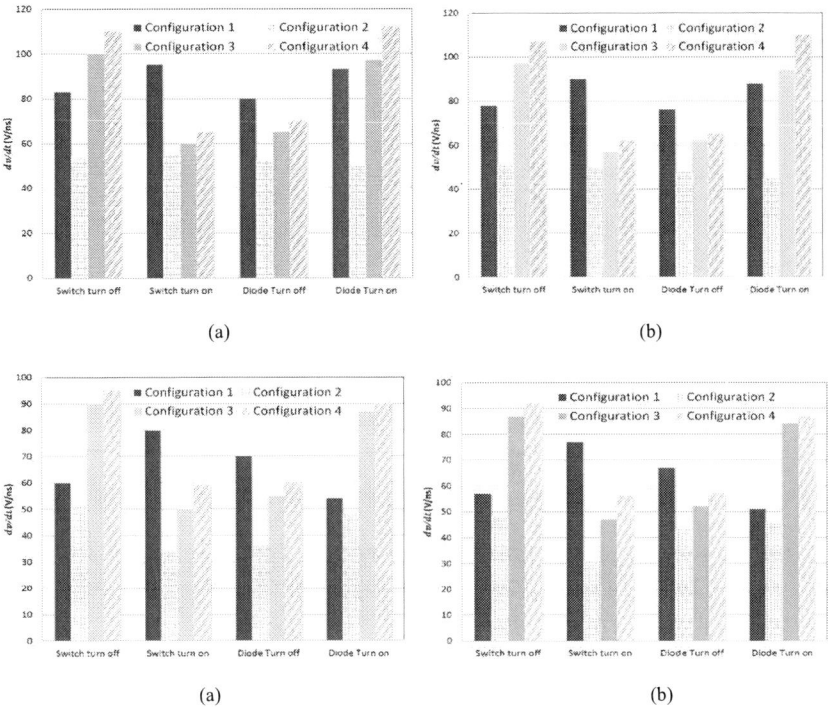

(a)   (b)

(a)   (b)

Figure 3: Simulated dv/dt across the switch and diode for various configurations of 200/380 V, 10 kW boost converter measured at (a) 5 MHz and 25°C, (b) 5 MHz and 150°C, (c) 1 MHz and 25°C, and (d) 1 MHz and 150°C. Configuration #1: GaN switch and SiC diode; Configuration #2: silicon CoolMOS and SiC diode; Configuration #3: silicon CoolMOS and silicon diode; Configuration #4: GaN transistor and silicon diode.

III.   Summary and Conclusions

High-voltage power switching devices made on Wide Band Gap (WBG) semiconductors are promising candidates for next-generation of electric utility and transportation energy conversion applications. Of particular interest at present are the commercially available SiC Schottky Barrier Diodes (SBD's) and lateral GaN power transistors. Both these devices have much lower on-state resistance, smaller capacitance, and reduced reverse recovery compared to commercial silicon diodes and CoolMOS transistors with similar ratings.

Simple physics-based circuit simulation models are used to calculate the voltage switching stresses in diodes and transistors in a boost (PFC) DC-DC power converter circuit. The model parameters are extracted from data sheet values. It is shown that high values of dv/dt stresses are imposed on SiC power diodes and GaN power transistors because of reduced capacitances. The high dv/dt switching stress is a serious concern that must be investigated from field-reliability considerations especially when designing compact power converters operating at higher junction temperatures.

References

1. K. Shenai, R. S. Scott and B. J. Baliga, "Optimum semiconductors for high-power electronics," IEEE Trans. Electron Devices, vol. 36, no. 9, pp. 1811-1823, Sept. 1989.
2. K. Shenai, "Potential Impact of Emerging Semiconductor Technologies on Advanced Power Electronic Systems," IEEE Electron Device Lett., vol. 11, no. 11, pp. 520-522, November 1990.
3. T. McDonald, "GaN based power technology stimulates revolution in conversion electronics," Electronics in Motion and Conversion, pp. 2-4, April 2009.
4. H. Zhang, L. M. Tolbert, and B. Ozpineci, "Impact of SiC Devices on Hybrid Electric and Plug-In Hybrid Electric Vehicles," IEEE Trans. Industrial Electronics, vol. 47, no. 2, pp. 912-921, March/April 2011.
5. G. Deboy, R. Rupp, R. Mallwitz and H. Ludwig, "New SiC JFETs Boost Performance of Solar Power Inverters," Power Electronics Europe, issue 4, 2011, pp. 29-33.
6. K. Shenai, P. G. Neudeck, M. Dudley, and R. F. Davis, "Rugged Electrical Power Switching in Semiconductors – What is Beyond Silicon Power Thyristor?," Digest of 2011 IEEE EnergyTech Conference, Cleveland, OH, July 12, 2011, IEEE Xplore Digital Object Identifier: 10.1109/EnergyTech.2011.5948505.
7. K. Acharya and K. Shenai, "On the dV/dt Rating of SiC Schottky Power Rectifiers," Proceedings Power Electronics Technology Conference, October 2002, pp. 672-677.
8. E. Abdoulin and A. Lidow, "High frequency 12V – 1V DC-DC converters – Advantages of using EPC's gallium nitride (GaN) power transistors vs. silicon-based power MOSFETs", Application Note, Efficient Power Conversion Corporation, www.epc-co.com
9. T. Uesugi and T. Kachi, "GaN Power Switching Devices for Automotive Applications," Proc. CS MANTECH Conference, May 18-21, 2009, Tampa, FL.
10. Ned Mohan, Tore M. Undeland, and William P. Robbins, Power Electronics: Converters, Applications and Design, Wiley: New York, 2003.
11. Robert W. Erickson, Fundamentals of Power Electronics, Chapman & Hall: New York, 1997.
12. K. Shenai, "A circuit simulation model for high-frequency power MOSFETs," IEEE Trans. Power Electronics, vol. 6, no. 3, pp. 539-547, July 1991.
13. S. Pendharkar, C. Winterhalter, and K. Shenai, "A Behavioral Circuit Simulation Model for High-Power GaAs Schottky Diodes," IEEE Trans. Electron Devices (TED), vol. 42, no. 10, pp. 1847-1854,October 1995.
14. K. Shah and K. Shenai, "A Simple and Accurate Circuit Model for GaN Power Transistors," IEEE Trans. Electron Devices (TED), vol. 59, n0. 10, pp. 2735-2741, Oct. 2012.

CHAPTER 5

POWER DEVICE RELIABILITY 1

GaN-based power HEMTs: Parasitic, Reliability and high field issues

G. Meneghesso[a], M. Meneghini[a], D. Bisi[a], R. Silvestri[a], A. Zanandrea[a], O. Hilt[b],
E. Bahat-Treidel[b], F. Brunner[b], A. Knauer[b], J. Wuerfl[b], E. Zanoni[a]

[a] Department of Information Engineering, University of Padova,
via Gradenigo 6/B, 35131 Padova, Italy
gaudenzio.meneghesso@dei.unipd.it
[b] Ferdinand-Braun-Institut Leibniz-Institut fuer Hoechstfrequenztechnik
Gustav-Kirchhoff-Strasse 4 12489 Berlin - Germany

This paper reviews the main mechanisms responsible for trapping
and breakdown in power HEMTs based on gallium nitride. With
regard to the trapping mechanisms, we describe the role of carbon
and iron buffer doping compensation in determining the dynamic
Ron. We also demonstrate how the use of double heterostructure
without doping or a single-heterostructure with proper buffer
doping compensation can effectively reduce trapping phenomena.
In addition, we investigate the breakdown limits of single and
double heterostructure (DH) HEMTs, by electrical and
electroluminescence characterization. Results indicate that, for the
devices adopting double heterostructure without doping or single-
heterostructure with proper buffer doping compensation, the
breakdown voltage linearly scales with the gate-drain distance, and
provides information on the origin of breakdown current
components for different bias levels and epitaxial structures.

Introduction

GaN-based High Electron Mobility Transistors (HEMTs) are expected to find soon
wide application in the power conversion field. Thanks to the wide bandgap, the high
breakdown field, and the high channel mobility, these devices can have breakdown
voltages in excess of 1000 V and switching frequencies of several MHz, thus being
suitable for replacing and surpassing conventional silicon devices for power application.
Despite the high theoretical performance of power transistors based on GaN, the
performance and reliability of these devices is still limited by a number of factors. More
specifically, the switching frequency is limited by the so-called dynamic-Ron (dynamic
on-resistance) collapse, i.e. a recoverable increase in the on-resistance induced by the
exposure to moderate-high drain bias levels [1-2]: in most of the cases, Ron collapse is
due to the trapping of carriers in the gate-drain access region (either at the surface or in
the semiconductor material) [3-4]. However, the nature of the traps responsible for Ron
collapse has not been clearly identified; furthermore, it is not clear how the structure of
the devices can influence the charge trapping process, e.g. by favoring/hindering the
injection of carriers at specific trap states, or by reducing the electric field responsible for
trapping. In addition, the breakdown voltage of GaN-based power transistors can be
lower than expected, due to several mechanisms: high gate-drain leakage currents [5],
source-drain conduction [6], and bulk leakage can increase the drain current at high

drain-source voltages, thus leading to a premature breakdown of the HEMTs [7-8]. Understanding the origin of the breakdown process is of fundamental importance for the optimization of power devices. Furthermore, the influence of device structure on the breakdown process must be analyzed in detail, with the aim of understanding how to fabricate high-robustness transistors.

The aim of this paper is to review the main mechanisms responsible for dynamic Ron and breakdown in high-power transistors based on GaN. In the first part of the paper, we describe the results of Ron transient measurements carried out on devices with carbon-doped buffer, with iron-doped buffer, with and without heterostructure. We show that from these measurements it is possible to extrapolate the properties (activation energy and cross section) of the traps responsible for dynamic Ron, and we demonstrate that it is possible to significantly reduce the Ron collapse by the adoption of a double-heterostructure epitaxy without any doping, or by the use of an iron-doped single-heterostructure.

In the second part of the paper, we summarize the results of breakdown measurements. The analysis was carried out on both single- (SH) and double-heterostructure (DH) devices: results indicate that (i) a single heterostructure without buffer doping compensation is not sufficient to achieve an acceptable breakdown voltages; (ii) the breakdown voltage can be significantly improved through the adoption of a double-heterostructure or buffer doping compensation in the single heterostructure; (iii) Breakdown voltage linearly increases with the gate-drain spacing. The results summarized within this paper provide important information towards the development of high performance and high reliability transistors based on gallium nitride.

Experimental Details

In this paper, we review the effect of the use of single and double heterostructure, and of carbon and iron doping on the dynamic characteristics of the transistors. The impact of the epitaxial structure on the dynamic Ron collapse was investigated by means of IV double-pulse characterization, and iso-thermal Ron transient analysis. The devices under test are grown on SiC substrate; the epitaxial-structures (Figure 1) was optimized for the normally-off and high-voltage breakdown operative conditions, which are mandatory features for high power switching applications. More specifically, four different wafers were analyzed: wafer #1, labeled SH:C, adopt a single-heterostructure configuration and employs a Carbon-doped GaN buffer layer, 13 nm-thick $Al_{0.23}Ga_{0.77}N$ barrier-layer, and a thin pGaN gate; wafer #2, labeled DH:C, is similar to the first structure, but features a 500nm-thick $Al_{0.05}GaN$ back-barrier layer; whereas Wafer #3 and Wafer #4, labeled SH:Fe and DH, employ respectively a single-heterostructure with Fe-doped buffer and a conventional double-heterostructure without GaN buffer. The devices have been grown and processed at FBH, Berlin. More detailed information on these samples can be found in [7].

The transient-characterization of the dynamic Ron was carried out by connecting a resistive load between the drain of the transistor and a fixed voltage supply (V_{DD}=50 V). Trapping was induced by keeping the devices in the off-state (V_{GS}=-3 V) for 100 s; devices were then switched to the on-state in the linear region (V_{DS}=5V;V_{GS}=1V), and the variation of the on-resistance 10us up to 100s was recorded. The measurements were repeated at several temperature levels, with the aim of identifying the signatures of the deep-levels responsible for Ron collapse, and the related defect-states which could be either intentionally introduced as doping species, or un-intentionally caused by the intrinsic defectiveness of the adopted materials and interfaces.

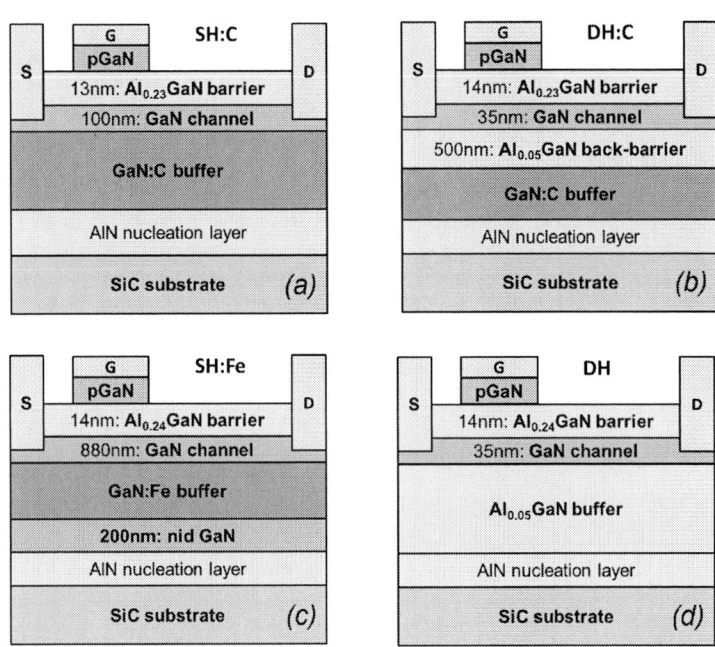

Figure 1: Device structures for (a) SH:C and (b) DH:C devices

Results of the dynamic-Ron measurements

Figure 2 reports the comparison between the normalized Ron-transient performed on one representative device for each series. As can be noticed, immediately after turn-on the dynamic on-resistance of the SH:C devices displays a strong increase (time constant in the order of milliseconds) followed by a slow decrease, which ends in the tens of seconds. On the other hand, the DH:C samples display a much lower initial dynamic Ron collapse, with no increase in the milliseconds-range, suggesting that the introduction the AlGaN back-barrier is beneficial for the suppression of the dynamic Ron collapse. A further improvement is obtained through the use of double heterostructure devices without any carbon doping (DH samples): these devices, after a trapping period, show only a +30% Ron increase, which is significantly smaller than what measured on the devices with carbon doping (SH:C and DH:C). It has to be noted that the trapping properties of C-doped SH and DH buffer layers significantly depend on the C-concentration as reported in [8]. An almost complete suppression of the Ron collapse was obtained by using the single-heterostructure devices with iron-buffer compensation (SH:Fe). These results suggest that, through the adoption of a proper buffer compensation doping scheme, even SH devices can have excellent performance in terms of dynamic Ron.

The observed trends have been investigated by means of the thermal characterization described in the previous section. The results of the current transients for the SH:C, DH:C, and DH samples, depicted respectively in Figure 3, Figure 4, and Figure 5 reveal the presence of three distinct processes, labeled H1 (detected only in the SH:C devices), E1 and E2 (detected in the SH:C, DH:C and DH devices), which are thermally-activated, and which display apparent activation energies and cross sections of $0.84eV/3x10^{-13}cm^2$, $0.85eV/4x10^{-14}cm^2$, and $0.83eV/1x10^{-15}cm^2$ respectively (see also the Arrhenius plot in Figure 6)

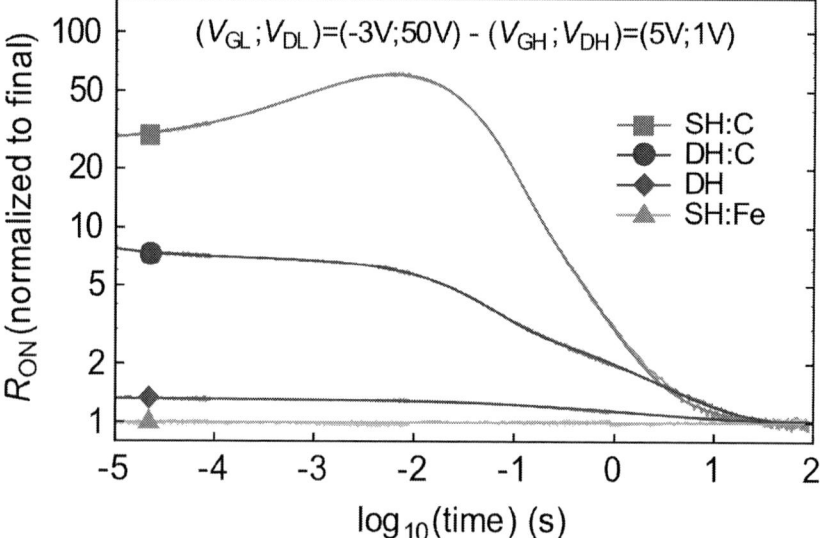

Figure 2: Comparison of the normalized-to-final Ron-transient performed on single- and double-heterostructure devices.

The process H1, which leads to the increase of dynamic Ron in initial phase of the Ron transient, is detected only in SH:C and can be modeled as a deep-acceptor state which captures carriers as the device is switched on. Hence, from the collected evidence, the deep-level H1 located at Ev+0.84eV within the band-gap, could be ascribed to the deep-acceptor states intentionally introduced by the Carbon-doping, in good agreement with reference experimental works [8] and simulations [9]; the comparison on the Arrhenius plot is depicted in Figure 6a.

On the other hand, E1 and E2 (detected in the SH:C, DH:C and DH samples), which compete with H1 by inducing a decrease in dynamic Ron, could be modeled as donor-like traps: the electrons, likely supplied either by the source- or the gate-leakage current, are captured by deep-levels during the high gate-drain voltage off-state trapping phase, causing a worsening of the dynamic Ron, and are thermally emitted to the conduction band during the measuring phase, causing the subsequent Ron decrease observed in the 100ms-range. As gate and/or drain leakage currents at off-state increase with temperature, more electrons can occupy trap states at higher temperatures during off-state biasing for a given time leading to an increased Ron immediately after switching. E1 and E2, whose Arrhenius plots are depicted in Figure 6b, reveal similar signature with both GaN and AlGaN-related defect-states, previously described in [10-18].

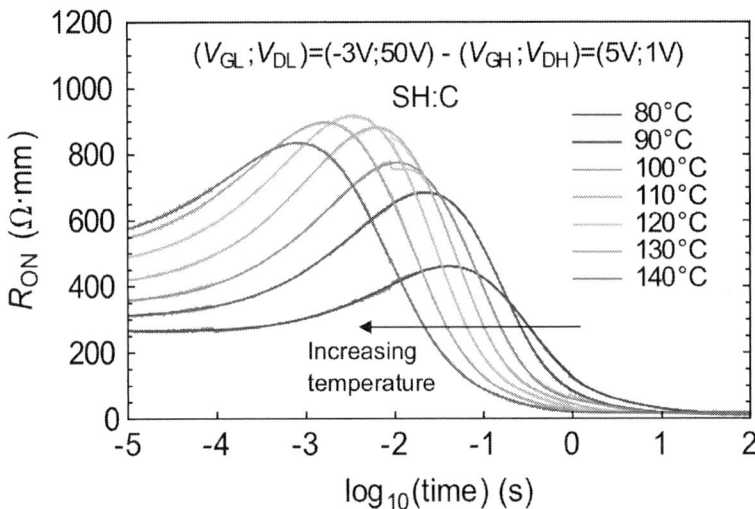

Figure 3: Thermal characterization of the Ron transient performed on single-heterostructure devices with carbon doping (SH:C).

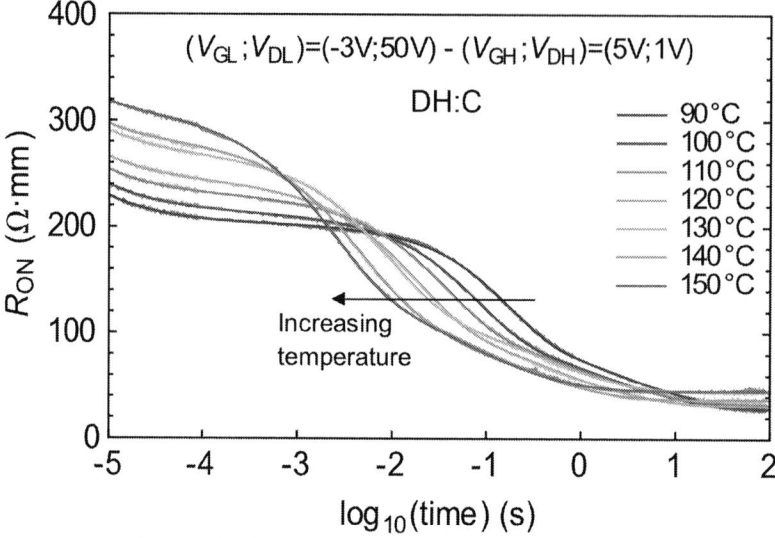

Figure 4: Thermal characterization of the Ron transient performed on double-heterostructure devices with carbon doping (DH:C).

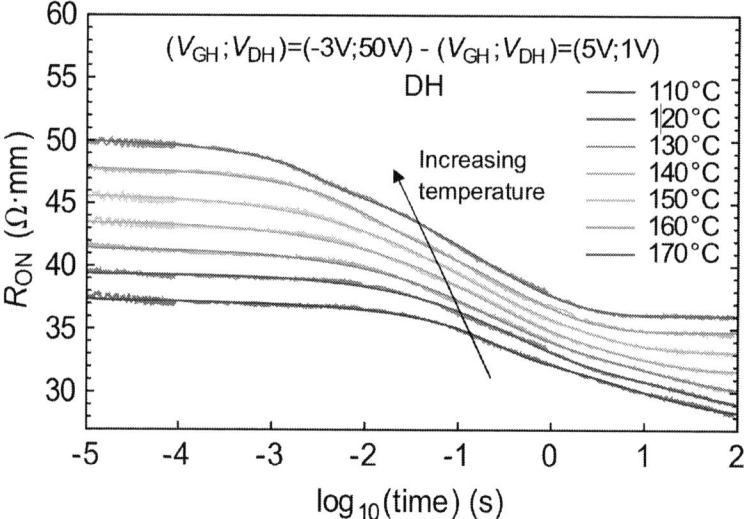

Figure 5: Thermal characterization of the Ron transient performed on double-heterostructure devices without doping

Interestingly, although the DH:C samples have a Carbon-doped buffer (similarly to the SH:C ones), these samples do not display the signature of the Carbon-related H1 trap. This result suggest that the 500nm-thick AlGaN back-barrier placed between the GaN channel and the Carbon-doped GaN buffer effectively reduces the possibility of trapping carriers at Carbon-related impurities, thus resulting in more stable dynamic Ron performance. Moreover, it can be noticed that also the parasitic effects caused by the electron-traps E1 and E2 are mitigated by the introduction of the double-heterostructure.

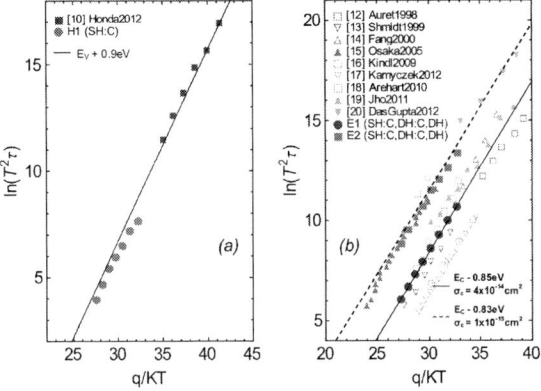

Figure 6: Arrhenius plots, and comparison with reference works, for (a) deep-acceptor-like, and (b) deep-donor-like defect states. τ represents the time constant of the detrapping processes

Results of the breakdown measurements

Since single (SH) devices with Fe-doping, and double (DH) heterostructure HEMTs (with no doping) showed the best characteristics in terms of Ron collapse, these devices were characterized also by means of breakdown measurements. For references, single heterostructure without buffer doping compensation has also been characterized. Tests were performed in closed channel conditions at room temperature in air ($V_{GS} = V_{TH}$ -2 V, where V_{TH} is the device threshold voltage) on devices with variable gate-drain distance.

Single-heterostructure devices without doping compensation in the buffer showed very poor breakdown values without any scaling with L_{GD} (Figure 7). On the other hand, both SH:Fe devices and DH HEMTs showed a pretty high breakdown voltage (BDV, Figure 8). BDV was found to scale almost linearly with gate-drain spacing (Figure 9), indicating that a very good isolation can be obtained both by using an iron compensation or a double heterostructure. A comparative analysis of the above described heterostructures can be found in Figure 9.

Furthermore, the p-gate contact was found to have a relatively low leakage current up to the source-drain off state device breakdown (see Figure 8), indicating the good electrical properties of the gate junction.

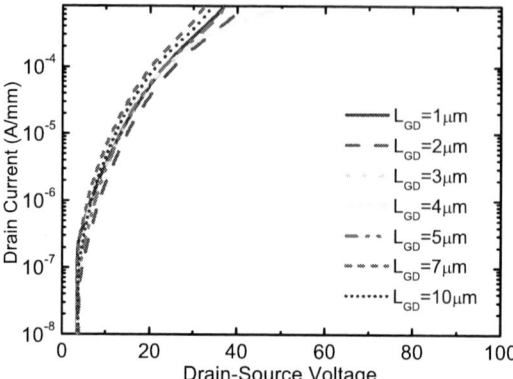

Figure 7: drain current measured on SH devices (without doping compensation) during a breakdown measurement ($V_{GS}=V_{TH} – 2V$). Devices with increasing L_{GD} were considered. These are normally-on devices

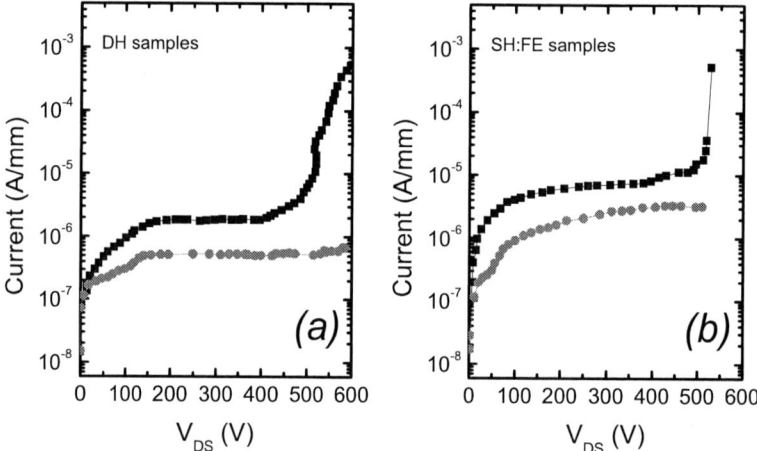

Figure 8: drain (black squares) and gate (red circles) current measured during and under breakdown conditions on the samples with double heterostructure (DH) and with Fe-doped single heterostructure (SH:Fe). $V_{GS}=V_{TH} - 2$ V

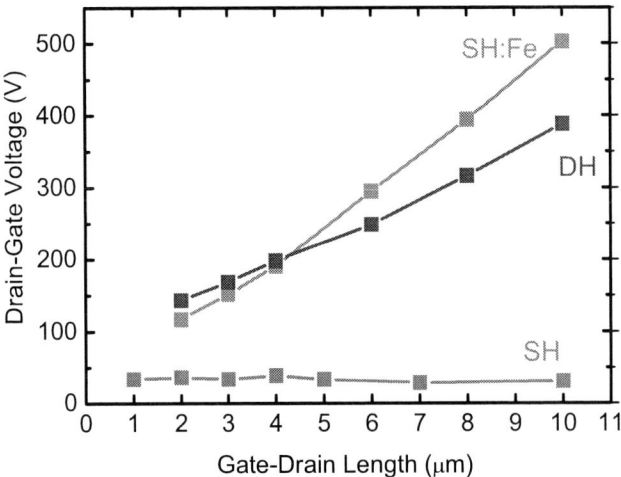

Figure 9: dependence of the breakdown voltage on the gate-drain spacing for devices with single and double heterostructure (V_{GS}=-2 V)

EL characterization can provide further insight into the origin of breakdown: results (Figure 10) indicate that the EL signal increases with increasing the drain-gate voltages. EL signal is localized in correspondence of preferential sites, thus indicating that breakdown does not occur uniformly along gate width, but is related to the presence of defective paths. Breakdown luminescence is supposed to originate from the deceleration of highly accelerated electrons injected towards the drain. The results in Figure 10 are in agreement with the breakdown tests in Figure 8.

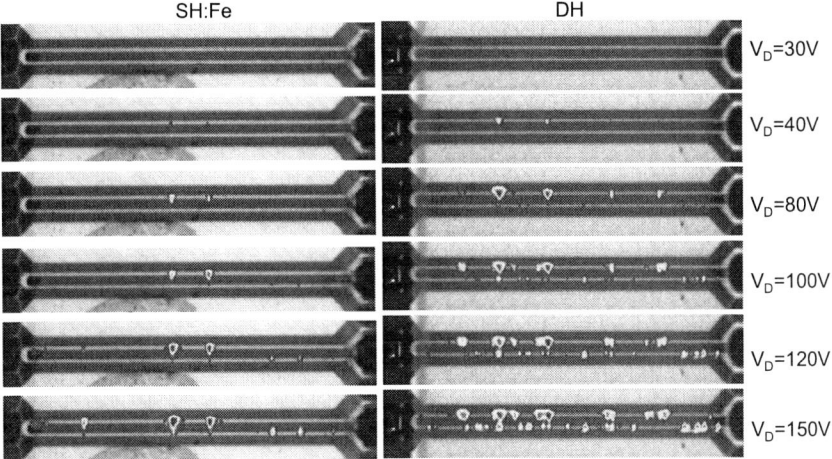

Figure 10: results of spatially-resolved EL measurements carried out under off-state conditions, for increasing drain voltages. V_{GS}=-3 V. Left: SH:Fe, Right DH

Conclusions

In summary, with this paper we have described an extensive analysis of the physical mechanisms responsible for the dynamic-Ron and for breakdown in GaN-based power transistors. The results of pulsed measurements provide information on the traps responsible for the increase in Ron: on the SH:C samples, we have been able to detect the deep-level related to the acceptor dopant used in the buffer (carbon), and to characterize its properties in terms of activation energy and cross section.

Moreover, we have demonstrated that the dynamic Ron collapse can be significantly reduced through the use of a double heterostructures (DH:C, and DH), possibly due to the difficulty of trapping electrons into the buffer, due to the additional band offset. SH devices with Iron compensation also present negligible Ron collapse.

The results of the breakdown measurements indicate that, for the devices analyzed within this work, a single heterostructure without any doping compensation is not sufficient for reaching high breakdown voltages, since SH devices show a very poor breakdown voltage, which does not significantly depend on the date-drain spacing. This can be ascribed to the existence of high gate-drain current components; poor carrier confinement, and the subsequent punch-through of electrons from the source to the drain may further decrease the breakdown voltage for these devices. On the other hand, single heterostructure samples with Fe-doping or devices with double heterostructure are found to largely improve the breakdown behavior: breakdown voltage increases linearly with gate-drain spacing.

As a general comment, however, the DH structures represent the most promising solution for the development of low dispersion, high breakdown voltages with good robustness levels, without critical negative side effects that can be present with the buffer doping compensation.

Acknowledgments

This work was partially supported by: the European Union Project, HiPoSwitch, GaN-based normally-off high power switching transistor for efficient power converters (Project Nr. 287602), the University of Padova (Project young researchers and Progetto di Ateneo 2010 "Development of normally off Gallium Nitride power devices for future green power applications"). Authors kindly thank Isabella Rossetto, Fabiana Rampazzo, Carlo de Santi, Marco Bertin and Antonio Stocco for their contribution.

References

1. O. Hilt, E. B.-Treidel, E. Cho, S. Singwald and J. Würfl, "Impact of Buffer Composition on the Dynamic On-State Resistance of High-Voltage AlGaN/GaN HFETs", Proceedings of the 24th International Symposium on Power Semiconductor Devices and ICs, IEEE 2012
2. D. Jin and J. A. del Alamo, "Mechanisms responsible for dynamic ON-resistance in GaN high-voltage HEMTs", Proceedings of the 24th International Symposium on Power Semiconductor Devices and ICs, IEEE 2012

3. C.-Y. Hu and T. Hashizume, "Non-localized trapping effects in AlGaN/GaN heterojunction field-effect transistors subjected to on-state bias stress", J. Appl. Phys. 111, 084504 (2012)

4. C. Ostermaier, P. Lagger, M. Alomari, P. Herfurth, D. Maier, A. Alexewicz, M.-A. di Forte-Poisson, S. L. Delage, G. Strasser, D. Pogany, E. Kohn, "Reliability investigation of the degradation of the surface passivation of InAlN/GaN HEMTs using a dual gate structure", Microelectronics Reliability 52 (2012) 1812-1815

5. W. S. Tan, P. A. Houston, P. J. Parbrook, D. A. Wood, G. Hill, and C. R. Whitehouse, Appl. Phys. Lett. 80 (2002) 3207.

6. M. Uren, K. J. Nash, R. S. Balmer, T. Martin, E. Morvan, N. Caillas, S. L. Delage, D. Ducatteau, B. Grimbert, and J. C. De Jaeger, IEEE Transactions on Electron Devices 53 (2006) 395.

7. D. Visalli, M. Van Hove, J. Derluyn, P. Srivastava, D. Marcon, J. Das, M.R. Leys, S. Degroote, Kai Cheng; E. Vandenplas, M. Germain, and G. Borghs, , IEEE Trans. Electron Devices, 57 (2), 3333(2010).

8. J. Würfl, O. Hilt, E. Bahat-Treidel, R. Zhytnytska, K. Klein, P. Kotara, F. Brunner, A. Knauer, O. Krüger, M. Weyersa and G. Tränkle, "Technological approaches towards high voltage, fast switching GaN power transistors" ECS Trans., vol. 52, no. 1, 979-989 (2013).

9. O. Hilt, P. Kotara, F. Brunner, A. Knauer, R. Zhytnytska, and J. Würfl, "Improved Vertical Isolation for Normally-off High Voltage GaN-HFETs on n-SiC Substrates", to appear on IEEE Transactions on Electron Devices

10. U. Honda , Y. Yamada, Y. Tokuda, and K. Shiojima, "Deep levels in n-GaN Doped with Carbon Studied by Deep Level and Minority Carrier Transient Spectroscopies", Japanese Journal of Applied Physics 51 (2012) 04DF04

11. J. L. Lyons, A. Janotti, and C. G. Van de Walle, "Carbon impurities and the yellow luminescence in GaN", Appl. Phys. Lett. 97, 152108 (2010)

12. F. D. Auret, S. A. Goodman, F. K. Koshnick, J.-M. Spaeth, B. Beaumont, and P. Gibert, "Electrical characterization of two deep electron traps introduced in epitaxially grown n-GaN during He-ion irradiation", Appl. Phys. Lett. 73, 3745, 1998

13. N. M. Schmidt et al., "Effect of Annealing on Defects in As-Grown and g-Ray Irradiated n-GaN Layers", phys. stat. sol. (b) 216, 533 (1999)

14. Z-Q. Fang, L. Polentah, J.W. Hemsky, and D. C. Look, "Deep centers in as-grown and electron-irradiated n-GaN", Semiconducting and Insulating Materials Conference, 2000. SIMC-XI. International, Canberra, ACT, pp. 35-45, 2000

15. J. Osaka, Y. Ohno, S. Kishimoto, K. Maezawa, and T. Mizutani, "Deep levels in n-type AlGaN grown by hydride vapor-phase epitaxy on sapphire characterized by deep-level transient spectroscopy", Appl. Phys. Lett. 87, 222112 (2005)

16. D. Kindl, P. Hubík, J. Krištofik, J. J. Mareš, Z. Výborný et al., "Deep defects in GaN/AlGaN/SiC heterostructures", J. Appl. Phys. 105, 093706 (2009); 10.1063/1.3122290

17. P. Kamyczek, E. Placzek-Popko, Vl. Kolkovsky, S. Grzanka, and R. Czernecki, "A deep acceptor defect responsible for the yellow luminescence in GaN and AlGaN", J. Appl. Phys. 111, 113105 (2012)

18. A. R. Arehart, T. Homan, M. H. Wong, C. Poblenz, J. S. Speck et al., "Impact of N- and Ga-face polarity on the incorporation of deep levels in ntype GaN grown by molecular beam epitaxy", Appl. Phys. Lett. 96, 242112 (2010)

19. J. Joh and J. A. del Alamo, "A Current-Transient Methodology for Trap Analysis for GaN High Electron Mobility Transistors", IEEE TRANSACTIONS ON ELECTRON DEVICES 58 132, 2011
20. DasGupta S, Sun M, Armstrong A, Kaplar R J, Marinella M J, Stanley J B, Atcitty S, Palacios T, "Slow Detrapping Transients due to Gate and Drain Bias Stress in High Breakdown Voltage AlGaN/GaN HEMTs", IEEE Transactions on Electron Devices 2012, 59, 2115
21. G. Meneghesso, A. Zanandrea, A. Stocco, I. Rossetto, C. De Santi, F. Rampazzo, M. Meneghini, E. Zanoni, E. Bahat-Treidel, O. Hilt, P. Ivo, J Wuerfl, "GaN-HEMTs devices with single- and double-heterostructure for power switching applications," Reliability Physics Symposium (IRPS), 2013 IEEE International , pp.3C.1.1,3C.1.7, 14-18 April 2013 doi: 10.1109/IRPS.2013.6531983
22. S. Bahl, and J. A. del Alamo, IEEE Transactions on Electron Devices 40 (1993) 1558

True "Figure of Merit (FOM)" of a Power Semiconductor Switch

Krishna Shenai

Energy Systems Division, Argonne National Laboratory
9700 South Cass Avenue, Bldg. 362, Argonne, IL 60439-4815

Abstract

Wide Band Gap (WBG) semiconductors such as Silicon Carbide (SiC) and Gallium Nitride (GaN) are known to be superior materials for the fabrication of power electronics switching devices compared to the industry work-horse silicon. This advantage primarily stems from wider energy bandgap, higher electric field strength for avalanche breakdown, and improved thermal conductivity of SiC and GaN semiconductors compared to silicon. This paper reviews the various "Figures of Merit (FOM)" proposed in the literature for the WBG power semiconductors. Unipolar power transistors with MOS channel conduction as well as transistors that employ two-dimensional electron gas (2DEG) channels are considered. A true FOM for a power semiconductor switching device needs to take into account all electrical power losses and thermal limitations. A new Safe Operating Area (SOA) limit that relates material defect density (D_{it}) to the switching current density (J_{on}) for a specified maximum junction temperature limit (T_{jmax}) is proposed as the true FOM for a power semiconductor switching device.

I. Introduction

It has been well-known for more than two decades that power electronic switching devices made from Wide Band Gap (WBG) semiconductors such as Silicon Carbide (SiC) and Gallium Nitride (GaN) offer much lower on-state resistance ($R_{DS(on)}$) than comparable silicon power devices because of their superior electrical and thermal conductivities, and breakdown field compared to silicon [1, 2]. Single-chip SiC JBS power diodes rated up to 1,700V/25A [3], and more recently, 1,700V vertical GaN Schottky power diodes [4] have been introduced into the commercial market in limited quantities. Although these devices are finding applications in computer/telecom power supplies, motor control, and smart grid [5-7], serious concerns pertaining to long-term reliability of these devices in compact power converters under stressful field operating conditions remain [8]. For example, majority of WBG commercial power diodes are not dv/dt- and avalanche-rated, and the data sheets rarely mention of the Safe-Operating Area (SOA), especially at elevated temperatures.

For high efficiency and high-frequency power conversion, MOS-controlled power semiconductor switches are needed [9]. Power MOSFET is the basic building block for other MOS-controlled high-power switches such as IGBT's. The "best-in-class" commercial 1,200V SiC power DMOSFET's [10] and trench-gate MOSFET's [11, 12] have a specific on-state resistance $R_{sp,on}$ of 3.7 mΩ-cm^2 and 2.6 mΩ-cm^2, respectively and a single-chip current rating of 50 amps. Figure 1 illustrates the unit cell cross-sections of these devices; the unit cell pitch is typically in the range of 10 to 12 microns for a 1,200V rated device. In the current state-of-the-art SiC power MOSFET's, the gate dielectric suffers from reliability problems, especially at elevated temperatures above 150°C. For example, gate MOS threshold voltage has been found to be unstable with prolonged gate voltage stress [13]. Low MOS inversion channel mobility and poor gate dielectric reliability have been largely attributed to a high density of interface states, especially in the upper half of the bandgap close to the conduction band edge. Recently, AlON/SiO$_2$ gate dielectric films deposited by CVD technique have been shown to improve both MOSFET performance and reliability of planar as well as trench-gate

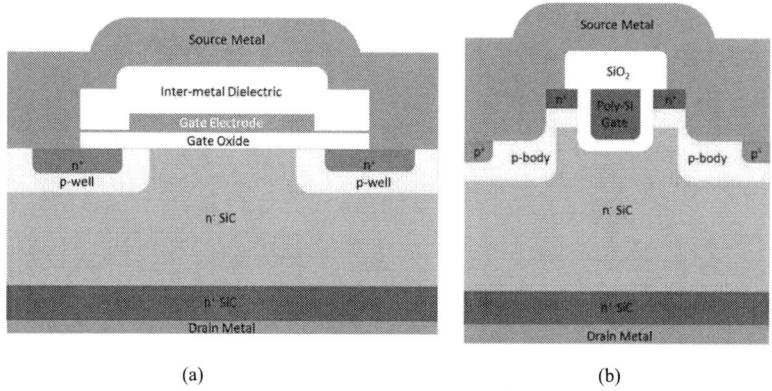

(a)   (b)

Figure 1: Cross-sections of "industry-best" normally-off vertical SiC power (a) DMOSFET and (b) trench-gate MOSFET unit cells [10-12].

devices [14]. For "industry-best" 1,200V SiC power MOSFET's, measured reverse leakage currents are typically high; and, devices are not rated for dv/dt and avalanche capabilities, especially at elevated temperatures. At the time of this writing, SiC power MOSFET's have not been proven reliable in power converter applications and their application is limited to junction temperatures below 150°C [15].

The best-in-class commercial GaN power transistors are rated at 600V and have single-chip current ratings of < 20 amps [16]. These gate-controlled normally-off GaN-on-Si power switches are lateral in configuration as shown in Figure 2, have a gate threshold voltage of ~ 1.5V, and a specific on-state resistance R$_{sp,on}$ = 2.5 mΩ-cm^2. The device utilizes hole-injection at the gate electrode from the p-AlGaN to the AlGaN/GaN hetero-junction, which simultaneously increases the electron density in the channel, resulting in a dramatic increase of the drain current due to conductivity modulation. A recent advance in this transistor technology pertains to an integrated Schottky diode that provides the reverse-conducting current path during switching [17]. Using GaN lateral enhancement-mode switches, a three-phase motor drive inverter rated at 900 Watts and switching at 6 kHz was demonstrated with a switching energy efficiency of 99.3% [18].

For "industry-best" lateral GaN power transistors, the measured reverse leakage currents are high, especially at elevated temperatures; device breakdown is primarily caused by breakdown at the buffer-substrate and/or at the device surface, and hence, these devices are not optimized for power switching. Furthermore, these devices are not rated for dv/dt and avalanche capabilities; and, are 3-5X more expensive than silicon power transistors with identical current ratings.

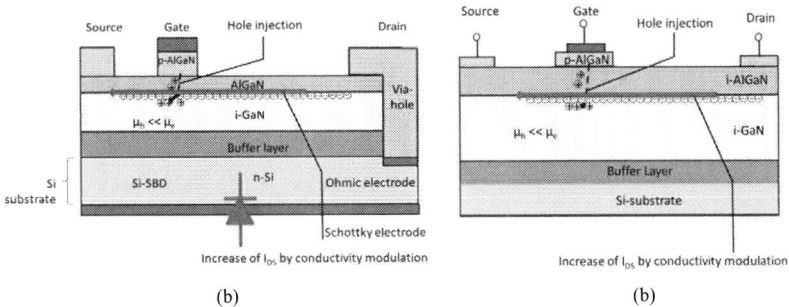

(b)   (b)

Figure 2: Cross-sections of "industry-best" normally-off lateral GaN-on-Si lateral power transistor (a) with and (b) without integrated body Schottky diode [16-18]

To the best of our knowledge, there is not a single silicon discrete lateral high-power switch available in the commercial market. This lack of market pull for a lateral power switch is largely due to the fact that lateral power devices are not scalable to higher voltages and currents, and are not cost-effective and reliable in power converter circuits [9]. Hence, vertical power transistor technology on free-standing bulk n$^+$GaN substrates must be developed.

Previous attempts to develop vertical power transistors in GaN using current aperture [19] and backside etching [20] technologies have not resulted in cost-effective and reliable and power switching devices. The CAVET device [19] shown in Figure 3(a) employs Mg-ion-implanted current blocking layer (CBL), and is similar in its operation

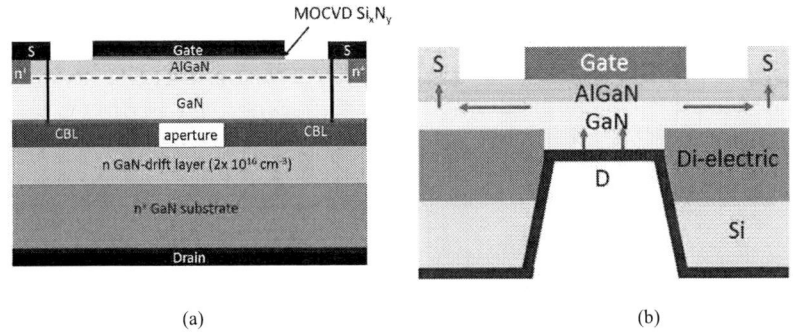

(a)   (b)

Figure 3: Cross-sections of state-of-the-art vertical GaN power transistor devices (a) normally-on CAVET transistor on free-standing GaN substrate and (b) normally-on GaN-on-Si power transistor fabricated using substrate etch-back technology

to the static induction transistor (SIT) in silicon [21]. Silicon SIT technology has not been commercially successful because of poor performance, reliability, and manufacturing issues. It is unlikely that GaN-based SIT-type devices can overcome the fundamental issues that have rendered silicon SIT's impractical for high-volume power electronics applications. The vertical GaN-on-Si power transistor structure shown in Figure 3(b) [20] suffers from the same issues as lateral GaN-on-Si power transistors currently deployed in the commercial market – severe manufacturing and reliability problems. The basic construction of the device structure does not allow for heat to diffuse easily from active regions of the device, and thus causes severe self-heating and leads to thermal destruction. Hence, a commercially viable and scalable vertical power transistor technology on free-standing GaN substrates is needed to adequately address these challenges.

II. Power Semiconductor "Figure of Merit (FOM)"

Traditionally, power semiconductor devices (and power electronics circuits) have been developed with a primary focus on achieving the highest power conversion efficiency [22]. This "performance-driven" practice is about to change as the end-customers are now demanding, in addition to energy-efficiency, improved field-reliability of power converters at reduced cost [23]. A paradigm shift in power converter design is needed in order to assure its field-reliability. A systematic approach to "reliability-driven" power technology development is also likely to lead to dramatic reduction in cost and significant performance gains; however, such a methodology is yet to be realized. On the other hand, for more than three decades, microprocessor (and other signal electronics) chips have been developed to guarantee at least 100,000 hours of active field-operation in a personal computer (PC) environment [24]. It is our contention that a similar "built-in" reliability design approach is exactly what is needed to further advance semiconductor power switching in order to meet the challenges of 21st century electric transmission and distribution infrastructure [25].

A power semiconductor switch operates quite differently from an analog or a digital semiconductor switch [9]. It is expected that when the power semiconductor switch is turned on, it should present the least electrical resistance in order to minimize the ohmic power loss; in the off-state, it should conduct zero leakage current with a very high avalanche breakdown voltage. Shenai first proposed the specific on-state resistance R_{sp} [1] and input switching charge Q_g [2] as the "Figures of Merit (FOM)" in order to assess the value of a given power semiconductor technology for power converter applications; these two FOM are respectively given by:

$$R_{sp} = \frac{4V_B^2}{\varepsilon_s \mu E_c^3} \qquad (1a)$$

$$Q_g = C_{iss} V_{gs} \qquad (1b)$$

where V_B is the breakdown voltage, ε_s is the permittivity, μ is the drift-region mobility, and E_c is the critical electric field strength at avalanche breakdown of the semiconductor material; the input switching charge assumes an MOS-controlled power switch where C_{iss} is the net input capacitance at zero gate bias voltage and V_{gs} is the gate-to-source voltage needed to fully control the "on" and "off" switching transitions. For a given power converter application, both $R_{sp,on}$ and Q_g must be minimized in order to obtain the highest electrical conductivity and the lowest input power loss. These two FOM have been adapted by the silicon power semiconductor industry to successfully optimize the power switching technology; since their first introduction more than two and half decades ago, $R_{sp,on}$ and Q_g have become "industry standards." A third FOM was also proposed by Shenai [1] which takes into account electrical and thermal conductivities in addition to avalanche breakdown field, and is given by:

$$Q_{F2} = \kappa \sigma_{sp} E_c \qquad (2)$$

where κ is the thermal conductivity and $\sigma_{sp} = 1/R_{sp}$ is the specific electrical conductance. Subsequently, a number of authors [26-28] have proposed various FOM to characterize power semiconductor switching devices. The high-frequency FOM proposed by Baliga [26] minimizes the chip area by considering the input switching power in addition to conduction power loss, and hence, it is really a combination of R_{sp} and Q_g first proposed by Shenai [1]. Note that BHFFOM does not include the output switching power. Recently, it has been shown that the output switching power is a significant portion of the total power loss, especially at higher switching frequencies using power switches with smaller capacitances [29, 30]. The FOM proposed by Kim et al [27] for high-frequency power switching, NHFFOM, although considers the output capacitance of the switch in addition to the on-state conduction power loss, it does not directly relate the output switching power loss to semiconductor material parameters, and there is no mention of the input switching power loss. Furthermore, the analysis by Kim et al optimizes the power switch die area by neglecting the input power loss. Huang [28] has proposed four FOM's; the first three FOM's (HDFOM, HMFOM, and HCAFOM) essentially convolve around specific on-state resistance and gate charge, and do not account for the output switching power loss or consider thermal limitations. The last FOM proposed by Huang, i.e., HTFOM, does refer to the semiconductor thermal conductivity; however, only heating due to on-state conduction and input switching power losses at the junction die attachment are included.

Thus, it is clear that there is a need for a "true" FOM for a power semiconductor switching device that considers all of the power loss mechanisms in addition to the thermal limitations of the semiconductor material and the package. This is particularly important given that superior high-voltage high-frequency SiC and GaN power devices capable of much higher temperature operation than silicon power devices are being introduced into the commercial market. At the time of this writing, both lateral and vertical power devices are being developed in SiC and GaN materials; lateral GaN power transistors are mostly based on two-dimensional electron gas (2DEG) formed at the AlGaN/GaN hetero-structure due to the polarization charge. Figure 4 illustrates the electric field profiles in the reverse-blocking mode for typical vertical and lateral gate-

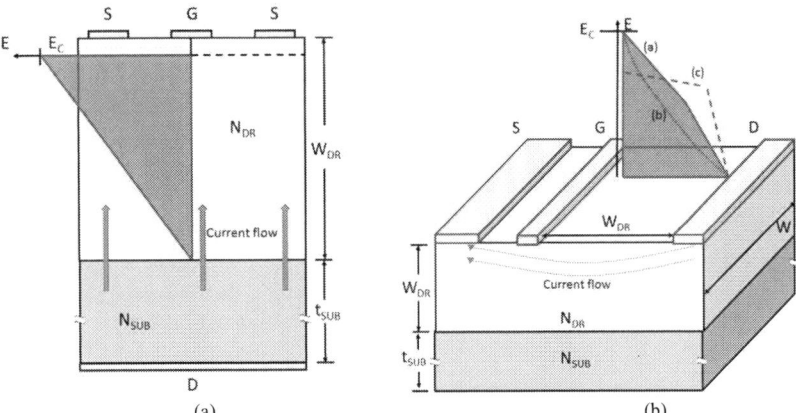

(a)   (b)

Figure 4: Schematic electric field profiles in the reverse blocking mode and current flow lines in typical (a) vertical-geometry and (b) lateral-geometry power switching devices.

controlled power switching devices; also shown are the current flow lines in the on-state. The peak electric field for a vertical power switch typically occurs near the top surface of the chip; for a lateral device, the electric field profile can be significantly adjusted using RESURF [31] and lateral charge control concepts [32]. The optimum breakdown and on-resistance performance is obtained when the drift-region parameters are adjusted such that the breakdown occurs by avalanche mechanism. The specific on-state resistance for a vertical device is given by eqn. (1a) and that of the lateral power switch with 2DEG conduction is given by:

$$R_{spl} = \frac{V_B^2}{q\mu_s n_s E_c^2} \qquad (3)$$

where μ_s and n_s are the mobility and sheet carrier density of the 2DEG, respectively in the lateral conduction channel. The results of calculations for various semiconductors are plotted in Figure 5 using the latest set of material parameters [1, 33-37]; for GaN, vertical devices that employ bulk material conduction are also evaluated. As the lateral devices

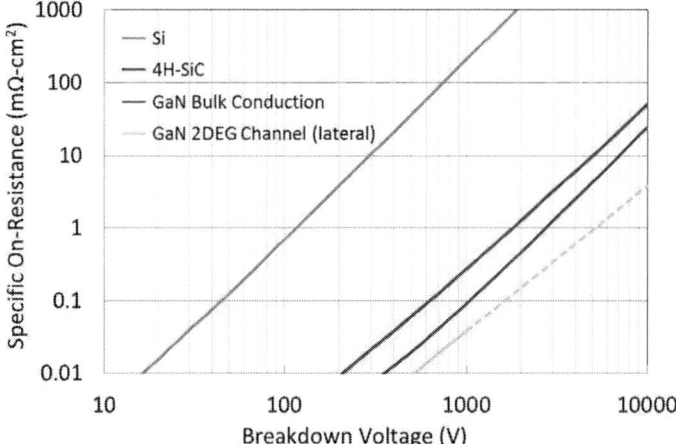

Figure 5: Calculated specific on-state resistance vs. avalanche breakdown for vertical and lateral power devices made in silicon, 4H-SiC and GaN materials.

are typically no scalable to higher voltages and higher currents, we expect GaN lateral power transistors to be limited to below 1,000 volts and 25 amps single-chip current ratings. The calculated values of various FOM are listed in Tables 1(a) – 1(c); also included are Johnson's [38] and Keye's [39] FOM although they are not relevant for power switching applications. In our calculations, accurate doping dependencies of electron mobility and critical field strength for avalanche breakdown are used.

Table 1(a) Calculated FOM's for various power semiconductor devices.

Material		Si	4H-SiC	GaN
BM $(\varepsilon_s \mu E_C^3)$		4.26E+20	1.83E+23	2.88E+23
BHM (μE_C^2)		1.22E+14	6.30E+15	9.80E+15
JM $(E_c v_{sat}/2\pi)^2$		9.09E+23	3.64E+26	6.87E+26
KM $\kappa(v_{sat}/\varepsilon_s)^{1/2}$		1.39E+03	6.46E+03	3.85E+03
Q_{F2} $(\kappa\sigma_{sp}E_C)$	V_B			
	1,000	1.76E+06	5.58E+10	4.51E+10
	10,000		2.32E+08	5.33E+08
	20,000		4.20E+07	1.34E+08
Q_{F1} $(\kappa\sigma_{spl}E_C)$	V_B			
	1,000			1.99E+11
	10,000			1.99E+09
	20,000			4.99E+08

Table 1(b) Specifc on-state conductance of lateral GaN power transistors using eqn. (3)

V_B	σ_{spl} $(q\mu_s n_s E_c^2/V_B^2)$
1,000	26283.015
10,000	262.83015
20,000	65.7075375

Table 1(c) Normalized values of FOM's for semiconductors.

Material		Si	4H-SiC	GaN
BM $(\varepsilon_s \mu E_C^3)$		1	429.88	674.98
BHM (μE_C^2)		1	51.85	80.67
JM $(E_c v_{sat}/2\pi)^2$		1	400.00	756.25
KM $\kappa(v_{sat}/\varepsilon_s)^{1/2}$		1	4.66	2.78
Q_{F2} $(\kappa\sigma_{sp}E_C)$	V_{BD}			
	1,000	1	3.17E+04	2.56E+04
Q_{F2} $(\kappa\sigma_{sp1}E_C)$	V_{BD}			
	1,000			1.13E+05

III. Summary and Conclusions

Various "Figures of Merit" (FOM's) proposed for power switching devices are critiqued and it is shown that there is a need to develop a "true" FOM that accounts for various forms of power losses and also considers thermal limitation of the semiconducting material and that of the package. Both vertical and lateral power devices that employ bulk charge conduction as well as 2DEG channel conduction are studied. It is shown that for voltages above 600V and single-chip current ratings > 20 amps, there is significant opportunity to develop vertical GaN power devices with 2DEG channels; vertical GaN 2DEG devices are shown to have nearly an order of magnitude improved overall power switching performance than the lateral GaN 2DEG power devices.

An important measure of robustness of a power semiconductor switch in power electronics switching is the Safe Operating Area (SOA). The SOA of a power semiconductor device refers to voltage and current limits within which the device can be safely switched. The SOA deviates from rectangular shape due to thermal heating as shown in Figure 6; the loss of SOA occurs when power dissipation is at its maximum and is a function of several parameters including switching frequency, duty ratio, dv/dt and di/dt among others. Provided bond wires and die-to-package interface remain intact, and

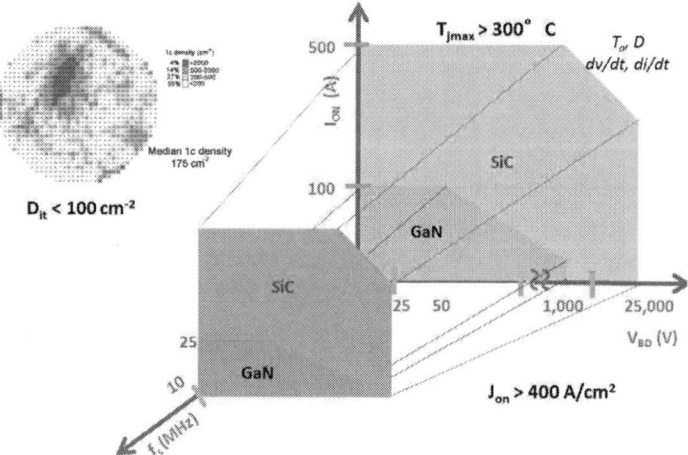

Figure 6: A proposed new "Figure of Merit (FOM)" for a power semiconductor switching device that relates the defect density, D_{it} in the drift-region of the device to the current density, J_{on} that can be reliably switched in given power converter under field-operating conditions at a specified junction temperature, T_{jmax} that the semiconductor chip can tolerate under "long-term" repetitive switching conditions

assuming isothermal die boundary conditions, it is known that "hot spots" occur locally during power switching and can lead to current filamentation and local "burn outs" primarily caused by Joule heating [40-42]. Maximum power dissipation occurs within the drift-region of the device where the electric field is high. The challenge is then to rapidly remove heat away from the local "hot spot" to avoid material burn-out; the thermal time constant must be small (typically less than a microsecond), and hence, substrate must be thinned down so that the cooling surface is brought closer to the drift-region of the device. As shown in Figure 6, the SOA of a GaN power device is expected to be smaller than that of a SiC power device primarily due to its relatively lower thermal conductivity. Also, because of the direct energy bandgap of GaN, minority carrier lifetime is small; hence, conductivity modulation of the drift-region is difficult; and, bipolar-mode power devices are not likely in GaN. Therefore, for higher voltage (above few kilo volts) and higher current (above 100 amps), it is our contention that the vertical SiC power devices hold the greatest promise.

As shown in Figure 6, the SOA of a semiconductor power switch is reduced as the power switching frequency is increased. This is largely due to the fact that at higher switching frequencies, switching power losses in the semiconductor power device increase. As is well-known in the literature [9, 22], SOA can be defined for single pulse as well as repetitive switching conditions. As the switch failure is caused by local temperature rise, separate SOA diagrams can be generated for the same semiconductor power switch for varying switching current densities, J_{on}. Hence, T_{jmax} must be specified as a function of J_{on} at a given ambient temperature, T_a of the power converter circuit. While this approach has had reasonable success with silicon power devices where the material crystal defect density is below 1 per cm^2, the same approach may not be valid in case of SiC and GaN power devices with several orders of magnitude higher defect densities. Assuming that crystal defects are uniformly distributed in the space-charge region of the device, SOA limits may have to be evaluated as a function of the net defect density, D_{it}. A revised SOA diagram shown in Figure 6 is proposed as a new "Figure of Merit (FOM)" that represents the amount of current density that can be reliably switched under field operating conditions at a specified T_{jmax} for a semiconductor power switching device with a space-charge region containing a net crystal defect density of D_{it}. It is therefore imperative that the revised SOA must be evaluated for SiC and GaN power devices and "bench-marked" against silicon power devices with identical voltage and current ratings, and packaging configurations.

References

1. K. Shenai, R. S. Scott and B. J. Baliga, "Optimum semiconductors for high-power electronics," IEEE Trans. Electron Devices, vol. 36, no. 9, pp. 1811-1823, Sept. 1989.
2. K. Shenai, "Potential Impact of Emerging Semiconductor Technologies on Advanced Power Electronic Systems," IEEE Electron Device Lett., vol. 11, no. 11, pp. 520-522, November 1990.
3. Power Products, Cree Inc., www.cree.com
4. I. C. Kizilyalli, and A. Edwards, D. Bour, H. Shah, H. Nie and D. Disney, "Vertical devices in bulk GaN drive diode performance to near-theoretical limits," HOW2POWER TODAY, pp. 1-7, March 2013.
5. SiC 2010: How SiC will impact electronics: A 10 year projection, Yole Development Market Report, Lyon, France, 2010.
6. T. McDonald, "GaN based power technology stimulates revolution in conversion electronics," Electronics in Motion and Conversion, pp. 2-4, April 2009.

7. G. Deboy, R. Rupp, R. Mallwitz and H. Ludwig, "New SiC JFETs Boost Performance of Solar Power Inverters," Power Electronics Europe, issue 4, 2011, pp. 29-33.
8. K. Shenai, "Switching megaWatts with power transistors," The Electrochemical Society Interface Magazine, vol. 22, no. 1, pp. 47-54, Spring 2013 (**invited paper**).
9. B. J. Baliga, Modern Power Devices, Wiley: New York, 1987.
10. S-H. Ryu, L. Cheng, S. Dhar, C. Capell, C. Jonas, R. Callanan, A. Agarwal, J. Palmour, A. Lelis, C. Scozzie, and B. Geil, "3.7 mΩ-cm^2, 1500 V 4H-SiC DMOSFET's for advanced high-power, high-frequency applications," Proc. 23rd *IEEE Int. Symp. Power Semiconductor Devices & IC's (ISPSD)*, pp. 227-230, May 23-26, 2011.
11. T. Nakamura, Y. Nakano, M. Aketa, R. Nakamura, S. Mitani, H. Sakairi, and Y. Yokotsuji, "High-performance SiC trench devices with ultra-low R_{on}," Digest of IEEE Int. Electron Devices Meeting (IEDM), pp. 599-601, Dec. 2011.
12. T. Nakamura, M. Sasagawa, Y. Nakano, T. Otsuka, and M. Miura, "Large current SiC power devices for automobile applications," Proc. 2010 IEEE Int. Power Electronics Conference, pp. 1023-1026, 2010.
13. A. J. Lelis, D. Habersat, R. Green, A. Ogunniyi, M. Gurfinkel, J. Suehle, and N. Goldsman, "Time dependence of bias-stress-induced SiC MOSFET threshold-voltage instability measurements," IEEE Trans. Electron Devices, vol. 55, no. 8, pp. 1835-1840, Aug. 2008.
14. T. Hosoi, S. Azumo, Y. Kashiwagi, S. Hosaka, R. Nakamura, S. Mitani, Y. Nakano, H. Asahara, T. Nakamura, T. Kimoto, T. Shimura, and H. Watanabe, "Performance and reliability improvement in SiC power MOSFET's by implementing AlON high-k gate dielectrics," Digest of IEEE Int. Electron Devices Meeting (IEDM), pp. 159-162, Dec. 2012.
15. K. Shenai, M. Dudley, and R. F. Davis, "Current status and emerging trends in wide bandgap (WBG) semiconductor power switching devices," ECS Journal of Solid State Science and Technology 2(8), N3055-N3063, July 2013 (**invited paper**).
16. Y. Uemoto, M. Hikita, H. Ueno, H. Matsuo, H. Ishida, M. Yanagihara, T. Ueda, T. Tanaka, and D. Ueda, "Gate Injection Transistor (GIT) – A normally-off AlGaN/GaN power transistor using conductivity modulation," IEEE Trans. Electron Devices, vol. 54, no. 12, pp. 3393-3399, Dec. 2007.
17. T. Morita, S. Ujita, H. Umeda, Y. Kinoshita, S. Tamura, Y. Anda, T. Ueda, and T. Tanaka, "GaN gate injection transistor with integrated Si Schottky barrier diode for high-efficient DC-DC converters," in Digest of IEEE Int. Electron Devices Meeting (IEDM), pp. IED12-151 to 12-154, 2011.
18. T. Morita, S. Tamura, Y. Anda, M. Ishida, Y. Uemoto, T. Ueda, T. Tanaka, and D. Ueda, "99.3% efficiency of three-phase inverter for motor drive using GaN-based gate injection transistors," in IEEE Applied Power Electronics Conference (APEC), pp. 481-484, 2011.
19. S. Chowdhury, M. H. Wong, B. L. Swenson, and U. K. Mishra, "CAVET on bulk GaN substrates achieved with MBE-regrown AlGaN/GaN layers to suppress dispersion," IEEE Electron Device Letters, vol. 33, no. 1, pp. 41-43, Jan. 2012.
20. B. Lu and T. Palacios, "AlGaN/GaN vertical power transistors," MIT Microsystems Technology Laboratories Annual Research Report, p. DE29, 2009.

21. J. Nishizawa, T. Terasaki, and J. Shibata, "Field effect transistor versus analog transistor (static induction transistor)," IEEE Trans. Electron Devices, vol. ED-22, pp. 185-197, 1975.
22. N. Mohan, T. M. Undeland, and W. P. Robbins, Power Electronics: Converters, Applications, and Design, 3rd ed., Wiley: New York, 2003.
23. K. Shenai, "Made-to-order Power Electronics," IEEE Spectrum, vol. 37, No. 7, pp. 50-55, July 2000 **(invited paper)**.
24. H. A. Schafft, D. L. Erhart, and W. K. Gladden, "Toward a building-in reliability approach," Microelectronic Reliability, vol. 37, no. 1, pp. 3–18, 1997.
25. National Electric Delivery Technologies Vision and Roadmap (2003), U.S. Department of Energy, November 2003.
26. B. Jayant Baliga, "Power semiconductor device figure of merit for high-frequency applications," IEEE Electron Dev. Lett., vol. 10, no. 10, pp. 455-457, Oct. 1989.
27. I. J. Kim, S. Matsumoto, T. Sakai, and T. Yachi, "New power device figure-of-merit for high-frequency applications," in Proc. IEEE Int. Symp. Power *Semiconductor Devices and IC's (ISPSD)*, pp. 309-314, 1995.
28. Alex Q. Huang, "New unipolar switching power device figures of merit," IEEE Electron Dev. Lett., vo. 25, no. 5, pp. 298-301, May 2004.
29. K. Shenai, "A circuit simulation model for high-frequency power MOSFETs," IEEE Trans. Power Electronics, vol. 6, no. 3, pp. 539-547, July 1991.
30. K. Shah and K. Shenai, "A Simple and Accurate Circuit Model for GaN Power Transistors," IEEE Trans. Electron Devices, vol. 59, n0. 10, pp. 2735-2741, Oct. 2012.
31. J. A. Appels and H. M. J. Vaes, "High voltage thin layer devices (RESURF) DEVICES)," IEEE International Electron Devices Meeting (IEDM) Digest, Abstract 10.1, pp. 238-241, Dec. 1979.
32. R. S. Wrathall, B. J. Baliga, K. Shenai,W. Hennessy, and T. P. Chow, "Charge Controlled 80-V Lateral DMOSFET with Very Low Specific On-Resistance Designed for an Integrated Power Process, " IEEE International Electron Devices Meeting (IEDM) Digest, pp. 954-957, Dec. 1990.
33. T. Hatakeyama, J. Nishio, C. Ota, and T. Shinohe, "Physical modeling and scaling properties of 4H-SiC power devices," in Proc. SISPAD, pp. 171–174, 2005.
34. S. Kagamihara, H. Matsuura, T. Hatakeyama, T. Watanabe, M. Kushibe, T. Shinohe, and K. Arai, "Parameters required to simulate electric characteristics of SiC devices for n-type 4H-SiC," J. Appl. Phys., vol. 96, no. 10, pp. 5601–5606, Nov. 2004.
35. N. D. Arora, J. R. Hauser, and D. J. Roulston, "Electron and hole mobilities in silicon as a function of concentration and temperature," IEEE Trans. Electron Devices, vol. 29, pp. 292–295, 1982.
36. S. J. Pearton, C. R. Abernathy, F. Ren, "Design and Fabrication of Gallium High-Power Rectifiers," Gallium Nitride Processing for Electronics, Sensors and Spintronics Engineering Materials and Processes, pp. 179-212, 2006.
37. T.T. Mnatsakanov, M.E. Levinshtein, L.I. Pomortseva, S.N. Yurkov, G. S. Simin, and M. A. Khan, "Carrier mobility model for GaN", Solid-State Electron., vol. 47, no. 1, pp. 111-115, Jan. 2003.
38. E. O. Johnson, "Physical Limitations on Frequency and Power Parameters of Transistors," RCA Review, vol. 26, pp. 163-177, 1965.
39. R. W. Keyes, "Figure of Merit for Semiconductors for High Speed Switches," Proc. IEEE, vol. 60, pp. 225-232, 1972.

40. Newman, R., W. C. Dash, R. N. Hall, and W. E. Burch, "Visible light from a Si p-n Junction," Physical Review, vol. 98, pp. 1536–1537, 1955.
41. Chynoweth, A. G., and K. G. McKay, "Photon emission from avalanche breakdown in silicon," Physical Review, vol. 102, no. 2, pp. 369–376, April 1956.
42. D. A. Dallmann and K. Shenai, "Scaling Constraints Imposed by Self-Heating in Submicron SOI MOSFETs," IEEE Trans. Electron Devices, vol. 42, no. 3, pp. 489-496, March 1995.

Progress in SiC MOSFET Reliability

D. R. Hughart, J. D. Flicker, S. D. DasGupta, S. Atcitty, R. J. Kaplar, and M. J. Marinella

Sandia National Laboratories, PO Box 5800, MS 1084, Albuquerque, NM 87185-1084

Bias-temperature stress experiments performed on two generations of SiC power MOSFETs from the same manufacturer show reductions in threshold voltage (V_T) shift at elevated temperatures from first- to second-generation. The negative V_T shift is reduced from a range of -1 V to -1.6 V to a range of -100 mV to -300 mV for temperatures from 125°C to 175°C. Plastic-packaged parts show a gate-bias-independent junction leakage current at temperatures above the rated temperature, suggesting that the plastic packaging introduces an extrinsic leakage path. Junction leakage in metal-packaged parts can be significantly reduced by applying a small negative gate bias at elevated temperatures. Switching gate bias temperature stresses show V_T shifts dependent on duty cycle, with a higher duty cycle resulting in a higher rate of V_T shift. Cumulative damage effects may be observed between switching gate bias stresses.

Introduction

SiC is a unique wide-bandgap material in that its native oxide is SiO_2 – an excellent insulator that is particularly useful in creating a field-effect transistor. SiC MOSFETs have been commercially available for over two years with blocking voltages of 1200 V and excellent on-state resistances (R_{DS-on}) as low as 80 mΩ. In addition, SiC devices can in theory operate at higher temperatures than competing Si devices. SiC MOS technology could offer major system-level improvements due to reduced size and weight for motor drives, hybrid electric vehicles, photovoltaic inverters, and even grid-level applications such as flexible AC transmission systems (FACTS) (1)-(2). Currently, the most significant hurdles to market penetration are reliability and cost, the latter of which is improving rapidly as the SiC technology becomes more widespread.

SiC MOSFET reliability challenges stem mainly from SiC/SiO_2 interface quality as well as interactions between devices and packaging. When operated under high gate electric fields and high temperatures, threshold voltage instabilities are often observed both in SiC MOSFET (3)-(4), and MOS capacitor (5)-(6) structures. Our constant-bias stress measurements comparing first- and second-generation commercially available SiC MOSFETs indicate significant progress on this front. Switching bias measurements performed on second generation parts simulate a more realistic operating condition and show that V_T degradation depends on duty cycle.

Junction leakage current at elevated temperatures is also a significant reliability concern. Comparisons of leakage measurements for first-generation SiC MOSFETs in plastic and metal packaging show increasing leakage current at high blocking voltages as temperature increases. However, the leakage current is worse for plastic packaging.

Applying a negative voltage on the gate can significantly reduce the leakage current for metal-packaged parts, but has no effect on plastic-packaged parts.

Experimental Details

First- and second-generation power MOSFETs from the same manufacturer were subjected to a variety of temperature and bias stresses. The first-generation devices that were tested had both plastic and metal packaging. The temperature rating for the first-generation devices is 125°C for the plastic-packaged parts and 225°C for the metal-packaged parts. The second-generation devices that were tested only had plastic packaging, and their temperature rating is 225°C (testing of metal-packaged devices is planned for future experiments). Current-voltage measurements were performed using a Keithley 2651A high-current sourcemeter coupled with a Keithley 2601A for gate control. For forward-blocking leakage measurements a Keithley 2410 high-voltage sourcemeter was used. First-generation parts were heated using a Corning ceramic hotplate for leakage measurements and a hot chuck for all other tests. Second-generation parts were heated using a VWR aluminum hot plate. Hot plate temperatures were verified using a temperature probe. Each part was allowed to stabilize at a given temperature for thirty to forty minutes after the desired temperature was reached, or until the gate sweep curve stopped shifting. After the gate bias stress at a given temperature, gate-sweep characterization curves were measured to ascertain changes in the MOSFET's V_T. Following this, in order to revert the device to its original condition, a gate bias of opposite polarity to the stress gate bias was applied in small time increments until the characterization gate sweep curve matched the initial gate sweep curve at that temperature. V_T is taken to be the voltage resulting in $I_D = 10$ mA, with the drain voltage at 100 mV. For leakage current measurements, the drain voltage was swept from 0 V to 900 V for gate biases of 0 V, -2 V, and -5 V at various temperatures.

Results

Threshold Voltage Shift

V_T decreases significantly for SiC MOSFETs compared to Si MOSFETs simply as a function of temperature, without considering gate bias stress. This is due to the higher interface trap density of the gate oxide for SiC devices, since at elevated temperature the surface potential under strong inversion is reduced along with the concentration of interface traps that must be charged or discharged to achieve that potential (7). This is demonstrated by the first-generation parts in Fig. 1. The low V_T at elevated temperatures makes further V_T shifts due to bias stress more concerning. This is particularly true for negative shifts.

In order to assess the V_T shift due to electron and hole injection in the oxide, SiC MOSFETs are stressed with positive and negative gate biases for thirty minutes at varying temperatures. Devices from both generations show varying amounts of V_T shift, depending on the magnitude and polarity of the bias, as well as temperature. Constant bias stress results using gate biases of 20 V and -20 V for plastic- and metal-packaged first-generation devices are shown in Fig. 2(a). ΔV_T is relative to the initial V_T at the stress temperature, not room temperature. Positive shifts are likely due to electron injection from the inverted SiC into the oxide (Fig. 2(b)) and negative shifts are likely due to hole injection from the accumulated p-type SiC into the oxide (Fig. 2(c)). Each type of packaging shows larger negative V_T shifts than positive V_T shifts, indicating that hole injection is a significant issue, similar to concerns in SiO_2 on Si where holes are trapped at oxygen vacancies (E' centers) (8) with analogous models being developed for SiO_2 on SiC (9). Negative V_T shifts appear to be slightly worse for plastic-packaged parts, but the differences are so small it is likely that packaging has no significant effect on ΔV_T. The negative V_T shift at 125°C (the rated temperature for plastic-packaged parts) is -1 V and increases to -4 V at 225°C. The positive V_T shift is below 100 mV at temperatures up to 150°C and reaches 1 V at 225°C.

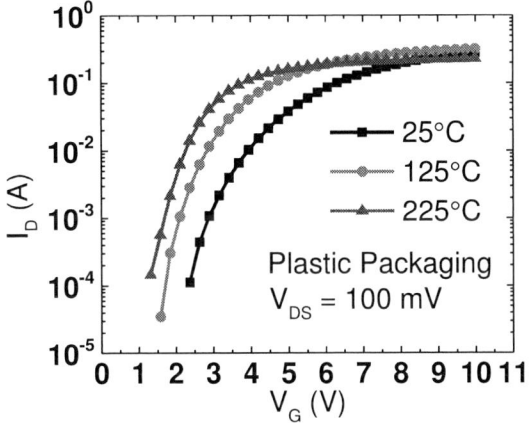

Figure 1. Drain current vs. gate voltage curves for first-generation 1200 V SiC MOSFETs in plastic packages, measured at various temperatures.

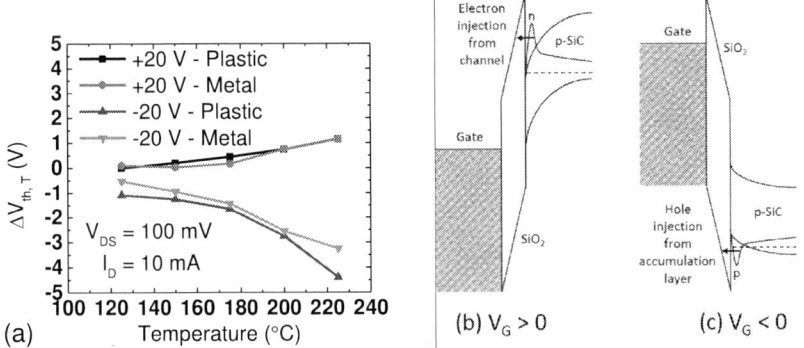

Figure 2. (a) ΔV_T for first-generation plastic- and metal-packaged parts stressed at ±20 V as a function of temperature. Schematic band diagrams illustrating (b) electron injection for $V_G > 0$ and (c) hole injection for $V_G < 0$.

The recommended off-state gate voltage is -5 V, so the devices are stressed with a gate bias of -5 V to evaluate them using more realistic operating voltages. Fig. 3 plots the results for plastic- and metal-packaged first-generation devices. No significant degradation is observed until a temperature of 250°C when the plastic-packaged parts show a V_T shift of -0.85 V. This degradation occurs at double the rated temperature of 125°C, demonstrating high resilience to bias and temperature stress at recommended operating voltages and temperatures.

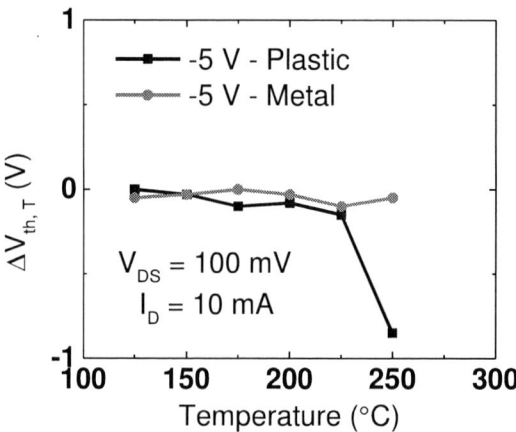

Figure 3. ΔV_T plotted vs. temperature for -5 V gate bias stress on plastic- and metal-packaged first-generation SiC MOSFETs.

Constant-bias-temperature stress experiments were repeated on second-generation plastic-packaged SiC MOSFETs. Gate biases of 20 V, -20 V, and -5 V were applied at temperatures of 125°C, 150°C, and 175°C. Fig. 4(a) plots the results and Fig. 4(b) compares the V_T shifts of first- and second-generation parts. Negative V_T shifts for a gate bias of -20 V were reduced significantly. Second-generation parts shifted by -300 mV at 175°C, compared to a shift of -1.5 V for first-generation devices. Positive V_T shifts due to a gate bias of 20 V were reduced as well, decreasing from 500 mV to 100 mV. The smaller V_T shifts indicate an improvement in the quality of the gate oxide from first-generation to second-generation devices.

In real-world applications, the power MOSFET will have a rapidly switching gate bias applied to it, which may result in different V_T shifts compared to constant bias stresses. Previous data using a switching gate bias stress have shown smaller V_T shifts compared to constant bias stress, even when tested for months (10). A second-generation SiC MOSFET stressed at 150°C with switching gate bias of +20 V / -5 V at a frequency of 100 Hz shows varying amounts of degradation depending on duty cycle. The part was stressed at a 50% duty cycle (50% of the time with 20 V on the gate, 50% of the time with -5 V on the gate) and 90% duty cycle (90% of the time with 20 V on the gate, 10% of the time with -5 V on the gate). Every half-hour the stress was stopped to perform a gate sweep. The device was recovered to its initial state using short stresses of positive bias after the first stress at 50% duty cycle. V_T rapidly drops for each stress, and then continues to shift negatively, but at a slower rate (Fig. 5). For a 50% duty cycle, after a fast initial shift of roughly -150 mV, the V_T shift recovers slightly and then gradually shifts at a rate of -2.5 mV/hr, with a cumulative V_T shift of -370 mV after 120 hours. For a 90% duty cycle, the fast initial shift is roughly -335 mV and after a slight recovery, V_T continues to shift negatively at a rate of -0.26 mV/hr, resulting in a final shift of -364 mV at 165 hours.

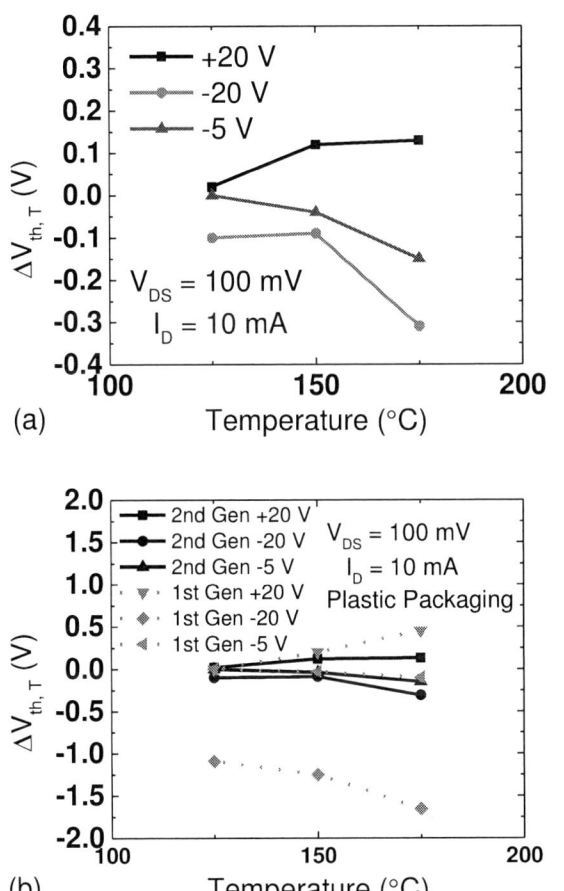

Figure 4. (a) ΔV_T for plastic-packaged second-generation SiC MOSFETs plotted vs. stress temperature, for ±20 V and -5 V gate bias stress conditions. (b) Direct comparison of first- and second-generation plastic-packaged parts for ±20 V and -5 V gate bias stress conditions.

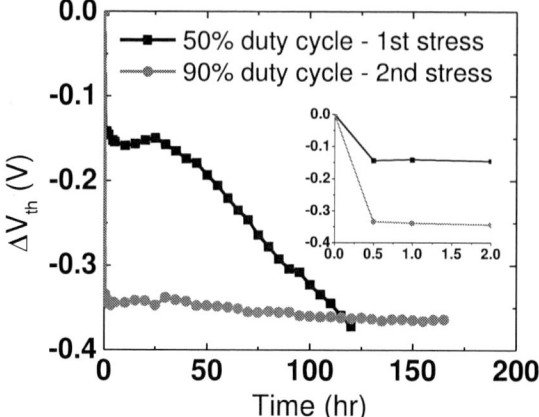

Figure 5. ΔV_T for a second-generation SiC MOSFET subjected to 50% and 90% duty cycle, +20 V / -5 V gate switching stress at 150°C. Inset shows expanded view of first two hours of stress (axis units same as main figure).

Leakage Current

Another key reliability concern is junction leakage current. This can be a significant concern at elevated temperatures, due to the negative V_T shifts observed at elevated temperatures that may prevent the device from being shut completely off with a gate bias of 0 V applied. Leakage current for a variety of temperatures was measured for first-generation SiC MOSFETs. Leakage current versus drain bias for a metal-packaged device is plotted in Fig. 6(a) and for a plastic-packaged device in Fig. 6(b). Both packages show increases in leakage current with increasing temperature, as expected. Note that the rated temperature for the plastic-packaged part is 125°C, so the temperatures at which these experiments were performed are above the rated temperature, indicating robust device performance within the recommended operating temperature range. For a drain bias up to 900 V, leakage remains below 1 μA for the metal-packaged part at 180°C and for the plastic-packaged part at 140°C. For temperatures of 140°C and 180°C, the leakage for the plastic-packaged part is roughly three times the leakage for the metal-packaged part, but at temperatures of 215°C and 250°C the leakage for the plastic-packaged part is larger by an order of magnitude.

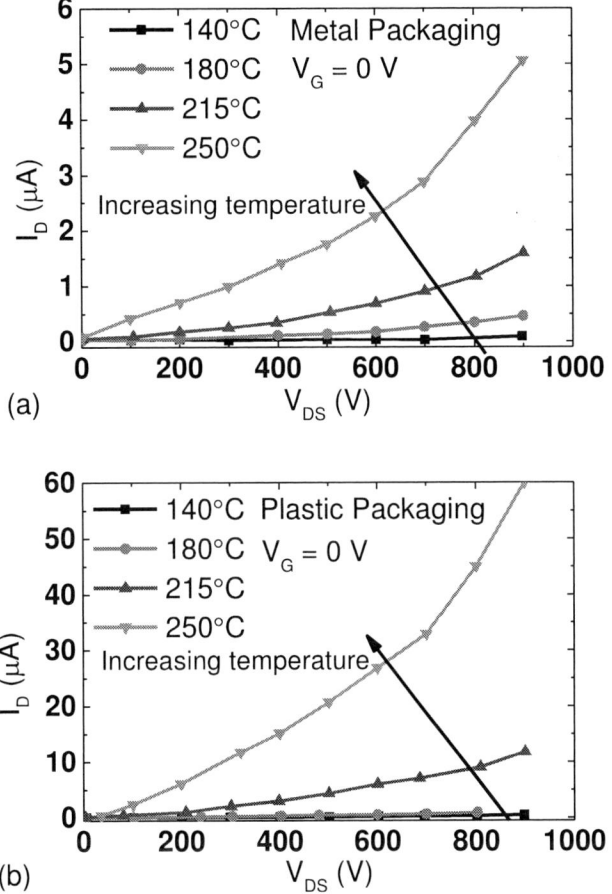

(a)

(b)

Figure 6. Drain current vs. drain voltage in the forward-blocking state ($V_G = 0$ V) for first-generation 1200 V SiC MOSFETs for the indicated temperatures, shown for (a) metal-packaged parts and (b) plastic-packaged parts.

Since V_T can be reduced at elevated temperatures, the experiments were repeated using gate biases of -2 V and -5 V. The results at 250°C, the highest temperature used in these experiments, are plotted for a metal-packaged part in Fig. 7(a) and for a plastic-packaged part in Fig. 7(b). A negative gate bias significantly reduced the leakage in the metal-packaged part, with a gate bias of -2 V resulting in less than 10 nA for a drain bias up to 900 V, compared to 5 μA with a gate bias of 0 V. However, there appeared to be virtually no effect of negative gate bias on leakage current for the plastic-packaged part. This result was obtained for multiple plastic-packaged parts. This suggests that the dominating leakage mechanism at these temperatures (which are above the rated temperature for this part) is related to the plastic packaging.

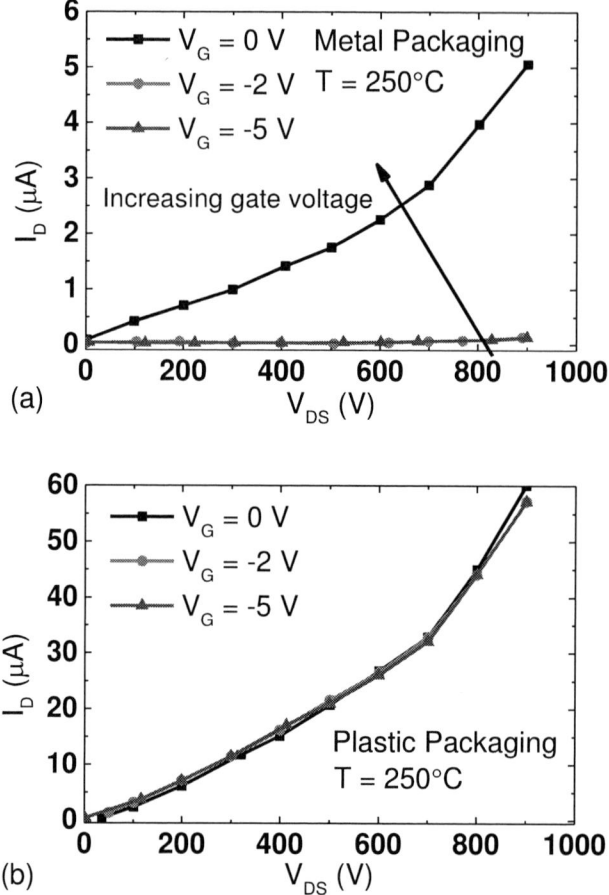

Figure 7. Drain current vs. drain voltage in the forward-blocking state for first-generation 1200 V SiC MOSFETs at 250°C and V_G = 0 V, -2 V, and -5 V for (a) metal-packaged parts and (b) plastic-packaged parts.

Discussion and Conclusions

SiC power MOSFETs are expected to endure high voltages and currents at elevated temperatures. The devices tested in this paper were assessed using two key reliability metrics, V_T shifts and junction leakage. V_T issues are likely a result of the quality of the oxide, with significant concentrations of interface traps contributing to larger V_T shifts with temperature, as well as additional V_T shifts with gate bias and stress due to charge trapping (more of an issue in this material system due to reduced band offsets compared to SiO_2/Si). However, packaging issues can also limit reliability (11), demonstrated in this experiment by the plastic-packaged parts having larger V_T shifts than metal-packaged parts after elevated temperature bias stresses and increased leakage current at elevated temperatures. Additionally, the plastic-packaged parts appear to have an extrinsic leakage path, since using negative gate bias does not reduce the leakage current. In contrast, metal-packaged devices show almost three orders of magnitude lower leakage current when using a gate bias of -2 V instead of 0 V. Constant bias temperature stress shows little effect of using a gate bias of -5 V up to at least 250°C (Fig. 3). The result for the metal-packaged part suggests that when not limited by packaging, applying small negative gate biases can reduce leakage current without a major impact on V_T.

However, initial switching gate bias experiments indicate that using a gate bias of -5 V can cause significant negative shifts in V_T. The V_T shift appears to be composed of a rapid change and slight recovery, followed by gradual degradation. The rate of degradation seen after the initial drop and recovery appears to decrease with increasing duty cycle, which corresponds to less time with a negative bias applied to the gate. Yet, even at 90% duty cycle the V_T shift is negative, indicating that hole injection is a greater concern than electron injection. This is especially surprising considering that second-generation parts did not show a significantly larger shift for -20 V compared to 20 V until 175°C. Some self-heating may occur (similar to (12)) during the switching stress that raises the temperature of the junction above the applied temperature of 150°C. This temperature increase may be significant, since even at an applied temperature of 175°C, when comparing the shift using 20 V on the gate with the shift using -5 V on the gate, they are nearly the same. Self-heating may also explain why the shift after half an hour of switching stress results in a shift of ~150 mV compared to a shift of ~50 mV for a constant bias stress of -5 V at 150°C.

The rapid initial change in V_T is different for the two duty cycles, which may be due to the different stress conditions. However, the initial change at the start of the second stress is similar to the final V_T shift at the end of the first stress, suggesting the possibility of cumulative damage between stresses, despite the recovery of the gate sweep curve. If the damage caused by worst-case conditions cannot be easily recovered, damage would persist for the lifetime of the device, which is particularly concerning for parts with parameters that vary over the course of their lifetime. Cumulative damage may occur if there are additional defects created during the first stress that remain in the oxide. Another possibility is that the holes trapped during the first stress are not actually annealed during the recovery. Electrons may be trapped at defects and form a dipole without recombining, so when a negative bias is applied again the electrons would be emitted and the resulting V_T shift would be similar to previous values. This is similar to the situation described previously for E' centers in SiO_2 on Si (13).

A variety of other factors may also affect the V_T shift observed for these devices. Using a higher frequency should increase the temperature of the junction further, likely increasing the V_T shift. Using a gate bias of -2 V may reduce the V_T shift while still helping to suppress leakage current. The effects these factors have on the rapid V_T shift and the slope of the gradual V_T shift, as well as how they interact with possible cumulative damage effects, are topics for further study. While reliability experiments are ongoing, the present results demonstrate that first-generation plastic-packaged parts must be stressed above the rated operating temperature before significant degradation is observed, and there is a significant reduction in negative V_T shifts due to negative gate bias on plastic-packaged second-generation parts. The improved V_T stability under negative gate bias allows more flexibility in using negative gate biases to reduce junction leakage at elevated temperatures and demonstrates the improving quality of gate oxides for SiC power MOSFETs.

Acknowledgments

The authors thank Dr. Imre Gyuk of the United States Department of Energy's Office of Electricity for his support of this work under the Energy Storage Program. Sandia National Laboratories is a multi-program laboratory managed and operated by Sandia Corporation, a wholly-owned subsidiary of Lockheed Martin Corporation, for the U.S. Department of Energy's National Nuclear Security Administration under contract DE-AC0494AL85000.

References

1. P. Friedrichs, Intl. Pwr. Electron. Conf. 3241, (2010).
2. M.J. Marinella et al., ECS Trans. **41**, 19 (2011).
3. A.J. Lelis, D. Habersat, R. Green, A. Ogunniyi, M. Gurfinkel, J. Suehle, and N. Goldsman, IEEE Trans. Elect. Dev. **55**, 1835 (2008).
4. A. J. Lelis, D. Habersat, R. Green, Mater Sci. Forum **679**, 599 (2011).
5. K. Matocha et al, IEEE Trans. Elect. Dev. **55**, 1830 (2008).
6. M.J. Marinella, D.K. Schroder, T. Isaacs-Smith, A.C. Ahyi, J.R. Williams, G.Y. Chung, J. W. Wan, and M. J. Loboda, Appl. Phys. Lett. **90**, 253508 (2007).
7. R. Schomer, P. Friedrichs, and D. Peters, IEEE Trans. Elec. Dev. **46**, 533 (1999).
8. P. M. Lenahan and P. V. Dressendorfer, Journal Appl. Phys. **55**, 3495 (1984).
9. C. J. Cochrane, P. M. Lenahan, and A. J. Lelis, Journal Appl. Phys. **109**, 014506:1-12 (2011).
10. M. K. Das, S. Haney, J. Richmond, A. Olmedo, J. Zhang, and Z. Ring, Materials Science Forum **717-720**, 1073 (2012).
11. T. Funaki, J. C. Balda, J. Junghans, A. S. Kashyap, H. A. Mantooth, F. Barlow, T. Kimoto, and T. Hikihara, "Power Conversion with SiC Devices at Extremely High Ambient Temperatures," IEEE Transactions on Power Electronics, Vol. 22, No. 4, pp. 1321-1329, 2007.
12. S. DasGupta, R. J. Kaplar, M. J. Marinella, M. A. Smith, and S. Atcitty, Proceedings of the 50th International Reliability Physics Symposium, 3D.3:1-5 (2012).
13. A. J. Lelis, T. R. Oldham, H. E. Boesch, and F. B. McLean, "The Nature of the Trapped Hole Annealing Process," IEEE Transactions on Nuclear Science, Vol. 36, No. 6, pp. 1808-1815, 1989.

Reliability of GaN HEMTs: Electrical and Radiation-Induced Failure Mechanisms

T. J. Anderson[a], A.D. Koehler[a], M. J. Tadjer[a], K. D. Hobart[a], P. Specht[b], M. Porter[c], T.R. Weatherford[c], B. Weaver[a], J.K. Hite[a], and F. J. Kub[a]

[a] Naval Research Laboratory, Washington, DC 20375 USA
[b] University of California, Berkeley, Berkeley CA 94720
[c] Naval Postgraduate School, Monterey, CA 93943

> The reliability of AlGaN/GaN HEMTs is currently a limiting factor in the development of next-generation power amplifier technology. In this work, we use atomic-resolution transmission electron microscopy (TEM) to directly image the defects associated with device failure. Furthermore, we attempt to induce defects through electrical and radiation-induced stress on the device, and compare the mechanism for failure by TEM analysis.

Introduction

The GaN-based materials system is known to be advantageous for fabrication of RF power amplifiers, high power amplifiers, and high breakdown voltage diodes. A fundamental understanding of the reliability and failure mechanisms in these devices is critical to further technology development and commercialization. Device degradation has typically been attributed to the inverse piezoelectric effect and the hot electron effect however the specific point defects and activation energy for these processes have yet to be determined [1,2]. Furthermore, the chemical inertness and strong bonding of the GaN crystal is intrinsically robust to harsh environments [3]. Given that the non-ionizing radiation hardness of a semiconductor material is inversely proportional to the lattice constant, GaN is predicted to have an activation energy for atomic displacement that is much higher than GaAs or Si. This makes GaN-based devices better candidates in for both displacement and ionizing radiation environments as well. In this work, we investigate the reliability of AlGaN/GaN HEMTs by employing atomic-resolution TEM to identify defects associated with radiation or electric field-induced device degradation, and understand the fundamental electrical stress and radiation-induced failure mechanisms.

Experimental

The HEMT devices were fabricated on Si substrates using standard processing steps, with a Ni/Au or Pt/Au gate metal and SiN$_X$ passivation. The devices were then cut into individual die and characterized by Hall, I-V, and C-V techniques to obtain the key metrics used to quantify device degradation: Hall mobility and 2DEG density, saturation current (I_{MAX}), off-state leakage (I_{MIN}), on-resistance (R_{ON}), gate leakage current (I_G), and threshold voltage (V_T). The devices were then subjected to either electrical stress or 2 MeV H$^+$ irradiation. After stressing, electroluminescence imaging was used to identify defective regions suitable for TEM analysis.

Electrical stressing was performed in a vacuum test station designed for high voltage electroluminescence (EL) measurements. The system employs a gate and drain SMU, both of which can be programmed to either hold a constant bias, perform a voltage step stress, or a step-recovery stress. A thermoelectric (TE) cooled CCD camera is mounted on the microscope and used to capture EL images as the voltage stress sequence is executed. A single bright spot was observed to appear at the gate edge relatively early in the stress program. While the intensity increased and more spots appeared during the stress, the location of the initial spot was the failure point when catastrophic breakdown occurred. Therefore EL imaging can be used to identify the defective regions of the material for TEM analysis. Proton irradiation was performed at Auburn University using a repeated test/irradiation cycle until the devices failed. EL was only performed on a virgin device and a failed device.

Results

Off-state drain voltage stress was applied in incremental voltage steps, allowing for simultaneous acquisition of the EL emission during measurement. A series of images showing the evolution of the EL emission is shown in Figure 1, with the accompanying I-V curves in Figure 2. The EL emission during off-state stress often initiates at one or few discrete locations between the gate and drain, near the gate edge, where the electric field is the highest. These spots increase in intensity as the drain voltage is increased. It is observed that the EL intensity is proportional to the current density. Also, under large off-state bias, EL emission is observed at the edges as well as off of the active device mesa, likely occurring at defects caused by plasma damage during the mesa etch processing step. The drain current increases as the drain voltage is increased. However, decreasing current transients are observed when the drain voltage is held at a specified drain voltage. This is likely due to charge trapping in the region between the gate and drain, acting as a virtual gate, reducing the current [4]. Eventually, irreversible breakdown occurs in the region between the gate and drain. Spatially, this breakdown is always correlated to the location of the initial EL spot. Reverse bias gate stress testing indicated the critical voltage was approximately 120V for both the Ni/Au and Pt/Au-gated samples.

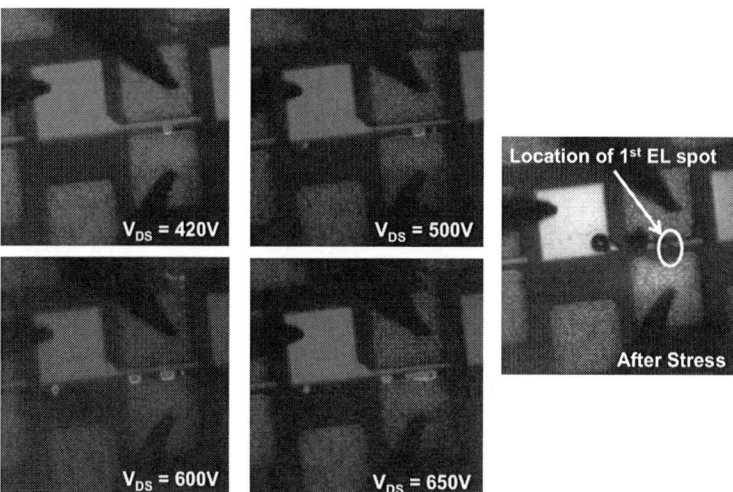

Figure 1. Sequence of EL images at increasing drain bias

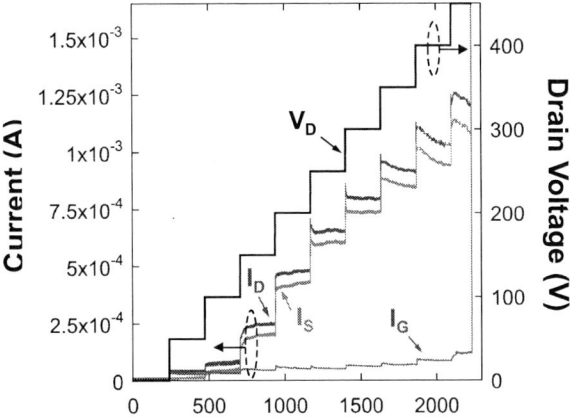

Figure 2. Current-voltage characteristics of the HEMT during the off-state step-stress program

To compare the impact of radiation to electrical stress, devices were irradiated at sufficiently high dose to induce failure. Initially, gamma ray irradiations were performed using a ^{60}Co source at doses up to 2 Mrad, however a less than 5% change in I_{DS} was observed, and the effects were comparable to current collapse, thus it was not possible to draw any conclusions from this study. In an ongoing experiment, devices have been irradiated a H^+ source at doses from $1x10^{12}$ cm^{-2} to $2x10^{14}$ cm^{-2}. At the final dose, the relevant HEMT parameters (I_{DS}, R_{ON}, μ, R_{SH}) had degraded approximately 20%. This is attributed to displacement damage at the AlGaN/GaN interface, causing increased scattering in the 2DEG. Given the destructive nature of EL and TEM measurements, the devices will only be imaged using these techniques after the final radiation dose, however recent work has suggested that irradiated devices will demonstrate an increased breakdown voltage [5].

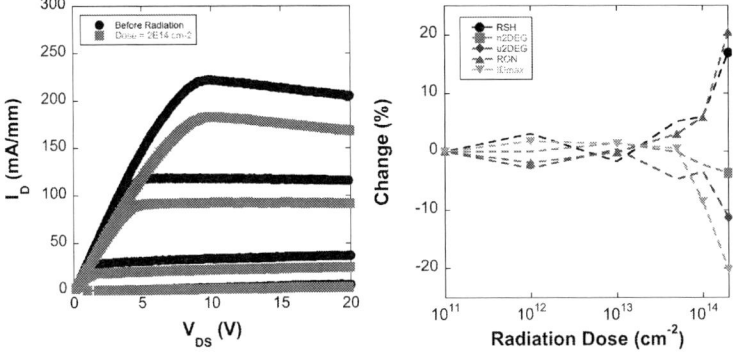

Figure 3. DC I-V curves before and after irradiation (left), and percent change in relevant HEMT parameters as a function of dose (right)

After either electrical stressing or irradiation, samples were retested, and devices were identified for TEM analysis. At present only virgin devices have been imaged, however several features are already visible. Notably, there is a feature visible at the gate edge that is indicative of metal movement during the passivation process. A second feature is a step on the surface, which appears to be associated with an incomplete layer spatially restricted by grain boundaries. Future analysis will be used to identify the nature of these defects, their chemical environment, and local strain. Future imaging of the stressed and irradiated devices will indicate whether these defects contribute to device failure.

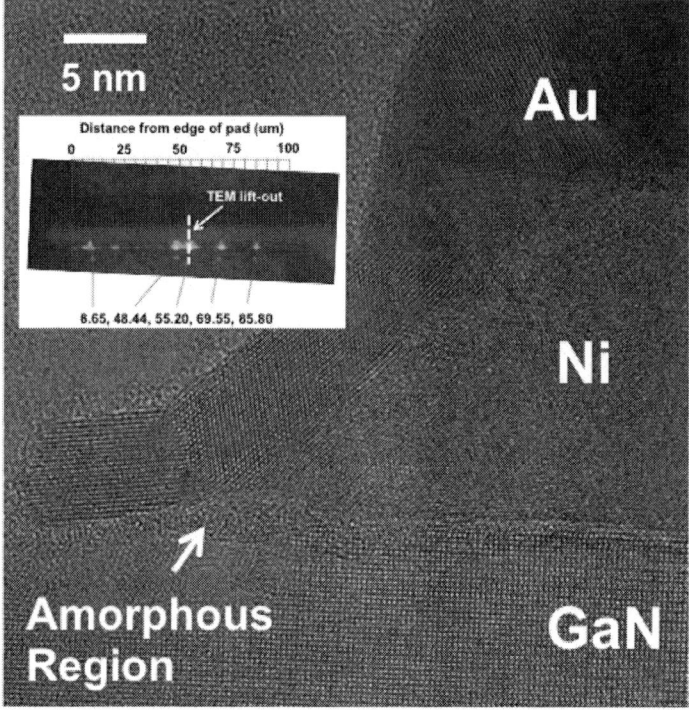

Figure 4. TEM image of process-induced defect at the gate edge of a HEMT. False-color EL image to identify defective region is shown as inset

Conclusion

Electroluminescence imaging has been used to quantify breakdown in GaN HEMTs due to electrical stress. It is possible to predict the failure point based on the position of the initial EL spot. Proton irradiation has also been used to degrade HEMTs by producing displacement damage around the 2DEG, and EL experiments will be performed to identify the failure point in these devices. TEM analysis has identified a number of process-induced defects that could be related to device failure. Experiments are in-progress to correlate the defects imaged in TEM to the failure points imaged in EL, and to further correlate the radiation-induced failure to the electrical stress-induced failure.

Acknowledgments

M. J. Tadjer thanks the American Society for Engineering Education – Naval Research Laboratory Postdoctoral Fellowship Program. The authors thank the NRL Institute for Nanoscience for equipment use and support.

References

1. J.A. del Alamo, J. Joh. *Microelectron. Reliab.* 49, 1200-1206 (2009)
2. G. Meneghesso, G. Verzellesi, F. Danesin, F. Rampazzo, F. Zanon, A. Tazzoli, M. Meneghini, E. Zanoni. *IEEE Trans. Device Mater. Reliab.* 8, 332-343 (2008)
3. A. Ionascut-Nedelcesu, C. Carlone, A. Houdayer, H.J. von Bardeleben, J.L. Cantin, S. Raymond. *IEEE Trans. Nucl. Sci.* 49, 2733-2738 (2002)
4. R. Vetury, N.Q. Zhang, S. Keller, U.K. Mishra. *IEEE Trans. Electron Devices* 48, 560-566 (2001)
5. C.F. Lo, L. Liu, F. Ren, H.Y. Kim, J. Kim, S.J. Pearton, O. Laboutin, Y. Cao, J.W. Johnson, I.I. Kravchenko. *J. Vac. Sci. Technol. B* 29, 061201 (2011)

226

CHAPTER 6

POWER SEMICONDUCTOR CURRICULUM

Systems-Driven Power Semiconductor Education

Krishna Shenai

Energy Systems Division, Argonne National Laboratory
9700 South Cass Avenue, Bldg. 362, Argonne, IL 60439-4815

Abstract

Today's power electronics curriculum practiced at most universities world-wide primarily focuses on power converters and power systems with little or no emphasis given to power electronics components and their interaction at the systems level. Consequently, graduates are trained to design power converters and power systems without adequate background on the physics and technology of power electronics components; graduate are often unable to understand and optimize component and systems technologies. The result is a widening disconnect that exists between the users and manufacturers of power electronics components. This paper discusses the challenges and approaches needed to bridge this educational gap and proposes a "systems-driven" power semiconductor educational curriculum in order to rapidly promote the power electronics industry.

I. Introduction

A typical power converter circuit consists of several nonlinear, reactive charge-storing circuit elements such as the semiconductor power switch, inductors, transformers, and capacitors; and, in most cases, the load also presents variable nonlinear reactive circuit impedance [1, 2]. Figure 1 illustrates the current approach used in the hierarchical design of a power converter where T_a and T_{jmax} are the ambient and maximum junction temperatures of the power chip, respectively; MTBF refers to the Mean-Time-Between-Failure of the power converter in the end-user application. At a given power switching frequency, all circuit power losses are minimized in order to reduce heating and increase the power conversion efficiency. Generally, power converter design including thermal management is performed as follows [3]:

- for a given set of system specifications, candidate power converter topologies are identified that appear to meet the system requirements;

- initial tradeoffs in converter size, weight, performance, cost, etc. are performed in order to select an appropriate converter topology to be used for the design;

- voltage and current ratings for the switching elements (semiconductor switches, inductors, capacitors, resistors, transformers, etc.) are determined, based on converter requirements and selected topology;

- from commercially available components, those components that meet the design criteria identified are selected;

- using circuit simulations, approximate power losses (conduction and switching) in power semiconductor devices are calculated for a chosen control strategy; circuit simulations typically employ first-order compact circuit models for all components including semiconductor switches and magnetic devices;

- the steady-state semiconductor chip junction temperature is estimated based on the module assembly stack-up of thermal impedances (as shown in Figures 8(a) and 8(b)) to the source of cooling, using the worst-case coolant temperature;

- if the estimated chip temperature is close to the allowable margin, then calculations are iterated considering thermal spreading in the stack-up;

- if the allowable chip temperature is exceeded, calculations are iterated considering the use of additional switching devices in parallel, or advanced cooling techniques (e.g., reduce the thermal impedances in the stack-up, lower temperature coolant, better heat sink, etc.);

- then transient circuit effects on the thermal management system are considered – for example, if it is a pulsed application vs. continuous power conduction, thermal management may require additional margin; also, peak temperature of the chips is often used as the junction temperature, and the entire chips is assumed to be isothermal; and,

- for the final power system configuration, circuit simulations are performed to determine the expected losses and estimate the final semiconductor chip junction temperature, and hence, final design of the thermal management and cooling system.

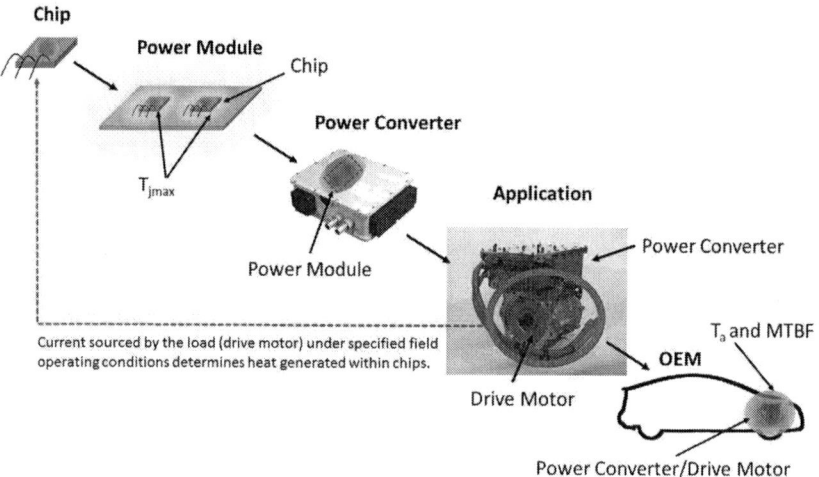

Figure 1: Schematic representation of hierarchical approach used for power converter design.

The approach described above is what is typically used in industry today, although some circuit designers take into account circuit parasitic elements, including those arising from packaging and interconnects. Such converter design approaches require the use of sophisticated computer-aided design (CAD) tools [4]. It is important to note that in the converter design process semiconductor power switch is treated to be an ideal power switch or at best represented using a compact circuit simulation model. In any case, the entire semiconductor power chip is considered isothermal. This methodology, although simplistic in nature, does not take into account non-linear switching charge dynamics within the semiconductor device, and hence, may not reveal the deleterious effects material defects may have on circuit switching. Additionally, circuit models used for power semiconductor devices do not contain reliability physics, especially physics related to long-term field-degradation mechanisms under circuit application stresses, and hence, are not suitable in predicting the field-reliability and lifetime of power converters [5-7].

II. Proposed Power Electronics Curriculum

Today, silicon is the industry work-horse for building power electronics switching devices. Although silicon has adequately addressed the needs of the power electronics industry until now, it may not be able to do so in the future. Next-generation power electronics systems will require the ability to efficiently handle very high power and operate at much higher temperatures. With silicon technology reaching its material limits, new material technologies, such as Wide Band Gap (WBG) semiconductors, including Silicon Carbide (SiC) and Gallium Nitride (GaN), become inevitable [8, 9]. A successful adoption of WBG technologies in power electronics systems will require new circuit topologies, new system design approaches, and efficient thermal management techniques. Reliability of power electronics systems must be addressed at the application-level from materials-to-converters including interactions with energy sources and non-linear loads as shown in Figure 2 [10]. This feature becomes particularly important when dealing with WBG power semiconductor devices because their performance and reliability in end applications are strongly dependent on the quality of the starting material which today consists of a high density of crystal defects [11].

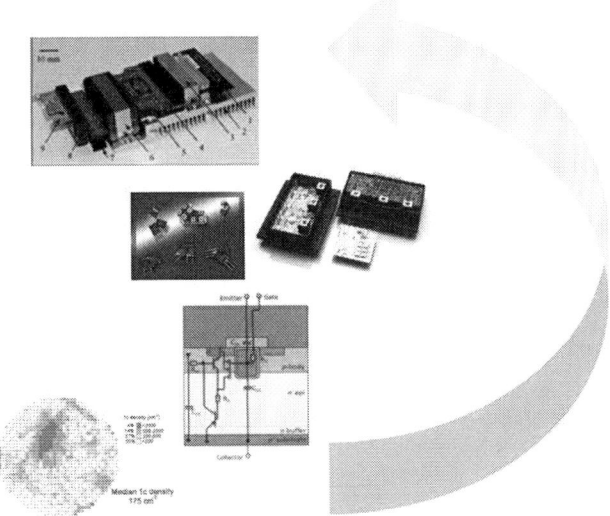

Figure 2: A pictorial representation of the interaction of material, device, module and converter parameters.

Material technology optimization must be pursued in concert with reducing the system cost. The current limited market for WBG power devices is largely due to the fact that commercial SiC and GaN power devices are prohibitively expensive, and are yet to be proven reliable in the field. In order to "unlock" the enormous potential of WBG power electronics, end-user Original Equipment Manufacturer (OEM) confidence on the field-reliability of WBG power devices in actual power converters is needed [11]. A fundamental understanding needs to be developed that links the basic material properties of commercially available WBG semiconductors to the reliability characteristics of power devices in power electronics converters when subjected to repetitive power switching conditions in the field. This problem, however is complicated by the fact that the internal charge dynamics within a typical power semiconductor switching device is highly nonlinear in nature, and is a strong function of instantaneous terminal electrical and thermal switching stresses [10]. As the power density of a power converter is increased, thermal issues play critical role on the reliability of the product. According to a study of the US Air Force Avionics Integrity Program (AVIP) [12], as illustrated in Figure 3(a), excessive temperatures of critical components, such as semiconductors and transformers, cause most equipment failures. As the junction temperature is increased, component failure rate increases rapidly as shown in Figure 3(b).

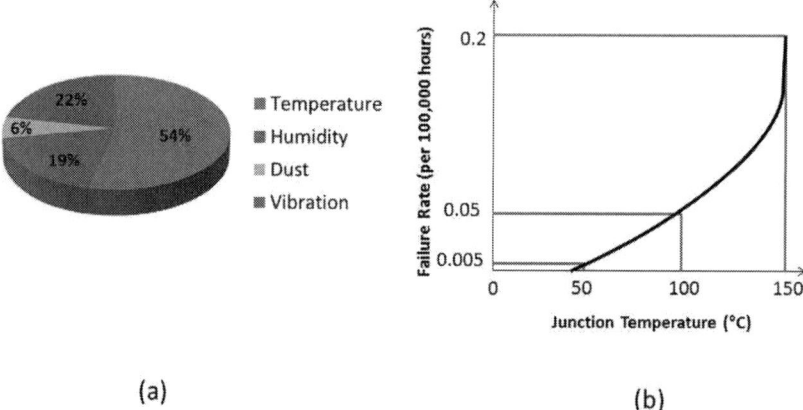

(a)   (b)

Figure 3: Importance of thermal management in electronics equipment: (a) breakdown of premature failure causes, and (b) equipment failure rate as a function of junction temperature (Source: US Air Force Avionics Integrity Program (AVIP) [12]).

An important measure of power electronics switching is the safe operating area (SOA) of a power semiconductor switch [13]. The SOA of a power semiconductor device refers to voltage and current limits within which the device can be safely switched. The SOA deviates from the rectangular shape due to thermal heating as shown in Figure 4; the loss of SOA occurs when power dissipation is at its maximum and is a function of switching frequency and duty ratio. Provided bond wires and die-to-package interface remain intact, and assuming isothermal die boundary conditions, it is well-known that "hot spots" occur locally during power switching and can lead to current filamentation and local "burn outs" caused by Joule heating [14, 15]. Maximum power dissipation occurs within the drift-region of the device as the electric field is high. The challenge is then to rapidly remove heat away from the drift-region; the thermal time constant must be small (typically less than a microsecond), and hence, the substrate must be thinned down so that the cooling surface is brought close to the drift-region of the device. As shown in Figure 4, the SOA of GaN power switching device is expected to be smaller than that of a SiC power device with similar rating primarily due to its lower thermal conductivity.

The skills needed to build next-generation of power semiconductor devices and integrate them into advanced power electronics systems require proper education and training of engineering students both at the undergraduate and graduate levels. Therefore, a curriculum that includes courses, modules, and laboratory sessions becomes an essential tool to acquire the needed skills and training. The curriculum must also integrate industry-relevant analysis and design techniques and computer-aided design (CAD) tools [16, 17]. A successful implementation of such advanced curricula is expected to position the US industry at a much higher competitive edge above the rest of the world.

There are very few educational programs in the US that offer courses on power electronics components and also on power electronic systems. The widely recognized energy conversion systems curriculum developed and offered through the Consortium of Universities for Sustainable Power (CUSP)™ by the University of Minnesota is primarily focused at the systems level with very little coverage on power electronics components and technologies [18].

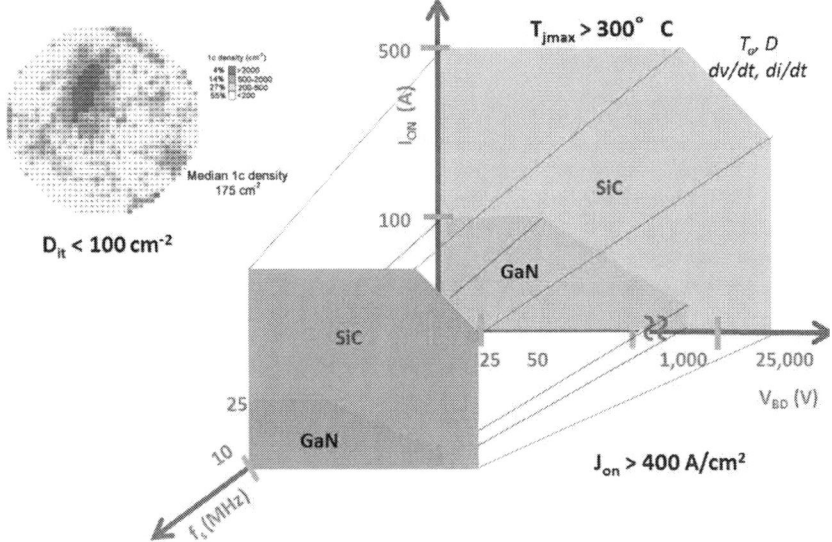

Figure 4: Typical defect density in the current state-of-the-art 4H-SiC wafers, and projected safe-operating-area (SOA) limits for GaN and SiC power devices.

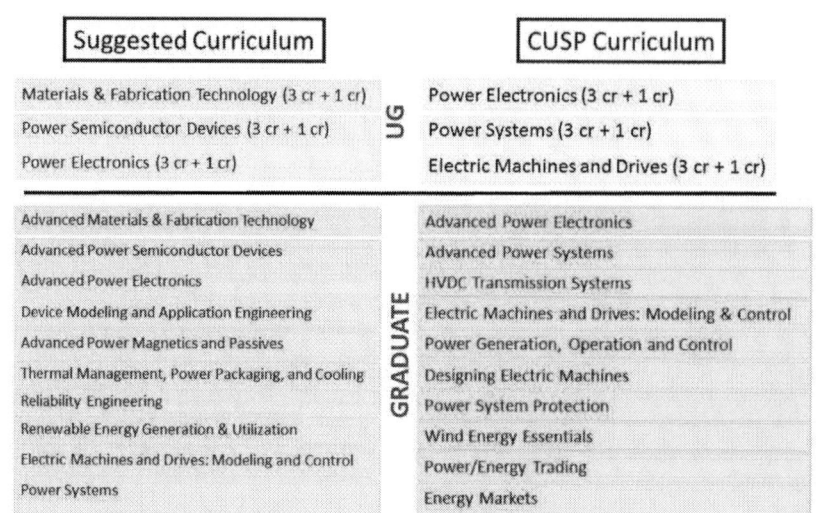

Figure 5: Proposed power-semiconductor driven power electronics curriculum in comparison with "systems-driven" CUSP curriculum.

To address science and technology related issues, not only component-level physics and fabrication should be covered, but also thermal management, temperature effects, reliability, power packaging, modeling, and systems integration should be addressed. Figure 5 illustrates the proposed power-semiconductor driven power electronics curriculum in comparison with "systems-driven" CUSP curriculum. A balance of focus between components and systems is needed in order to develop and promote power electronics components and power electronics systems industries. No single academic institution typically has the resources or the expertise to develop all elements of such an advanced power electronics curriculum. Only a collaborative effort among educational institutions in this technical field can achieve this goal. To foster such collaboration, meeting at a premiere conference such as this one is needed in order to bring together leading experts from educational institutions, industry and research laboratories to vigorously exchange ideas and propose a plan of action. Discussions on future course of actions, curriculum structures, course contents, pre-requisites, ABET requirements, laboratory components, course delivery methods, interdisciplinary topics, and system integration approach will be needed. These deliberations are expected to generate discussions on how to proceed to ensure the adoption of a strongly integrated curriculum in power electronics components, technologies, and systems.

References

1. Ned Mohan, Tore M. Undeland, and William P. Robbins, Power Electronics: Converters, Applications and Design, Wiley: New York, 2003.

2. Robert W. Erickson, Fundamentals of Power Electronics, Chapman & Hall: New York, 1997.

3. H. A. Schafft, D. L. Erhart, and W. K. Gladden, "Toward a building-in reliability approach," Microelectronic Reliability, vol. 37, no. 1, pp. 3–18, 1997.
4. U. Drofenik, D. Cottet, A. Müsing, and J. W. Kolar, "Design tools for power electronics: trends and innovations," Proceedings of the 2nd International Conference on Automotive Power Electronics (APE '07), Paris, France, September 26-27, 2008.
5. K. Shenai, "A circuit simulation model for high-frequency power MOSFETs," IEEE Trans. Power Electronics, vol. 6, no. 3, pp. 539-547, July 1991.
6. S. Pendharkar, C. Winterhalter, and K. Shenai, "A Behavioral Circuit Simulation Model for High-Power GaAs Schottky Diodes," IEEE Trans. Electron Devices (TED), vol. 42, no. 10, pp. 1847-1854,October 1995.
7. K. Shah and K. Shenai, "A Simple and Accurate Circuit Model for GaN Power Transistors," IEEE Trans. Electron Devices (TED), vol. 59, n0. 10, pp. 2735-2741, Oct. 2012.
8. K. Shenai, R. S. Scott, and B. J. Baliga, "Optimum semiconductors for high-power electronics," IEEE Trans. Electron Devices, vol. 36, no. 9, pp. 1811-1823, September 1989.
9. K. Shenai, "Switching megaWatts with power transistors," The Electrochemical Society Interface Magazine, vol. 22, no. 1, pp. 47-54, Spring 2013 (**invited paper**).
10. K. Shenai, "Made-to-order Power Electronics," IEEE Spectrum, vol. 37, No. 7, pp. 50-55, July 2000 (**invited paper**).
11. K. Shenai, M. Dudley, and R. F. Davis, "Current status and emerging trends in wide bandgap (WBG) semiconductor power switching devices," ECS Journal of Solid State Science and Technology 2(8), N3055-N3063, July 2013 (**invited paper**).
12. Avionics Integrity Program (AVIP), MIL-STD-1796A (USAF), October 13, 2011.
13. K. Shenai, "Power Semiconductor Manufacturer SOA Specifications Require Updating," PCIM, vol. 26, No. 12, pp. 26-32, December 2000 (**invited paper**).
14. Newman, R., W. C. Dash, R. N. Hall, and W. E. Burch, "Visible Light from a Si p-n Junction," Physical Review, vol. 98, pp. 1536–1537, 1955.
15. Chynoweth, A. G., and K. G. McKay, "Photon Emission from Avalanche Breakdown in Silicon," Physical Review, vol. 102, no. 2, pp. 369–376, April 1956.
16. E. McShane, M. Trivedi, and K. Shenai, "An Improved Approach to Application-Specific Power Electronics Education–Curriculum Development," IEEE Trans. Education, Volume 44, Issue 3, pp. 282-288, August 2001.
17. M. Trivedi, E. McShane, R. Vijayalakshmi, A. Mulay, S. Abedinpour, S. Atkinson, and K. Shenai, "An Improved Approach to Application-Specific Power Electronics Education—Part II: Switch Characterization and Modeling," IEEE Trans. Education, Vol. 45, Issue 1, pp. 57-64, Feb. 2002.
18. http://cusp.umn.edu/

236

ECS Transactions, 58 (4) 237-244 (2013)
10.1149/05804.0237ecst ©The Electrochemical Society

Power Semiconductor Device Education: Which Topics and What Depth?

William P. Robbins and Ned Mohan
Dept. of Electrical and Computer Engineering, University of Minnesota
Minneapolis, Minnesota 55455, USA

Power electronics is a significant enabling technology in many areas of electrical engineering and the introduction of wide bandgap power devices utilizing silicon carbide and gallium nitride will make it even more important and pervasive in the future. In developing the power electronics workforce going forward, two major questions arise concerning power semiconductor devices. First what device topics should be covered and second what should be the depth of the topical coverage? One example discussed in the paper are the power semiconductor device topics are taught at the undergraduate and graduate level at the University of Minnesota and to what depth. The topical details of a graduate level device physics and device design course at the graduate level are also considered.

1. Introduction

In the past few decades, power electronics applications have expanded to such an extent that power electronics is considered an enabling technology (some would argue even a foundational technology) in the broad field of electrical engineering. A major contributor has been the extraordinary development of power semiconductor devices. The development of new device concepts or geometries and the use of new semiconductor materials have resulted in significant improvements in the capabilities of power semiconductor devices.

Many universities and colleges now offer or are planning to offer one or more courses in power electronics, usually as technical electives. A major curricular concern in planning these courses is the level of discussion of the power semiconductor devices. The two basic questions are what power devices should be discussed and what is the depth or details of the coverage. The natural tendency is to minimize the discussion of device topics in order to cover more power electronics topics and applications. However skimping too much on the device discussion will leave students with too little knowledge of devices to effectively design power electronic converters.

2. Power Semiconductor Devices in the Undergraduate Curriculum

In most undergraduate EE curricula across the country, the discussion of power semiconductor devices is minimized. The devices that are covered include pn junction and Schottky diodes, MOSFETS, and IGBTs. SCRs and GTOs are usually discussed only if the power electronics courses consider power system applications. BJTs (bipolar junction transistors) are no longer being used in new designs and so are left out of most power device discussions. The level of the device discussion is limited to a black box description of the devices, i.e. circuit symbols, device ratings and terminal characteristics. Discussion of device physics is reserved to the graduate curriculum.

The major reason for the limited discussion of power devices is the crowded nature of the undergraduate curriculum. The first three undergraduate years are filled with required math and physics courses , liberal arts requirements, and core electrical engineering courses (circuits, electronics, logic design and microprocessors, signals and systems, fields and transmission lines, and semiconductor devices). The senior year is the only place where elective courses that go into more depth in a particular area like power electronics can be offered.

Within the senior year, it is unrealistic to have more than a couple of courses in a particular area such as electric energy systems. Most undergraduate EE programs have a breadth and depth requirement in their senior technical elective offerings. Thus students must take at least one course from several different areas (typically 4-6) for the breadth requirement and at least two courses in one area for the depth requirement. The University of Minnesota senior technical program shown below in Fig. 1 illustrates just how crowded the senior year can be. A top down organization of the power electronic courses in which basic circuit topologies and applications are discussed with the power devices being idealized (i.e. terminal characteristics only) as much as possible makes the most efficient use of the limited time available. Only after the students have taken such a course will they then get significant benefit from a detailed discussion of power device physics. Consider the fact that in most undergraduate programs, required core courses in electronics cover transistor circuits from a circuit viewpoint and only consider the terminal characteristics of the transistors. Only after taking the required electronics course do students take the required semiconductor device course.

Senior Design Project (all students) – 4 credits

Senior Electives – 26 credits required

Controls
- Linear Controls & lab
- Space State Controls & lab

Electric Energy Systems
- Power electronics & lab
- Electric drives & lab
- Power systems & software lab

RF/Wireless Systems
- Digital signal integrity
- Wireless systems design

Computer & Logic Systems
- Digital Design with FPGAs & lab
- Microprocesser System Design & lab
- Computer Architecture
- Intro to Predictive Learning

Communications & Signal Processing
- Communications systems & lab
- Digital signal processing

Electronics and Devices
- Advanced analog electronics design
- Energy conversion and storage & lab

Figure 1. University of Minnesota senior technical elective program.

It has been argued that a required undergraduate course in semiconductor devices could include discussion of power semiconductor devices. However this is unrealistic for several reasons. The typical topical coverage is basic semiconductor physics, signal level devices including pn junctions, Schottky diodes, BJTs, and MOSFETs., Covering these topics in a one semester course does not leave time to discuss power devices. Moreover the terminal characteristics of the devices covered in the course are already familiar to the students from the preceding electronics courses thus providing motivation to the students to learn the material. However until a student takes a power electronics course or a related course such as electric drives, the student has no exposure to power devices. Discussing power devices before the students see the applications will not elicit much interest or motivation from the student to learn about power devices. Finally a course required of all students should not include topics that are specialized to an elective topic that not all students will take.

If instructional laboratories are offered in power electronics related courses such as done in many undergraduate programs students will acquire significant practical experience with power semiconductor devices. For example, the power electronics instructional laboratory at the University of Minnesota has the experiments listed in Table 1. The experiments and hardware used in the laboratory give the students the needed exposure to power devices, especially diodes and MOSFETs without overcrowding the senior year with too many courses.

Table I. UMN power electronics laboratory experiments

Exp. #	Title of Experiment
1	Lab and Equipment Familiarization
2	Buck Converter
3	Switching Characteristics of MOSFETs and Diodes
4	Boost Converter
5	Buck-Boost Converter
6	Voltage-Mode Control
7	Peak Current Mode Control
8	Flyback Converter
9	Forward Converter

3. Power Semiconductor Devices in the Graduate Curriculum

Two different types of power device courses are needed at the graduate level. The first is a survey type course that discusses the devices from a qualitative viewpoint and does not attempt to cover topics in any significant quantitative detail.,. The second is a power device physics course which discusses power devices in quantitative detail. Each course is described in detail in the following sections.

3a. Power Semiconductor Device Survey Course

In order to design a well-functioning power electronics converter, an engineer needs to have a good understanding of the physical operation of power devices, their terminal characteristics and limitations. Otherwise he will not be able to optimize the inevitable tradeoffs that must be made between desired converter characteristics and power device capabilities. The idealized black box description of power devices from an undergraduate program is not sufficient. Conversely the ability to quantitatively design a power device, while useful, is more information than is needed for designing power electronic converters.

An example of the survey power device course is the device section of a first year graduate power electronics course at the University of Minnesota which utilizes the textbook by Mohan, Undeland, and Robbins. [1] The major topics are:

- Semiconductor Physics Review
- Power Diodes (including Schottky diodes)
- MOSFETS (including superjunction MOSFETs)
- IGBTs
- Thyristor and GTO Devices
- Wide Bandgap Materials and Devices

Device topics that are left out include normally-on device (JFETS), legacy devices such as BJTs, and other seldom used devices such as MCTS and triacs.

The semiconductor physics review is summary of material covered in an undergraduate semiconductor device course. It considers only silicon and utilizes classical physics to describe the properties of silicon and step pn junctions culminating in the standard exponential I-V characteristic based on the charge control model. No quantum mechanics or Fermi-Dirac statistics are utilized. This topic is the most quantitative of the major topics.

For each of the device topics listed, only silicon devices are discussed in a first pass in order to avoid the different properties of other materials like silicon carbide complicating the presentation, Once the basic device concepts are understood in the context of silicon, describing wide bandgap versions of the same devices is straightforward. For each different device type, the following topics are discussed:

- Device geometry
- Terminal characteristics
- Physics of device operation
- Breakdown voltage considerations
- On-state losses
- Switching characteristics
- Safe operating area

Figure 2 which shows cross-sectional views of a pn junction diode and a GTO illustrates the level of detail possible in a survey type presentation. Another example is Figure 3 which shows the origins of MOSFET on-state losses.

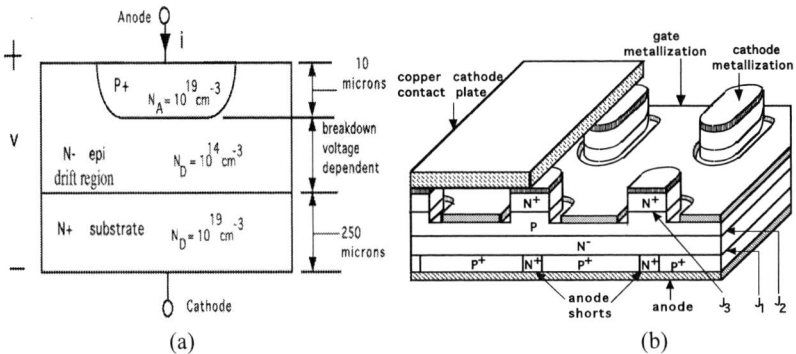

(a)   (b)

Figure 2. Examples of power device geometries. a) PN junction diode cross-section and b) GTO cross-section.

Using only the step junction quantitative relationships, several important device parameters can be estimated in a semi-quantitative manner. These include breakdown voltages, drift region lengths and doping levels. Approximate quantitative tradeoffs between on-state losses and switching speed in bipolar devices can be derived. Allowable current density versus breakdown voltage rating of both bipolar and unipolar devices can be shown. Such relationships show that the breakdown electric field strength is the most important semiconductor parameter in determining device performance and explains why wide bandgap devices are of such interest.

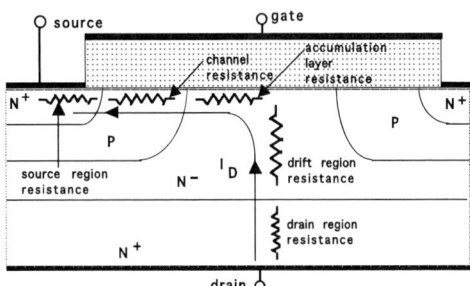

Figure 3. Contributions to MOSFET on-state losses.

Our experience with this type of device coverage is that it takes only about 8-9 weeks of a three lectures per week 15 week semester course. We combine the device discussion with other component considerations such as heat transfer fundamentals and heat sinks, design of high frequency transformers and inductors for power electronics, snubbers for hard-switched converters, and general drive circuit design considerations. With some prior familiarization with the device topics, a faculty member with power electronic circuit design interests but no background in devices other than an undergraduate device course can teach the device material effectively.

3b. Power Semiconductor Device Physics and Design Course(s)

The students who will design the next generation power devices will need detailed quantitative device physics and design training which are not provided by the survey course described previously. A different course organized around the same topics listed in the previous section but delving deeply into the quantitative details of each topic is needed. An example of the level of coverage needed is the recent textbook by Baliga. [2], Figure 4 illustrates the analytical modeling of the turn-on behavior of a MOSFET-based buck converter and is indicative of the level of detail desired in the device physics and design course. The level of detail should be sufficient that simulation models of the devices can be discussed in the course that would be used in circuit simulation programs to accurately simulate the nonlinear time-dependent behavior of the converter circuits.

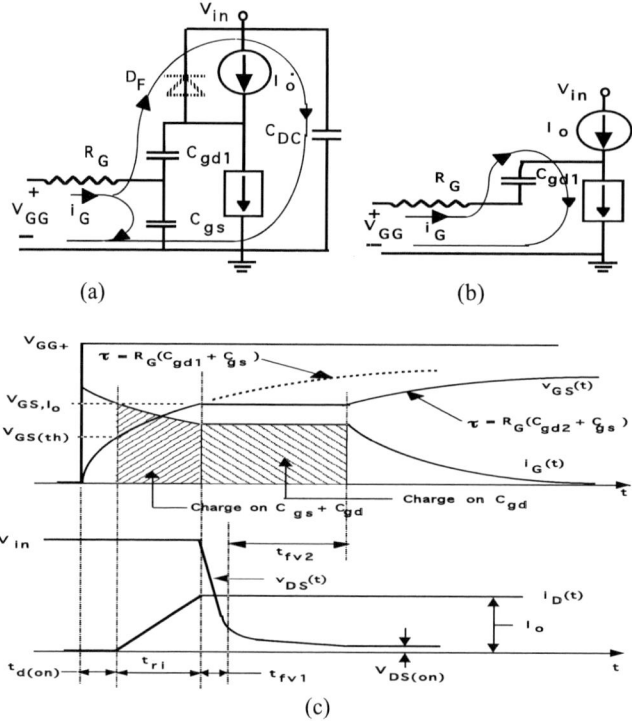

(a)   (b)

(c)

Figure 4. Analytical modeling of the turn-on waveforms of a MOSFET-based buck converter with (a) the equivalent during the current rise time interval, (b) the equivalent circuit during the initial voltage fall time, and (c) the complete turn-on waveforms.

Silicon-based devices should be discussed first for the reasons cited in the previous section followed by a discussion of same devices in wide bandgap materials. To avoid overloading the course with too many topics, device concepts that are not widely used such as normally on devices (JFETs) and legacy devices such as BJTs should not be covered unless there is a special need for that topic or time available at the end of the course.

Other ancillary graduate courses need to be taken in addition to the device physics course in order to give the student a comprehensive background in power device design. Such courses might include:

- Instructional laboratory experimentally characterizing power devices
- Semiconductor fabrication course with laboratory
- Solid state physics
- Theory of semiconductor materials
- Heat transfer and heat sink design

Offering a quantitative power device physics course requires considerable resources in terms of facilities, faculty numbers, and availability of ancillary supporting courses discussed in the previous paragraph. In particular, there should be one or more faculty actively engaged in power device research available to teach the course. Only a few schools will have these resources compared to the much larger number who can offer graduate level power electronic courses that concentrate on converter design.

This should not be considered a serious problem for two reasons. First the need for power device designers is much less than the need for circuit/converter designers. Second the technology available today for remote video conferencing via the internet makes it possible for institutions to share their course offerings. If institution A needs to have a device physics course available to their students, but they do not have the capability to offer it, it is possible to make arrangements with another institution that does have the needed course so that the students at institution A can participate in the course over the internet.

Other types of interactions are possible. A group of universities with complementary course offerings in the device physics area could form a consortium and work together to develop the full range of needed courses without each participant having to offer the full slate of courses on their own. Each consortium member would make their unique courses available to the rest of the consortium members. This consortium approach is being used to respond to the workforce development part of a Dept. of Energy FOA. The goal is to design a broad-based curriculum in power electronics with specific emphasis on the device topics and depth of coverage to meet the needs of all personnel involved in the wide bandgap power device manufacturing supply chain ranging from material growth to OEM manufacturers utilizing the wide bandgap devices. Another approach is exemplified in the CUSP (Consortium of Universities for Sustainable Power) initiative. [3] In this approach courses are developed as a series of online video lectures and all other needed course materials such as online homework problems, quizzes, exams, etc. The course materials are freely available online to CUSP members (membership is simple and free) to use as they see fit.

4. Conclusions

The questions of which power semiconductor device topics should be discussed and to what depth or detail in a power electronics curriculum depends on whether the students are undergraduates or graduate students. At the undergraduate level, diodes (pn junction and Schottky) MOSFETS, IGBTs, and possibly SCRs and related devices should be discussed only in terms terminal characteristics i.e. a black box presentation. In a graduate program two different types of coverage are needed. For graduates who intend to be power electronic circuit/system designers, a survey type device course is needed that discusses semiconductor physics, diodes, MOSFETS, IGBTs, thyristor-based devices, and wide bandgap devices in a qualitative manner. Graduate students interested in developing new power devices need a power device course or courses that cover the devices in a quantitative manner including device models appropriate for circuit simulation programs.

References

1. Power Electronics: Converters, Applications, and Design, 3rd Edition, Ned Mohan, Tore Undeland, and William P. Robbins, John Wiley and Sons, ISBN: 0471226939, (2002)

2. Fundamentals of Power Semiconductor Devices, B . Jayant Baliga, Springer US, ISBN-13: 978038743130, (2008)

3. www.cusp.umn.edu

ECS Transactions, 58 (4) 245-252 (2013)
10.1149/05804.0245ecst ©The Electrochemical Society

Power Semiconductor Devices, Course Contents Revisited

I. M. Abdel-Motaleb

Department of Electrical Engineering, Northern Illinois University, DeKalb, IL 60115, USA

The skills needed to build the next-generation of power semiconductor devices and to integrate them into power electronic systems require proper education and training of engineering students, both at the undergraduate and graduate levels. To assess the current state of power electronics education, a study of curricular offering in power electronics in the U.S. universities is conducted. The study shows that there is a lack of course offering in semiconductor power devices and almost nonexistence of critical topics in the curriculum, especially thermal management, temperature effects, reliability, power packaging, modeling, systems integration, and failure analysis. To remedy these deficiencies, new power electronics tracks/concentrations for undergraduate and graduate programs are proposed. The proposed curricular modifications are designed to be rigorous to meet the future needs of the power electronic industry, but they remained flexible enough to be easily adopted by many institutions.

Introduction

The skills needed to build next-generations of power semiconductor devices and integrate them into power electronic systems require proper education and training of engineering students both at the undergraduate and graduate levels. Therefore, a curriculum that includes courses, modules, and laboratory sessions becomes an essential tool to acquire the needed skills and training. The curriculum must also integrate industry-relevant analysis and design techniques and computer-aided design (CAD) tools. A successful implementation of such advanced curricula is expected to position the U.S. industry at a higher competitive edge above the rest of the world.

Proposing viable curricular activities, to meet the needs of industry, requires studying existing curriculum activities in the U.S. and assessing their strengths and weaknesses. Based on the results of this study, one can propose modifications, change, or addition to both undergraduate and graduate programs.

Implementing curricular modification to the existing ones faces some constraints. First, the modification must be within the allowed number of credit hours for the program. Increasing the number of credit hours to accommodate these changes would face fierce resistance from the institutions, the accrediting agencies, the students, and the parents. Second, the proposed changes must have enough flexibility to be adapted by the different programs. Proposing rigid models would not be successful, since the curricular structures of the engineering programs are not identical. A close look at the engineering programs at different U.S. institutions would reveal that some programs follow a liberal arts model, where students have more flexibility in choosing the courses. Other programs adopt the model of tracks starting from the freshman or the sophomore years. Third, programs can be fundamentally different, where some are straight electrical engineering programs, while others can be a combination of two or more subprograms, such as electrical, electronics, and computer engineering. Such fundamental difference in program structures results in a big variation in course emphasis and topics covered. Adding to that the fact that some programs use the quarter system, while others use the semester system, one would realize that one model will not fit all.

This paper reports on the results of a study conducted to appraise the curricular activities in power semiconductor devices and systems that are offered by U.S. educational institutions. Modifications to existing undergraduate and graduate programs are proposed and discussed.

Assessment of Current Power Electronics Curricula

The electrical engineering programs offered by the 200 universities ranked by the U.S. News and World Report were studied (1). The study indicates that the number of programs that offer some courses in power engineering is small. This is a result of the elimination of power engineering subjects in favor of the newly developed electronics and digital subjects in the seventies and eighties of the 20th century. The study shows also that, even with the small number of power engineering programs, many of them do not offer all the required courses to ensure viable programs. Most undergraduate programs with power engineering emphasis offer the following required courses: (a) Basic Circuit Analysis, with laboratory components, (b) Electronic Circuits, with laboratory components, (c) Power Systems, and (d) Physics of Semiconductor Devices; see Figure 1. These courses are required for all students, and they are normally offered during the sophomore and the junior years.

For students who wish to specialize in power engineering, there are courses that can be taken as electives. Depending on the institution, some or all of the following additional courses can be available to senior students: (e) Fabrication of Devices and ICs, (f) Power Electronics, and (g) Power Semiconductor Devices; see Figure 1. Some of these courses are offered for both undergraduate and graduate credits.

The topics covered in the first set of courses (a-d) are shown in Table I; while the topics covered by the second set (e-g) are shown in Table II. The study shows that the topics covered by the Basic Circuits, Electronic Circuits, and Semiconductor Devices are almost the same for all programs. However, these topics may be covered in more than the identified three separate courses (a-c), depending on the curriculum structure and whether a semester system or a quarter system is used. However, in all cases, the topics, shown in Table I, for these courses are normally covered.

The topics of the Basic Circuits course are normally offered in one course (2). Institutions that offer combined sub-programs cover these topics in more than one course. The Electrical and Computer Engineering program at the University of Illinois at Urbana-Champaign covers these topics in three courses ECS 110, ECS 205, and ECS 210 (3).

Similar to the Basic Circuits course, the Electronic Circuits and the Physics of Semiconductor Devices courses are almost identical in all institutions.

The topics covered in the Electric Power Systems course differ from one institution to the other. This course can be a combination of topics such as: motors, generators, power transmission, distribution systems, power electronic circuits, or renewable energy systems. An example of the topics covered by this course is the MIT course, Introduction to Electrical Power Systems (4). In other institutions, these topics may be covered in more than one course.

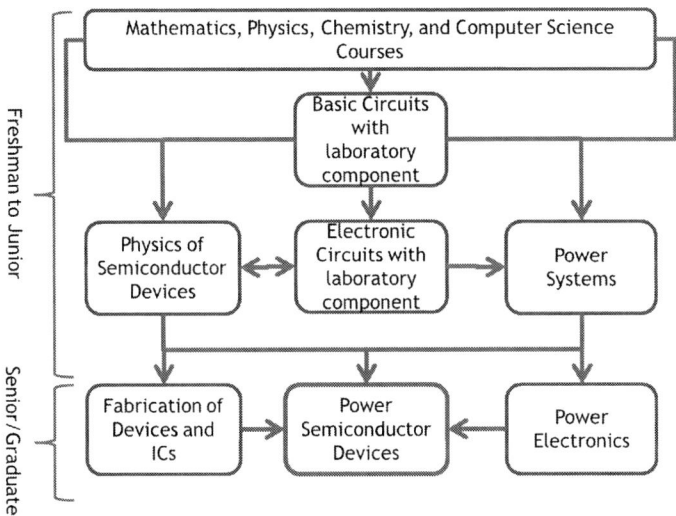

Figure 1. An example of current curriculum for power electronics.

The second group of courses, namely, Fabrication of Devices and ICs, Power Electronics, and Power Semiconductor Devices, are offered as part of the elective courses or as part of a specific track of study. Device and ICs fabrication laboratory courses are offered by almost all major institutions that have micro-fabrication facilities (2-16). The topics that are typically covered in a device fabrication course are shown in Table II (2-16). Some programs offer laboratory fabrication courses for both undergraduate and graduate credits. Others restrict laboratory experience to graduate students, while offering lectures-only-courses for undergraduate credit. Small institutions with no fabrication facility, such as Bradley University, do offer laboratory courses (17).

Institutions with strong power programs offer the Power Electronics topics as a full course (2, 4, 8, 11, 13, 15, 18). Others introduce power electronics subjects as few topics in one of the Electronic Circuits or Power Systems courses (17, 19). In general, there is a fewer number of institutions that offer a course in power electronics and even fewer that offer a laboratory component. The topics that are typically covered in the first course in power electronics is shown in Table II.

TABLE I. Topics of Required/Basic Courses.

Course	Topics
Basic Circuit	Dc circuit analysis, RLC circuits, ac circuit analysis, operational amplifier circuits, phasor, and a minimum of two hours/week laboratory
Electronic Circuits	Characteristics of diodes, Bipolar Junction Transistors (BJT), Junction Field Effect Transistors (JFETs), amplifier circuits, Metal Oxide FETs (MOSFETs), biasing, feedback, stability, operational amplifiers, and a minimum of two hours/week laboratory.
Physics of Semiconductor Devices	Lattice structure, energy band theory, current conduction in solids, physical structure, operation of semiconductor devices: diodes, BJTs, and MOSFETs, and optoelectronic phenomena in semiconductors.
Introduction to Electric Power Systems (4)	Electric circuit theory with application to power handling electric circuits. Modeling and behavior of electromechanical devices, including magnetic circuits, motors, and generators. Operational fundamentals of synchronous, induction and DC machinery. Interconnection of generators and motors with electric power transmission and distribution circuits. Power generation, including alternative and sustainable sources.

Topics on power semiconductor devices are normally introduced as sections in other courses, such as Power Electronics, Semiconductor Devices, or Electronics Circuits. These topics can also be covered as part of the traditional Advanced Semiconductor Devices courses. In this case, power devices are treated as one of the many compound semiconductor devices, not as devices with unique characteristics. The study shows that it is rare to find an engineering program offering power semiconductor device topics as a separate course. Cornell University is among the rare institutions that offer such a course for graduate and undergraduate credits (20). The topics of the Cornell course are shown in Table II. As can be seen from the Table, the course covers aspects related to Si as well as wide band gap semiconductors. Some institutions cover power devices in seminars, in special topics courses, or as a research topic for thesis or dissertation.

TABLE II. Topics of Power Courses.

Course	Topics
Semiconductor Device Fabrication Laboratory (2)	Design and fabrication of active semiconductor devices. Laboratory exercises include artwork and pattern generation, mask making, oxidation, photolithographic processing, diffusion, metallization, and device testing.
Power Electronics (14)	Introduces power semiconductor devices and power electronic converters, including single-phase and three-phase ac/dc rectifiers, ac voltage controllers, dc/dc converters and dc/ac inverters.
Power Semiconductor Devices (20)	Basic power electronic devices, basic power electronic circuits, and their suitability for power applications spanning a large range of currents and voltages. Device operation, design, fabrication, and power electronic circuit issues are discussed. Emphasis is on the device aspects. Examples are drawn from current Si device technology as well as emerging power devices technologies developed from wide band gap semiconductors.

Currently, silicon is considered the industry workhorse for building power electronic devices. Although silicon has adequately addressed the needs of the power electronic industry until now, it may not be able to do so in the future. Next generation systems will require the ability to handle very high power applications and operate at much higher temperatures. With silicon technology reaching its physical limits, new material technologies, such as wide energy band gap (WBG) semiconductors, become inevitable. A successful adoption of WBG technologies in power electronics systems will require new circuit topologies, new system design approaches, and efficient thermal management techniques. Reliability of the power electronics systems must be addressed at the application level, including interactions with energy sources and non-linear loads. Current educational programs lack adequate coverage of the application of WBG semiconductor devices in the design of power systems. It should be noted here that the widely recognized energy conversion systems curriculum developed and offered through the consortium of universities for sustainable power (CUSP)™ by the University of Minnesota is primarily focused at the systems level with very little coverage on power electronics components and technologies (21, 22).

Restructuring Power Electronics Curricula

Undergraduate Program

The proposed Power Electronics track for undergraduate programs is shown in Figure 2. In this track, three regular courses are offered: Fabrication of Devices, Power Semiconductor Devices, and Power Electronics. The track is then sealed by the Senior Design Project course, where the students are expected to design and build power electronic systems or design and fabricate power devices or circuits.

The proposed fabrication course can be offered as lectures with or without laboratory component, depending on the facilities available. If laboratory components could not be offered, device and process simulators can be used instead (23-26). Some of these tools not only have circuit simulators, but also have PCB design tools to build complete systems. Using simulation tools can emulate the real devices, since these tools utilize numerical techniques such as finite elements. A multi-physics simulation tool such as COMSOL can be used to investigate thermal stress and heat dissipation issues (27).

The topics of the Semiconductor Power Devices course shown in Table II and Figure 1 are suitable for the proposed power device course, since it addresses Si and wide band gap devices. However, temperature effect, thermal stress, heat transfer, thermal management, reliability, and power packaging need to be included. Many of these issues can be addressed using a muliphysics CAD tool (27).

Regarding the Power Electronics course, the topics shown in Table II are the standard topics covered. However, thermal effects, system reliability, and failure analysis need to be addressed. A laboratory component for this course is essential, since engineers at the BS level will be involved in power system design and maintenance more than designing and fabricating individual power components or devices.

The Senior Design Project course will be used to provide the practical experience of designing and constructing a power electronic system. The course will be used to test the validity of the knowledge acquired in the previous courses, especially reliability and thermal effect. This course ensures that the students have adequate knowledge about the design of reliable and efficient power electronic systems.

The courses can be delivered either as elective courses or as a new track. They can also be used to issue a certificate for students or working engineers. The proposed courses form a framework; however, each institution may need to modify the topics covered to fit into their existing program. Industry, through Industrial Advisory Boards, should be consulted during the development of the proposed program. The dissemination of this knowledge will increase the number of institutions that participate in this program.

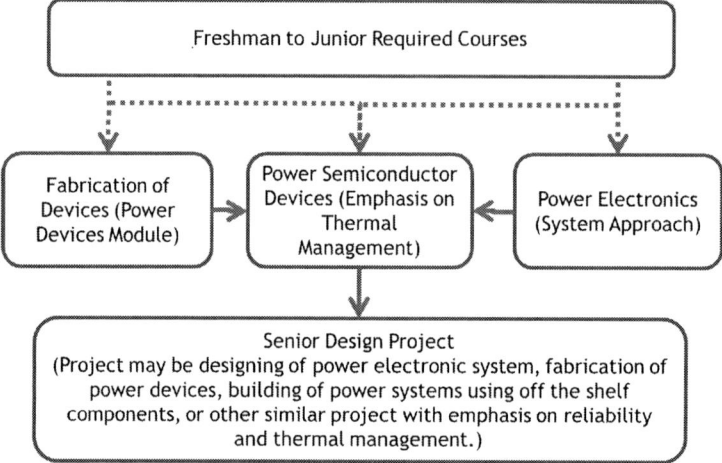

Figure 2. Proposed undergraduate concentration in power semiconductor devices and systems.

Graduate Programs

Creating a graduate program in power electronics is much easier. This is because courses can be more advanced, students have more freedom in taking courses, and research and development can be more advanced. The proposed program is shown in Figure 3. The first three courses are the same courses shown in Figure 2 and Table II, where it is assumed that these courses are senior/graduate courses. However, more advanced open only for graduate credit can be offered, if the program prefers that.

Conclusions

The proposed program changes provide strong background in the area of power electronics and devices that can meet the needs of industry. The proposed courses can be taken as elective courses or as a separate track without exceeding the maximum limit of credit hours. This proposal will help the educational community and industry to begin developing a road map for advancing this area. This shall include faculty training, curriculum development, and textbook publishing to accelerate the full development of this area.

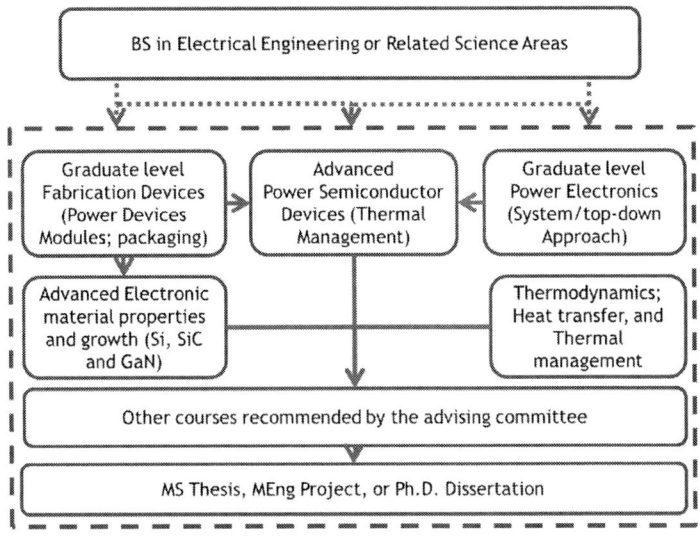

Figure 3. Proposed graduate track for power semiconductor devices and systems.

Acknowledgments

The author would like to acknowledge Dr. Krishna Shenai of Argonne National Laboratory for his help, advice, and intellectual discussions without which this work would not be complete.

References

1. U.S. News and World Report: http://colleges.usnews.rankingsandreviews.com/best-colleges/rankings/national-universities.
2. Northern Illinois University: http://catalog.niu.edu/content.php?catoid=20&navoid=612
3. University of Illinois at Urbana-Champaign: http://www.ece.illinois.edu/courses/
4. Massachusetts Institute of Technology: http://ocw.mit.edu/courses/electrical-engineering-and-computer-science/6-061-introduction-to-electric-power-systems-spring-2011/index.htm
5. Harvard: http://www.registrar.fas.harvard.edu/courses-exams/course-catalog/engineering-sciences-177-microfabrication-laboratory
6. Princeton: http://www.princeton.edu/ee/courses/course-catalog/
7. California Institute of Technology:

http://catalog.caltech.edu/courses/listing/aph.html

8. University of California Berkeley: http://general-catalog.berkeley.edu/catalog/gcc_list_crse_req?p_dept_name=Electrical+Enginee ring&p_dept_cd=EL+ENG&p_path=1

9. Carnegie Mellon University: http://www.ece.cmu.edu/courses/

10. University of California Los Angeles: http://www.ee.ucla.edu/academics/courses/catalog-description-of-electrical-engineering-courses

11. University of Michigan-Ann Arbor: http://www.engin.umich.edu/bulletin/eecs/courses.html

12. University of Virginia: http://www.ece.virginia.edu/curriculum/ugrads/handbook.pdf

13. Boston University: http://www.bu.edu/ece/undergraduate/courses/

14. Georgia Institute of Technology: http://www.ece.gatech.edu/academics/courses/course_listing.php?r_course=3&sel _tig=45

15. University of Wisconsin: http://courses.engr.wisc.edu/ece/

16. University of Minnesota: http://onestop2.umn.edu/courses/courses.jsp?designator=EE&submit=Show+the+ courses&campus=UMNTC

17. Bradley University: http://www.bradley.edu/academic/undergradcat/20122013/egt-eecourses.dot

18. Notre Dame: http://www.ee.nd.edu/

19. University of Dayton: http://catalog.udayton.edu/allcourses/ece/

20. Cornell University: http://courses.cornell.edu/content.php?catoid=12&navoid=2156&filter%5Bcpage %5D=35&filter%5Bonly_active%5D=1&filter%5B3%5D=1&filter%5Bitem_typ e%5D=3&expand=1

21. Consortium of Universities for Sustainable Power(CUSP)™ http://cusp.umn.edu/

22. K. Shenai, Workshop on Electric Energy Systems Curriculum for Sustainability, Napa, California, February 8-9, 2013. http://cusp.umn.edu/Napa_2013/Saturday/WBG_Curriculum_Shenai.pdf

23. Mentor Graphics: http://www.mentor.com/

24. MIcroTec: http://www.siborg.ca/microtec.html

25. Cadence: http://www.cadence.com/us/Pages/default.aspx

26. Synopsys: http://www.synopsys.com/tools/tcad/Pages/default.aspx

27. COMSOL: http://www.comsol.com/

Power Electronic Module Packaging at UA

S. S. Ang, H. A. Mantooth, and J. C. Balda

High Density Electronics Center & Department of Electrical Engineering
University of Arkansas, Fayetteville, Arkansas 72701, USA
siang@uark.edu

Power electronic modules incorporating silicon carbide (SiC) and gallium nitride (GaN) power semiconductors enable many power electronic system applications otherwise not possible for their silicon counterparts. These power electronic modules are desired to operate at junction temperatures of greater than 200°C. As such, the design and fabrication of power electronic modules become more challenging and involve several multi-disciplinary engineering expertise. This paper addresses the educational and research efforts in power electronic module packaging at the University of Arkansas.

Introduction

Power electronic modules are power packages containing several power semiconductor devices on a common packaging substrate. The advent of wide bandgap power semiconductor devices (e.g., SiC and GaN) has changed tremendously the design, material, and fabrication aspects of conventional silicon-based power electronic modules. The desire to operate at junction temperatures of 200°C and higher makes the selection of packaging materials and fabrication processes of these power electronic modules very challenging. The ability of the SiC power devices to withstand tens of thousands volts of breakdown voltage has made high-voltage applications attractive since series connections of power semiconductor devices are not necessary. The high-temperature and high-voltage requirements make the design and fabrication of these power electronic modules becoming a multi-disciplinary approach involving electrical, mechanical and material engineering.

At the University of Arkansas (UA), educational efforts in power electronic module are undertaken through course work and research projects. Two courses in microelectronic packaging have been offered since 1994. As the results of these educational efforts, a multi-author edited textbook entitled "Advanced Electronic Packaging" was published in 1997 [1] and its second edition was published in 2006 [2]. Several universities offer courses in microelectronic packaging in their undergraduate and graduate curricula. Table I lists some examples of microelectronic packaging related courses offered by several universities in the United States. The Georgia Tech's ECE 4754 course provides hands-on instruction in electronics packaging, including assembly, reliability, thermal management, and test of next-generation microsystems [3]. Another Georgia Tech's ECE 4755 course provides hands-on instruction in basic packaging substrate fabrication techniques, including interconnect design and testing, dielectric deposition, via formation, and metallization [3].

TABLE I. Examples of Packaging–related Courses.

Course No.	Course Name	University	Course Level
ELEG 5273	Electronic Packaging	University of Arkansas	Undergraduate/ graduate
ELEG 6273	Advanced Electronic Packaging	University of Arkansas	Graduate
ECE4754 ME4754 MSE4754	Electronics Packaging Assembly, Reliability, Thermal Management, and Test	Georgia Institute of Technology	Undergraduate
ECE4755 CHE4755	Electronic Packaging Substrate Fabrication	Georgia Institute of Technology	Undergraduate
ENME 473	Mechanical Design of Electronic Systems	University of Maryland	Graduate
ENME690	Mechanical Fundamentals of Electronic Systems	University of Maryland	Graduate
ENME693	High Density Electronic Assemblies and Interconnects	University of Maryland	Graduate

The Maryland's ENME 473 course delivers design considerations in the packaging of electronic systems, production of circuit boards and design of electronic assemblies, including vibration, shock, fatigue and thermal considerations [4]. The Maryland's ENME 690 course gives an understanding of the fundamental mechanical principles used in the design of electronic devices and their integration into electronic systems. Focus will be placed on the effect of materials compatibility, thermal stress, mechanical stress, and environmental exposure on product performance, durability and cost [4]. The Maryland's ENME 693 course presents the mechanical fundamentals needed to address reliability issues in high-density electronic assemblies. Design methods to prevent failures within the life cycle are developed in this course [4].

The remaining sections of the paper describe the UA electronic packaging courses and power electronic module research program. The research program consists of a research agenda that includes transportation [5], grid-connected [6], energy-exploration, and aeronautical power electronic systems. Section II describes the pedagogical approach behind the electronic packaging courses and some details of the courses. Section III describes several aspects of the research programs in power electronic module. This paper is concluded with a description of the training of researchers in power electronic modules that the UA microelectronic packaging education has enabled.

II. UA Pedagogical Approach

The first course in microelectronic packaging (ELEG 5273) is an undergraduate/graduate introductory course covering topics such as packaging materials, processing, electrical, thermal, mechanical, and design aspects of electronic packaging, integral passives, assembly technology, multichip/3D packaging, RF, power, and low-temperature co-fired ceramics packaging, reliability/analytical, and cost/evaluations. Some of these topics have a laboratory component to reinforce the students' learning experience. For example, the laboratory component for the power packaging lectures involves a laboratory demonstration of the power semiconductor die attachment using a solder alloy in an annealing furnace. Students are exposed to die attachment process, including the solder reflow profile and practical operation of the reflow furnace. This course is frequently taken by electrical engineering, mechanical engineering, and microelectronic-photonics students over the years. The pre-requisite for this first course in microelectronic packaging is senior or graduate standing in engineering and science. Over the last 20 years, many highly motivated undergraduate engineering students have successfully completed this course.

The second course in microelectronic packaging (ELEG 6273) is graduate level and offers opportunities for individual students to specialize in a particular packaging technology such as power electronic module packaging, high-frequency packaging, or flexible electronic packaging. In the power electronic module packaging, substrate and material selections, the attachment strategy, design/processing/ assembly issues for the development of high temperature and high power silicon, silicon carbide (SiC), and gallium nitride (GaN) power electronic modules are discussed. These power semiconductor devices have been shown capable of operating at temperatures of 350-400°C. Students learn that satisfying the requirements for operating at higher temperatures for power electronic modules require substantial changes to conventional power module fabrication techniques and material selections. These operating environments typically desire an operating junction temperature greater than 200°C, coupled with the need to increase power density at both the device and system level consistently extending the limits of the available power semiconductor devices, module packaging, passive and driver technology, and thermal management systems to ensure a balance of electrical, thermal, and mechanical properties over the proposed temperature range to minimize electrical power losses, minimize thermal impedance of the package and module, and provide mechanical reliability.

III. Relation to Research

The need for high power, high temperature device packages and modules has led to significant research and development efforts in power electronic module packaging. At these high operating temperatures, some failure modes manifest themselves more readily. For example, passivation and encapsulations utilized for voltage isolation at both the device and module levels break down more readily at high temperatures presenting a major reliability problem for power electronic modules. The requirements for high voltage breakdown in power electronic module applications require series connection of several power semiconductor devices to increase their high voltage handling capability. The most common high-voltage press-pack power semiconductor packages are adapted from the traditional "hockey puck" packages [7]. This rigid pressure contact packaging technology, intended for the rugged silicon thyristors, is not optimized for the sensitive microstructures on the surface of wide-bandgap power semiconductor devices [7]. As a consequence, a great deal of care is required during module assembly process. The issue is further aggravated when the press-pack module size is increased to increase current handling capabilities. There is a significant impact on system production cost as a result of these shortcomings [7]. Recent advancements of high-voltage power semiconductor devices such as silicon carbide and gallium nitride power semiconductor devices require a different high-voltage power module packaging approach. Currently silicon carbide (SiC) devices are achieving breakdown voltages above 10 kV and are at the limits of traditional high-voltage packaging solutions. Cree and Powerex have developed high voltage power modules utilizing 10kV SiC MOSFETs in a wire bonded module [8]. However, as these device breakdown voltages increase, new approaches in module package design, passivation, and encapsulation become necessary.

The UA approach examines several wire-bonded and wire-bondless module packaging architectures including the use of 8kV rated devices with a direct solder attachment hierarchy for interconnection [9]. This involves the use of two direct-bond copper (DBC) substrates, one acting as the power substrate connected to a copper base plate and the second as an interconnection lead-frame between the power semiconductor devices. This module configuration was chosen to maximize power density of the module while also achieving the potential for double-sided cooling [10]. Naturally, voltage breakdown between the topside and backside of the devices is inevitable due to the small distances between them. In order to eliminate this voltage breakdown potential, UA researchers have developed a two-step passivation technique using a nano silica embedded polyamide imide layer [11].

IV. Conclusion

The UA electronic packaging courses have been taught since 1994 and have undergone several major revisions due to the fast-changing microelectronic packaging field. These courses have benefited the power electronic module research and development program through more team-based research projects. A typical project might include a device modeling student, a power packaging student, and a power electronic design student all focused on the research and development of a new integrated power module [12].

References

[1] W. D. Brown, "Advanced Electronic Packaging – With emphasis on Multichip Modules," Wiley-Interscience, 1997.

[2] R. K. Ulrich and W. D. Brown, "Advanced Electronic Packaging," 2nd Edition, Wiley-Interscience, 2006.

[3] http://www.ece.gatech.edu/academics/courses/course_listing.php

[4] http://www.gradschool.umd.edu/catalog/courses/enme.htm

[5] Atanu Dutta, Shijie Wang, Jinchang Zhou, Simon S. Ang, June-Chien Chang, and Chang-Sheng Chen," The Design and Fabrication of a 50KVA 450A Silicon Carbide Power Electronic Module," *The 4th International Symposium on Power Electronics for Distributed Generation Systems*, July 8-11, 2013, USA.

[6] H. Zhang, S. S. Ang, C. Farnell, Y. Liu, J. C. Balda, H. A. Mantooth, "A SiC SGTO/PIN Diode Power Electronic Module for a Fault Current Limiter," *The 4th International Symposium on Power Electronics for Distributed Generation Systems*, July 8-11, 2013, USA.

[7] S. Kaufmann, T. Lang, and R. Chokhawala,"Innovative press pack modules for high power IGBTs," pg. 59-62, *Proceedings of the 13th International Symposium on Power SemiconductorDevices and ICs*, ISPSD'01 Osaka, Japan (2001)..

[8]] J. Richmond, S. Leslie, B. Hull, M. Das, "Roadmap for megawatt class power switch modules utilizing large area silicon carbide MOSFETs and JBS diodes," pg.106 – 111, *Proceedings of IEEE Energy Conversion Congress and Exposition* (2009).

[9] Hao Zhang, Simon Ang, Alan Mantooth and Juan Balda, "A 6.5kV, Wire-Bondless, Double-Sided Cooling Power Electronics Module," *2012 IEEE Energy Conversion Congress and Exposition*, September 16-20, 2012, Raleigh, North Carolina, USA.

[10] H. Zhang, S. S. Ang, H. A. Mantooth, and S. Krishnamurthy, "A Double-side Cooling Power Electronic Module Using a Low-Temperature Co-fired Ceramic Device Carrier," *2013 IEEE Energy Conversion Congress and Exposition*, Denver, Colorado, USA in September 15-19, 2013.

[11] Jinchang Zhou, Simon Ang, Alan Mantooth, and Juan C. Balda, "A Nano-Composite Polyamide Imide Passivation for 10 kV Power Electronics Modules," *2012 IEEE Energy Conversion Congress and Exposition*, September 16-20, 2012, Raleigh, North Carolina, USA.

[12] H. A. Mantooth, S. Ahmed, and S. S. Ang, "Power Semiconductor Device Modeling and Simulation," *Electrochemical Society Meeting Abstract Paper 25364*, October 2013.

CHAPTER 7

GaN POWER DEVICES 2

260

III-N High-Power Bipolar Transistors

Russell D. Dupuis[a], Jeomoh Kim[a], Yi-Che Lee[a], Zachary Lochner[a], Mi-Hee Ji[a], Tsung-Ting Kao[a], Jae-Hyun Ryou[b], Theeradetch Detchphrom[a], and Shyh-Chiang Shen[a]

[a] School of Electrical and Computer Engineering, Georgia Institute of Technology, 777 Atlantic Drive NW, Atlanta, GA 30332-0250, USA.
[b] Now at University of Houston, Houston, USA

We report state-of-the-art d.c. and RF performance of GaN/InGaN npn DHBTs grown by the MOCVD technology on sapphire substrates. The fabricated GaN/InGaN HBTs achieved a collector current density greater than 50 kA/cm^2 and a d.c. current gain of 60 at the collector voltage of 13 V. A open-base collector-emitter breakdown voltage of greater than 90 V was measured. Using the same layer structure and device fabrication techniques, we also demonstrated state-of-the-art RF performance of GaN/InGaN HBTs with f_T greater than 8 GHz and f_{max} greater than 1.8 GHz at the collector current density of 11 kA/cm^2. These d.c. and RF measurement results demonstrated the capability and feasibility of III-N HBTs for high-power circuit applications.

Introduction

In the recent decade, wide-bandgap III-Nitride (III-N) transistor technologies have been actively researched and developed. The device focus of this work has largely been in the area of field-effect transistors which exploit the piezoelectric properties of the AlGaN-GaN heterojunction and the high electron saturation velocity properties of the corresponding two-dimensional electron gas. These HFET-based III-N transistor technologies have been proposed as the basis for next-generation electronic systems with a wide variety of applications for both civilian and military use. In the past few years, reports of III-N high-electron mobility transistors (HEMTs) with ultra-high voltage operation in the kV-range have been made (1,2). III-N HEMTs with high cut-off frequencies greater than 200 GHz were also reported (3,4).

The development of III-N heterojunction bipolar transistor (HBT) technologies has been less successful due to many technological barriers in the growth and the fabrication of these devices. However in spite of these difficulties, a number of npn AlGaN/GaN HBTs were demonstrated with high current gains and high breakdown voltages (58). GaN-based HBTs have also shown the capability of operating at temperatures up to 300C (9). However, few of these devices can achieve high collector current density (J_C) combined with meaningful RF characteristics (10). To overcome some of the limitations, npn GaN/InGaN HBTs have been developed to take advantage of the higher free-hole concentration in a p-InGaN base layer (11). These devices demonstrated high J_C combined with good current gain by employing either base-regrowth, emitter-regrowth, or single-pass growth (i.e., direct growth) techniques. (1216) Recently, direct-growth *npn* GaN/InGaN HBTs have been demonstrated with a DC current gain > 100 and J_C > 6.5 kA/cm^2 (17). Also, *pnp* GaN/InGaN HBTs were reported with good DC current gain (86) and J_C = 7.3 kA/cm^2 (18).

However, in spite of these advancements, very little meaningful RF data has been reported on III-N HBTs to date. Recently, we reported the first RF performance of *npn* GaN/InGaN DHBTs with a direct-growth technique (19). The devices are fabricated on sapphire substrates using high-quality epitaxial growth and optimized processing techniques to achieve high current drive and RF power gain. The fabricated RF-DHBTs show a d.c. common–emitter current gain (β) of 24, J_C >16 kA/cm^2, and the open-base common-emitter breakdown voltage (BV_{CEO}) > 95 V. At J_C = 4.7 kA/cm^2, the DHBT with an emitter area (A_E) of 4×20 μm^2 shows a cut-off frequency (f_T) > 5 GHz and a maximum oscillation frequency (f_{max}) > 1.3 GHz. To the best of our knowledge, this report is the first RF demonstration of GaN/InGaN HBTs with f_T > 5 GHz. We also reported a ultra-high d.c. power density of > 3 MW/cm^2 for an InGaN HBTs built on a free-standing (FS) GaN substrate (20).

In this paper, we will report the device performance of GaN-based *npn* double-heterojunction bipolar transistors (DHBTs) grown by metalorganic chemical vapor deposition (MOCVD). Through further device processing improvement and material growth optimization, we demonstrate that, when the GaN/InGaN HBTs are grown on a sapphire substrate, high current gain (> 70), high J_C (> 50 kA/cm^2) can be achieved. We also demonstrated the state-of-the-art RF amplification performance with f_T greater than 8 GHz and f_{max} greater than 1.8 GHz. These results for GaN HBTs on sapphire indicate that high-performance III-N power HBTs can be achieved without the need for complex re-growth schemes as reported earlier.

GaN/InGaN HBT Growth and Fabrication

Epitaxial Layer Structure and Growth Optimization of NpN-GaN/InGaN HBT

The epitaxial layer structures developed for this study were grown by metalorganic chemical vapor deposition (MOCVD) in an AIXTRON 6x2 MOCVD reactor system equipped with a Close-Coupled^{TM} showerhead. EpiPure^{TM} (21) trimethylgallium (TMGa, Ga(CH$_3$)$_3$), trimethylindium (TMIn, In(CH$_3$)$_3$), and high purity ammonia (NH$_3$) were used as precursors for GaN and InGaN layers. Silane (SiH$_4$) and bis-cyclopentadienylmagnesium (Cp$_2$Mg, Mg(C$_5$H$_5$)$_2$) were used for the *n*- and *p*-type dopants, respectively. The epitaxial structures of HBTs were grown on GaN templates, which is composed of a c-plane (0001) sapphire substrate, ~20 nm-thick low temperature (T$_g$=550 °C) GaN nucleation layer, and ~2.5 μm-thick high-temperature (T$_g$=1050 °C) GaN buffer layer. The epitaxial layer structure developed in this study is described in Table I.

The collector and sub-collector layers were grown at 1065 °C with different free electron concentrations. The *p*-In$_{0.03}$Ga$_{0.97}$N:Mg base layer was grown at 850 °C. The InGaN layer is suitable for the *p*-type base region of a GaN/InGaN HBTs because of lower bandgap energy and a corresponding lower Mg acceptor activation energy of InGaN layers than those of similar Mg-doped *p*-type GaN layers, resulting in higher hole concentrations of the *p*-InGaN:Mg base layer followed by lower base resistance and higher d.c. gain in a GaN/InGaN HBTs. To avoid the lattice strain between the GaN and InGaN layer, resulting in degradation of device performance via strain relaxation and defect formation, a graded heterojunction has been developed for the base-collector (BC) and base-emitter (BE) junctions. In addition to the strain relief and improved crystal quality, these grading layers also serve to eliminate the effects of conduction band discontinuities at each heterojunction, enhancing transport of minority carriers across the junction interfaces.

The growth temperature for base-collector (BC) graded junction was the same as the temperature for the base layer, while the TMIn molar flow rate was ramped to accomplish grading junction. The base-emitter (BE) grading layer was grown by ramping the TMIn flow rate and temperature, from the base layer growth temperature to emitter layer growth temperature. The emitter layer was grown at 950 °C, which is well below the optimal growth temperature for high quality GaN (~1050 °C), but this lower temperature is required to prevent desorption of indium from the p-InGaN:Mg base layer during emitter layer growth.

Table I. Epitaxial layer structures of the npn GaN/InGaN HBTs.

Epitaxial layer	Material:doping	Thickness
Emitter	n-GaN:Si^{++}	70 nm
Base-Emitter Grading	$In_{0.03}Ga_{0.97}N$:Si → GaN:Si	30 nm
Base	p-$In_{0.03}Ga_{0.97}N$:Mg	100 nm
Collector-Base Grading	GaN:Si → $In_{0.03}Ga_{0.97}N$:Si	30 nm
Collector	n-GaN:Si	500 nm
Sub-collector	n-GaN:Si^{+}	1000 nm
Buffer	GaN:ud	2.5 μm
Substrate	Sapphire	

III-N HBT device fabrication

The device fabrication processing of III-N HBTs in this study starts with three sequential mesa etching steps using low-damage inductively coupled plasma (ICP) etching to form the base pedestal, the collector pedestal and the device isolation, respectively. After the ICP mesa etching, the wafer is treated in a diluted $KOH/K_2S_2O_8$ solution under ultraviolet light illumination to remove the dry-etching-induced surface damage (22). Ti/Al-based metal stacks are used to form the collector and the emitter ohmic contacts. The base contact is formed using Ni/Au-based metal stacks. After the device-level processing is completed, the HBTs are passivated by benzocyclobutene (BCB) and via accessing holes are subsequently opened by ICP etching. NiCr films are deposited by an e-gun evaporation tool for the on-chip resistor and for the on-wafer radio-frequency (RF) calibration standards. Finally, a 1-micron-thick Ti/Au metal layer is deposited for the device interconnection.

Device Performance Summary

In this study, the DC and RF characteristics of fabricated GaN/InGaN DHBTs are measured at room temperature. The DC performance of fabricated devices is characterized using a Keithley SCS-4200 semiconductor characterization system. The RF small-signal S-parameter measurements of the transistors are carried out in an Agilent E8364B PNA

Microwave Network Analyzers. On-wafer short-open-load-through (SOLT) calibrations are also adopted in the small-signal RF characterization to remove the pad parasitic components. The devices under test are biased in the common-emitter configuration.

Shown in Figure 1 is a Gummel plot of a fabricated GaN/InGaN DHBT with $A_E = 11.7$ μm^2. The device is grown on a sapphire substrate and the measurement was taken at $V_{CB} = 0$ V. The I_B-I_C crossover point (i.e., the unit current gain condition) is ~ 0.8 A/cm^2 at $V_{BE} = 7$ V. The biasing voltage at the crossover point is higher than a typical GaN-based pn junction because of high sheet resistance in the base region and the non-ohmic contact property in the base metal contact. Nevertheless, the differential current gain (h_{fe}) increases to > 70 at $J_C > 20$ kA/cm^2 (V_{BE} = 11 V) and starts to roll off at higher current region ($J_C > 25$ kA/cm^2) due to the series resistance effect. It is also shown that $h_{fe} > 55$ can still be achieved at $J_C > 65$ kA/cm^2 ($V_{BE} = 13$ V).

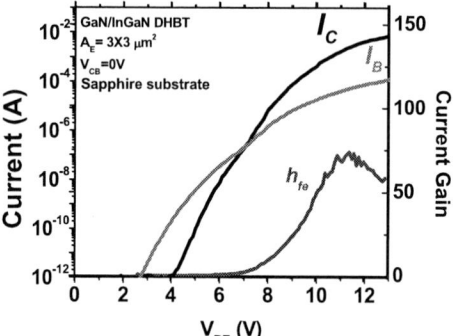

Figure 1. A Gummel plot of a fabricated GaN/InGaN DHBT with $A_E = 11.7$ μm^2.

Shown in Figure 2 is a set of common-emitter family curves of the same GaN/InGaN DHBT discussed in the previous section. Compared to other GaN-based HFETs, the device shows a sharper knee-voltage transition. In this set of curves, $J_C > 51$ kA/cm^2 ($I_C = 6$ mA) and β ($\triangleq I_C/I_B$) = 60 are achieved at $I_B = 100$ μA and the measured BV_{CEO} is 90 V at $I_C = 0.5$ μA. When compared to the previously reported highest J_C for III-N HBTs on free-standing GaN substrate, this operable J_C is about 40% of that for devices grown on FS-GaN substrates ($J_C > 140$ kA/cm^2 on FS-GaN (20) v.s. $J_C > 50$ kA/cm^2 on sapphire in this report for the same device design). The reduction in the highest operable J_C could come from (1) the increased recombination centers arising from higher dislocation and point defects for HBTs grown on sapphire and (2) the poor heat conduction of the substrate under high J_C operation. The elevated junction temperature not only limits the achievable J_C but may also lead to premature device degradation for high-power transistor operation. A proper choice of substrates and proper heat management would be an important consideration for the implementation of these high-power vertical carrier transport III-N bipolar transistors.

The RF characteristics of the GaN/InGaN DHBTs are measured in an Agilent E8364B PNA from 40 MHz to 20 GHz using an on-wafer SOLT calibration at room temperature. As shown in

Figure 3, the frequency responses of short-circuit current gain ($|h_{21}|^2$), the Mason's unilateral gain (U), and the maximal available gain (MAG) were measured at $V_{CE} = 13$ V and $J_C = 11$ kA/cm^2. The fabricated device has A_E of 5×20 μm^2. A value of $|h_{21}|^2$ of 30 dB can be achieved for the operation frequency of < 200 MHz. An asymptotic 20 dB/decade however can be clearly seen on the measured $|h_{21}|^2$ curve beyond a corner frequency around 1 GHz. The frequency response represents a typical single-pole frequency response in a well-behaved transistor and the f_T value of 8 GHz can be unambiguously determined.

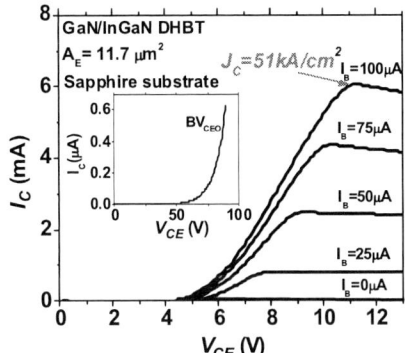

Figure 2. The common-emitter characteristics of a fabricated GaN/InGaN DHBT with $A_E = 3 \times 3$ μm^2 grown on a sapphire substrate.

It should be noted that the tailing of the $|h_{21}|^2$ frequency response at the high-frequency region that shows an erroneous f_T of 20GHz at dB($|h_{21}|^2$) = 0 dB is due to the reactive coupling between the collector and the emitter through the lightly doped underlying buffer layer ($n \sim 10^{16}$ cm^{-3}), which should be discounted from the f_T assessment. The S-parameter measurement also shows that the device is unconditionally stable for frequencies up to 20 GHz. The f_{max} of 1.8GHz is determined either by U at 0 dB.

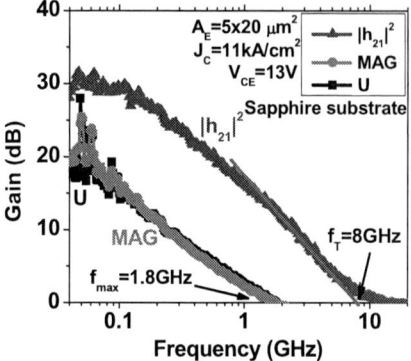

Figure 3. The frequency response of the measured $|h_{21}|^2$, MAG, and U of a GaN/InGaN DHBT with $A_E = 5 \times 20$ μm^2.

Conclusions

In conclusion, we report state-of-the-art high-power d.c. and RF performance of GaN-based *npn* DHBTs grown by the MOCVD technology. Through growth condition optimization and improved processing technology for III-N HBT fabrication, we successfully demonstrated high current density operation with J_C 50 kA/cm² J_C for InGaN HBTs grown on a sapphire substrate. Using the same layer structures and fabrication technology, we also demonstrate the RF performance of GaN/InGaN HBTs with $f_T > 8$ GHz and $f_{max} > 1.8$ GHz. To the best of our knowledge, the measured f_T and f_{max} values are the highest RF performance on III-N bipolar transistors reported to date. These d.c. and RF results clearly demonstrated the capability and feasibility of III-N HBTs for future high-power circuit applications.

Acknowledgments

R. D. Dupuis acknowledges the support of the Steve W. Chaddick Endowed Chair in Electro-Optics and the Georgia Research Alliance. S. C. Shen also acknowledges financial support from the Georgia Tech Foundation. The work was supported in part by NSF under contract ECCS-0725736. We also acknowledge the support of SAFC Hightec.

References

1. T.-T. Kao, C.-Y. Wang, S.-C. Shen, D. A. Girdhar and F. Hebert, *2011 CSMANTECH Conference*, Palm Springs, CA, May 16-19, 2011.

2. K. Ota, K. Endo, Y. Okamoto, Y. Ando, H. Miyamoto, and H. Shimawaki, *Proc., IEEE Intl. Elect. Dev. Meeting*, Piscataway NJ, USA, 2009.

3. H. Sun, A. R. Alt, H. Benedickter, E. Feltin, J.-F. Carlin, M. Gonschorek, N. R. Grandjean and C. R. Bolognesi, *IEEE Elect. Dev. Lett.*, **31**(9), 957, (2010).

4. M. Higashiwaki, T. Mimura and T. Matsui, *Appl. Phys. Express*, **1**, 021103 (2008).

5. L. McCarthy, P. Kozodoy, M. Rodwell, S. Denbaars, and U. Mishra, *Compd. Semicond.*, **4** (8), 16 (1998).

6. J. Han, A. G. Baca, R. J. Shul, C. WIllison, L. Zhang, F. Ren, A. Zhang, G. Dang, S. Donovan, X. Cao, H. Cho, K. Jung, C. Abernathy, S. Pearton, and R. Wilson, *Appl. Phys. Lett.*, **74**(18), 2702 (1999).

7. J. J. Huang, M. Hattendorf, M. Feng, D. Lambert, B. Shelton, M. Wong, U. Chowdhury, T. Zhu, H. Kwon, and R. D. Dupuis, *IEEE Elect. Dev. Lett.*, **22** (4), 157 (2001).

8. H. G. Xing, and U. K. Mishra, *Int. J High Speed Electron. Syst.*, **14** (3), 819 (2004).

9. D. Keogh, P. Asbeck, T. Chung, J. Limb, D. Yoo, J.-H. Ryou, W. Lee, S.-C. Shen, and R. D. Dupuis, *Electron. Lett.*, **42** (11), 661 (2006).

10. L. S. McCarthy, I. P. Smorchkova, P. Fini, et al. *Electron. Lett.*, **38** (3), 144 (2002).

11. K. Kumakura, T. Makimoto, and N. Kobayashi, *Phys. Stat. Sol. (A)*, **188** (1), 363, (2001).

12. T. Makimoto, Y. Yamauchi, and K. Kumakura, *Appl. Phys. Lett.*, **84** (11), 1964 (2004).

13. B. F. Chu-Kung, C. Wu, G. Walter, M. Feng, N. Holonyak, Jr., T. Chung, J. Ryou, and R. Dupuis, *Appl. Phys. Lett.*, **91** (23), 2321141, (2007).

14. A. Nishikawa, K. Kumakura, and T. Makimoto, *Appl. Phys. Lett.*, **91**(13), 1 (2007).

15. S.-C. Shen, Y. Lee, H.-J. Kim, Y. Zhang, S. Choi, R. D. Dupuis, and J.-H. Ryou, *IEEE Elect. Dev. Lett.*, **30** (11), 1119 (2009).

16. T. Makimoto, K. Kumakura, and N. Kobayashi, *Jpn. J. Appl. Phys.*, **43**, 4B,1922, (2004).

17. Y.-C. Lee, Y. Zhang, H.-J. Kim, S. Choi, Z. Lochner, R. D. Dupuis, J.-H. Ryou, S.-C. Shen, *IEEE Trans. Electron Dev.*, **57**(11), 2964 (2010).

18. K. Kumakura, and T. Makimoto, *Appl. Phys. Lett.*, **92**(15), 153509 (2008).

19. S.-C. Shen, R. D. Dupuis, Y.-C. Lee, H. J. Kim, Y. Zhang, Z. Lochner, P. D. Yoder, and J.-H. Ryou, *IEEE Electron Device Lett.* **32** (8), 1065 (2011).

20. Y. Lee, Y. Zhang, Z. Lochner, H.-J. Kim, J.-H. Ryou, R. D. Dupuis, and S.-C. Shen, *Physica Status Solidi(a)*, **209** (3), 497 (2012).

21. Purchased from SAFC Hitech, Haverhill, MA, USA.

22. Y. Zhang, J.-H. Ryou, R. D. Dupuis, and S.-C. Shen, *Proc. Int. Conf. Compd. Semicond. Manuf. Technol. Dig. Papers*, Chicago, IL, Apr. 2008.

AlGaN/GaN MIS-HEMT Gate Structure Improvement Using Al₂O₃ Deposited by
PEALD and BCl₃ Gate Recess Etching

R. Meunier[a], A. Torres[a], M. Charles[a], E. Morvan[a], M. Plissonier[a], and F. Morancho[b]

[a] CEA-Leti, MINATEC Campus, 17 Rue des martyrs, 38054 Grenoble Cedex 9, France
[b] LAAS-CNRS, 7 Avenue du Colonel Roche, 31400 Toulouse, France

The enhancement of electric properties of MIS structures on an AlGaN/GaN heterostructure using Al_2O_3 a gate dielectric are investigated using C(V) and $I_d(V_g)$ measurements. The Al_2O_3 layer was deposited using two types of atomic layer deposition (ALD) techniques: thermal ALD and plasma enhanced ALD. Using PEALD over thermal ALD led to an increase of the threshold voltage V_{th} of 4V, and the suppression of non-uniform C(V) behavior by reducing traps at the Al_2O_3/AlGaN interface. Gate leakage current was also reduced by 6 decades and an I_{on}/I_{off} ratio of 10^9 was achieved, with a subthreshold slope of 81mV/decades. Further improvements were achieved by gate recess etching before the high-k deposition through BCl_3 reactive ion etching (RIE). We were able to further increase V_{th} by 4V while reducing gate leakage current, achieving a 10^{10} I_{on}/I_{off} ratio, without degrading the subthreshold slope and the abruptness of the transition.

Introduction

AlGaN /GaN heterostructures are very promising for the elaboration of high-power and high frequency devices because of their excellent electrical properties such as a high breakdown voltage, a high electron saturation velocity and a high mobility of the 2D electron gas (2DEG) (1-2). Furthermore, the possibility of growing high quality AlGaN/GaN heterostructures on large diameter Si substrates represent a good step forward in order to reduce the production costs. In terms of device realization, the Metal Insulator Semiconductor (MIS) gate structure with the introduction of high dielectric constant (high-κ) materials as a gate dielectric represents one of the most promising ways to achieve viable power electronic devices (3-4). Such structures can lead to effective gate leakage reduction compared to non-insulated Schottky gate structures. Among various insulators commonly selected in the world of microelectronics, Al_2O_3 is mostly used for its deposition simplicity and offers advantages of a large band gap (9eV), a high dielectric constant (κ~10) and a high breakdown field. It has already brought very good results, though it often needs post deposition treatments and surface pre-conditioning (5). Recently, plasma-enhanced atomic layer deposition (PEALD) has been given considerable interest since it can lead to MIS-gate structures with better performances compared to a Schottky-gate device (6).

In order to achieve the best performances, it is crucial to understand the different mechanisms taking place at the interface between the high-κ and the AlGaN. While all those different interactions are not yet fully understood, previous papers already stated that the oxidation as well as the carbon contamination on the surface may induce the presence of surface traps, which in turn have a drastic influence on RF dispersion as well as breakdown voltage, decreasing the overall efficiency of the devices (7). In this context, it is very important to have a good interface quality between the high-k and the AlGaN surface.

This work is focused on capacitance/voltage C(V) and drain-current/gate-voltage Id(Vg) measurements analysis of MIS structures realised using two types of Al_2O_3 as gate dielectric. In both cases, tri-methyl aluminium (TMA) was used as a precursor, but in one case water is used as oxidizer while oxygen plasma is used in the other. We will respectively refer to these techniques as thermal and plasma-enhanced ALD (PEALD).

Due to the properties of the AlGaN/GaN heterostructure, the devices realised are in a normally-off configuration. In order to increase the threshold voltage V_{th}, rapid thermal annealing (RTA) were carried out just after the gate dielectric deposition, and MIS structures with a gate recess etching prior to ALD were also realised. The aim is to increase V_{th} as much as possible by reducing positive charge trapping in the Al_2O_3 in case of the RTA, and by reducing the thickness of the AlGaN barrier under the gate in case of the recess etching, without degrading the good performances achieved through the PEALD MIS structure.

Experimental procedure

Device processing

The AlGaN/GaN heterostructure was grown on silicon using a metal-organic chemical vapor deposition process. The epitaxy consists of a 2µm thick buffer and GaN bilayer over which a 25nm AlGaN barrier is then grown.

Device processing is carried through multiple lithography steps. Isolation is first made by MESA etching. Ti/Al Ohmic contacts are then formed at 900°C. A 10nm Al_2O_3 layer is then deposited at 250°C by either thermal ALD or PEALD, and the gate is deposited using a Cr/Au stack. Figure 1 represents a typical MIS-HEMT structure.

Regarding post-PEALD RTA, a 400°C and a 650°C for 1 minute were performed just before gate stack deposition on a 6nm thick Al_2O_3 layer. As for the gate recess etching, it was performed through a BCl_3 RIE etching. Gate recesses of 10, 15 and 20nm were realized, followed by the deposition of 6nm thick Al_2O_3 layer deposition through PEALD. All the ALD processes were carried out on a Cambridge Nanotech Fiji.

The C(V) measurement were performed on 400µm diameter diodes and Id(Vg) measurements were performed on 1mm width circular transistors with a 100µm gate length.

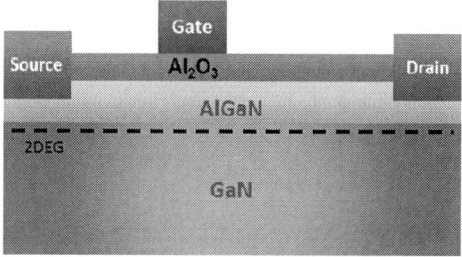

Figure 1. MIS-HEMT on an AlGaN/GaN heterostructure.

Results

Thermal ALD vs PEALD:

As we can see in Figure 2, two distinct behaviors appeared depending on the oxidation process used during the ALD. The one using H_2O showed a stepped C(V) curve while the one using oxygen plasma led to a smooth and steep non-stepped on/off transition. The threshold voltage (Vth) was also increased from ~-9V to ~-5V. In the latter case, the same sharp behavior and steady capacitance below Vth was also obtained for frequencies as low as 0.1kHz, while the H_2O samples led to negative capacitance below 50kHz.

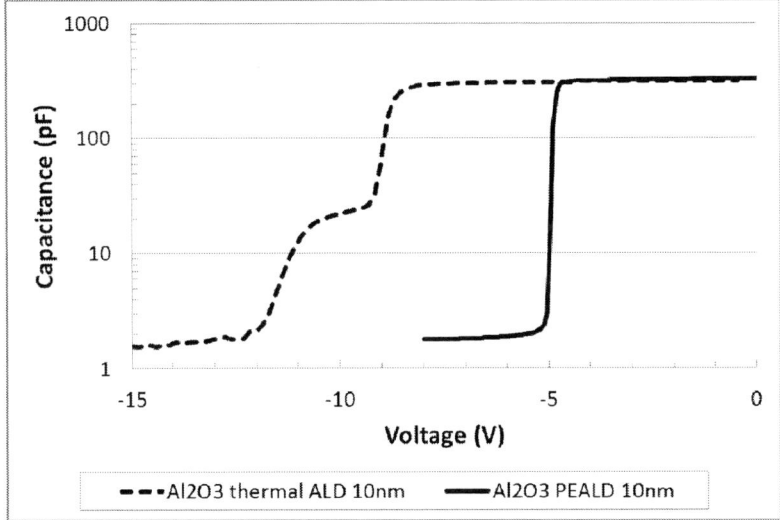

Figure 2. C(V) measurements for MIS diodes with 10nm Al_2O_3.

Regarding Id(Vg) measurements, we see in Figure 3 the same increase in Vth as before, as well as a drastic gate leakage current (I_{leak}) reduction for the PEALD sample. Gate leakage current as low as 1e-12A/mm was measured, for an average of 1e-11A/mm for a drain/source bias Vds=5V. The I_{on}/I_{off} ratio was increased from 10^3 to 10^9, with an I_{leak} reduction of more than 6 orders of magnitude. Speed of the transition between the on and off state was also greatly increased, and we were thus able to obtain a sub-threshold slope of 80mV/decades.

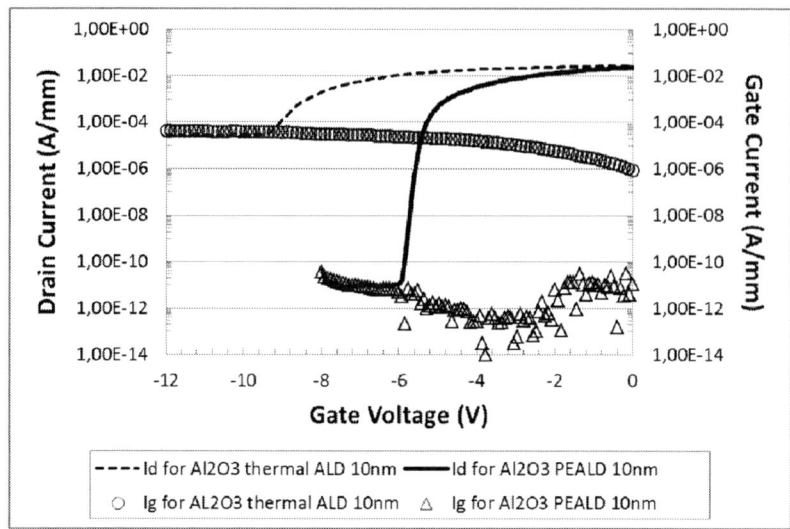

Figure 3. $I_d(V_g)$ measurements for MIS diodes with 10nm Al_2O_3.

Regarding C(V) and Id(Vg) results, the V_{th} improvement can be linked to a reduction of trapped charges through the O_2-plasma ALD deposition technique compared to thermal ALD. Furthermore, the better quality of the O_2-plasma oxide is confirmed by stable low frequency measurements, while the negative capacitance obtained with H_2O deposition is characteristic of a leaky behavior. Overall reduction of traps can also be confirmed by the sharper transitions between the on and off states.

Those traps can be associated to the carbon contamination of the AlGaN surface. Thus the improvement of the results between the two deposition techniques may come from a better carbon removal at the surface during the first cycles of plasma assisted ALD. XPS studies confirmed that an oxygen plasma treatment could greatly reduce carbon contamination (8)

In case of thermal ALD, we also observed through TEM images non-uniformity of the Al_2O_3 deposition, with a detachment of the oxide layer at the gate periphery as visible in Figure 10. This is due to a degradation of the MIS structure though the deposition of thermal ALD and could explain the stepped behavior of the device. It can thus be assimilated as having two capacitances in parallel (9).

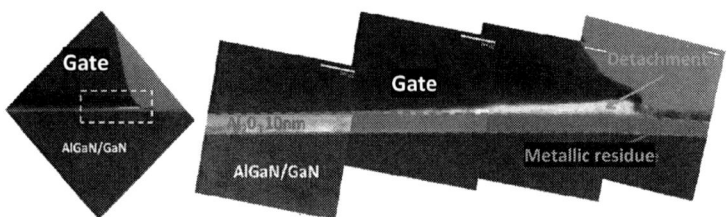

Figure 10. TEM image of the gate detachment at the gate periphery for thermal ALD

Rapid thermal annealing:

As previously stated, the aim of the introduction of a post-PEALD RTA was to see if it was possible to further improve the electrical properties of our PEALD MIS structures. The objective here is to anneal the high-k at different temperature in order to reduce positive charge trapping inside the Al_2O_3 and thus increase the V_{th}.

As we can see in Figure 4, the RTA didn't have the expected effect. Compared to a MIS structure without any RTA, it decreased the Vth and slightly improved the minimum capacitance C_{min}. And the higher the RTA temperature, the more the V_{th} was decreased and the C_{min} increased.

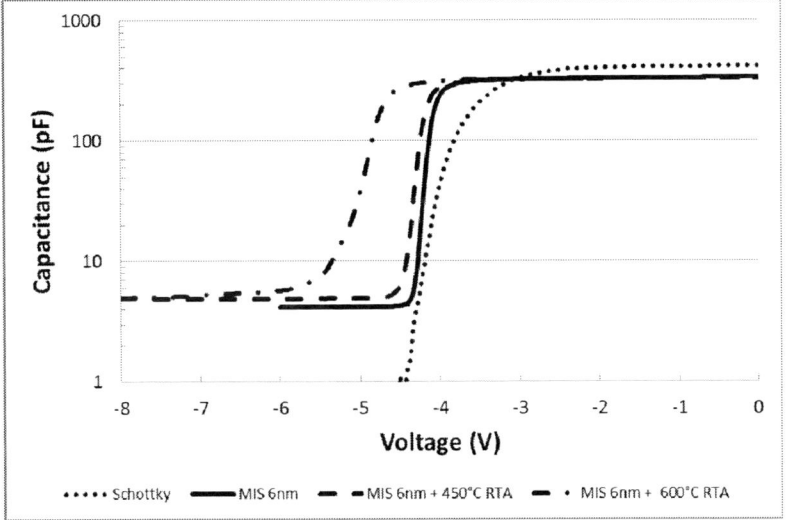

Figure 4. C(V) measurements for MIS diodes with 6nm Al_2O_3 with and without RTA, with a Schottky reference.

As for $I_d(Vg)$ measurements, we see in Figure 5 that RTA led to lower V_{th} and higher gate leakage current in both cases, with again higher degradations the higher the RTA temperature was.

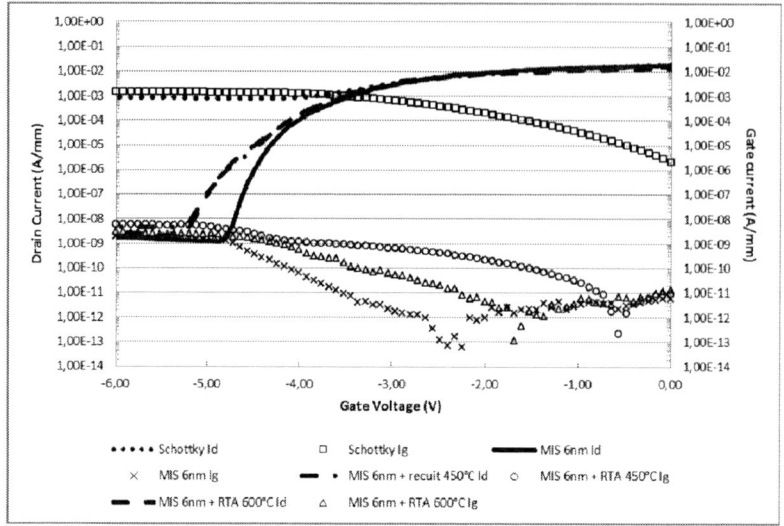

Figure 5. $I_d(V_g)$ measurements for MIS diodes with 6nm Al_2O_3 with and without RTA, with a Schottky reference.

After RTA, one possible explanation regarding the overall decrease of electrical performances could be due to the fact that there is a formation of a bad quality Ga_2O_3 at the interface. This is confirmed by the fact that the C(V) transitions are less abrupt and with a higher C_{min}, and that the gate leakage current levels are higher. As with the thermal ALD, this behavior is characteristic of bad quality interface between the oxide and the AlGaN surface.

Gate recess etching:

Numerous papers have already stated that it is possible to increase V_{th} in AlGaN/GaN heterostrctures by reducing the thickness of the AlGaN barrier. If we look at the band diagram in figure 6 and equation [1], assuming the Schottky barrier height ϕ_B is constant, theory predicts a linear behavior of the V_{th} with the remaining AlGaN thickness t_{RA}, N_{2D} being the density of the 2D electron gas. Figure 7 represents a typical MIS-HEMT transistor with a partially recessed gate structure.

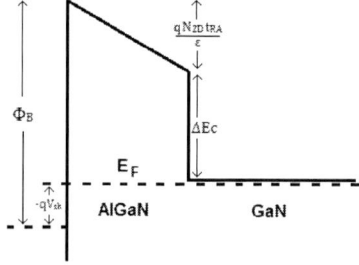

Figure 6. Band diagram for an AlGaN/GaN heterostructure with a gate recess.

$$qV_{th} = \Phi_B - \Delta E_C - \frac{qN_{2D}t_{RA}}{\varepsilon}$$

[1]

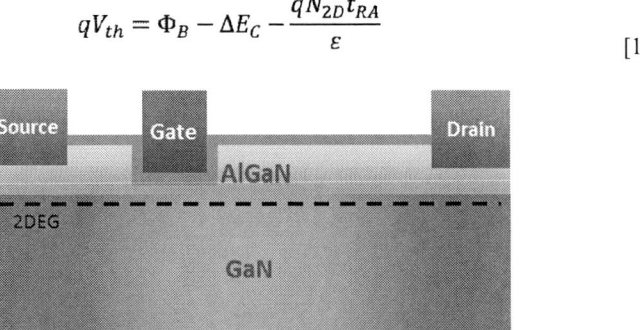

Figure 7. MIS-HEMT transistor on AlGaN/GaN on AlGaN/GaN heterostrcture with partial gate recess.

As we can see in Figure 8, the deeper the recess the higher the V_{th} increase. With a recess of 20nm, it was possible to increase V_{th} by 4V, without sacrificing the abrupt transition between the on and off state. Furthermore, measurements were stable again for frequencies as low as 0.1kHz, which indicates a good preservation of the low leakage properties introduced via the PEALD MIS structure. However, depletion was slightly degraded with a higher C_{min}, and a small hysteresis was introduced between the depletion and accumulation regime. Regarding the Schottky reference sample, the depletion appears to be better, but it is due to the fact that we obtained negative capacitances in the off state, even at 100kHz. This is characteristic of a leaky behavior of the structure.

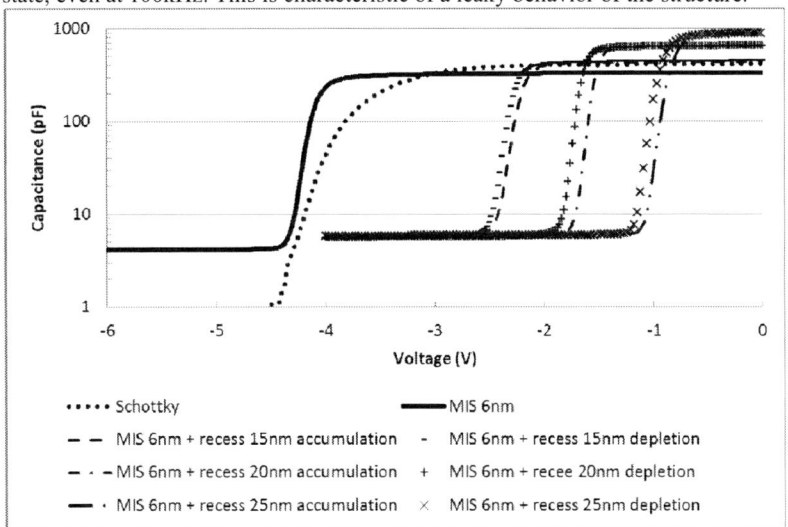

Figure 8. C(V) measurements for MIS diodes with 6nm Al_2O_3 with and without RTA, with a Schottky reference.

Regarding $I_d(V_g)$, we can see in Figure 9 a nice V_{th} increase, as well as gate leakage current reduction. Again we find that the further the recess, the higher the V_{th}, and the lower the gate leakage current. We were thus able to achieve gate leakage currents as low as 10^{-12}A.mm^{-1} in the case of the 20nm recess, for a record I_{on}/I_{off} ratio of 10^{10}. We can see that the reference Schottky sample is again very leaky, thus confirming what we already obtained on the C(V) characteristic.

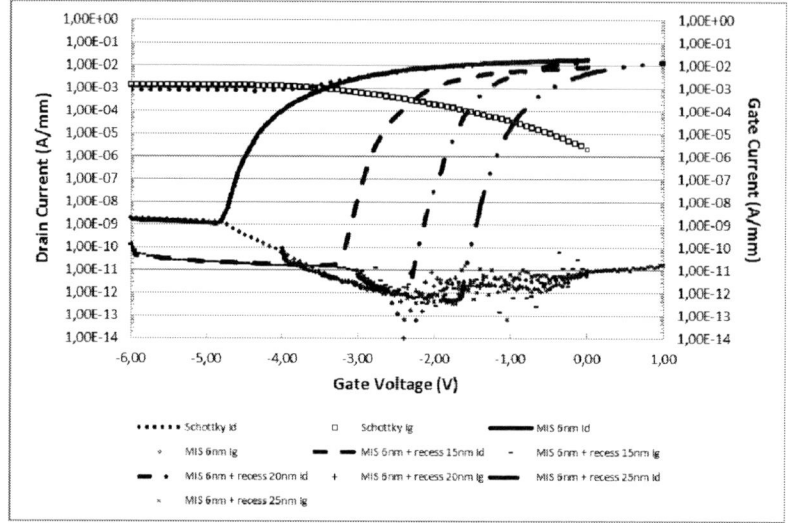

Figure 9. $I_d(V_g)$ measurements for MIS diodes with 6nm Al$_2$O$_3$ with and without RTA, with a Schottky reference.

Table 1 is a summary of all the different results obtained through RTA and gate recess processing.

TABLE I. Summary of the different results obtained through Schottky and MIS structures, with and without RTA or gate recess etching.

Samples	V_{th} (accumulation / depletion)	Electron density N_S (cm^{-2})	I_{on}/I_{off} ratio
Schottky	-4.21V	7.12e^{12}	1.59e^1
MIS 6nm	-4.24V	6.65e^{12}	1.55e^7
MIS 6nm + RTA 450°C	-4.36V	6.76e^{11}	6.98e^6
MIS 6nm + RTA 600°C	-5.00V	7.55e^{11}	2.74e^6
MIS 6nm + recess 10nm	-2.33V/-2.38V	4.79e^{12}	5.72e^8
MIS 6nm + recess 15nm	-1.73V/-1.94V	5.06e^{12}	1.68e^{10}
MIS 6nm + recess 20nm	-0.96V/-1.00V	3.67e^{12}	2.85e^{10}

While it not fully linear when going from the reference MIS structure to the one with a 10nm recess, the increase in V_{th} can be considered linear when looking at the three recessed structure with a 0.7V increase every 5nm recessed. Regarding the C_{min} augmentation, it can be related to the roughness increase after the etching. Previous work has already proven that a roughness increase could hinder the canal depletion (8). Regarding the hysteresis generation, it might be due to the overall interface traps increase because of a slightly poorer Al$_2$O$_3$/AlGaN interface, despite the use of PEALD. This could be remedied to by using a high temperature TMAH treatment after the BCl$_3$ etching in order to smooth the surface out(10). As for the lower gate leakage current and the conservation of the abrupt transition between the on and off state, this could come from the fact that we are closer to the canal under the gate, thus achieving a better control of our devices.

Conclusion

In regard to this study, it appears clearly that PEALD is vastly superior to classic thermal ALD in terms of electrical stability, rapidity of transition and gate leakage reduction. An average increase of 4V in the threshold voltage was obtained while the gate leakage current was reduced by 7 orders of magnitude. The I_{on}/I_{off} ratio was also greatly increased to 10^9 with a subthreshold slope of 80mV/dec. Furthermore, those very good results can be achieved without any prior surface preparations.

While RTA proved ineffective in order to further increase our results, gate recess etching represent a viable solution to increase the V_{th} towards positive voltage, without degrading in a significant way the overall performances achieved through the PEALD MIS structure. We were thus able to further increase the threshold voltage by 4V, while reducing again the gate leakage current level, leading to a record 10^{10} I_{on}/I_{off} ratio.

Regarding further perspectives, a post-etching surface regeneration could be achieved through a hot TMAH chemical treatment, while a complete recess of the AlGaN barrier combined with a good quality PEALD oxide could lead to normally-off devices.

Acknowledgments

We would like to thank the III-V Lab which provided some of the substrates on which we conducted the experiments and the people in charge at the PTA (Plateforme Technologique Amont) where devices were processed.

References

1. O. Ambacher, B. Foutz, J. Smart, J. R. Shealy, N. G. Weimann, K. Chu, M. Murphy, A. J. Sierakowski, W. J. Schaff, L. F. Eastman, R. Dimitrov, A. Mitchell, and M. Stutzmann, *Journal of Applied Physics, AIP*, **87**, 334-344, (2000).
2. R. Oberhuber, G. Zandler and P. Vogl, *Applied Physics Letters, AIP*, **73**, 818-820 (1998).
3. T. Hashizume, S. Anantathanasarn, N. Negoro, E. Sano, H. Hasegawa, K. Kumakura and T. Makimoto, *Japanese Journal of Applied Physics*, The Japan Society of Applied Physics, **43**, L777-L779 (2004).
4. N. Maeda et al., *Microelec. Proceedings of SPIE*, vol. 7216 (2009)
5. O. Saadat and T. Palacios, *Solid-State Device Research Conference*, (2011) 287-290
6. R. Lossy et al., *Journal of Vacuum Science and Technology A*, **31** (2013)
7. R. Vetury, N. Zhang, S. Keller and U. Mishra, *Electron Devices, IEEE Transactions*, **48**, 560-566 (2001).
8. R. Meunier et al., *ECS Meeting Abstracts*, MA2012-02 (2012) 2571
9. R. Escoffier et al., *SOI Conference (SOI)*, 2012 IEEE International (2012)
10. K.-S. Im, K.-W. Kim, D.-S. Kim, H.-S. Kang, D.-K. Kim, S.-J. Chang, Y.-H. Bae, S.-H. Hahm, S. Cristoloveanu and J.-H. Lee, *International Journal of High Speed Electronics and Systems*, **21**, No. 1 (2012) 1250007

278

GaN Power Transistors with Integrated Thermal Management

C.R. Eddy, Jr.[a], T.J. Anderson[a], A.D. Koehler[a], N. Nepal[b], D.J. Meyer[a], M.J. Tadjer[b], R. Baranyai[c], J.W. Pomeroy[c], M. Kuball[c], T.I. Feygelson[a], B.B. Pate[a], M.A. Mastro[a], J.K. Hite[a], M.G. Ancona[a], F.J. Kub[a] and K.D. Hobart[a]

[a] U.S. Naval Research Laboratory, 4555 Overlook Ave., SW, Washington, DC 20375, USA
[b] American Society for Engineering Education, 1818 N Street, NW, Washington, DC 20036, USA(residing at NRL)
[c] University of Bristol, H.H. Wills Physics Laboratory, Bristol, BS8 1TL, United Kingdom

The concept for and design/fabrication of a GaN power transistor with integrated thermal management is presented. Key elements of the design, including those supporting enhancement-mode operation, high breakdown voltages and low on-resistance are described in detail. The importance of surface preparation and growth of high-κ gate dielectrics using atomic layer deposition is summarized. Aspects of barrier design and surface passivation to promote low access resistance including lattice matched barriers and novel low-temperature AlN passivations are discussed. Finally, the integration of nanocrystalline diamond coatings on both the top-side and the back-side of the device for thermal management are described. Initial performance assessments of each of these components, as measured in the device operation, are presented and future efforts are highlighted.

Introduction

GaN power transistors offer great promise for applications in compact, high efficiency power systems simply based upon the properties of III-N materials. These properties result in a Baliga power figure-of-merit that is almost 700X greater than silicon and 160X that of 4H silicon carbide. Despite this promise there remain several challenges to reliable GaN power switches including reducing the thermal limit to enable ultimate performance, reducing gate and forward blocking leakage and reducing the on-resistance of the device. This paper presents recent efforts to address GaN power switch limitations through the development of an enhancement-mode (e-mode) lateral HEMT with integrated nanocrystalline diamond coatings for thermal management – a schematic of which is shown in Figure 1.

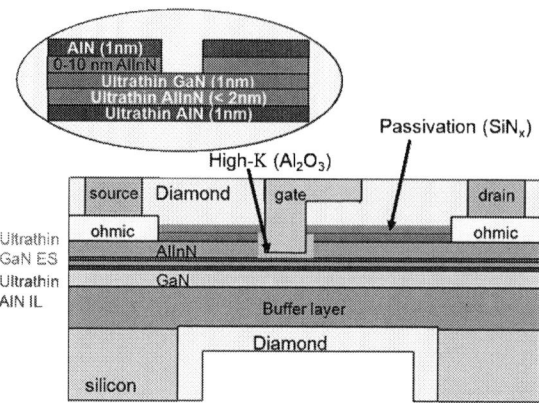

Figure 1.   Schematic cross-section of GaN power transistor with integrated nanocrystalline diamond coatings.

Device Design Concept

The device concept presented schematically in the cross section of Figure 1 aims to achieve a higher overall power density and preserve enhancement-mode (e-mode) operation all while keeping high reliability in mind. High current density is promoted through the use of an AlInN barrier that provides both a lattice matched barrier (improved reliability) and high sheet charge density, >2X conventional AlGaN/GaN(1) as well as high temperature stability. E-mode operation is achieved through a recessed gate process that brings the gate dielectric and metal within nanometers of the channel. Additional threshold voltage engineering can be achieved through implementation of a back-barrier to confine the GaN channel (2). Finally, nanocrystalline diamond layers are integrated on both the top-side and the back-side of the device to aid in efficient heat extraction from the drain edge of the gate where heat generation is known to occur (3).

Normally-off Operation

Normally-off operation is made possible through a combination of barrier design, recess processing and, to avoid gate leakage under forward bias, development of a gate dielectric. As shown in Figure 1, the barrier for the device involves a stack of very thin III-N layers that are designed to allow recess processing while remaining nearly-lattice-matched (to GaN) and, therefore, reducing built-in strain-related reliability concerns. The ultrathin, psuedomorphic AlN interlayer is employed to reduce scattering of carriers in the 2DEG, while the ultrathin $Al_{0.83}In_{0.17}N$ lattice-matched barrier has been shown to result in enhanced 2DEG sheet carrier density. The ultrathin GaN layer above the barrier enables a selective wet-chemical gate etch recess process we've previously demonstrated (4), while the top-most AlInN layer is designed to keep the channel sheet carrier density in the access regions high for low on-resistance performance. Growth of these layers will be challenged in both stoichiometry control of ternaries and thickness control. For these reasons we a developing novel growth technique in addition to conventional metalorganic chemical vapor deposition (MOCVD), namely, atomic layer epitaxy (ALE). Our initial results clearly demonstrate the potential of this method to grow crystalline quality III-N layers (5,6). Demonstrations of the complex barrier are expected soon.

With the recessed barrier demonstrated, there remains a key technical hurdle to be cleared, namely the high quality integration of a high-κ gate dielectric. Current state-of-the-art efforts employ high-k dielectrics deposited by atomic layer deposition (ALD) (7). We are investigating two ALD dielectrics – aluminum oxide (Al_2O_3) and hafnium oxide (HfO_2). Our early work has highlighted the critical need to prepare the III-N surface for high performance as will be discussed below.

High Power Operation

As the device is being developed for high power conversion applications, it is important that it both block high voltages in the off-state and be able to conduct high currents in the on-state. To support this objective it is essential to promote high sheet charge density and mobility through the gate stack design as discussed above and to ensure maximum breakdown between the gate and drain (where the highest electric fields are sustained in particular near the gate contact). As the III-N semiconductors have high breakdown fields, such failure is not expected in the barrier itself, but along the surface of the device instead. Traditional coatings in this region of rf devices have primarily involved silicon nitride layers aimed at passivating surface traps that limit frequency performance. These coatings perform well at low voltages, but it is felt that alternative coatings may be necessary for higher and higher voltage applications. In our approach we are examining solutions that address both breakdown concerns and thermal management as both AlN coatings and nanocrystalline diamond (NCD) coatings are being explored.

Integrated Thermal Management

A primary objective of our design is to address the thermal limitations experienced in rf HEMT devices and which are expected to be more challenging in high power switching HEMTs. To this end we employ an NRL developed diamond coating with a nanocrystalline structure that is highly insulating (10^{12} Ω-cm), has low dielectric loss and possesses a thermal conductivity (12 W/cm-K) that is 3X greater than copper. A key challenge is to integrate the NCD growth process (750°C hydrogen plasma for several hours) with the e-mode HEMT fabrication process. A crucial component to that integration is the nature of the surface upon which the NCD will be grown. Top-side NCD integration efforts have focused on seeded growth on silicon nitride (good HEMT passivation layer) and directly on the top of the barrier (to minimize interface thermal resistance). Early efforts showed that this approach provided good spreading of the heat from its generation source (drain edge of the gate) to contact pads on the top surface, but there remained a need to remove the heat from the device. To this end, efforts to integrate backside diamond are now being explored.

Highlights of Device Fabrication Progress

Fabrication of the device of Figure 1 involves a sophisticated fabrication process that includes a number of unique processing steps. Upon determination of an optimal process flow sequence, unit processing steps are being individually developed to be inserted in the process flow appropriately. Progress of several of these key steps is highlighted below.

High-κ Gate Dielectric Integration

High–κ gate dielectrics of Al_2O_3 and HfO_2 have been developed using ALD on both GaN and AlGaN surfaces as initial steps toward implementation on AlInN(8). The goal of these efforts is to develop a dielectric/semiconductor stack that exhibits the lowest gate leakage levels possible with minimal hysteresis and interface charge. For both semiconductor surfaces, surface pre-treatment is essential with each requiring a specific *ex situ* treatment (piranha etch for GaN, HF for AlGaN). Such dependence is important to understand as the III-N surfaces that will ultimately be in play are likely to have been subjected to surface altering processes that will need to be corrected. In addition, various post-deposition anneals were investigated (with the optimal for dielectrics on GaN being 600°C in N_2 for 60 sec) and found to enable improvements in the performance of the dielectric/semiconductor interface. Using these conditions hysteresis as small as 0.03V (Figure 2), interface trap densities below 10^{11} cm^{-2} and leakage current densities of $<10^{-7}$ A/cm^2 (Figure 3) have been achieved for gate biases less than +5V.

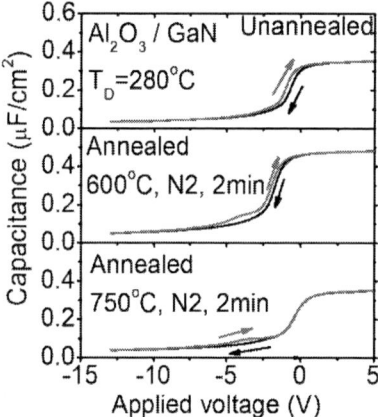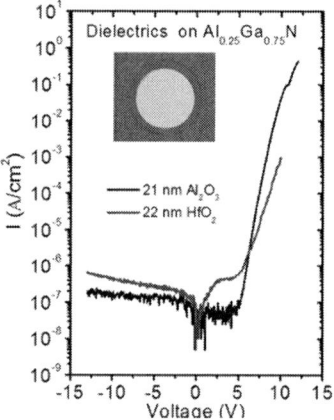

Figure 2. Capacitance-Voltage characteristics for Al_2O_3 dielectric capacitors on optimally treated GaN surfaces and then subjected to post-deposition anneals as described.

Figure 3. Current-voltage characteristics for high-κ dielectrics on optimally treated $Al_{0.25}Ga_{0.75}N$ surfaces demonstrating leakage current densities of $< 10^{-7}A$ cm^{-2} for biases below 5V.

Access Region Engineering

In order to maintain a low on-resistance in an e-mode device it is essential to keep the conductivity of the access regions as low as possible. Further, because GaN power switches offer the advantage of a high switching frequency, it is important to minimize any dispersion in the device's operation due to surface states. Both of these issues must be addressed through passivation approaches in the gate-drain and gate-source surface regions of the device. In the case of our diamond integrated HEMT, it is also important that the passivation solution be compatible with the diamond growth process. We have evaluated a number of passivation approaches including silicon oxides and silicon nitrides (PECVD and MOCVD), but have found interesting recent results using AlN passivation layers deposited by atomic layer epitaxy (ALE) (9). Using these layers we demonstrate lower sheet resistance (10%), lower off-state leakage (100X) and reduced dynamic on-resistance degradation (50%) compared to conventional SiN_x passivations. Key parameters of the device with this new passivation approach are compared with unpassivated and conventionally passivated devices in Table I.

TABLE I. HEMT parameters for HEMTS that are unpassivated and with 100 nm SiN_x, 4nm AlN Grown at 300C and 500C passivation.

Parameter	Units	Unpassivated	SiN_x	AlN 300 °C	AlN 500 °C
μ_{2DEG}	$cm^2/(V{\cdot}s)$	1703	1567	1790	1765
n_s	$10^{12}\ cm^{-2}$	7.12	8.60	8.30	8.55
R_{SH}	$\Omega/\square$	523	463	418	414
S	mV/dec	–	131	133	86
V_T	V	–	-3.51	-3.12	-3.25
$R_{ONDYN,0}$	Ω-mm	–	12.0	13.2	11.4
I_{ON}/I_{OFF}	–	–	4.5×10^3	8.3×10^4	3.0×10^5

Complex Barrier Development

In order to simultaneously ensure normally-off and high on-state current density operation, the design of the HEMT barrier becomes complex involving AlN interlayers, lattice-matched AlInN ternaries and ultrathin GaN etch stop layers, as shown in Figure 1. Growth of such layers is challenged using conventional MOCVD and MBE approaches. Recent efforts have focused on employing ALE to grow the barrier stack with first experiments focused on growing individual layers. Early results show much promise through the growth of all three barrier components (AlN, GaN and AlInN) (5,6). These layers are grown at temperatures that are 50% of conventional growth methods and result in crystalline materials that demonstrate x-ray rocking curve (XRC) full-width-at-half-maximum (FWHM's) that are comparable to much thicker films grown by conventional methods. For example, 36nm thick AlN layers demonstrate a 670 arc-sec XRC-FWHM, only 40% larger than 1-2 micrometer thick AlN grown by MOCVD and MBE. Even more importantly, the significantly reduced growth temperature permits realization of the full range of stoichiometries (Figure 4) of the AlInN ternary to be grown through careful use of the pulse sequence during ALE. Current efforts are aimed at assessing the quality of the 2DEG that forms when such layers are grown on GaN and will be the subject of reports in the near future.

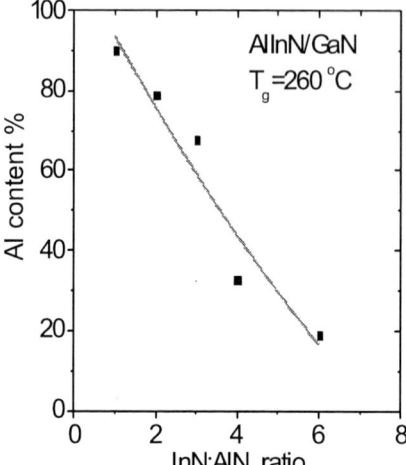

Figure 4. Variation of aluminum content in ALE-grown $Al_xIn_{1-x}N$ ternaries grown with the InN/AlN cycle ratio as determined by XPS.

Diamond Integration for Thermal Management

Aiming for high power operation is only worthwhile if the thermal limitations of even a high efficiency GaN power switch are mitigated. The primary objective is to effectively remove the heat from its source (drain edge of the gate) to heat sinks. We have employed top-side NCD (in our gate-after-diamond) approach to achieve 20% (25-50°C) reductions in device operating temperature (Figure 5) while actually <u>enhancing</u> the device's operation (both DC and RF), Figures 6 and 7. This is achieved through improved heat spreading from the generation point to contact pads with simultaneous improvement in sheet resistance and surface passivation. These approaches are quite effective is spreading the heat over a larger area on the front-side of the device and lowering the internal operating temperature, but the result also make clear that additional approaches will be needed to remove the heat from the device entirely. To that end a backside diamond-filled via approach is being investigated that will bring a significant volume of diamond within a couple microns of the heat generation point with minimal thermal boundaries. Shown schematically in Figure 1 and in scanning electron microscopy images in Figure 8, we demonstrate the feasibility of coating vias etched in silicon uniformly with diamond – a critical first step toward a high performance thermal management solution.

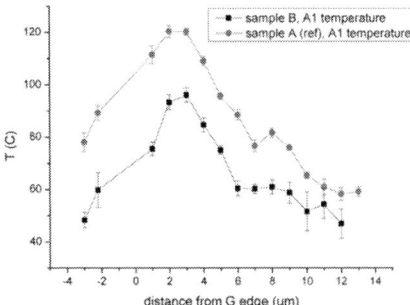

Figure 5. Local temperature as measured by Raman thermography of a NCD-coated HEMT (sample B) and a reference HEMT (sample A) as a function of distance along the gate-drain region of the device, using the Al Raman modes for the temperature measurement. A reduction in device temperature of up to 20-30% is demonstrated.

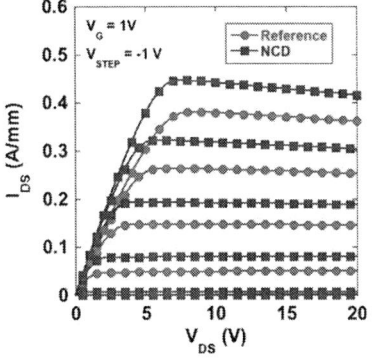

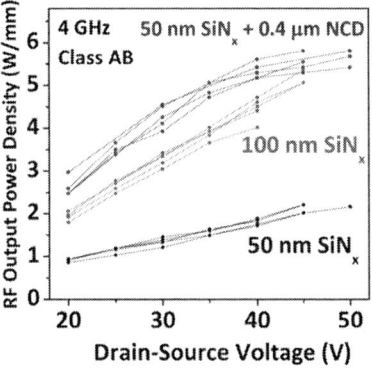

Figure 6. Comparison of DC I_D vs. V_{DS} curves for HEMT with (NCD) and without (Reference) NCD coating showing enhanced I_D^{max}.

Figure 7. RF output performance for GaN HEMT with various passivation coatings showing the enhanced RF output power performance for NCD.

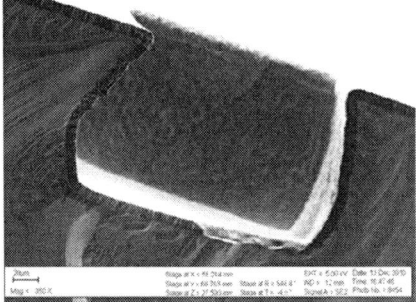

Figure 8. Demonstration of conformal coating of via in silicon with NCD – a first step toward backside thermal management strategies.

The Next Steps

These advances have resulted in critical demonstrations for the envisioned high power switch, but there remain more hurdles to be solved as efforts continue forward toward its realization. Current efforts are focused on increasing the threshold voltage which will also require increases in the performance of the gate dielectric. Of course, demonstrating the appropriate dielectric on the required III-N (AlInN) is essential and optimization will be required. Finally, continuing to refine the diamond filled via process to ensure maximum fill fractions while managing any induced strains will be an area of increasing effort.

Summary

These early developmental results to realize a GaN power switch with integrated thermal management offer great promise for GaN power switches that can perform at near-theoretical levels and, therefore, deliver on the promise of GaN power switches for high-power, high-efficiency, compact power conversion systems.

Acknowledgments

N. Nepal and M.J. Tadjer acknowledge the support of the American Society for Engineering Education NRL Postdoctoral Fellowship Program. Research conducted at the U.S. Naval Research Laboratory is supported by the Office of Naval Research.

References

1. S. Guo, et al., *Phys. Status Solidi. A* **207**, 1348 (2010).
2. T. Oka and T. Nozawa, *IEEE Electron Dev. Lett.* **29**, 668 (2008).
3. A. Wang, M.J. Tadjer and F. Calle, Semicon. Sci Technol. 28, 055010 (2013).
4. T.J. Anderson, M.J. Tadjer, M.A. Mastro, J.K. Hite, K.D. Hobart, C.R. Eddy, Jr. and F.J. Kub, *ECS Trans.* **28**, 65 (2010).
5. N. Nepal, N.A. Mahadik, L.O. Nyakiti, S.B. Qadri, M.J. Mehl, J.K. Hite and C.R. Eddy, Jr., *J. Cryst. Growth & Design* **13**, 1485 (2013).
6. N. Nepal, S.B. Qadri, J.K. Hite, N.A. Makadik, M.A. Mastro and C.R. Eddy, Jr., accepted for publication in *Appl. Phys. Lett.*, July 2013.
7. J.A. Kittl et al., *Microelectronic Eng.* **86**, 1789 (2009).
8. N. Nepal, J.K. Hite, N.Y. Garces, M.A. Mastro, D.J. Meyer and C.R. Eddy, Jr., *Appl. Phys. Exp* **4**, 055802 (2011).
9. A. Koehler, N. Nepal, T.J. Anderson, M.J. Tadjer, K.D. Hobart, C.R. Eddy, Jr. and F.J. Kub, accepted for publication in *IEEE Elect. Dev. Lett.*, July 2013.

Ammonothermal Bulk GaN Substrates for Power Electronics

M. P. D'Evelyn, D. Ehrentraut, W. Jiang, D. S. Kamber, B. C. Downey, R. T. Pakalapati, and H.-D. Yoo

Soraa, Inc., Goleta, California 93117, USA

Soraa has developed a novel ammonothermal approach for growth of high quality, true bulk GaN crystals at a greatly reduced cost, known as SCoRA (Scalable Compact Rapid Ammonothermal). SCoRA GaN growth has been performed on seed crystals with diameters between 5 mm and 2" to thicknesses of 0.5-4 mm. The highest growth rates are greater than 40 μm/h and rates in the 10-30 μm/h range are routinely observed. Two-inch diameter, crack-free, free-standing, n-type bulk GaN crystals have been grown. The crystals have been characterized by a range of techniques, including x-ray diffraction rocking-curve (XRC) analysis, optical microscopy, cathodoluminescence (CL), optical spectroscopy, and capacitance-voltage measurements. The crystallinity of the grown crystals is very good, with FWHM values of 15-80 arc-sec and average dislocation densities below 5×10^5 cm^{-2}.

Introduction

Although most attention in GaN-based power devices has focused on lateral, GaN-onSi or GaN-on-SiC devices, vertical, GaN-on-GaN power device technology is widely believed to be extremely promising. Recently, Disney et al. demonstrated vertical GaN-on-GaN diodes with near-theoretical specific on-resistance as a function of breakdown voltage, substantially better than the best SiC-on-SiC or heteroepitaxial lateral GaN devices [1]. In addition, early-generation vertical GaN switches have been reported by a number of groups [2], [3], [4], [5], [6], [7], [8].

It is clear that one of the largest obstacles to further development of GaN-on-GaN power electronics is the substrate. Virtually all commercial bulk GaN substrates today are fabricated by hydride vapor phase epitaxy (HVPE) after nucleation on foreign substrates and have wafer-averaged dislocation densities in the range of 3×10^6 cm^{-2} – 1×10^7 cm^{-2}. These substrates became the standard for GaN-based laser diode (Blu-Ray^{TM}) manufacturing in about 2005 [9] and are currently being used by Soraa for GaN-on-GaN^{TM} high-performance LED manufacturing [10]. While HVPE GaN substrates represent a considerable improvement over heteroepitaxial GaN, they have a number of limitations that limit their usefulness for power switches, including diameter, cost, and undesirably-high dislocation density. Threading dislocations act as vertical current-leakage paths in GaN device structures [11], accelerate degradation of AlGaN/GaN high electron mobility transistors (HEMTs) [12], [13], and have a well-documented negative impact on laser diode lifetime [14]. Consequently, dislocation densities below 10^4 cm^{-2}, as are available in GaAs and should be available in true bulk GaN, are often regarded as being necessary for vertical GaN-on-GaN power devices.

The ammonothermal method has attracted considerable attention as a potential bulk GaN manufacturing method [15], [16], [17], [18], [19], [20], [21], [22], [23], [24]. Ammonothermal crystal growth is analogous to hydrothermal crystal growth, which is employed commercially to grow thousands of tons of α-quartz per year at low cost. The AMMONO company has demonstrated 1" and 2" diameter crystals with a dislocation density of about 5×10^3 cm^{-2}, a (0002) x-ray rocking curve with a full width at half maximum (FWHM) of about 17 arcsec, and a radius of curvature in the range of 100-1000 m using an ammonothermal technique [25], [26]. The crystalline quality of this material is believed to be fully adequate for GaN-on-GaN power devices. However, unfortunately, the growth rates in Ammono's process and in similar processes being investigated by other groups [23], [27], [28] are small, generally about 1-4 μm/h, despite more than a decade of development. The very low growth rates impact both the cost (currently much higher than HVPE) and the rate at which substrate size can be increased. HVPE can in principle be used to grow ultralow-defect crystals, given ultralow-defect seed crystals, but it seems likely that the cost will remain relatively high.

Other techniques have been proposed for true bulk GaN crystal growth, including growth from a flux comprising Na-Ga alloys [29]. However, while very respectable growth rates, ca. 30 μm/h, and very low defect levels [30] have been demonstrated, the solubility of N and/or Ga in all fluxes investigated to date is at least an order of magnitude lower than that for other crystals that are successfully manufactured by flux-based methods, raising questions about the manufacturability of these methods.

Experimental

Soraa's novel ammonothermal approach, known as SCoRA (Scalable Compact Rapid Ammonothermal) and shown schematically in Fig. 1, utilizes internal heating rather than external heating as in conventional ammonothermal reactors. Raw material, including seed crystals, polycrystalline GaN nutrient, a mineralizer, and ammonia, are placed inside a capsule and sealed. The capsule is surrounded successively by a heater, a ceramic shell

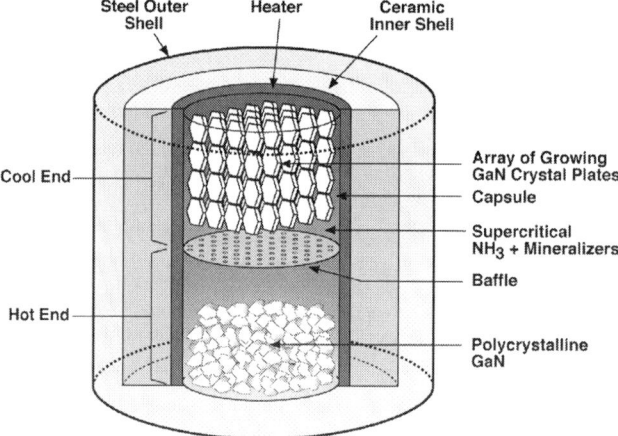

Figure 1. Schematic of internally-heated high pressure SCoRA apparatus.

providing structural support and thermal insulation, and an externally-cooled steel outer shell that provides mechanical confinement. The use of steel rather than a nickel-based superalloy for the pressure apparatus greatly reduces the cost, including both raw material and machining costs, and enables much larger volumes. The low thermal conductivity of the ceramic enables the steel shell to remain below 200 °C even at an operating temperature of 750 °C, maintaining high creep resistance. The SCoRA reactor has demonstrated capability for temperatures and pressures as high as 750 °C and 600 MPa, respectively, enabling higher growth rates than conventional ammonothermal techniques, but is less expensive and more scalable than conventional autoclaves fabricated from nickel-based superalloys.

The SCoRA reactor may be used with a range of mineralizer chemistries. The configuration shown in Fig. 1 is appropriate for a mineralizer with a positive solubility coefficient, e.g., NH_4Cl. With mineralizers showing retrograde solubility, e.g., $NaNH_2$, the positions of the seeds and nutrient are reversed.

In the present work we used either specially-prepared HVPE bulk GaN substrates or previously-grown SCoRA crystals or wafers as seeds. After hanging the seeds on a rack, placing the rack and the desired quantities of GaN nutrient, mineralizer, and ammonia into the capsule, the capsule was welded closed and placed in the SCoRA reactor. The SCoRA reactor was then closed and the capsule heated to the desired temperature distribution for the desired length of time. After cooling and removal of the ammonia from the capsule, the capsule was removed from the SCoRA reactor, opened, and the grown crystals collected and characterized. For the present study, growth temperatures and pressures ranged between 650-750 °C and 100-600 MPa, respectively.

The crystals were characterized by a combination of techniques, including optical microscopy, x-ray diffraction rocking-curve (XRC) analysis, and etch-pit density measurements of dislocation density. XRC was performed using a PANalytical MRD PRO high-resolution four-circle X-ray diffractometer operating in a receiving slit mode utilizing Cu Kα radiation (1.5405 Å) and a hybrid two-bounce Ge(220) monochromator. A 0.5° source slit and a 1 mm vertical detector slit were used throughout the analysis. Both on-axis and off-axis ω rocking curves were measured for the (002) and (201) GaN reflections, respectively, for c-plane oriented GaN substrates.

In some cases the crystals were sliced into wafers, lapped or ground, polished, and chemical-mechanically polished.

Results and Discussion

Some representative 2" c-plane SCoRA crystals, either as-grown or partially wafered, are shown in Fig. 2. Macro-steps, attributed to high supersaturation, have been observed occasionally [Fig. 2(a)]. Growth rates as high as 40 μm/h have been observed, although rates in the 10-30 μm/h range are more typical. The crystals typically are n-type, with a carrier concentration in the mid 10^{16} - mid 10^{18} cm^{-3} range.

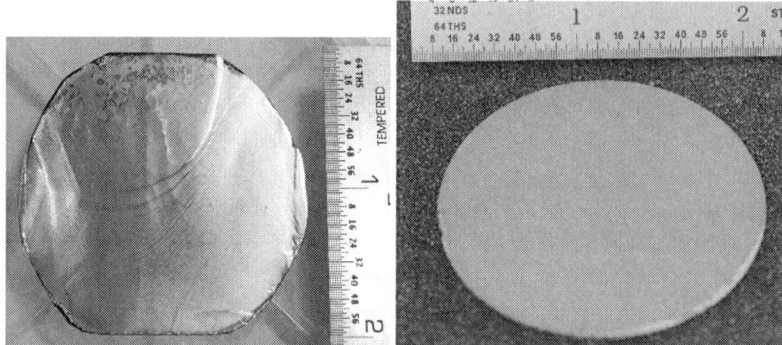

Figure 2. (a) As-grown; (b) cored, partially-polished 2" c-plane SCoRA crystals.

Dislocation densities in the as-grown or wafered crystals were evaluated by etch pit density measurements, being careful not to over- or under-etch. Depending on processing conditions, the values ranged from 5×10^4 cm^{-2} to 5×10^6 cm^{-2}. We are optimistic that the dislocation densities can be further reduced below 1×10^4 cm^{-2}.

XRC data for some c-plane crystals are shown in Fig. 3. Our XRD measurements where performed with a $1/2°$ source slit and a 1 mm detector slit, producing a beam approximately 1 mm wide and 12 mm in length, so a considerable area of the surface was illuminated for these measurements. These values indicate excellent crystallinity.

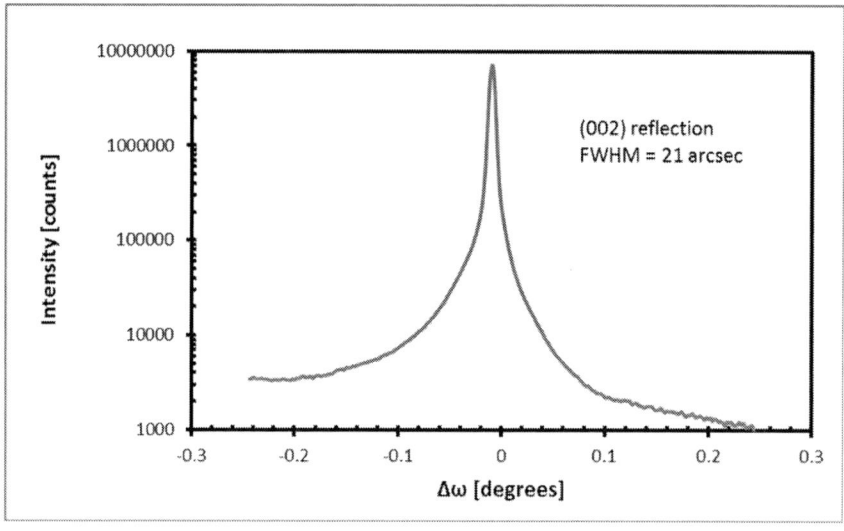

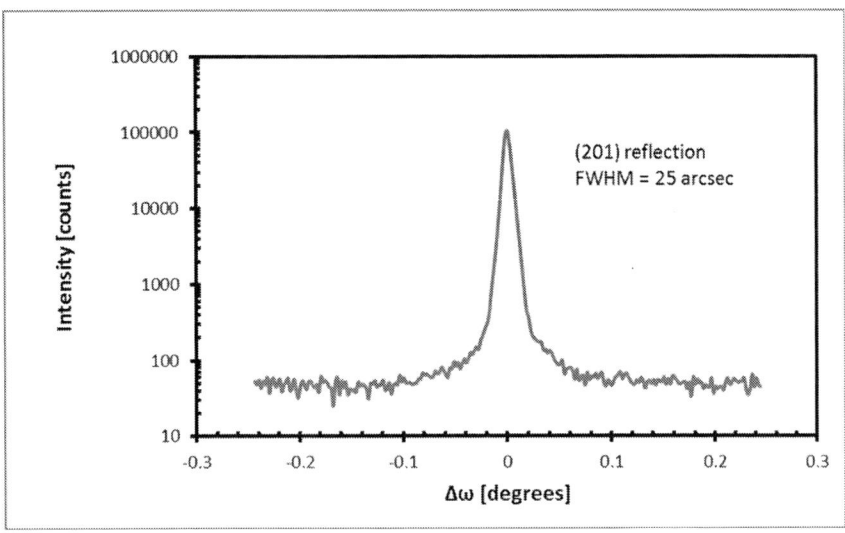

Figure 3. XRC data for (a) symmetric (002) reflection and (b) asymmetric (201) reflection of a c-plane SCoRA crystal;.

We believe that the rapid progress we have made synthesizing 2" bulk GaN crystals with a quality approaching requirements for power electronics is very encouraging. Current work focuses on further improvements in crystal quality, productivity, and wafering.

Acknowledgments

The authors gratefully acknowledge partial support for this work from the U.S. Advanced Research Projects Agency—Energy (Cooperative Agreement No. DE-AR0000020) and the U.S. National Science Foundation (Grant No. IIP-1026896). The authors also thank K. Hughes, M. Bagwell, E. Limon, J. Nink, D. Pocius, R. Vargas, and J. Cook for assistance with the crystal growth and characterization.

References

[1] D. Disney, H. Nie, A. Edwards, D. Bour, H. Shah, and I. C. Kizilyalli, "Vertical Power Diodes in Bulk GaN," *Proceedings of the 25th International Symposium on Power Semiconductor Devices*, pp. 59–62, 2013.

[2] M. Kanechika, M. Sugimoto, N. Soejima, H. Ueda, O. Ishiguro, M. Kodama, E. Hayashi, K. Itoh, T. Uesugi, and T. Kachi, "A Vertical Insulated Gate AlGaN/GaN Heterojunction Field-Effect Transistor," *Japanese Journal of Applied Physics*, vol. 46, no. No. 21, pp. L503–L505, May 2007.

[3]   T. Uesugi and T. Kachi, "GaN Power Switching Devices for Automotive Applications," *Paper presented at CS Mantech Confererence, Tampa, Florida, http://mantech.org/Digests/2009/2009 Papers/2.1.pdf*, 2009.

[4]   H. Otake, K. Chikamatsu, A. Yamaguchi, T. Fujishima, and H. Ohta, "Vertical GaN-Based Trench Gate Metal Oxide Semiconductor Field-Effect Transistors on GaN Bulk Substrates," *Applied Physics Express*, vol. 1, no. 1, p. 011105, Jan. 2008.

[5]   T. Fujishima, H. Otake, and H. Ohta, "Reduction in threshold voltages in GaN-based metal oxide semiconductor field effect transistors," *Applied Physics Letters*, vol. 92, no. 24, p. 243505, 2008.

[6]   S. Chowdury, B. L. Swenson, and U. K. Mishra, "Enhancement and Depletion Mode AlGaN / GaN CAVET With Mg-ion-implanted GaN as Current Blocking Layer," *IEEE Electron Device Letters*, vol. 29, no. 6, pp. 543–545, 2008.

[7]   S. Chowdhury, M. H. Wong, B. L. Swenson, and U. K. Mishra, "Dispersion-free AlGaN/GaN CA VET with low Ron achieved with plasma MBE regrown channel with Mg-ion-implanted current blocking layer," in *Device Research Conference - Conference Digest, DRC*, 2010, pp. 201–202.

[8]   S. Chowdhury, M. H. Wong, B. L. Swenson, and U. K. Mishra, "CAVET on bulk GaN substrates achieved with MBE-regrown AlGaN/GaN layers to suppress dispersion," *IEEE Electron Device Letters*, vol. 33, no. 1, pp. 41–43, 2012.

[9]   S. Uchida, S. Agatsuma, T. Hashizu, T. Yamamoto, and M. Ikeda, "Short wavelength lasers based on GaAs and GaN substrate for DVD and Blu-ray technology," in *Technical Digest - International Electron Devices Meeting, IEDM*, 2006.

[10]   M. J. Cich, R. I. Aldaz, A. Chakraborty, A. David, M. J. Grundmann, A. Tyagi, M. Zhang, F. M. Steranka, and M. R. Krames, "Bulk GaN based violet light-emitting diodes with high efficiency at very high current density," *Applied Physics Letters*, vol. 101, no. 22, p. 223509, 2012.

[11]   S. W. Kaun, M. H. Wong, S. Dasgupta, S. Choi, R. Chung, U. K. Mishra, and J. S. Speck, "Effects of Threading Dislocation Density on the Gate Leakage of AlGaN/GaN Heterostructures for High Electron Mobility Transistors," *Applied Physics Express*, vol. 4, no. 2, p. 024101, Jan. 2011.

[12]   F. A. Marino, N. Faralli, T. Palacios, D. K. Ferry, S. M. Goodnick, and M. Saraniti, "Effects of threading dislocations on AlGaN/GaN high-electron mobility transistors," *IEEE Transactions on Electron Devices*, vol. 57, no. 1, pp. 353–360, 2010.

[13]   M. Tapajna, S. W. Kaun, M. H. Wong, F. Gao, T. Palacios, U. K. Mishra, J. S. Speck, and M. Kuball, "Influence of threading dislocation density on early

degradation in AlGaN/GaN high electron mobility transistors," *Applied Physics Letters*, vol. 99, no. 22, pp. 223501–223503, Nov. 2011.

[14]   S. Tomiya, T. Hino, S. Goto, M. Takeya, and M. Ikeda, "Dislocation Related Issues in the Degradation of GaN-Based Laser Diodes," *IEEE Journal of Selected Topics in Quantum Electronics*, vol. 10, no. 6, pp. 1277–1286, Nov. 2004.

[15]   R. Dwiliński, R. Doradziński, J. Garczyński, L. P. Sierzputowski, a. Puchalski, Y. Kanbara, K. Yagi, H. Minakuchi, and H. Hayashi, "Excellent crystallinity of truly bulk ammonothermal GaN," *Journal of Crystal Growth*, vol. 310, no. 17, pp. 3911–3916, Aug. 2008.

[16]   R. Dwiliński, R. Doradziński, J. Garczyński, L. Sierzputowski, R. Kucharski, M. Zając, M. Rudziński, R. Kudrawiec, J. Serafińczuk, and W. Strupiński, "Recent achievements in AMMONO-bulk method," *Journal of Crystal Growth*, vol. 312, no. 18, pp. 2499–2502, Sep. 2010.

[17]   D. R. Ketchum and J. W. Kolis, "Crystal growth of gallium nitride in supercritical ammonia," *Journal of Crystal Growth*, vol. 222, no. 3, pp. 431–434, Jan. 2001.

[18]   A. P. Purdy, R. J. Jouet, and C. F. George, "Ammonothermal Recrystallization of Gallium Nitride with Acidic Mineralizers," *Crystal Growth & Design*, vol. 2, no. 2, pp. 141–145, Mar. 2002.

[19]   B. Wang and M. J. Callahan, "Ammonothermal Synthesis of III-Nitride Crystals," *Crystal Growth & Design*, vol. 6, no. 6, pp. 1227–1246, Jun. 2006.

[20]   T. Hashimoto, F. Wu, J. S. Speck, and S. Nakamura, "Ammonothermal growth of bulk GaN," *Journal of Crystal Growth*, vol. 310, no. 17, pp. 3907–3910, Aug. 2008.

[21]   T. Fukuda and D. Ehrentraut, "Prospects for the ammonothermal growth of large GaN crystal," *Journal of Crystal Growth*, vol. 305, no. 2, pp. 304–310, Jul. 2007.

[22]   D. Ehrentraut, Y. Kagamitani, C. Yokoyama, and T. Fukuda, "Physico-chemical features of the acid ammonothermal growth of GaN," *Journal of Crystal Growth*, vol. 310, no. 5, pp. 891–895, Mar. 2008.

[23]   M. Saito, D. S. Kamber, T. J. Baker, K. Fujito, S. P. DenBaars, J. S. Speck, and S. Nakamura, "Plane Dependent Growth of GaN in Supercritical Basic Ammonia," *Applied Physics Express*, vol. 1, p. 121103, Dec. 2008.

[24]   Y. Nojima, M. Ikari, E. Letts, and T. Hashimoto, "Improvement of structural quality in the initial stage of GaN growth by basic ammonothermal method," *Journal of Crystal Growth*, vol. 317, no. 1, pp. 132–134, Feb. 2011.

[25]   R. Dwiliński, R. Doradziński, J. Garczyński, L. P. Sierzputowski, a. Puchalski, Y. Kanbara, K. Yagi, H. Minakuchi, and H. Hayashi, "Excellent crystallinity of truly

bulk ammonothermal GaN," *Journal of Crystal Growth*, vol. 310, no. 17, pp. 3911–3916, Aug. 2008.

[26] R. Dwiliński, R. Doradziński, J. Garczyński, L. Sierzputowski, R. Kucharski, M. Zając, M. Rudziński, R. Kudrawiec, J. Serafińczuk, and W. Strupiński, "Recent achievements in AMMONO-bulk method," *Journal of Crystal Growth*, vol. 312, no. 18, pp. 2499–2502, Sep. 2010.

[27] D. Ehrentraut, Y. Kagamitani, T. Fukuda, F. Orito, S. Kawabata, K. Katano, and S. Terada, "Reviewing recent developments in the acid ammonothermal crystal growth of gallium nitride," *Journal of Crystal Growth*, vol. 310, no. 17, pp. 3902–3906, Aug. 2008.

[28] B. Wang, D. Bliss, M. Suscavage, S. Swider, R. Lancto, C. Lynch, D. Weyburne, T. Li, and F. a. Ponce, "Ammonothermal growth of high-quality GaN crystals on HVPE template seeds," *Journal of Crystal Growth*, vol. 318, no. 1, pp. 1030–1033, Mar. 2011.

[29] Y. Mori, M. Imade, K. Murakami, H. Takazawa, H. Imabayashi, Y. Todoroki, K. Kitamoto, M. Maruyama, M. Yoshimura, Y. Kitaoka, and T. Sasaki, "Growth of bulk GaN crystal by Na flux method under various conditions," *Journal of Crystal Growth*, vol. 350, no. 1, pp. 72–74, 2012.

[30] M. Imanishi, K. Murakami, H. Imabayashi, H. Takazawa, Y. Todoroki, D. Matsuo, M. Maruyama, M. Imade, M. Yoshimura, and Y. Mori, "Coalescence growth of dislocation-free GaN crystals by the Na-flux method," *Applied Physics Express*, vol. 5, no. 9, 2012.

ECS Transactions, 58 (4) 295-298 (2013)
10.1149/05804.0295ecst ©The Electrochemical Society

1000V Vertical JFET Using Bulk GaN

Q. Diduck, H.Nie, B. Alvarez, A. Edwards, D.Bour, O. Aktas, D. Disney,
and I. C. Kizilyalli

Avogy Inc.,677 River Oaks Parkway, San Jose, California 95134, USA

Bulk GaN substrates with low defect density are now commercially available. This material enables the fabrication of vertical GaN devices with high breakdown voltage, and excellent electrical performance and power device figure of merit. In this paper, we present a 1000V Vertical JFET that has a positive threshold of 1V. The normally-off FET integrates a hetero-junction in the design to improve the transconductance and enable efficient switching operation into the 10s of MHz.

Introduction

There is a great interest in developing power devices based on wide bandgap materials, such as silicon carbide (SiC) and gallium nitride (GaN), as silicon power devices are approaching their physical limits. The theoretical material property based Power device-figure-of-merit (PFOM) of GaN is at least 5X better than SiC and nearly 1000X better than Si. To date, the majority of GaN power device development effort has been directed toward lateral devices, such as high-electron mobility transistors (HEMTs), fabricated using thin layers of GaN that are grown on foreign substrates. However, the performance and reliability of these lateral GaN devices have fallen well short of expectations. Some well-known issues include current-collapse, dynamic on-resistance, and inability to support avalanche breakdown are limiting their potential. By fabricating vertical transistors on bulk GaN substrates, it is possible to realize the full potential of GaN including true avalanche breakdown capability. The fundamental benefits of GaN, coupled with the many shortcomings of existing lateral GaN structures, provide ample motivation to develop vertical power devices in utilizing bulk GaN substrates. This paper reports on the progress we have made in our vertical transistor devices.

Lateral HEMT based power devices suffer from a design limitation that makes it difficult to support a large breakdown field. This is due to large fields being applied to lateral gates near the surface. This exposes high fields to passivation layers and creates high field regions near the gate that can cause electro-migration and even semiconductor cracking near the gate area, resulting in unreliable operation [1]. To compensate for this a variety of field plate methods have been employed to limit the field near the gate. Also surface modification methods have been used local to the gate to increase breakdown. All of these methods improve performance at the cost of increasing device complexity. Handling the surface and defect states at high fields has also been challenging. While low voltage microwave devices have overcome most of these limitations, the higher energies involved with high voltage power switches significantly increases the challenge.

295

To realize the full potential of GaN and to achieve normally-off operation we utilize a combined lateral and vertical structure in our JFET design. Additionally we incorporate the benefits of a heterojunction, but control the fields by incorporating a secondary vertical structure. The net result is a device that has good trans-conductance, low capacitance and supports large breakdown potential.

Vertical JFET Device Structure - Measurement Results and Performance

We utilize bulk GaN substrates to reduce our active area defect density to below 1×10^4 /cm^2. This significantly reduces the defect challenges presented to the lateral portion of the device, but is also necessary for operation of the vertical structure. The device structure has been optimized such that the gate depletion achieves a 1V threshold and thus the device operates in enhancement mode. A range of breakdown voltage and on-resistance design can be realized by adjusting the drift region doping and thickness. The elimination of lattice mismatch supports the fabrication of large vertical drift regions that allow for multi-kV operation.

Several device design splits were fabricated on multiple wafers, sweeping a variety of process and dimensional parameters. The net result was a design that performed with a 1000V breakdown voltage measured at less than 0.1mA of leakage as well as devices with a specific on resistance of less than 4.8mΩ-cm^2. It should be noted that over 50% of this on resistance can be attributed to the substrate not being thinned prior to back metalization. While this result is not beyond the silicon carbide limit, even without substrate thinning it is well beyond the silicon limit. Current device research is focusing on reducing the on-resistance to demonstrate the full potential of GaN. Shown in Figure 1 is a typical current-voltage response and transfer curve.

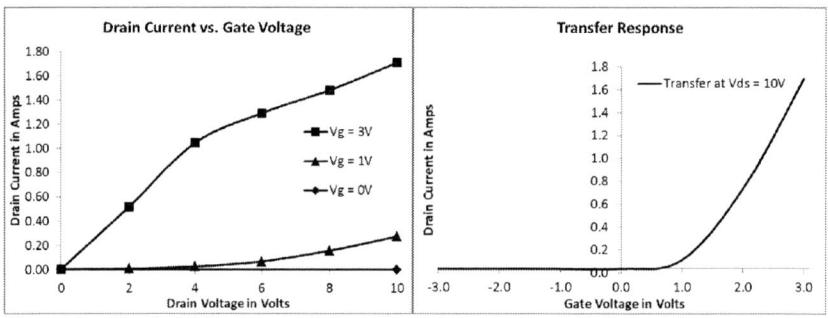

Figure 1.   Typical IV characteristics of a 1A component (left) and its transfer curve (right).

As can be observed from the transfer curve in Figure 1, the device has a threshold voltage of approximately 1V. The linear operating region of the device extends to over 1A with a knee voltage of approximately 4.5V suggesting good ohmic contact performance. Improvements from wafer thinning and drift region doping are expected to bring future devices close to the GaN BFOM limit. Published results from vertical SiC power devices indicate that SiC devices are already approaching the theoretical limit for this material system and GaN is so far the only path forward to an improved power switch [2-5].

Several devices were packaged and utilized in power switching applications. The devices were operated in switch mode at frequencies in excess of 10 MHz with very sharp transitions. This suggests that these devices will enable a new class of power supplies that operate at very high switching frequencies. This will lead to much more compact power conversion systems. Shown in Figure 2 is a die photo of a packaged part.

Figure 2. Die photo of GaN Vertical JFET.

Conclusion

In this paper we presented a Vertical JFET using bulk GaN. The device supported over 1000V breakdown, with less than 0.1mA of leakage. By utilizing bulk GaN it is possible to achieve a 5x performance improvement over Silicon Carbide and these initial device results suggests that it is realistic to expect that level of performance from GaN. The vertical nature of these devices supports the development of multi-kilovolt components. These initial results suggest that optimized devices using GaN are likely to displace silicon carbide device as the component of choice for high performance power switching applications.

Acknowledgments

The authors would like to acknowledge the support of this work by ONR under contract number N0001413M0034 and project monitor Daniel Green.

References

1. R. Quay, *Gallium Nitride Electronics*, Springer-Verlag, Berlin Heidelberg (2008)
2. D. Sheridan, A. Ritenour, V. Bondarenko, P. Burks, and J. Casady, *J. Proc. of Intl. Symp. Power Semiconductors (ISPSD)*, pp. 335-338, (2009).

3. A. Furukawa, S. Kinouchi, H. Nakatake, Y. Ebiike, Y. Kagawa, N. Miura, Y. Nakao, M. Imaizumi, H. Sumitani, and T. Oomori, *Proc. of Intl. Symp. Power Semiconductors (ISPSD)*, pp. 288-291, (2011).
4. T. Nakamura, Y. Nakano, M. Aketa, R. Nakamura, S. Mitani, H. Sakairi, and Y.Yokotsuji, *IEDM Tech. Dig.*, pp. 26.5.1 – 26.5.3, (2011).
5. S. Balachandran, C. Li , P.A. Losee, I.B. Bhat, and T.P. Chow, *Proc. of Intl. Symp. Power Semiconductors (ISPSD)*, pp. 293-296, (2007).

CHAPTER 8

MANUFACTURING CHALLENGES

300

Manufacturing Challenges in Wide Band Gap (WBG) Power Electronics

Krishna Shenai

Energy Systems Division, Argonne National Laboratory
9700 South Cass Avenue, Bldg. 362, Argonne, IL 60439-4815

Abstract

Wide Band Gap (WBG) semiconductors, Silicon Carbide (SiC) and Gallium Nitride (GaN), have enormous potential for transformative impact on electric utility and transportation infrastructures. However, commercialization of WBG power electronics switching devices has been slow primarily due to high cost and unproven field-reliability. This paper reviews the manufacturing challenges facing the WBG power electronics industry in order to "unlock" the true potential of this "game changing" solid-state energy conversion technology.

I. Introduction

It has been well-known for more than two decades that power electronic switching devices made from Wide Band Gap (WBG) semiconductors such as Silicon Carbide (SiC) and Gallium Nitride (GaN) offer much lower on-state resistance ($R_{DS(on)}$) than comparable silicon power devices because of their superior electrical and thermal conductivities, and breakdown field compared to silicon [1, 2]. Single-chip SiC JBS power diodes rated up to 1,700V/25A [3], and more recently, 1,700V vertical GaN Schottky power diodes [4] have been introduced into the commercial market in limited quantities. Although these devices are finding applications in computer/telecom power supplies, motor control, and smart grid [5-7], serious concerns pertaining to long-term reliability of these devices in compact power converters under stressful field operating conditions remain [8]. For example, majority of WBG commercial power diodes are not dv/dt- and avalanche-rated, and the data sheets rarely mention of the Safe-Operating Area (SOA), especially at elevated temperatures.

For high efficiency and high-frequency power conversion, MOS-controlled power semiconductor switches are needed [9]. Power MOSFET is the basic building block for other MOS-controlled high-power switches such as IGBT's. The "best-in-class" commercial 1,200V SiC power DMOSFET's [10] and trench-gate MOSFET's [11, 12] have a specific on-state resistance $R_{sp,on}$ of 3.7 mΩ-cm^2 and 2.6 mΩ-cm^2, respectively and a single-chip current rating of 50 amps. In the current state-of-the-art SiC power MOSFET's, the gate dielectric suffers from reliability problems, especially at elevated temperatures above 150°C. For example, gate MOS threshold voltage has been found to be unstable with prolonged gate voltage stress [13]. Low MOS inversion channel mobility and poor gate dielectric reliability have been largely attributed to a high density of interface states, especially in the upper half of the bandgap close to the conduction band edge. Recently, AlON/SiO$_2$ gate dielectric films deposited by CVD technique have been shown to improve both MOSFET performance and reliability of planar as well as trench-gate devices [14]. For "industry-best" 1,200V SiC power MOSFET's, measured reverse leakage currents are typically high; and, devices are not rated for dv/dt and avalanche capabilities, especially at elevated temperatures. At the time of this writing, SiC power MOSFET's have not been proven reliable in power converter applications and their application is limited to junction temperatures below 150°C [15].

The best-in-class commercial GaN power transistors are rated at 600V and have single-chip current ratings of < 20 amps [16]. These gate-controlled normally-off GaN-on-Si power switches are lateral in configuration, have a gate threshold voltage of ~ 1.5V, and a specific on-state resistance $R_{sp,on} = 2.5$ mΩ-cm^2. The device utilizes hole-injection at the gate electrode from the p-AlGaN to the AlGaN/GaN hetero-junction, which simultaneously increases the electron density in the channel, resulting in a dramatic increase of the drain current due to conductivity modulation. A recent advance in this transistor technology pertains to an integrated Schottky diode that provides the reverse-conducting current path during switching [17]. Using GaN lateral enhancement-mode switches, a three-phase motor drive inverter rated at 900 Watts and switching at 6 kHz was demonstrated with a switching energy efficiency of 99.3% [18]. However, the measured reverse leakage currents are high, especially at elevated temperatures; device breakdown is primarily caused by breakdown at the buffer-substrate and/or at the device surface, and hence, these devices are not optimized for power switching. Furthermore, these devices are not rated for dv/dt and avalanche capabilities; and, are 3-5X more expensive than silicon power transistors with identical current ratings. To the best of our knowledge, there is not a single silicon discrete lateral high-power switch available in the commercial market. This lack of market pull for a lateral power switch is largely due to the fact that lateral power devices are not scalable to higher voltages and currents, and are not cost-effective and reliable in power converter circuits [9]. Hence, vertical power transistor technology on free-standing bulk n$^+$GaN substrates must be developed.

II. Market Drivers

As shown in Figure 1, WBG power semiconductor switches have enormous potential to miniaturize power electronics converters (higher frequency and higher temperature operation) with much higher energy efficiency compared to today's silicon-based power converters. The result is dramatic cost reduction and energy savings of power processing equipment. Figure 2 illustrates industry trend in the miniaturization of power processing equipment and the need for SiC-based power converters in transportation, grid, and other applications [19].

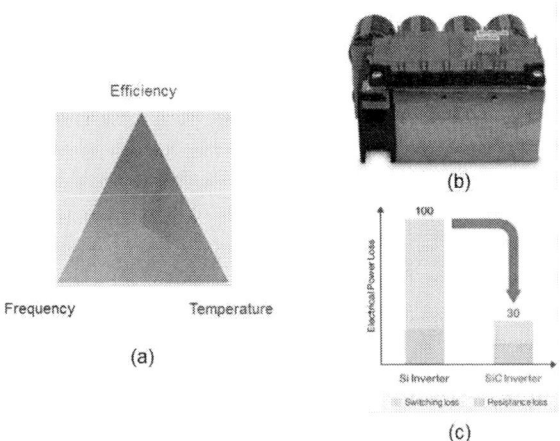

Figure 1: (a) Power electronics converter design criteria, (b) A 11 kW all-SiC inverter from Mitsubishi Electric, and (c) Energy savings in the 11 kW all-SiC inverter compared to all-silicon inverter.

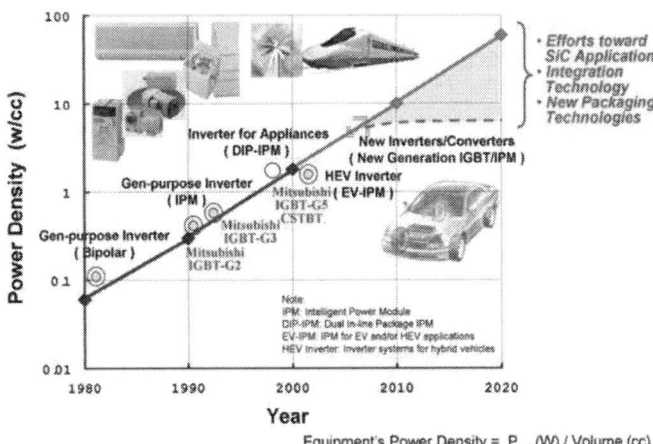

Figure 2: Growth of power density in power electronics systems and its projection by Mitsubishi Electric. SiC power devices, high-density packaging, and system integration are needed for further miniaturization of power converters and enable next generation transportation, grid, and other applications.

III. Economic Opportunity

Key applications of WBG power devices are shown in Figure 3 [19]. Vertical SiC power devices rated up to 1.7kV and lateral GaN/Si power transistors rated up to 200V are commercially available today. A few companies have started to provide samples of 600V lateral GaN/Si power transistors. At the time of writing of this report, only SiC power diodes are used in select power converter applications. The global market share of US companies for SiC power diodes in 2012 is about $100 million per year which is about half of world-wide sales. Most market analysts [20] predict a world-wide market size of about $3 billion per year by 2020 (about 9% of the total market of power switching devices) for SiC power devices. Figure 4(a) illustrates the SiC device market share, split by voltage range. This analysis clearly suggests that bulk of SiC power device market in the next 5 years is for medium-voltage (1.2 – 1.7kV). On the other hand, GaN power device market size is expected to be less than $1 billion per year in the same time frame, and these devices are expected to be rated blocking capabilities of < 600V. From cost considerations, at least for the near future, GaN power devices are likely to be lateral in construction and will be fabricated on low-cost silicon substrates. As illustrated in Figure 4(b), GaN-on-Si lateral power transistors will find applications in low-voltage (< 200V) point-of-load (POL) power converters mostly used in battery-powered wireless OEM devices and PFC/power supplies requiring power semiconductor switches with blocking voltages below 600V.

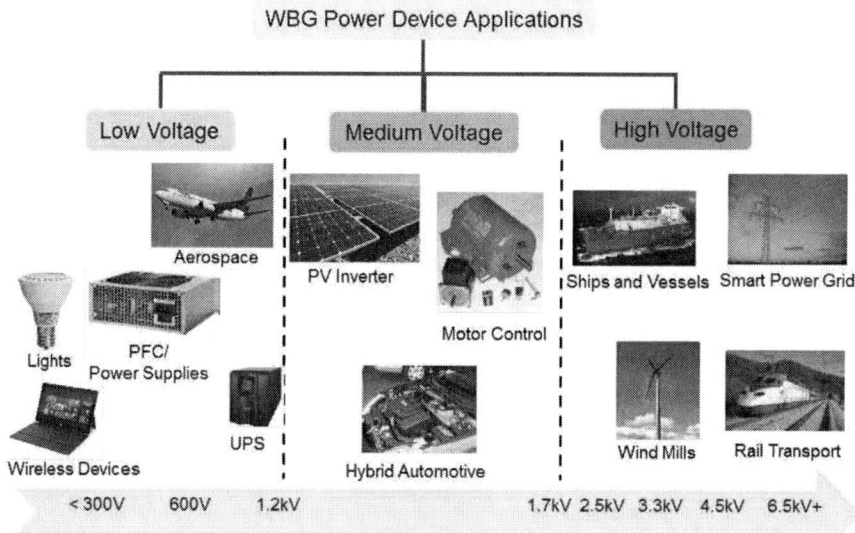

Figure 3: Key applications of SiC and GaN power devices.

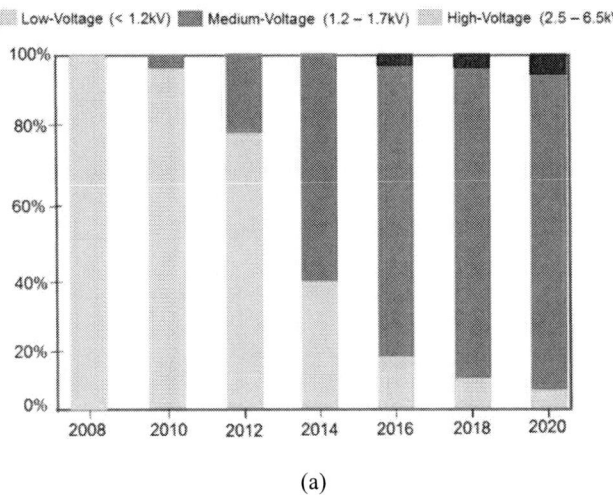

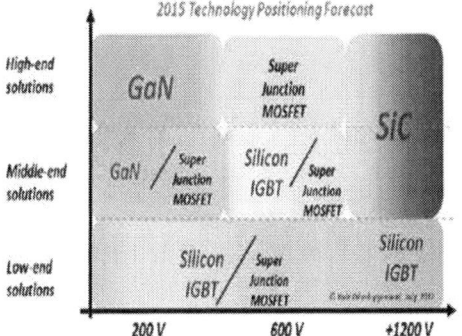

(b)

Figure 4 (a) SiC device market share in %, split by voltage range (data based on expected sales of devices in $ millions per year), and (b) projected application space of GaN power devices (Source: Yole Development).

IV.   Manufacturing Challenges

In order to realize the full market potential of both SiC and GaN power devices in low- and medium-voltage power converter applications, significant manufacturing challenges must be overcome:

- Develop low defect-density, large-area (6 inch diameter) starting wafers: The key challenges are to increase the wafer throughput (and reduce cost/wafer) and reduce the defect density. For SiC, native defects in the substrate (needed to seed the crystal) propagate into the epitaxial layer; additional defects are also introduced during the high-temperature growth of lightly-doped epitaxial layers needed to support high voltage. A recent study (see Figure 5) reports breakdown voltage yield on 76 mm 4H-SiC wafers as low as 40% for 0.5 – 2.0 kV JBS power diodes [21] suggesting a possible breakdown voltage degradation due to defect-induced leakage currents.

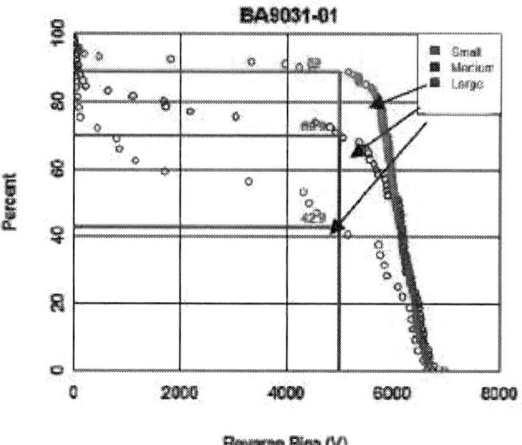

Figure 5: Distribution of diode blocking voltages measured on 76 mm diameter 4H-SiC epitaxial wafer, averaging 52 micron thick, with average doping density of 8.5×10^{14} cm^{-3} [21].

For GaN-on-Si wafers, key challenges are to improve the quality of the interface at the buffer layer and substrate, improve the quality of the two-dimensional electron gas (2DEG), and reduce the defect density (and trapping centers) in the drift-region that supports high voltage in lateral power devices [22-24]. For vertical GaN power device manufacturing, high-quality (low defect density and high thermal conductivity) homo-epitaxial n⁻GaN layer on bulk n⁺GaN substrate (GaN/GaN) is needed; significant improvement in the thermal conductivity of n⁻GaN epitaxial material has been reported with a reduction in the crystal defect density [25].

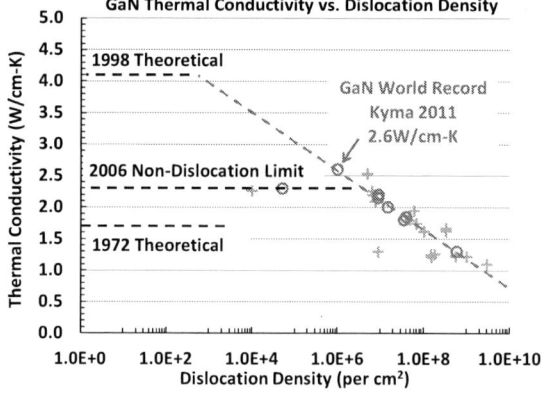

Figure 6: Non-dislocation density limited thermal conductivity of HVPE GaN substrates [25].

- Develop rugged device structures with > 100 amps/chip current-handling capability: Important technical challenges need to be addressed in order to develop rugged high-voltage and high-power switching devices. Perhaps the immediate challenge is to develop 600V – 650V power MOSFET's with single-chip current-handling capability exceeding 100 amps as shown in Figure 7 [15].

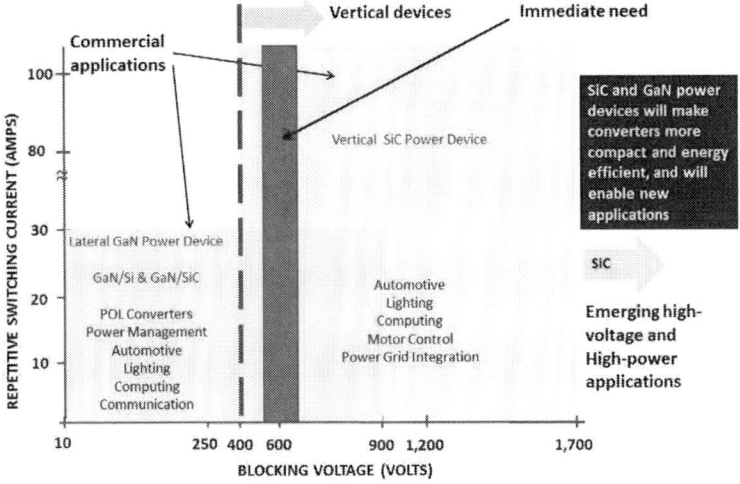

Figure 7: Suggested technology roadmap for WBG power switching devices.

Other issues that need to be addressed are:

- Demonstrate manufacturable high-quality vertical trench metal-semiconductor oxide (MOS) technology in SiC with high channel mobility and long-term reliability.
- Demonstrate vertical channel insulated-gate 2DEG in GaN for the manufacture of high-voltage vertical GaN power devices.
- Develop techniques to control the minority carrier lifetime in the drift-region of a SiC bipolar device.
- Demonstrate thermally stable and reliable low-resistance contact metallization to both n^+ and p^+ layers in SiC and GaN.
- Demonstrate manufacturable and efficient high-voltage edge termination technology.
- Demonstrate reliable and stable high-voltage surface passivation technology for both GaN and SiC power devices.
- Develop power device structures that provide the best on-state and switching performance for a specified blocking voltage that are field-reliable and cost-effective.

- Develop manufacturable high-density, low-cost, high-performance packaging technology that is reliable in field-application environments for ambient temperatures in excess of 200°C.

- Demonstrate field-reliability of power devices and modules in high-volume commercial applications.

- Develop the metrology needed for yield assessment at various facets of power semiconductor manufacturing including bulk substrates, epitaxial wafers, fully processed devices, and packaged modules.

- Develop power semiconductor device and module product data sheet formats that can be used by power converter designers to assess both performance and reliability of application circuits.

- Develop a systematic approach to assess system-level benefits from a given power switching technology. The primary benefits include energy savings, cost reduction, reliability improvement, and size/weight reduction.

- Develop a systematic approach for cost reduction strategy in various facets of WBG power device manufacturing.

- Develop continuously evolving technology roadmaps and standards for power switching devices and modules that can be used by the power semiconductor manufacturing industry as the guideline.

IV. Summary and Conclusions

Formidable manufacturing challenges exist before WBG power switching technology can be widely accepted by end-user power electronics community. Arguably, cost and quality of starting wafers remain the foremost among these challenges. A clear understanding of the role of crystal defects on power device wafer yield and long-term field reliability must be understood; performance optimization and cost-reduction must be pursued synergistically with a "defect reduction" roadmap.

References

1. K. Shenai, R. S. Scott and B. J. Baliga, "Optimum semiconductors for high-power electronics," IEEE Trans. Electron Devices, vol. 36, no. 9, pp. 1811-1823, Sept. 1989.

2. K. Shenai, "Potential Impact of Emerging Semiconductor Technologies on Advanced Power Electronic Systems," IEEE Electron Device Lett., vol. 11, no. 11, pp. 520-522, November 1990.

3. Power Products, Cree Inc., www.cree.com

4. I. C. Kizilyalli, and A. Edwards, D. Bour, H. Shah, H. Nie and D. Disney, "Vertical devices in bulk GaN drive diode performance to near-theoretical limits," HOW2POWER TODAY, pp. 1-7, March 2013.

5. SiC 2010: How SiC will impact electronics: A 10 year projection, Yole Development Market Report, Lyon, France, 2010.

6. T. McDonald, "GaN based power technology stimulates revolution in conversion electronics," Electronics in Motion and Conversion, pp. 2-4, April 2009.

7. G. Deboy, R. Rupp, R. Mallwitz and H. Ludwig, "New SiC JFETs Boost Performance of Solar Power Inverters," Power Electronics Europe, issue 4, 2011, pp. 29-33.

8. K. Shenai, "Switching megaWatts with power transistors," The Electrochemical Society Interface Magazine, vol. 22, no. 1, pp. 47-54, Spring 2013 (**invited paper**).

9. B. J. Baliga, Modern Power Devices, Wiley: New York, 1987.

10. S-H. Ryu, L. Cheng, S. Dhar, C. Capell, C. Jonas, R. Callanan, A. Agarwal, J. Palmour, A. Lelis, C. Scozzie, and B. Geil, "3.7 $m\Omega$-cm^2, 1500 V 4H-SiC DMOSFET's for advanced high-power, high-frequency applications," Proc. 23rd IEEE Int. Symp. *Power Semiconductor Devices & IC's (ISPSD)*, pp. 227-230, May 23-26, 2011.

11. T. Nakamura, Y. Nakano, M. Aketa, R. Nakamura, S. Mitani, H. Sakairi, and Y. Yokotsuji, "High-performance SiC trench devices with ultra-low R_{on}," Digest of IEEE Int. Electron Devices Meeting (IEDM), pp. 599-601, Dec. 2011.

12. T. Nakamura, M. Sasagawa, Y. Nakano, T. Otsuka, and M. Miura, "Large current SiC power devices for automobile applications," Proc. 2010 IEEE Int. Power Electronics Conference, pp. 1023-1026, 2010.

13. A. J. Lelis, D. Habersat, R. Green, A. Ogunniyi, M. Gurfinkel, J. Suehle, and N. Goldsman, "Time dependence of bias-stress-induced SiC MOSFET threshold-voltage instability measurements," IEEE Trans. Electron Devices, vol. 55, no. 8, pp. 1835-1840, Aug. 2008.

14. T. Hosoi, S. Azumo, Y. Kashiwagi, S. Hosaka, R. Nakamura, S. Mitani, Y. Nakano, H. Asahara, T. Nakamura, T. Kimoto, T. Shimura, and H. Watanabe, "Performance and reliability improvement in SiC power MOSFET's by implementing AlON high-k gate dielectrics," Digest of IEEE Int. Electron Devices Meeting (IEDM), pp. 159-162, Dec. 2012.

15. K. Shenai, M. Dudley, and R. F. Davis, "Current status and emerging trends in wide bandgap (WBG) semiconductor power switching devices," ECS Journal of Solid State Science and Technology 2(8), N3055-N3063, July 2013 (**invited paper**).

16. Y. Uemoto, M. Hikita, H. Ueno, H. Matsuo, H. Ishida, M. Yanagihara, T. Ueda, T. Tanaka, and D. Ueda, "Gate Injection Transistor (GIT) – A normally-off AlGaN/GaN power transistor using conductivity modulation," IEEE Trans. Electron Devices, vol. 54, no. 12, pp. 3393-3399, Dec. 2007.

17. T. Morita, S. Ujita, H. Umeda, Y. Kinoshita, S. Tamura, Y. Anda, T. Ueda, and T. Tanaka, "GaN gate injection transistor with integrated Si Schottky barrier diode for high-efficient DC-DC converters," in Digest of IEEE Int. Electron Devices Meeting (IEDM), pp. IED12-151 to 12-154, 2011.

18. T. Morita, S. Tamura, Y. Anda, M. Ishida, Y. Uemoto, T. Ueda, T. Tanaka, and D. Ueda, "99.3% efficiency of three-phase inverter for motor drive using GaN-based

gate injection transistors," in IEEE Applied Power Electronics Conference (APEC), pp. 481-484, 2011.

19. G. Majumdar and T. Oomori, "Some key researchers on SiC device technologies and their predicted advantages," Power Electronics Europe, issue 6, pp. 18-22, 2009.

20. SiC 2010: How SiC will impact electronics: A 10 year projection, Yole Development Market Report, Lyon, France, 2010.

21. M. J. Loboda, G. Chung, E. Carlson, R. Drachev, D. Hansen, E. Sanchez, J. Wan, and J. Zhang, "Advances in SiC substrates for power and energy applications," Compound Semiconductor MANTECH Conference Digest, presented at Compound Semiconductor MANTECH Conference, May 16th-19th, 2011, Palm Springs, California, USA.

22. T. Paskova and K. R. Evans, "GaN substrates-progress, status, and prospects," IEEE J. Sel. Topics Quant. Elect. vol. 15, pp. 1041 – 1052, 2009.

23. K. Fujito, S. Kubo, H. Nagaoka, T. Mochizuki, H. Namita, and S. Nagao, "Bulk GaN crystals grown by HVPE," Journal of Crystal Growth, vol. 311, pp. 3011–3014, 2009.

24. M.P. D'Evelyn, H.C. Hong, D. S. Park, H. Lu, E. Kaminsky, R.R. Melkote, P. Perlin, M. Lesczynski, S. Porowski, and R.J. Molnar, "Bulk GaN crystal growth by the high-pressure ammonothermal method," Journal of Crystal Growth, vol. 300, pp. 11–16, 2007.

25. C. Mion, J. F. Muth, E. A. Preble, and D. Hanser, "Accurate dependence of gallium nitride thermal conductivity on dislocation density," Appl. Phys. Lett. 89, 092123 (2006).

Production Readiness of AlGaN/GaN HEMT on 6"/8" Si

Dong S. Lee, Jie Su, Balakrishnan Krishnan, George D. Papasouliotis and Ajit Paranjpe

Veeco MOCVD Operations, 394 Elizabeth Avenue, Somerset, NJ 08873, USA

Growth of crack-free AlGaN/GaN heterostructures on 6 and 8 inch Si(111) by 5x6" and 3x8" multi-wafer K465i production MOCVD system is presented. The two-dimensional-electron-gas is formed at the AlGaN/GaN interface with average Hall mobility values more than 1800 cm^2/v.s and sheet resistance less than 400 Ohm/sq. Run to run repeatability of AlGaN/GaN structural qualities, wafer bow, and 2DEG properties show the potential manufacturing possibility with the epitaxial process stability and longevity.

Introduction

AlGaN/GaN High Electron Mobility Transistors (HEMTs) on Si have emerged as a promising solution for high frequency power amplification and high voltage power switching applications, as they can utilize CMOS fab lines and the availability of low cost Si substrates in larger diameters (> 6 inch) compared to SiC and GaN substrates. Encouraged by the recent device performance results achieved for AlGaN/GaN HEMTs on Si aided by the advance of high quality crack-free GaN growth on large substrates (6-8 inch), the power electronics industry is currently engaged in pilot production of devices. With the large mismatch of lattice constant and thermal expansion coefficients between Si and GaN, epitaxy on Si leads to problems such as cracks, high-density misfit and threading dislocations. Many techniques have been utilized to relieve this stress and create crack-free GaN, such as using a low-temperature AlN layer [1], graded AlGaN buffer layers [2-3], AlN/GaN superlattices [4], and a SiN interlayer [5]. In addition, large wafer bow caused by the compressive stress from GaN during growth hinders the uniform temperature control across the wafer, which results in the non-uniform composition/thickness, layer stress and device performance. With the growth challenges of heteroepitaxy of GaN on Si, device quality GaN and manufacturability have to be demonstrated for the potential mass production of AlGaN HEMTs on Si.

We have tested GaN/Si and HEMT repeat runs on 6" and 8" Si (111) substrates to confirm the repeatability and stability in large scale production MOCVD. Run to run repeatability of AlGaN/GaN structural qualities, wafer bow, and 2DEG properties show the potential manufacturing possibility with the epitaxial process stability and longevity.

Experiment

The epitaxy process is carried out in 5x6" and 3x8" multi-wafer Veeco TurboDisc K465i production MOCVD systems. The Si wafer is first annealed at 1050 °C under Hydrogen ambient for about 5 minutes, which is followed by the pre-flow of TMAl without the presence of ammonia. The buffer layers consist of a 150nm AlN nucleation layer and three step-graded $Al_xGa_{1-x}N$ intermediate layers with x=0.75, 0.5, and 0.25. The thickness of the three $Al_xGa_{1-x}N$ layers is 200 nm, 250 nm and 300nm, respectively. The growth conditions for AlN and AlGaN buffer layer are above 1000 °C at a pressure of 75-100Torr. A GaN layer, about 2-3 µm, is grown on top of the AlGaN buffer layers with one optional stain-relieving low temperature (LT) AlN interlayer. The unintentionally doped $Al_{0.25}Ga_{0.75}N$ barrier layer is grown without a GaN cap in order to inspect the AlGaN surface morphology. A 1-nm thick AlN spacer is inserted between the AlGaN barrier and GaN to improve the performance of the two-dimensional electron gas (2DEG). The crystalline quality of GaN layer is performed by high resolution X-Ray Diffraction (HRXRD) for (002) symmetrical and (102) asymmetrical Bragg reflections. The thickness of epilayers is measured by white light reflectance and Al composition by photoluminescence and HRXRD. The surface morphology of the AlGaN HEMT barrier is studied by Atomic Force Microscope (AFM), and the electrical properties of the 2DEG are evaluated by Van der Pauw-Hall measurement using indium dots as the ohmic contact without heat treatment.

AlGaN/GaN HEMT structures exhibit good thickness uniformity (<2% in 1σ) and are free of cracks from the edge to the wafer center. The Al composition variation across the wafer is less than 1.5%. XRD FWHM (full width at half maximum) average values for (002) and (102) planes are less than 600" and 800" respectively for GaN layers with Carbon concentration of $6.0x10^{18}cm^{-3}$. (102) XRD FWHM values for HEMTs without Carbon doped GaN are less than 700". HEMT samples with 25 nm $Al_{0.25}Ga_{0.75}N$ barriers exhibit an average 2DEG mobility ($µ_H$) greater than $1500cm^2/V.s$, and sheet carrier concentrations (n_s) higher than $6.0x10^{12}cm^{-2}$. [6] With the insertion of AlN spacer, the average mobility $µ_H$ is greater than 2000 $cm^2/V.s$ and n_s higher than $8.0x10^{12}cm^{-2}$. All these values are valid for 6" and 8" runs. The enhancement of 2DEG performance compared to those without AlN spacer could be due to the produced large effective conduction band offset between AlGaN and GaN (ΔEc), reduction of the alloy disorder scattering, and improved polarization. [7]

Table1. 8" AlGaN/GaN HEMT Repeatability Runs

Run #	Wafer Bow (µm)	Thickness (µm)	(002) XRD FWHM (arc sec)	(102) XRD FWHM (arc sec)
HEMT_8inSi #1	-6.95	2.94	525	659
HEMT_8inSi #2	-11.13	2.98	539	670
HEMT_8inSi #3	-12.19	2.98	520	644
HEMT_8inSi #4	-13.39	2.97	522	644
HEMT_8inSi #5	-16.77	3.04	525	654
HEMT_8inSi #6	+2.30	3.02	536	643
HEMT_8inSi #7	-13.17	3.00	545	669
HEMT_8inSi #8 PM	-6.53	3.07	539	649
HEMT_8inSi #9	-15.79	3.03	533	666
HEMT_8inSi #10	-11.22	2.96	539	666
HEMT_8inSi #11	-17.01	3.02	551	655
HEMT_8inSi #12	-15.08	3.04	538	646
HEMT_8inSi #13 LED_8inSi	-14.57	3.06	541	667
HEMT_8inSi #14	-17.37	3.08	539	668

Table 1 shows the run to run (RtR) repeatability for the growth of 8" AlGaN/GaN HEMT structures. The growth rates of all the layers in the repeat runs show <0.5% in 1σ (not individual layers in the table). The average GaN XRD FWHM of (002) and (102) diffraction planes are 535 and 660 arc-second, with <2% in 1σ (within measurement limit), respectively. The post growth wafer curvatures at room temperature are consistently less than 50μm, for ~3μm stacks. The repeat runs were interrupted by system maintenance after more than 50 GaN/Si runs and LED growths to establish the system stability and flexibility.

Table2. 6" AlGaN/GaN HEMT Repeatability Runs

Run #	Wafer Bow (μm)	Thickness (μm)	(002) XRD FWHM (arc sec)	(102) XRD FWHM (arc sec)
HEMT_6inSi #1-1	17.6	3.0	588	814
HEMT_6inSi #1-2	17.6	3.0	597	814
HEMT_6inSi #1-3	16.2	3.01	582	812
HEMT_6inSi #1-4	16.2	3.0	597	818
HEMT_6inSi #2	18.0	3.07	596	801
HEMT_6inSi #3	17.2	3.07	592	814
HEMT_6inSi #4	20.4	3.0	599	819

Table2 shows the wafer to wafer (WtW) and run to run (RtR) repeatability for the AlGaN/GaN HEMT on 6" Si(111) with carbon doped GaN. The growth rates of AlN, 25% AlGaN, and GaN show <0.5% standard deviation from the 4 repeat runs. The GaN crystal quality is analyzed by HRXRD. The full width at half maximum (FWHM) for rocking curves of (002) and (102) diffraction planes of GaN show an average of 595" and 812", with <1.5% standard deviation. The wafer bows are consistently <50um, measured at room temperature for the ~3μm stack.

The HEMT surfaces have no micro-cracks and exhibited a step-flow feature comparable to the surface of the underlying GaN layer, indicating pseudomorphic growth of AlGaN with less strain relaxation. The root-mean-square (RMS) roughness for RtR and WtW is around 0.2-0.4 nm in a 5μmx5μm scanned area as shown in Figure 1

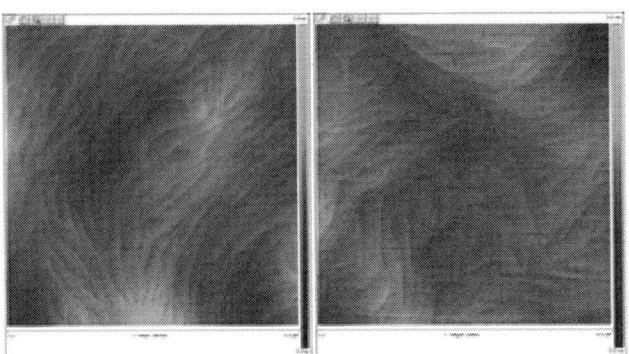

Figure 1. Identical surface morphology (RMS < 0.5nm for 5μm x 5μm scan) with pseudomorphic growth mode for run-to-run and wafer-to-wafer.

The in-situ measurement shows reproducible wafer curvature for 6" HEMT repeat runs, as shown in Figure 2 (a) and (b). Figure 2(a) shows identical in-situ curvature data, and 2(b) shows the identical data for 4 repeat runs.

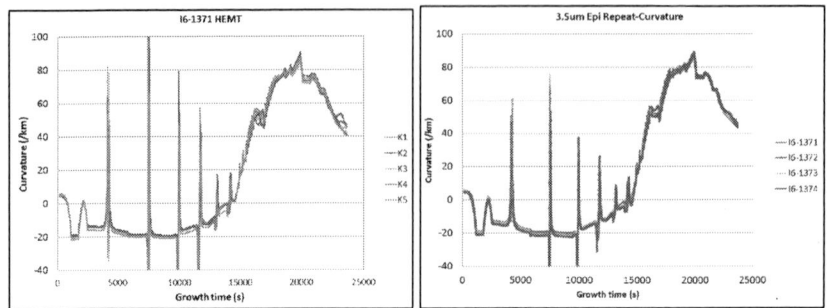

Figure 2. (a) wafer-to-wafer reproducibility and (b) run-to-run repeatability/stability for 6" HEMT repeat runs.

Conclusions

In summary, we report here the repeated growths of crack-free AlGaN/GaN HEMT on 6" and 8" silicon by large scale production MOCVD. The HEMT runs show good wafer-to-wafer and run-to-run repeatability, demonstrating the readiness of using Si substrates for the production of AlGaN/GaN HEMT epi.

Acknowledgements

The authors would like to thank the characterization lab at Veeco MOCVD operation for all the measurements. Special thanks to the night and weekend personnel for their continuous process support.

References

1. A. Dadgar, J. Blasing, A. Diez, A. Alam, M. Heuken, and A. Krost, Jpn. J. Appl. Phys. **39**, L1183, 2000
2. H. Marchand, L. Zhao, N. Zhang, B. Moran, R. Coffie, U. Mishra, J. Speck, S. DenBaars, and J. Freitas, J. Appl. Phys. **89**, pp7846, 2001
3. A. Able, W. Wegscheider, K. Engl, and J. Zweck, J. Crystal Growth, **276**, pp415, 2005
4. E. Feltin, B. Beaumont, M.Laugt, P. de Mierry, P. Vennegues, H. Lahreche, M. Leroux, and P. Gibart, Appl. Phys. Lett. **79**, pp3239, 2001
5. K.Cheng, M. Leys, S. Degroote, M. Germain, and G. Borghs, Appl. Phys. Lett. **92**, pp192111, 2008
6. S. Tripathy, Vivian K. X. Lin, S. B. Dolmanan, Joyce P. Y. Tan, R. S. Kajen1, L. K. Bera, S. L. Teo, M. Krishna Kumar, S. Arulkumaran, G. I. Ng, S. Vicknesh, Shane Todd, W. Z. Wang, G. Q. Lo, H. Li, D. Lee, and S. Han,, Appl. Phys. Lett. **101**, 082110, 2012
7. S. Tan, T. Suzue, S. Selvaraj, and T. Egawa, Jpn. J. Appl. Phys. **48**, pp111002, 2009

Synchrotron X-Ray Topography Studies of the Evolution of the Defect Microstructure in Physical Vapor Transport Grown 4H-SiC Single Crystals

M. Dudley[a], B. Raghothamachar[a], H. Wang[a], F. Wu[a], S. Byrappa[a], G. Chung[b], E. K. Sanchez[b], S. G. Mueller[b], D. Hansen[b], M. J. Loboda[b]

[a] Department of Materials Science & Engineering, Stony Brook University, Stony Brook New York 11794, USA
[b] Dow Corning Compound Semiconductor Solutions, Midland, Michigan, 48686, USA

Synchrotron White Beam X-ray Topography studies are presented of dislocation behavior and interactions in a new generation of 100mm diameter, 4H-SiC wafers grown using Physical Vapor Transport under specially designed low stress conditions. Such low stress growth conditions have enabled reductions of dislocation density by two or three orders of magnitude from lowest previously reported levels of 10^4-10^5cm^{-2} down to current levels of 10^2-10^3 cm^{-2}. This provides a unique opportunity to discern the details of dislocation configurations and interactions which were previously precluded due to complications of image overlap at higher dislocation densities. Detailed topography analysis have revealed dislocation multiplication by the hopping Frank-Read source mechanism, interactions between threading c, a and $c+a$ dislocations and deflections of threading dislocations resulting in stacking fault formation. These insights greatly aid in eliminating such undesirable defects or engineering their structures to minimize their impact leading to improved performance in device applications.

Introduction

Silicon carbide (SiC), chiefly the 4H polytype, is a wide bandgap semiconductor highly suited for electronic and optoelectronic devices operating under high temperature, high power, high frequency and/or strong radiation conditions, where conventional semiconductor materials like silicon, GaAs and InP are considered to have reached their limits (1). The primary requirement for SiC-based devices is the production of high quality thin films, which in turn require high quality substrates. Commercial SiC wafers are obtained from bulk crystals predominantly grown by physical vapor transport (PVT, also called seeded modified Lely growth) (2). During bulk growth, management of the various growth parameters including thermal profile of the growth system and resultant thermal gradient stresses within the growing crystal are critical to control the nucleation and propagation of various defects. Crystalline imperfections such as growth dislocations of screw character with closed-cores and hollow-cores (micropipes), deformation induced basal plane dislocations, planar defects (stacking faults, small angle boundaries), etc. affect device performance to different extents (3-6) and the lowering of their densities is a primary goal of the SiC crystal growth community. Synchrotron white beam X-ray topography (SWBXT) (7) has played a key role in revealing the detailed configurations of these defects and has been able to shed much light on their origins (8-9) thereby enabling the development of strategies for their elimination. Reduction of thermal shear stress during growth to levels below the critical resolved shear stress of 1 MPa (10) has enabled the lowering of basal plane dislocations (BPD) densities by up to three orders of magnitude from between 10^4 and a few times 10^5 cm^{-2} to between $10^2 - 10^3$ cm^{-2}. This has provided a unique opportunity to discern details of BPD behavior as well as threading dislocations which were previously mostly unresolvable. In this paper, recent observations of defect configurations and behavior in 4H-SiC crystals, carried out primarily using X-ray topography (both white beam and monochromatic) are discussed.

Experimental

Synchrotron X-ray topography studies were carried out on 4H-SiC wafers grown by the PVT technique in transmission, reflection and grazing-incidence geometries. Images were recorded on Agfa Structurix D3-SC film at specimen-to-film distances ranging from 10–40cm. White beam imaging was carried out at the Stony Brook Synchrotron Topography Station, Beamline X-19C, at the National Synchrotron Light Source at Brookhaven National Laboratory while monochromatic imaging was carried out at 33-BM-C, at the Advanced Photon Source at Argonne National Laboratory.

Primary Defects in 4H-SiC

Micropipes (MPs) and closed-core threading screw dislocations (TSDs). Screw dislocations lying along the [0001] axis have Burgers vectors equal to nc with hollow cores becoming evident with $n \geq 2$ for 6H and SiC $n \geq 3$ for 4H SiC (11). Using back-reflection geometry in SWBXT, both closed- and hollow-core (micropipes) screw dislocations (Fig. 1(a)) are revealed (12). Recent improvements in the PVT growth process have effectively eliminated the nucleation of micropipes in 4H-SiC but 1c TSDs are always present.

Basal Plane Dislocations (BPDs). BPDs are glissile dislocations with both line directions and Burgers vectors (i.e. 1/3<11-20>) in the basal plane (0001) (Fig. 1(b)). These are deformation induced and nucleated both at the crystal edges and at the sites of micropipes/TSDs (13).

Threading Edge Dislocations (TEDs). TEDs are dislocations with line directions roughly parallel to c-axis and Burgers vectors in basal plane i.e 1/3<11-20>. TEDs are one of the major components of low angle grain boundaries (LAGBs). The six different directions of the TED Burgers vector in 4H-SiC (Fig 1(c)) can be unambiguously identified by comparing their image features to their corresponding ray tracing simulated grazing-incidence topograph images (14).

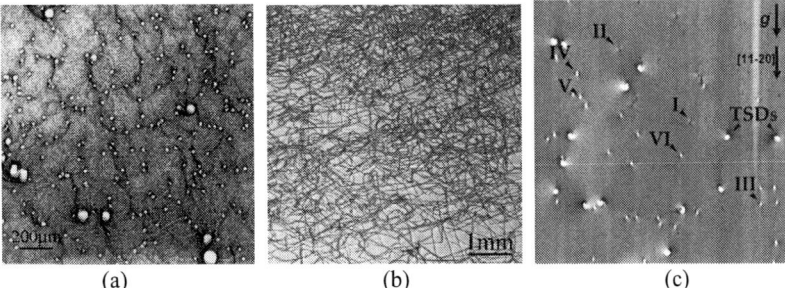

| (a) | (b) | (c) |

Figure 1. (a) Back-reflection X-ray topograph images of close-core (smaller white spots) and hollow-core (large white spots) screw dislocations in a (0001) 6H SiC wafer; (b) Transmission x-ray topograph of 4H-SiC substrate showing network of BPDs; (c) Monochromatic beam X-ray topograph (g = 11-28) showing the images of six types of TEDs in 4H-SiC; I: b=1/3[-12-10]; II: b=1/3[-2110]; III: b=1/3[-1-120]; IV: b=1/3[1-210]; V: b=1/3[2-1-10]; VI: b=1/3[11-20].

Interactions between Primary Defects in 4H-SiC

Hopping Frank-Read Source mechanism of BPD multiplication

Studies of the configurations and behavior of BPDs in wafers cut from PVT-grown SiC boules have enabled identification of the operation of single-ended Frank-Read sources which are interconnected on different basal planes (15). Fig. 2(a) shows an example of a final configuration of a BPD with Burger vector 1/3[11-20] which has developed into a sequence of diamond-shaped loops with loop segments along <1-100> line directions on different basal planes connected by TED segments. When a macrostep deflection of the surface outcrop of a TED onto the basal plane creates a segment of BPD (Fig. 2(b)) and the BPD segment is in screw orientation, further advancement of the macrostep will replicate the BPD in the direction of step flow (Fig.2(c)). If the advancing macrostep encounters a step advancing in the opposite direction, for example a spiral step from a screw dislocation, the dislocation will be re-deflected into the threading direction (Fig. 2(d)). At a later stage, if the outcrop of the re-deflected TED encounters another advancing macrostep, the whole process can be repeated (Fig. 2(e)).

As soon as the TED is deflected onto the basal plane, a single-ended Frank-Read (marked as No.1) source will begin to operate (Fig 3(a)). The deflected BPD segments moves away from screw orientation and becomes more susceptible to conversion into a TED segment. When the BPD is re-deflected into threading orientation, a second single-ended source (marked as No.2) will initiate at that point. As this new TED segment is deflected again onto the basal plane, a third single-ended source (marked as No.3) will initiate (see Fig. 3(c)). For these latter two sources to be able to operate independently the BPD segments on parallel basal planes must be separated by a distance large enough that their mutual repulsive forces can be overcome. Thus the resultant BPD configuration effectively behaves like a series of single-ended Frank-Read sources which "hop" from one slip plane to other parallel slip planes through a process of double deflection and this behavior is termed as a Hopping Frank-Read source mechanism (15).

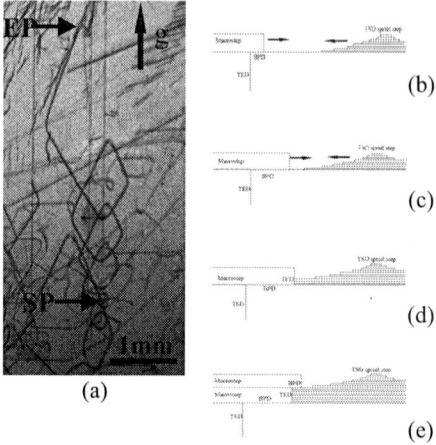

Figure 2. Transmission X-ray topograph (g=$\bar{1}\bar{1}20$) showing a diamond shaped BPD loop (SP – starting point; EP – ending point). (b)-(e) Schematic cross-sectional view of the deflection of a TED onto the basal plane by a macrostep followed by re-deflection into threading orientation through the encounter between macrostep and a TSD spiral advancing in the opposite direction.

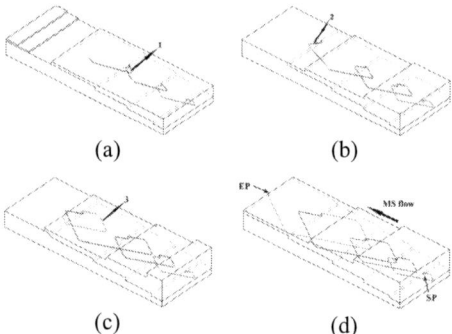

Figure 3. (a)-(c) Schematic of the formation of a single-ended Frank-Read source through the deflection of a TED into a BPD and back again. The TED segments act as pinning points for the BPD glide; (d) The final configuration of the loops produced by the Hopping Frank-Read Source. "SP" indicates the starting point, and "EP" end point.

Interactions between threading dislocations in 4H-SiC

In addition to micropipes and pure screw dislocations, 4H-SiC crystals also contain threading dislocations with both c and a components (16,17). Fig. 4 shows an SWBXT image (g=0004) from an axial slice of 4H-SiC crystal where segments of six threading dislocations labeled **1** thru **6** are visible (dislocations **3** and **4** appear to have interacted such that parts of their length have annihilated). By carrying out detailed **g.b** contrast analysis on multiple reflections (see Table I), the Burgers vectors of these dislocations are first estimated. Then the exact Burgers vectors are revealed through the interactions between the threading dislocations. Two closed-core threading dislocations with opposite c-component experience a strong attractive force but the reaction is difficult via slip and can be achieved by interaction with non-equilibrium concentration of vacancies (i.e. climb). These reactions are summarized below and shown in Figure 5.

(a) Reaction between threading dislocations with Burgers vectors of $-c+a$ and $c+a$ wherein the opposite c-components annihilate leaving behind the two a-components. In Fig. 4, segments of dislocation lines **3** and **4**, marked with arrows, appears to have annihilated and shows no contrast. As illustrated in Fig. 5(a), dislocation **3** and **4** have opposite sign c-components and these c-components annihilate while the two a-components are left behind as see in Fig. 5(b). Fig. 5(c) shows a section topograph superimposed onto a 0004 projection topograph. From the sense of the mutual shift between the bimodal image components, it can be seen that dislocation **3** and **4** have right-handed and left-handed screw components respectively and thus annihilate when they meet.

(b) Reaction between threading dislocations with Burgers vectors of $-c$ and $c+a$ leaving behind the a-component. Fig. 5(d) shows a reaction between $-c$ and $c+a$ dislocations. The $-c$ TSD annihilates with the opposite sign c-component of the $c+a$ dislocation leaving a single a-component behind. TSD dislocation **7** is visible in Fig. 5(f) and out of contrast in Fig. 5(e) while dislocation **8** is visible in both images, which suggest that it has both c- and a-components. The annihilated segment, marked by arrows, is invisible in Fig. 5(e) and visible in Fig. 5(f), which indicates that the annihilated segment has only a-component of Burgers vector.

(c) Reaction between opposite-sign threading screw dislocations with Burgers vectors c and $-c$. Fig. 5(h) shows a typical helical morphology of a TSD and when two dislocations like this are close to each other, the interaction between TSDs with opposite signs of Burgers vectors will occur, as illustrated in Fig. 5(g) and shown in Fig. 5(i) where some segments annihilate leaving others in the form of trails of stranded loops comprising closed dislocation dipoles.

TABLE I. **g.b** values for dislocations from **1** to **6** in Fig. 4.

b \ g.b \ g	0004	0-110	1-210	01-11	0-111	10-1-5
1&6:1/3[1-213], **c+a**	4	1	2	0	2	-5
3: 1/3[-1-123], **c+a**	4	1	1	0	2	-5
4: 1/3[-1-12-3], **-c+a**	-4	1	1	-2	0	5
2&5: <0001> , **c**	4	0	0	1	1	-5
Arrow:2/3[-1-120], **2a**	0	2	2	-2	2	-2

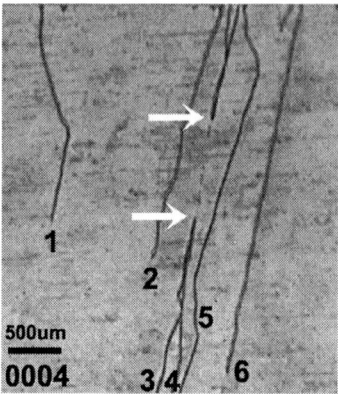

Figure 4. Transmission X-ray topograph (g = 0004) of axial-cut 4H-SiC showing threading dislocations numbered 1 to 6.

Deflection of threading dislocations and creation of stacking faults

Three types of stacking faults have been observed to date in 4H-SiC: Shockley faults, Frank faults and those which comprise some kind of combination of these two. Such faults can arise from the deflection of c-axis threading dislocations of Burgers vector c and $c+a$ on to the basal plane (18,19). If the spiral step risers of such dislocations divide into $c/4$, $c/2$ and $3c/4$ increments, overgrowth can be facilitated by the simultaneous dragging of one of the Shockley partials associated with the core structure of the original dislocation by the overgrowing macrostep and this adds the Shockley component to the Frank component of fault vector. Stacking faults vectors can be determined by X-ray topography by analyzing the contrast from stacking faults on multiple x-ray topographs which arises from the phase shift experienced by the x-ray wavefields as they cross the fault plane. This phase shift has been computed to be equal to $\delta = (-2\pi \, g.R)$, where g is the active reciprocal lattice vector for the reflection and R is the fault vector. Four fault vectors have been observed in 4H-SiC wafers:

(i) Frank fault. These are "in-grown" stacking faults and result from the overgrowth of c-axis screw dislocations whose surface growth spiral steps are separated into $c/2$ increments [20] (Fig 6(e)). Detailed analysis of the contrast from different reflections shows the fault vector is ½[0001] (Fig. 6(a)).

(ii) Shockley faults. Fig 6(b) is a X-ray topograph showing stacking fault regions with one straight edge corresponding to the sessile partial and the other being curved corresponding to the glissile partial. The fault vector is revealed to be 1/3[01-10] which corresponds to a Shockley fault. Fig. 6(f) is an illustration showing the integral number of step height, where the glissile partial lying below the step has created faults while the sessile partial at the top gets locked.

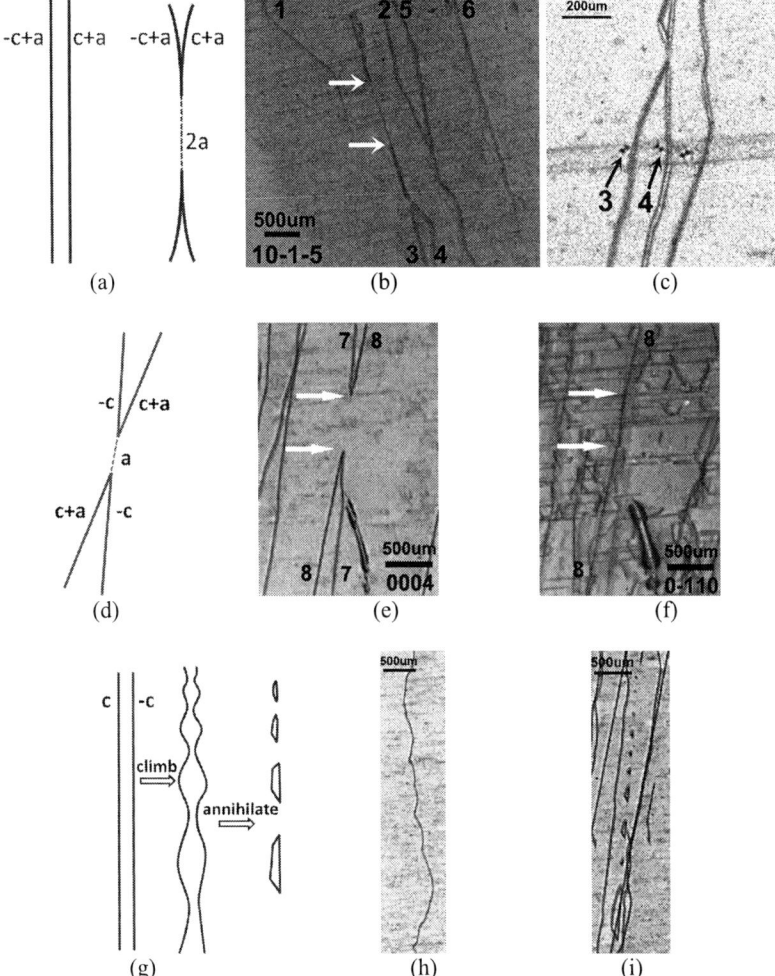

Figure 5. Reactions between threading dislocations with Burgers vectors: (a)-(c) **c+a** (**3**) and **-c+a** (**4**) leaving behind **2a** dislocation - (a) Illustration (b) 10-15 topograph showing the partially annihilated segment, marked by arrows, and the rest of dislocations **3** and **4** (c) superimposed 0004 section topograph on projection topograph showing opposite signs of dislocation **3** and **4**; (d)-(f) **-c** (**7**) and **c+a** (**8**) leaving behind **a** dislocation - (d) Illustration (e) 0004 topograph showing the annihilation of -c and c+a dislocation (f) 0-110 reflection showing strong contrast of dislocation **8** as well as "the annihilation segment", and absence of dislocation **7**; (g)-(i) **c** and **-c** - (g) Illustration (h) 0004 topograph showing curved slightly helical morphology of **c** dislocation indicating climb; (i) 0004 topograph showing annihilation of **c** and **-c** dislocations at some segments leaving remaining segments in form of loops comprising closed dislocation dipoles.

(iii) Combination of Shockley and Frank fault with fault vector S+c/2. The overgrowth of c-axis TSD whose spiral steps have divided into two c/2-high demisteps and **c+a** dislocations which has protruded on the terrace created by the demisteps results in both **c**/2 Frank fault and Shockley fault associated with **c+a** to lie on the same basal plane [24] leading to the formation of this fault (see Figs. 6(c) for X-ray topograph & Fig. 6(g) for illustration of fault formation).

(iv) Combination of Shockley plus a Frank fault with fault vector S+c/4. This is also a combination of Shockley and a Frank fault but of the type S+c/4, where the c-axis TSD has divided itself into c/4 and 3c/4 step heights. Overgrowth results in the simultaneous dragging of one of the Shockley partials and adds the Shockley component to the c/4 Frank component of fault vector. Figure 6(d) shows X-ray topographs showing the fault area marked C with R= 1/12[4-403] which is a combination of Shockley and c/4 Frank fault and area marked D with R=1/12[-4043] which is due to a second c+a dislocation protruding the terrace.

Such deflection process generally leads to exit of the threading dislocations through the crystal sidewalls and provide a mechanism by which the density of threading dislocations in the boules can be lowered.

Conclusions

New insights into behavior of BPDs and threading dislocations in bulk 4H-SiC materials have been obtained by detailed characterization of defect configurations in basal and axial-cut wafers of low BPD density ($10^2 - 10^3$ cm^{-2}) using SWBXT in conjunction with other characterization techniques. Dislocation multiplication by the hopping Frank-Read source mechanism, interactions between threading c, a and $c+a$ dislocations and deflections of threading dislocations resulting in stacking fault formation in bulk crystals has been observed. These results will provide valuable information to either eliminate defects or engineer their structures to minimize their impact.

Acknowledgments

Work supported by Dow Corning. Topography experiments were carried out at the Stony Brook Synchrotron Topography Facility, Beamline X-19C, at the NSLS (DOE Office of Basic Energy Sciences Contract no. DE-AC02-76CH00016) and at beamline 33-BM at the APS (DOE Office of Basic Energy Sciences Contract No. DE-AC02-06CH11357).

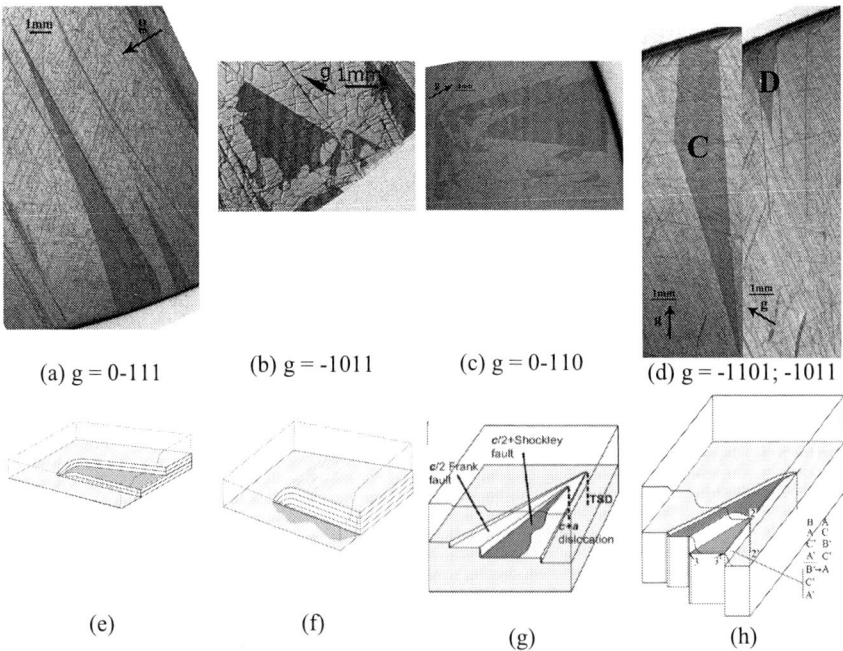

(a) g = 0-111 (b) g = -1011 (c) g = 0-110 (d) g = -1101; -1011

(e) (f) (g) (h)

Figure 6. Transmission topographs from 4H-SiC crystals showing stacking faults with fault vector: (a) R = 1/2[0001] (Frank fault); (b) R = 1/3[1-100] (Type I Shockley fault); (c) R = 1/6[20-23] vector (S+c/2 fault); (d) R = 1/12[-4043] (C) & R =1/12[-4403] (D) (vector S+c/4 fault). Illustrations of the stacking faults: (e) Frank type stacking fault (shaded area); (f) Type I Shockley fault with the glissile partial lying below the step and the sessile at the top which is locked; (g) Overgrowth of c dislocation and formation of the c/2 Frank type fault within which a c+a dislocation with one sessile Shockley partial and one glissile Shockley partial converts some part of the fault to c/2 plus a Shockley type; (h) Overgrowth of c+a dislocation with a c-height step converting it into a Frank dislocation plus two Shockleys with one sessile and the other glissile. Overgrowth of c+a dislocation 11' with c/4 and 3c/4- height steps which has second 22' with c-height spiral step protruding onto the terrace between these two step risers. "Interfacial Shockley" converts A layer into B', allowing overgrowth by the A' layer at the bottom of the macrostep. Following overgrowth the Shockley associated with the deflected 22' dislocation located at 23 glides under stress until it reaches the edge of the step 23' creating the fault of type C shown in (d).Transmission X-ray topograph (g = 0004) of axial-cut 4H-SiC showing threading dislocations numbered 1 to 6.

References

1. A. A. Lebedev, V. E. Chelnokov, *Semiconductors*, **33**, 999 (1999).
2. Y. M. Tairov, V. F. Tsvetkov, *J. Crystal Growth,* **43**, 209 (1978).
3. P. G. Neudeck, J. A. Powell, *IEEE Electron. Dev. Lett.,* **15**, 63 (1994).

4. P. Neudeck, W. Huang, and M. Dudley, *IEEE Trans. Electron. Dev.*, **46**, **478** (1999).
5. H. Chen, B. Raghothamachar, W. Vetter, M. Dudley,Y. Wang, B.J. Skromme, *Mater. Res. Soc. Symp. Proc.*, **911**, 169 (2006).
6. H. Lendenmann, F. Dahlquist, N. Johansson, R. Soderholm, P. A. Nilsson, J. P. Bergman and P. Skytt, *Mater. Sci. Forum*, **353–356**, 727 (2001).
7. Raghothamachar, G. Dhanaraj, M. Dudley in Handbook of Crystal Growth, G. Dhanaraj, K. Byrappa, V. Prasad, M. Dudley, p. 1425, Springer (2010).
8. M. Dudley, Y. Chen and X. R. Huang: *Mater. Sci. Forum*, **600-603**, 261 (2009).
9. M. Dudley and X. Huang, *Mater. Sci. Forum*, **338-342**, 431 (2000)
10. M. Selder, L. Kadinski, F. Durst, T.L. Straubinder, P.L. Wellmann and D. Hofmann, *Mater. Sci. Forum*, **353-356**, 65 (2001).
11. W. Si, M. Dudley, R. Glass, V. Tsvetkov, C. H. Carter, Jr., *Mater. Sci. Forum*, **264-268**, 429 (1998).
12. M. Dudley, W. Huang, S. Wang, J. A. Powell, P. Neudeck, C. Fazi, *J. Phys. D: Appl. Phys.*, **28**, A56 (1995).
13. M. Dudley, Y. Chen and X. R. Huang, *Mater. Sci. Forum*, **600-603**, 261 (2009).
14. I. Kamata, M.Nagano, H.Tsuchida, Yi.Chen, M.Dudley, *J. Cryst. Growth*, **311**, 1416 (2009).
15. H. Wang, F. Wu, S. Byrappa, S. Sun, B. Raghothamachar, M. Dudley, E. K. Sanchez, D. Hansen, R. Drachev, S. G. Mueller, and M. J. Loboda, *Appl. Phys. Lett.*, **100**, 172105
16. F. Wu, H. Wang, S. Byrappa,.B. Raghothamachar, M. Dudley, E.K. Sanchez,.D. Hansen, R. Drachev, S.G. Mueller and M.J. Loboda, *Mater. Sci. Forum,* **717-720**, 343 (2012).
17. F Wu., S. Byrappa., H. Wang, Y. Chen, B. Raghothamachar, M. Dudley, E.K. Sanchez, G. Chung,, D. Hansen, S.G.Mueller and M. J. Loboda, *Mater. Sci. Forum*, **740-742**, 217 (2013).
18. M. Dudley, F. Wu, H. Wang, S. Byrappa, B. Raghothamachar, G. Choi, S. Sun, E. K. Sanchez, D. Hansen, R. Drachev, S. G. Mueller, and M. J. Loboda, *Appl. Phys. Lett.*, **98**, 232110 (2011).
19. M. Dudley, S. Byrappa, H. Wang, F. Wu, B. Raghothamachar, G. Choi, S. Sun, E. K. Sanchez, D. Hansen, R. Drachev, S. Mueller, and M. J. Loboda, *Mater. Sci. Forum,* **679-680**, 269 (2011).
20. M. Dudley, S. Byrappa, H. Wang, F. Wu, Y. Zhang, B. Raghothamachar, G. Choi, E.K. Sanchez, D. Hansen, R. Drachev, and M.J. Loboda, *Mater. Res. Soc. Symp. Proc.*, **1246**, 1246-B02-02 (2010).

Basal Plane Dislocation Mitigation using High Temperature Annealing in 4H-SiC Epitaxy

N. A. Mahadik[a], A. Nath[b], E. A. Imhoff[a], R. E. Stahlbush[a], and R. Nipoti[c]

[a] Naval Research Laboratory, Washington, DC, USA
[b] George Mason University, Fairfax, Va, USA
[c] CNR-IMM Bologna, Bologna, Italy

Basal plane dislocations (BPD) were mostly eliminated in 4H-SiC epitaxy using post growth high temperature annealing in the range of 1600 °C - 1950 °C for 30s - 2 mins. The samples annealed at temperatures >1700 °C showed the best BPD reduction. However, surface morphology was degraded for samples annealed >1850 °C, and new BPDs were generated. A better capping technique was developed to improve the surface morphology and avoid generation of new BPDs, while significantly reducing the existing BPDs in the SiC epitaxial layers.

Introduction

Basal plane dislocations (BPD) in 4H-silicon carbide (SiC) epilayers cause formation of stacking faults (SF) during device operation, which leads to forward voltage degradation in minority carrier devices (1, 2). More than 90% of the BPDs in the substrate are converted to benign threading edge dislocations (TED) at the beginning of the epigrowth, but the remaining BPD density is too high for power device fabrication (3). For epitaxial growth on 4 degree offcut substrates, most of the BPDs are converted to TEDs during the first few microns of growth (4). However, BPDs in many areas of the epilayers can be greater than 100 cm^{-2}, which adversely affects the device yield and reliability. Post growth high temperature annealing at 1800 °C was shown to cause small BPD shortening in 8 degree offcut SiC epilayers (5), when annealed without a protective cap only. This results in poor surface morphology, which is detrimental to any device fabrication. In this paper, we report on post growth high temperature annealing to eliminate or significantly shorten BPDs in the both 4 degree and 8 degree epilayers that had propagated from the substrate. In all our anneals we used protective capping layers to preserve the surface morphology, while attaining the BPD elimination. Additionally, during the device fabrication process, for most of the devices, implantation is used in selective areas to obtain required doping levels. This is followed by high temperature annealing, typically greater than 1600 °C to activate the implanted species and remove the crystalline damage due to implantation. It has been shown that high temperature annealing in SiC can generate extended defects such as BPDs, which is also detrimental to device yield (6). We also investigated generation of extended defects upon high temperature annealing and devised a better capping and annealing technique to mitigate the generation of new defects.

Experimental procedure

For this study, we used 15 μm, n-type 6E15 4H-SiC epilayers commercially grown on either 4 or 8 degree offcut substrates. The as-grown samples were imaged using ultraviolet photoluminescence (UVPL) imaging, which shows the BPDs and several extended defects in the epitaxial layers (7). Carriers were excited with the 334 nm line from an Ar-ion laser, and images were collected in the emission range of 600-1000 nm. Some of the samples were implanted with phosphorus for a uniform doping concentration of $\sim 2 \times 10^{20}$ /cm^3 up to 0.2μm from the sample surface. All the samples were capped with graphitic carbon. All the high temperature annealing were performed using a microwave annealing technique described elsewhere (8). Table I below provides the annealing schedule followed. All the post-annealed samples were also imaged again using the UVPL technique to observe the BPDs in the epilayers.

TABLE I. Annealing schedule for the 4H-SiC samples.

Sample	Temperature (oC)	Time (s)
Un-implanted 4° offcut	1750, 1850 , and 1950	30
Un-implanted 4° offcut	1850, and 1950	60
P⁻ implanted 4° offcut	1750, 1850, and 1950	30
P⁻ implanted 4° offcut	1850, and 1950	60
Un-implanted 8° offcut	1600 and 1700	30 and 60
Un-implanted 8° offcut	1600	120
P⁻ implanted 8° offcut	1600 and 1700	30 and 60
P⁻ implanted 8° offcut	1600	120
Un-implanted 4° offcut/ change conditions	1600	60

Results and Discussion

The UVPL images of the as grown samples showed an average BPD density ~ 1000 cm^{-2} in the 4 degree offcut epilayers and ~ 2000 cm^{-2} in the 8 degree offcut epilaters . Upon high temperature annealing the UVPL images of the samples had a complex structure, where most of the BPDs that were previously present in the epilayers had disappeared, for annealing >1700 °C, or were greatly shortened. Figure 1 shows the UVPL images of (a) as grown and (b) 1750 °C / 30 s annealed un-implanted sample. The bright (white) lines are the BPDs. The left end point of the line is where the BPD entered the epilayer from the substrate, and the right end point can be where it hits the surface or converts to a TED during the growth. The thickness the BPD propagates in the epilayer is the product of the length of the bright line and tangent of the offcut angle of the sample.

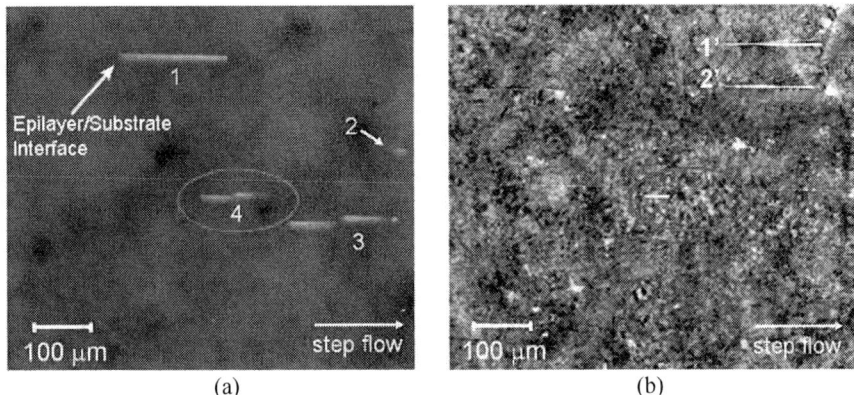

(a)   (b)

Figure 1. UVPL images of SiC epilayers (a) before and (b) after annealing at 1750 °C/30s

It can be seen that BPDs labeled 1-3 disappear upon annealing. BPD #4 (encircled) that had propagated 9.3 μm before converting to a TED gets shortened to 4 μm after annealing. It is proposed that upon annealing the BPDs that had not yet converted to a TED during epigrowth forms a TED near the surface. Then, during the annealing process, the TEDs at the end of all the BPDs glide in a prismatic plane causing the BPD to shorten. This effect of BPD removal was observed in all the annealed samples, with no observable difference for implanted samples.

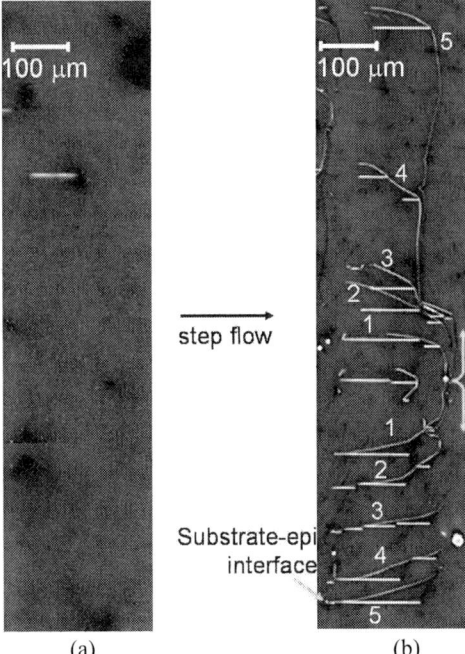

(a)   (b)

Figure 2: UVPL images of (a) before annealing and (b) new BPDs that were formed after high temperature annealing.

However, new BPDs, poor surface morphology and other extended defects such as in-grown faults and hexagonal pits were observed to have formed during the annealing process. This was more severe for the samples annealed at 1850 °C and higher, and was more extensive for the implanted samples probably due to higher microwave coupling and heating of the implanted layer. Figure 2 shows the (a) as grown and (b) new BPDs that were generated upon annealing at the higher temperatures. In the 8° offcut samples, these BPDs were also formed as low as 1600°C. The new BPDs were observed to form at the sample surface and create a dislocation loop that glides in the basal plane towards the substrate. During the high temperature annealing, the graphite cap appears to delaminate first in small region of the surface causing Si sublimation, which would induced strain in the sample and create a dislocation, which glides during the annealing process and forms the dislocation loops. We then developed another graphite capping process, which results in a thicker capping layer, which is also densified at a higher temperature. The annealing process was further optimized by changing to a slower ramping rate of ~55 °C/s from the initial 170 °C/s. We also used an overpressure of nitrogen in the annealing chamber during annealing to further prevent oxidation of the graphite cap from residual oxygen. This resulted in obtaining almost no new BPDs, while shortening or eliminating the existing BPDs as shown in Figure 3.

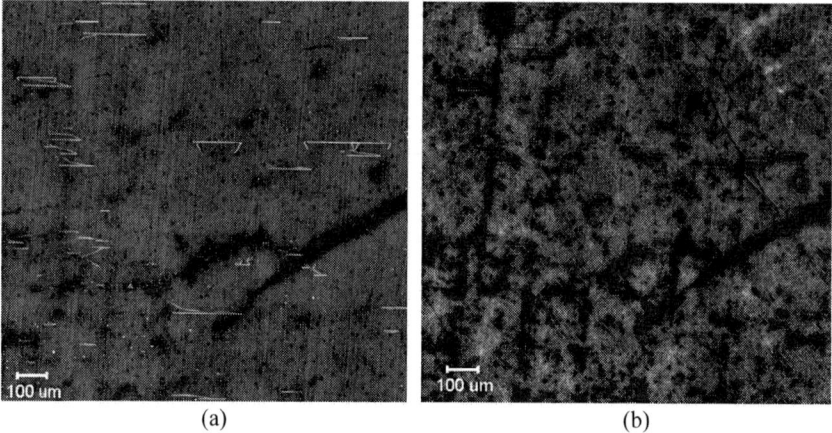

(a)   (b)

Figure 3: UVPL images of (a) as grown and (b) after annealing at 1600 °C/ 60 s using new annealing conditions.

The sample shown in Fig. 3 is a epilayer grown on 4° offcut substrate, un-implanted and was annealed at 1600 °C for 60s. It can be seen in Fig. 3(b) that after annealing, almost all the BPDs either disappear or are shortened significantly, and no new BPDs appear to form at the surface. In order to further enhance BPD elimination in the epilayers, higher temperature anneals have to be optimized to protect the surface, to limit degradation of the sample surface or generation of new BPDs.

Conclusions

We achieved elimination of almost all BPDs that had propagated from the substrate to the epilayers in 4H-SiC grown on either 4° or 8° offcut substates using high temperature microwave annealing. The anneals with the most efficient BPD mitigation was performed in temperatures greater than 1700 °C for 30s or higher. New BPDs that formed dislocation loops were generated at temperatures greater than 1850 °C in 4° offcut samples and all 8° offcut samples - 1600 °C and above). These were formed due to non-optimized surface protection and annealing conditions. After developing a better capping technique and optimizing the annealing conditions, the new BPDs were not generated and almost all the BPDs were either eliminated or significantly shortened. This technique of post-growth elimination of BPDs, would greatly improve the device performance and yield fabricated in 4H-SiC epilayers.

Acknowledgments

This work was supported by the Office of Naval Research.

References

1. J. P. Bergman, H. Lendenmann, P. A. Nilsson, and U. Lindefelt, Mater. Sci. Forum **353**, 299 (2001)
2. R. E. Stahlbush, J. B. Fedison, S. D. Arthur, L. B. Rowland, J. W. Kretchmer, and S. Wang, Mater. Sci. Forum **389**, 427 (2002)
3. S. Ha, and M. Skowronski, Phys. Rev. Lett. **92**, 175504 (2004)
4. R. L. Myers-Ward, B. L. VanMil, R. E. Stahlbush, S. L. Katz, J. M. McCrate, S. A. Kitt, C. R. Eddy, Jr., and D. K. Gaskill, Mat. Sci. Forum **615-617**, 105 (2009)
5. X. Zhang, and H. Tsuchida, MRS Proc. **1433**, mrss12-1433-h05-03 (2012)
6. N. A. Mahadik, R. E. Stahlbush, J. D. Caldwell, M. O'Loughlin, and A. Burk, Mater Sci. Forum, **717-720**, 297 (2012)
7. R.E. Stahlbush, K.X. Liu, Q. Zhang and J.J. Sumakeris, Mater. Sci. Forum **556-557**, 295 (2007)
8. S. G. Sundaresan, Y. L. Tian, M. C. Ridgway, N. A. Mahadik, S. B. Qadri, and M. V. Rao, Nucl. Inst. Meth. B, **261**, 616, (2007)

330

ECS Transactions, 58 (4) 331-339 (2013)
10.1149/05804.0331ecst ©The Electrochemical Society

3D TCAD Simulations for More Efficient SiC Power Devices Design

L. V. Phung[a], D. Planson[a], P. Brosselard[c], D. Tournier[a], and C. Brylinski[b]

[a] Ampère Laboratory, Université de Lyon, INSA de Lyon, Villeurbanne, France
[b] Laboratoire des Multimatériaux et Interfaces, Université de Lyon, Villeurbanne, France

SiC devices become more and more prominent in the power semiconductor industry. Thanks to a technology that seems to be mature enough, SiC devices are becoming more and more sophisticated. Therefore, they can be serious competitors to existing silicon devices in not so distant future at least for high temperature and high power applications. In addition to undoubtedly better electrical and thermal properties, SiC devices still require attention regarding their design. Indeed, the material is still more expensive than silicon and some limitations such as the inability to create deep p-n junctions prevent from re-using existing silicon design. Therefore, SiC devices should be designed so that the best trade-off between active area and breakdown voltage is achieved. In such a context, 3D TCAD can become an interesting approach mostly thanks to modern computer farms.

Towards 3D TCAD Simulations

During these past decades, power semiconductor industries have been drawn by a relentless need to achieve always more power efficient devices that target high voltage and high temperature applications. To deal with power hungry applications (railway or aeronautic applications...), wide bandgap materials such as SiC are progressively more and more appealing since significant milestones regarding device fabrication were reached. Therefore, SiC devices are becoming more and more widespread.

However, device design still plays a major role in achieving better electrical performances. For instance, superjunctions introduced at the very end of the 90s reinvigorate silicon power MOSFETs. SiC devices still do not benefit from this concept yet but this example shows the importance of device design.

Typically, devices such as transistors, whether made of silicon or SiC, opt for an interdigitated design (Figure 1). While such a design is satisfying for low-to-medium power applications, it becomes less suitable for the ones that are very power hungry. Indeed, those devices need to carry significant amounts of current leading to important die sizes. Nonetheless, the resistance of metallic fingers is no longer negligible at this scale. The consequence of such a resistance is that some device areas become less effective. Device designs were rethought for thyristors and cannot solely rely on 2D simulations. As a consequence, large thyristors opt for concentric designs so as to maximize their active area.

331

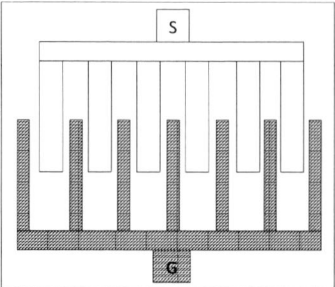

Figure 1. Schematic top view of an interdigitated FET. G and S refer respectively to the Gate and the Source terminals.

Breakdown voltages are also a concern. While SiC allows fabricating very high voltage devices, some material properties such as a poor diffusion coefficient forbid the fabrication of devices with deep PN junctions. Semiconducting wells are very thin and their bending radius is very small leading to electric field crowding (Figure 2). The breakdown voltage can be far smaller than expected. PiN diodes are subject to this issue since their layouts are rounded or circular in order to minimize the resistivity of metallic layers. Oversizing the device is not an option due to material cost. Junction edges must then be "protected". One way to mitigate the crowding effect is to rely on Junction Termination Extension (JTE) or guard rings [1-4].

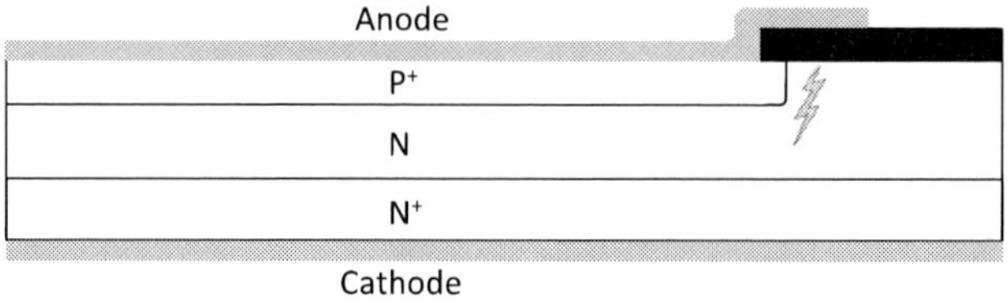

Figure 2. Schematic cross-section of a PiN diode. The low bend radius at the right edge of the junction leads to a premature breakdown.

The two present cases are better dealt with 3D simulations. Thanks to modern computer farms and adaptive meshing, running times can be kept reasonable. Following simulations results below are obtained from finite elements method thanks to Sentaurus^{TM} TCAD software [5].

Two typical cases where 3D simulations are helpful

PiN Diodes periphery protection: Junction Termination Extension (JTE)

Junction Termination Extension (JTE) is the most common technique to protect junction edges of SiC devices from the electric field crowding phenomenon by adding a p-layer very close to critical junction edges. When the device is reverse biased, the JTE becomes fully depleted. Consequently, the electric field peak that would result from crowding is lower. This protective surrounding ring is realized thanks to a boron or aluminum implantation (p-type). The dose must be carefully controlled in order to get the maximum efficiency from this kind of periphery protection. An insufficient dose will reduce the JTE effectiveness to almost nothing while too important doses will shift the unwanted electric field peak to the periphery. An appropriate JTE will produce electric peaks with similar magnitudes at the edges of the pn junction and the JTE (Figure 3).

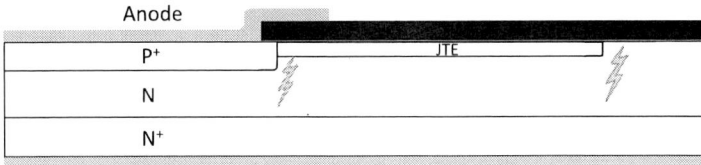

Figure 3. Schematic cross-section of a PiN diode that features a JTE periphery protection.

However, in the 3D context, the edge of this protection is still subject to electric field crowding to a lower extent [6]. Figure 4.b depicts a quarter of a PiN mesa diode with its JTE while Figure 4.a shows the corresponding schematic cross-section view along the yz-plane [7]. This diode structure should sustain 12000V provided by a semi-infinite plane parallel junction. Two bend radiuses R_B are chosen: 120µm and 450µm. Devices with low bend radiuses tend to be rectangular shaped and have a more important active area. However, they are more prone to premature breakdowns.

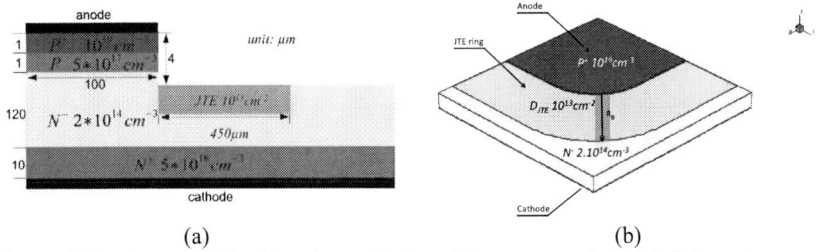

(a)   (b)

Figure 4. PiN mesa diode featuring a JTE periphery protection. (a) Schematic cross-section along the yz-plane. (b) 3D top view of a quarter of a structure.

Figure 5.a plots the holding voltage capability of these diodes from experimental data [8] and simulations while Table 1 summarizes breakdown voltages that are obtained from the very same two bend radiuses and from a theoretical 1D diode. The crowding effect is more pronounced in the case where the bend radius is small hence a lower breakdown voltage than expected.

TABLE I. *Comparison between breakdown voltages obtained from 3D structures of two different JTE bend radiuses and 1D ideal SiC diode.*

JTE bend radius (µm)	Simulated breakdown voltage (V)	Experimental breakdown voltage (V)	Efficiency from experimental results (%)
120	8700	9000	75
450	9800	10200	82

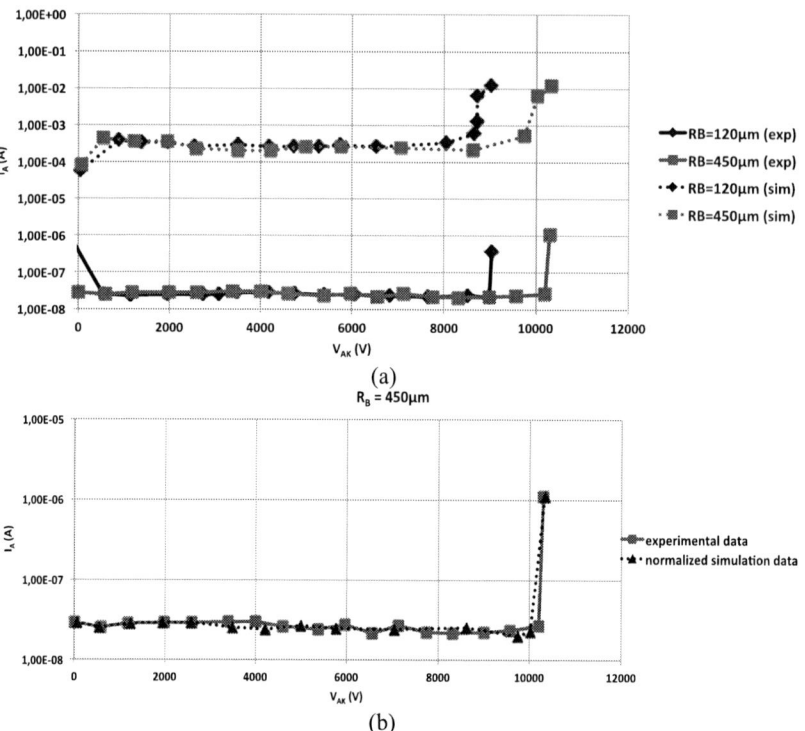

Figure 5. Comparison between simulated and experimental reverse electrical characteristics of 3D SiC diodes. (a) For different bend radiuses R_B. (b) Between normalized simulation data and experimental data for $R_B = 450\mu m$.

Nonetheless, the main noticeable difference between these experimental and simulated data comes from the leakage current value. Indeed, the default accuracy setting that defines the relative error convergence criterion must be tighten due to the small intrinsic carrier density in SiC. However, increasing accuracy can slow down simulations especially when the computer farm does not support double precision natively. The speed can then be improved by increasing leakage current level hence the difference between simulations and experimental data. One way to increase this current is to increase the temperature. An optical generation can also be used instead. Therefore, the running time can be significantly reduced while not altering breakdown voltage. Figure 5.b shows that simulation data normalized with respect to thermal generated current almost overlap the experimental ones confirming the accuracy of such an approach.

These simulation and experimental results confirm that higher efficiency can be achieved if a bigger bend radius is chosen. In this context, 3D simulations could provide a better view of the structure to device designers. Indeed, different types of termination can be combined allowing reaching even higher efficiency. For example, single or multiples JTE can be used in conjunction with guard rings [9].

Thyristors active area design

Contrary to transistors, thyristors only need a sufficient current pulse on the gate terminal to be switched on. Due to a bistable behavior, the switch-on process is more complicated since it involves more external inputs such as dV/dt ramps. Junction capacitances are subject to inject currents under high dV/dt, which can accidentally turn on the device. Thyristor is a device of choice for railway applications since it can carry high amount of current. Chips can be large and, as a consequence, design plays a major role in order to efficiently exploit the large available active area.

Several designs are encountered for SiC thyristors. The conventional circular design (Figure 6.a) can be deal with cylindrical simulations. However, this design does not maximize the gate current injection efficiency since the "anode-gate" pn junction is very close to the center of the chip. An "involute" design (Figure 6.b) would be better suited for a reasonable trade-off between the device anode area and the gate efficiency [10]. While such a design guarantees a constant bend radius, the device periphery is susceptible to behave differently.

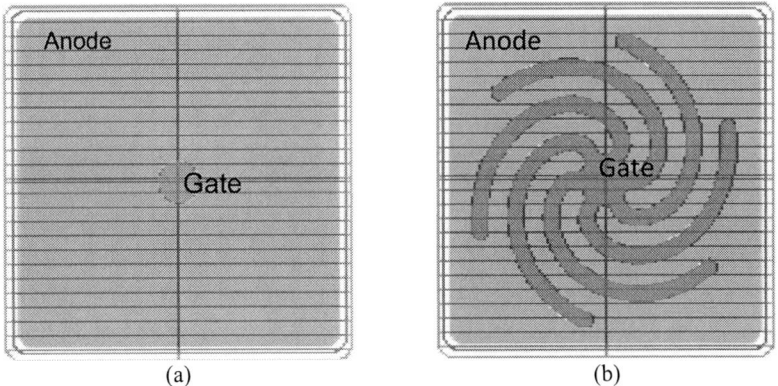

(a)   (b)

Figure 6. Two thyristor mask designs without periphery protection. (a) Circular. (b) Involute.

Rather than modeling the whole involute of a circle, two simpler portions were retained. Both are designed following the same way: the substrate and the epilayer are formed in a quarter of a "cylinder". An etching is then defined in order to form the mesa. Depending on which area of the chip is chosen, two 3D structures are possible (Figure 7).

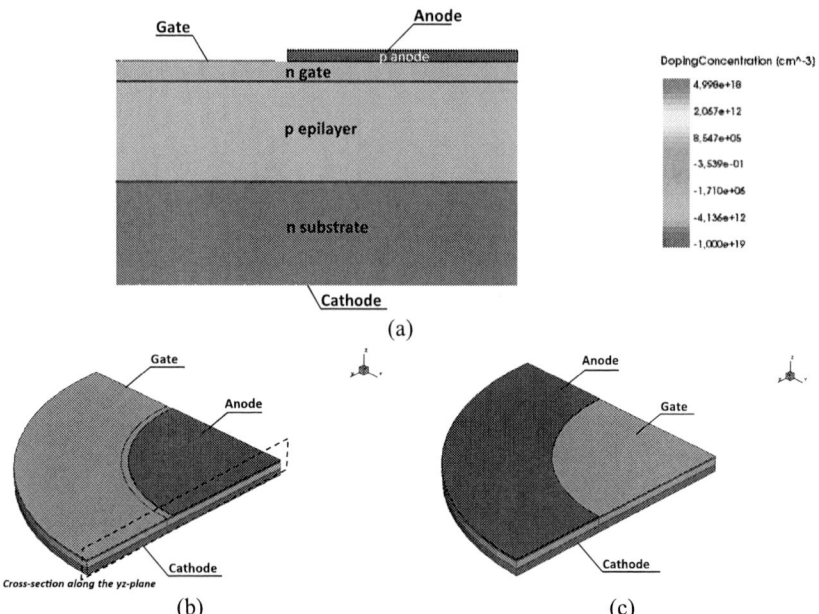

Figure 7. (a) Schematic cross-section of an involute thyristor. (b) 3D structure 1. (c) 3D Structure 2.

Figure 7.a shows what the structure in Figure 7.b would look like in 2D along the yz-plane. In this cross-section, the gate length is 150μm while the anode length is 200μm. The ratio between the areas of the outer and inner electrodes stays the same between Figure 7.b and Figure 7.c. For the sake of simplicity, these structures will be respectively named "structure 1" and "structure 2". The anode always sits on the mesa while the bend radius is 350μm and kept unchanged between the two structures. Identical arbitrary voltage and current are applied on the two structures respectively between the anode and cathode terminals and on the gate terminal. Figure 8 shows the obtained waveforms when such a thyristor turns on.

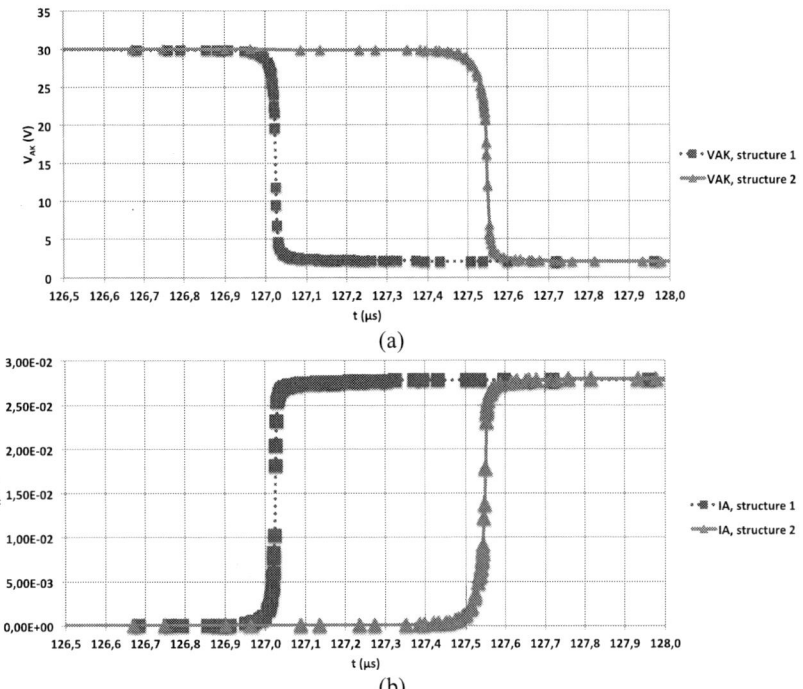

(a)

(b)

Figure 8. Waveforms of the voltage drop V_{AK} (a) and the load current I_A (b) across the thyristor at turn-on. $dI_G/dt = 3A.s^{-1}$.

Structure 1 tends to switch on slightly before structure 2 by 1.5µs. While the electrodes areas are inverted between the two structures, there is no noticeable difference between the I_A current levels. Since the outer electrode area is larger than the inner one, the gate current injection efficiency would be slightly better in structure 1 than in structure 2. Figure 9 depicts the total current density in the cross-section along the xy-plane in the n gate layer and along the yz-plane during the whole turn-on process. Regions in the xy-plane named "anode" refers to the n-gate layer just beneath the p anode mesa. Snapshots have been taken before, at the beginning ($I_A = I_{Amax}/4$) and after the turn-on process. Each pair of snapshots has been taken for same current levels. The upper color scale has been compressed on purpose so as to display more accurately low gate current density.

Figures 9.a-d show that the total current density gradient in structure 2 injected from the gate to the anode is more important than the one found in structure 1 for the same input current value. This gap can be attributed to the geometry. The consequence is that structure 1 is easier to turn-on since the current distribution, which is mostly an electron current before turn-on, is more uniform.

Figures 9.e-f depict the total current distribution at the very end of the simulation after the gate has been disengaged. Both structures still carry current, as one would expect from thyristors.

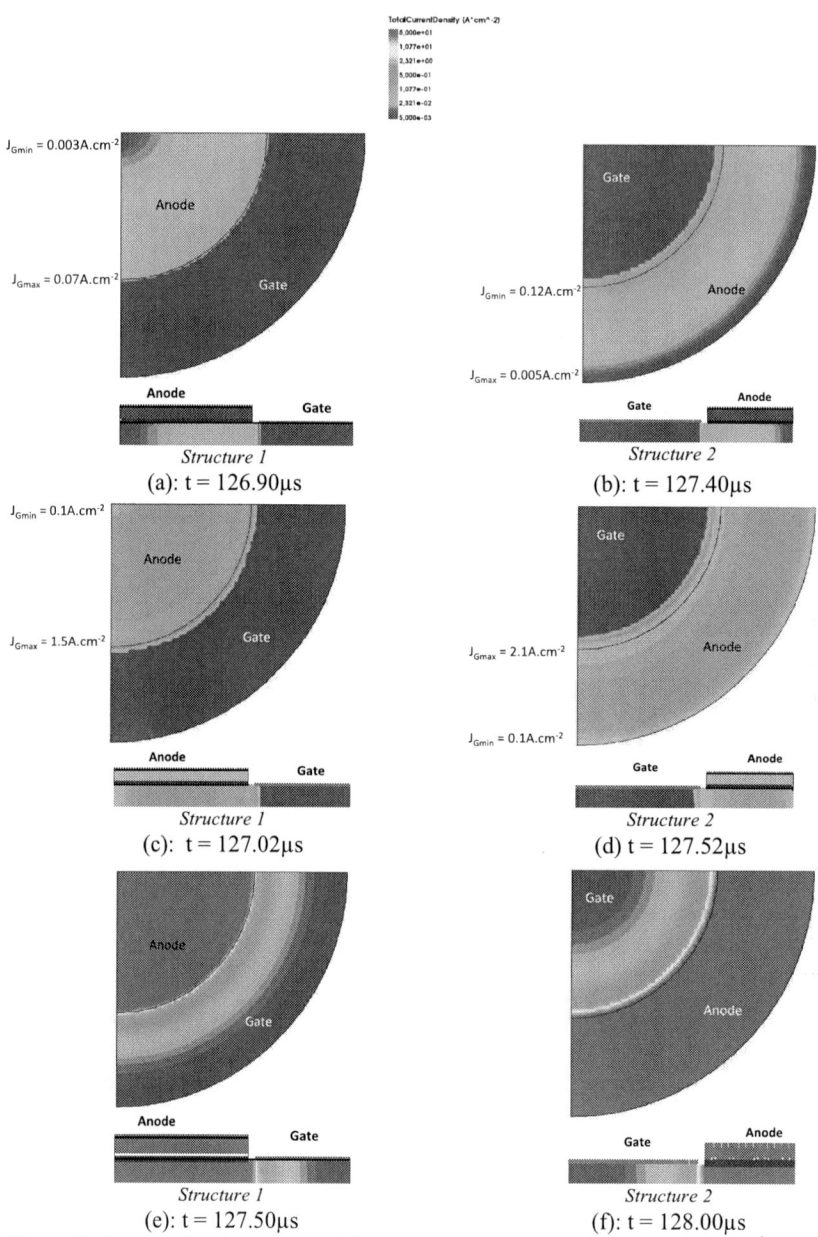

Figure 9. Current density extracted from cross-sections along the xy-plane in the n-gate layer (beneath the mesa) and along the yz-plane. The same scale of current density is used for all figures. (a-b): before turn-on. (c-d): at the beginning of turn-on. (e-f): after turn-on.

These simulations confirm the benefit of a fully 3D involute structure over a circular one. Areas where electron injection is better can compensate less effective areas, for example, in the inner side of the involute of a circle. The outer periphery of the structure might also not work properly but this should remain a relatively minor drawback.

Prospects for 3D simulations for SiC

In order to maximize their efficiency, devices design tends to be more sophisticated. One structure that could serve as an example is the involute thyristor that tries to merge the advantages of interdigitated and circular designs. Even if 3D challenges can be tackled more easily by studying 2D cross-sections or by performing cylindrical simulations, a fully 3D simulation can help designers to grasp the different mechanisms that occur within such a structure. For example, light triggered thyristors could be studied and optimized by this way. The study could also be extended to interdigitated designs (JFET, HEMT, BJT) where fingers geometry can be optimized in term of depolarization.

References

1. M. K. Das et al., *Mater. Sci. Forum* 483, 965 (2005).
2. D. Peter et al., *Mater. Sci. Forum* 645, 901 (2010).
3. B. A. Hull et al., *Proceedings of 18th ISPSD*, Naples (2006).
4. M. Berthou et al., *Mater. Sci. Forum 711*, 124 (2012).
5. Sentaurus, TCAD simulation tool by Synopsys Inc., ver. F (2011).
6. D. Planson et al, *223th ECS Meeting*, Hawai (2012).
7. R. Huang, *PhD thesis*, INSA de Lyon (2011).
8. D.M. Nguyen et al., Mater. Sci. Forum 740, 609 (2013).
9. G. Feng. *IEEE Trans. Elect. Dev.*, 99, 1 (2012).
10. H. F. Storm and J. G. Clair. *IEEE Trans. Elect. Dev.*, 21, 520, (1974).

340

CHAPTER 9

POWER DEVICE RELIABILITY 2

342

Heat Dissipation in GaN Based Power Electronics

Zonghui Su and Jonathan A. Malen[a]

[a] Department of Mechanical Engineering, Carnegie Mellon University, Pittsburgh, PA 15213, USA

Gallium Nitride (GaN) possesses superior electronic properties for RF power electronics that play critical roles in various wireless communication technologies and military applications [1]. Heat generated as a byproduct of operation in these devices, increases their operating temperature and degrades their performance and lifetime. While bulk GaN has a high thermal conductivity (k) approaching 250 W/m-K, [2, 3] GaN thin films and devices experience a much lower k due to the presence of additional phonon scattering mechanisms and departures from Fourier transport [4, 5]. We will review thermal transport in GaN based devices, broadly addressing the impact of heat source dimensions, film thicknesses, interfaces, and defects.

Introduction

Contemporary GaN HEMTs utilize a high electron mobility 2-dimensional electron gas (2-DEG) that exists at the interface of AlGaN and GaN thin films. SiC or sapphire substrates with AlN nucleation layers are the platform for epitaxial growth of the HEMT heterostructure, because GaN and AlN substrates are not economically viable. Joule heat generated in the 2-DEG must pass through a series of thermal resistances before its ultimate disposition to the environment. While package-level thermal management strategies have been called upon to mitigate these concerns, substantial thermal resistance exists within the device itself [6]. The size of the devices and the presence of defects and interfaces leads to thermal resistances that are higher than expected.

Previous experimental studies of thermal conductivity (k) have shown that single crystal GaN, AlN, and SiC have very high k that is dominated by phonons. Room temperature values of k exceed 200 W/m-K in GaN [7] and AlN [8] and are reported as high as 490 W/m-K in SiC [9]. Yet achievement of such high values is more challenging in practice, when the mean free paths (MFPs) of phonons are reduced by added scattering from interfaces, grain boundaries, and dislocations in real devices [10]. Since the distribution of MFPs spans nearly four orders-of-magnitude, some phonons remain diffusive, while others are scattered by such defects.

Theoretical Models

To better understand thermal transport in GaN based electronics it is important to understand thermal conductivity in terms of phonon transport. Simple kinetic theory can be used to approximate thermal conductivity as $k = C v_s \bar{l}/3$ where C is the volumetric specific heat, v_s is the sound velocity, and $\bar{l}$ is the average phonon mean free path (a.k.a. the gray mean free path). However, phonons exist over a range of frequencies, and each frequency phonon contributes differently to thermal conductivity. A more accurate expression of thermal conductivity is hence,:

$$k = \sum_s \int_0^\infty \frac{1}{3} C_\omega(\omega) v(\omega) l(\omega) d\omega \quad [1]$$

where ω is the phonon frequency, l is the phonon MFP, v is the phonon group velocity, C_ω is the volumetric heat capacity per unit frequency, and s indexes the polarization of phonons. Since thermal conductivity suppression occurs when the phonon mean free paths are commensurate to the device size, it is helpful to recast equation [1] as an integral over l as follows,

$$k_{accum} = \sum_s \int_0^{l^*} \frac{1}{3} C_l(l) v(l) l \, dl \quad [2]$$

where k_{accum} is the thermal conductivity accumulation function and C_l is the volumetric heat capacity per unit phonon mean free path. When $l*$ is set to infinity this predicts the bulk k because all phonon MFPs are included. The value of k_{accum} when $l*$ is set to the film thickness or defect spacing can be used as an estimate of the thermal conductivity suppression when some MFPs are reduced. Due to non-diffusive effects, it has also been shown that heat sources with dimension d perceive a reduced k that can be approximated as k_{accum} when $l*=d$ [11].

To truly understand experimental observations, simple analytical models based on kinetic theory and interface scattering must be complemented by detailed simulation of phonon properties. Molecular dynamics (MD) simulations have been applied to better understanding thermal conductivity and thermal boundary resistance in range of material systems[12-20]. A set of interatomic potentials that describe the interactions between atoms are required to perform an MD simulation. Interatomic potentials for nitrides have been developed using Stillinger-Weber (SW)[21-23], Tersoff[24-26], and modified embedded atom method[27] formulations. Potentials do not exist for SiC-nitride or Al_2O_3-nitride interactions. Hence, due to the lack of a complete set of potentials and the complexity of nitride devices, much progress to date has been based on experiments.

Thermal conductivity accumulation in GaN, AlN, and SiC

We have made direct experimental measurements of k_{accum} in GaN, AlN, and SiC using a novel technique known as broadband frequency domain thermoreflectance (BB-FDTR). [28, 29] Here we use high frequency surface temperature modulation that generates non-diffusive phonon transport to probe k_{accum} of GaN, AlN, and 4H-SiC at temperatures near 80K, 150K, 300K, and 400K. [29] In Figure 1 the values of k_{accum} for GaN are normalized by the bulk values of thermal conductivity (see legend), and the value at any particular MFP represents the thermal conductivity due to phonons having a MFP less than or equal to that value. In stark contrast to sub-100 nm estimates of the average phonon MFP based on kinetic theory, we find that phonons with MFPs greater than 1000±200 nm, 2500±800 nm, and 4200±850 nm contribute 50% of the bulk thermal conductivity of GaN, AlN, and 4H-SiC near room temperature. Hence, localized suppression of k will increase the operating temperature of HEMTs, built from thin, highly defective films having small localized heat sources.

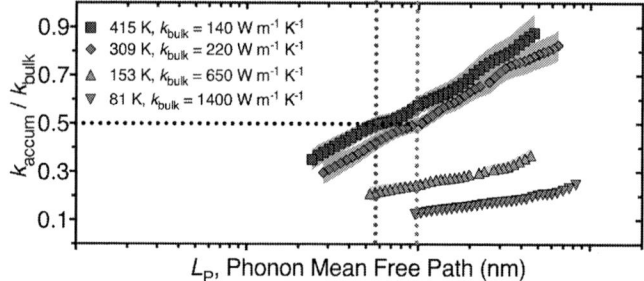

Figure 1. Normalized k_{accum} of GaN at different temperatures. This figure is modified from reference [29].

Prior experiments have shown that when heat sources are smaller than the mean free paths of phonons, a suppression in thermal conductivity is experienced. An example is in the TDTR measurement of Si at low temperature where smaller laser spot sizes yielded smaller thermal conductivity [30]. Qualitatively, if we make the heater size or thermal penetration depth smaller than a certain value, then all phonons having MFP longer than this length will ballistically escape this region without being sensed. In this case, we would only measure the thermal conductivity due to phonons with MFP smaller than the specific length. While this behavior enables experimental interrogation of k_{accum}, it also leads to increased operating temperatures in devices. To capture this effect in GaN we have calculated the ratio of the temperature rise with and without thermal conductivity suppression (based on k_{accum} in Fig. 1) for a circular heat source, as a function of heat source diameter at temperatures near 300 K and 400 K. The temperature rise due to a circular heat source is $\Delta T = A/(\pi k d)$, where A is the heater power and d is its diameter. For heat sources smaller than $d=1$ μm, the temperature rise is significantly higher than our bulk expectations. The effect is more profound at lower temperatures because long mean free path phonons contribute more to thermal transport.

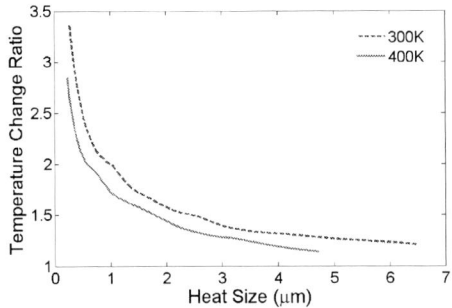

Figure 2. Ratio of the temperature rise with and without thermal conductivity suppression, due to a circular heat source of varied size (x-axis) in a GaN half-plane.

Thermal Boundary Resistance in GaN Devices

Another consideration is the effect of interfaces, which induce increased thermal resistance as phonons transmit from one material to another. Our prior studies on similar GaN-based LED structures show that the AlN-substrate interface is the dominant thermal resistance in the system. Kuball and his colleagues have also shown that the AlN layer and its interfaces with substrate and the GaN layer are the dominant thermal resistance in GaN based electronics [6, 35]. Data shown in Figure 3 prove that interface roughness plays a significant role in reducing the thermal boundary resistance (TBR) at the AlN-substrate interface. Here we plot the thermal resistance (*thermal conductivity/thickness=k/L*) of AlN films vs. film thickness for AlN on mechanically polished SiC (MP; RMS roughness=1.2nm), chemomechanically polished SiC (CMP; RMS roughness=0.2nm), and sapphire (RMS roughness=0.2nm) substrates.[30, 31] Since the data are only weak functions of film thickness, we conclude that thermal boundary resistance (TBR) dominates over film resistance.

The TBR, which includes contributions from top and bottom interfaces, is determined by extrapolating to $L=0$ (the y-intercept) in Figure 3. After subtracting the TBR of the top contact, we estimate that the TBR of the AlN/CMP-SiC interface is 5.1±2.8 m²-K/GW—equal to 1450 nm of single crystal AlN! The TBR of AlN/CMP-SiC is ~10 times lower than that of AlN/MP-SiC, regardless of whether the AlN film is grown by organometallic vapor phase epitaxy (industry standard) or plasma vapor deposition. This effect is corroborated by TEM images of the AlN/SiC interface, which are shown in Figure 4. While the CMP interface is clean and the AlN is highly crystalline within a few atomic layers, the MP interface has planar defects and strain fields that we hypothesize scatter phonons and lead to high TBR.

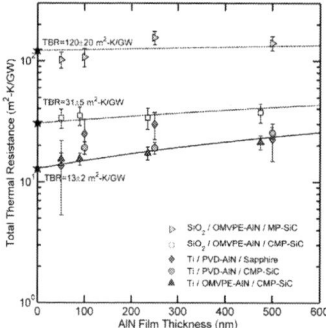

Figure 3. Thermal boundary resistance of AlN films vs. film thickness. Legend indicates top and bottom contacts. This figure is modified is from reference [31].

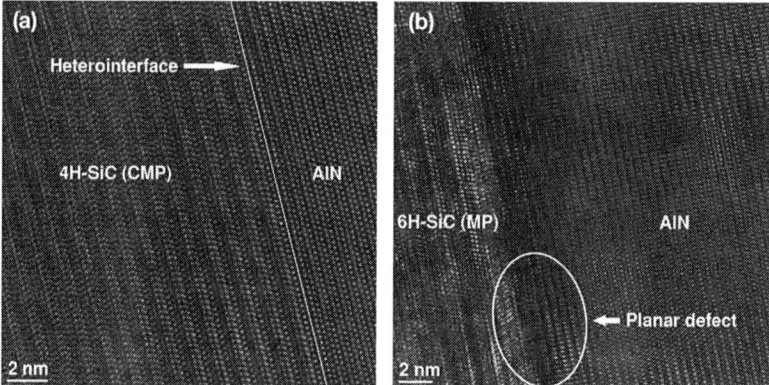

Figure 4. Comparison of the interfaces between AlN/CMP SiC and AlN/MP SiC. (a) AlN grown on CMP 4H-SiC (0001) has a clearly defined interface with no apparent additional strain in the film. (b) By contrast, microstructural roughness of the surface of the MP 6H-SiC (0001) generated stresses and resultant strains that caused planar defects within a few nanometers of the interface in both films, supporting our hypothesis that interface defects contribute to the thermal boundary resistance of the heterostructures. This figure is modified from reference [31].

Interface conductance has been of great interest to many researchers in the thermal transport field; multiple experimental investigations have measured the interface between GaN and substrates.[32] Raman spectroscopy was first used by Sarua et al.[6] to study the temperature profile in the GaN/SiC structure, and to extract the thermal boundary resistance due to the AlN nucleation layer. Based on a 3D analysis, Sarua et al. determined that TBR for GaN/SiC and GaN/Si is ~33m^2-K/GW, while TBR for GaN grown on sapphire is estimated to be 120 m^2-K/GW. In the TDTR measurements performed by Cho et al.[32], both the intrinsic thermal conductivity of GaN film and GaN-SiC thermal boundary resistance (TBR) were extracted. In this case, thermal resistance coming from the intrinsic thermal conductivity of the nucleation layer of 28nm AlN between SiC and GaN was incorporated into the TBR. Two methods were used to analyze the TDTR data, a simultaneous fit of the TDTR temporal traces for all experimental samples gives us a result of 157±11W/m-K for GaN conductivity and TBR$_{GaN-SiC}$ = 4.2±0.6m^2-K/GW. In the second approach, extrapolation of the thermal resistances was determined separately for each sample. This method reveals a result of k_{GaN} = 182±33W/m-K and TBR$_{GaN-SiC}$ = 3.6±1.6m^2-K/GW at room temperature. Compared with previous work, our values of TBR (5.1±2.8 m^2-

346

K/GW) are similar to Cho et al. [34], but much lower than the Raman studies from refs. [6, 33]. Such differences can result from experimental methods and uncertainties, but also from differences in the AlN transition layer due to growth conditions and substrate preparation.

The effect of defects on thermal conductivity in nitride semiconductors.

Defects and dislocations can also greatly reduce phonon transport within the GaN layer or across the interface. Our previous work [30] has shown that over the 100K to 400K temperature range, a dislocation density of 1.2×10^8 cm^{-2} can cause GaN's intrinsic thermal conductivity to drop by ~40%. We used the truncated Callaway model (TC), as described in Koh et al.[34], to predict thermal conductivity reduction in GaN films due to dislocation scattering. Phonon relaxation times were estimated based frequency independent scattering by phonons traveling between dislocations separated by l_D at their group velocity as,

$$\frac{1}{\tau_D} = \frac{v}{l_D} = v\sqrt{\frac{3^{0.5}\rho_D}{2}} \quad [3]$$

where ρ_D is the areal dislocation density and the factor of $3^{0.5}/2$ results from the assumption that the dislocations are areally close packed. This additional relaxation time is added to the phonon-phonon relaxation time τ_{P-P} and the defect scattering relaxation time τ_{def}. The effective relaxation time for the parallel scattering channels can be calculated by the Matthiessen's Rule:

$$\frac{1}{\tau} = \frac{1}{\tau_D} + \frac{1}{\tau_{def}} + \frac{1}{\tau_{P-P}} \quad [4]$$

This prediction of phonon relaxation times is implemented using a truncated Callaway model for thermal conductivity prediction, and is compared with historical data measured by Mion et al [3] in Figure 5. Good agreement is established, suggesting that this basic frequency independent model for τD may be capturing the phonon properties well. The plan view TEM image of GaN shown in Figure 6 indicates that high dislocation densities, of order 10^{10}, are typical in nitride devices. Other studies looking into the impurities and dislocations vs. thermal conductivity were performed by multiple groups [35, 36]. In Kotchetkov et al.'s work[35], phonon scattering on dislocations is described by a combined term that includes scattering from core dislocation, edge dislocation and screw dislocation

$$\frac{1}{\tau_D} = \frac{1}{\tau_{core}} + \frac{1}{\tau_{screw}} + \frac{1}{\tau_{edge}} \quad [5]$$

With the combined phonon relaxation time, Kotchetkov et al. calculated lattice thermal conductivity at 300K. If the dislocation line density drops from 10^{12} to 10^{10} cm^{-2}, thermal conductivity would increase from 131 to 197W/m-K, which is in agreement with recent experimental observations [37-39].

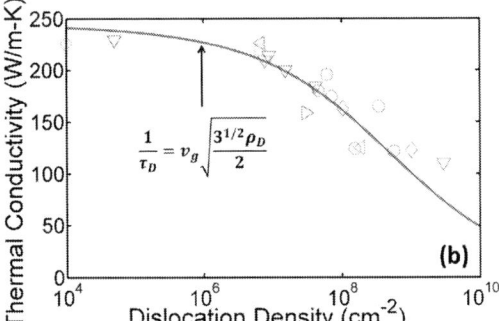

Figure 5. Thermal conductivity of GaN vs. dislocation density at 300K. The new relaxation time based

on dislocation density was added to the TC model and compared with data from Mion et al. (see Ref. 3). This revised TC was used to calculate the GaN thermal conductivity vs. temperature. This figure is modified from reference [30].

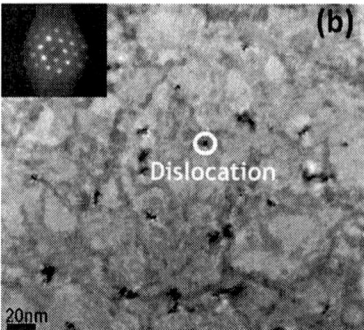

Figure 6. A plan view TEM image of the AlN layer acquired along the [0001] zone axis. The TEM image reveals a high dislocation density of $4\times10^{10}cm^{-2}$. Inlet is the associated diffraction pattern. This figure is modified from reference [30].

Acknowledgments

We would like to thank partial support of the NSF GOALI grant (CBET 1133394) and the CMU ICES-Dowd Fellowship.

References

1. Mishra, U.K., et al., *GaN-Based RF power devices and amplifiers.* Proceedings of the Ieee, 2008. **96**(2): p. 287-305.
2. Shibata, H., et al., *High thermal conductivity of gallium nitride (GaN) crystals grown by HVPE process.* Materials Transactions, 2007. **48**(10): p. 2782-2786.
3. Mion, C., et al., *Accurate dependence of gallium nitride thermal conductivity on dislocation density.* Applied Physics Letters, 2006. **89**(9).
4. Liu, W.L. and A.A. Balandin, *Temperature dependence of thermal conductivity of AlxGa1-xN thin films measured by the differential 3 omega technique.* Applied Physics Letters, 2004. **85**(22): p. 5230-5232.
5. Daly, B.C., et al., *Optical pump-and-probe measurement of the thermal conductivity of nitride thin films.* Journal of Applied Physics, 2002. **92**(7): p. 3820-3824.
6. Sarua, A., et al., *Thermal boundary resistance between GaN and substrate in AlGaN/GaN electronic devices.* Ieee Transactions on Electron Devices, 2007. **54**(12): p. 3152-3158.
7. Jezowski, A., et al., *Thermal conductivity of GaN crystals in 4.2-300 K range.* Solid State Communications, 2003. **128**(2-3): p. 69-73.
8. Slack, G.A., et al., *The Intrinsic Thermal-Conductivity of Aln.* Journal of Physics and Chemistry of Solids, 1987. **48**(7): p. 641-647.
9. Slack, G.A., *Thermal Conductivity of Pure + Impure Silicon Silicon Carbide + Diamond.* Journal of Applied Physics, 1964. **35**(12): p. 3460-&.
10. Mion, C., et al., *Accurate dependence of gallium nitride thermal conductivity on dislocation density.* Applied Physics Letters, 2006. **89**(9): p. 092123.
11. Minnich, A.J., et al., *Thermal Conductivity Spectroscopy Technique to Measure Phonon Mean Free Paths.* Physical Review Letters, 2011. **107**(9).
12. Zhou, X.W., et al., *Towards more accurate molecular dynamics calculation of thermal conductivity: Case study of GaN bulk crystals.* Physical Review B, 2009. **79**(11).
13. Turney, J.E., et al., *Predicting phonon properties and thermal conductivity from anharmonic lattice dynamics calculations and molecular dynamics simulations.* Physical Review B, 2009. **79**(6).
14. McGaughey, A.J.H., et al., *Phonon band structure and thermal transport correlation in a layered diatomic crystal.* Physical Review B, 2006. **74**(10).
15. Landry, E.S. and A.J.H. McGaughey, *Thermal boundary resistance predictions from molecular dynamics simulations and theoretical calculations.* Physical Review B, 2009. **80**(16).

16. Landry, E.S. and A.J.H. McGaughey, *Effect of interfacial species mixing on phonon transport in semiconductor superlattices.* Physical Review B, 2009. **79**(7).

17. Landry, E.S., M.I. Hussein, and A.J.H. McGaughey, *Complex superlattice unit cell designs for reduced thermal conductivity.* Physical Review B, 2008. **77**(18).

18. Sellan, D.P., et al., *Size effects in molecular dynamics thermal conductivity predictions.* Physical Review B, 2010. **81**(21).

19. Schelling, P.K., S.R. Phillpot, and P. Keblinski, *Phonon wave-packet dynamics at semiconductor interfaces by molecular-dynamics simulation.* Applied Physics Letters, 2002. **80**(14): p. 2484-2486.

20. Schelling, P.K., S.R. Phillpot, and P. Keblinski, *Comparison of atomic-level simulation methods for computing thermal conductivity.* Physical Review B, 2002. **65**(14).

21. Aichoune, N., et al., *An empirical potential for the calculation of the atomic structure of extended defects in wurtzite GaN.* Computational Materials Science, 2000. **17**(2-4): p. 380-383.

22. Bere, A. and A. Serra, *Atomic structure of dislocation cores in GaN.* Physical Review B, 2002. **65**(20).

23. Bere, A. and A. Serra, *On the atomic structures, mobility and interactions of extended defects in GaN: dislocations, tilt and twin boundaries.* Philosophical Magazine, 2006. **86**(15): p. 2159-2192.

24. Benkabou, F., M. Certier, and H. Aourag, *Elastic properties of zinc-blende GaN, AlN and InN from molecular dynamics.* Molecular Simulation, 2003. **29**(3): p. 201-209.

25. Kioseoglou, J., P. Komninou, and T. Karakostas, *Interatomic potential calculations of III(Al, In)-N planar defects with a III-species environment approach.* Physica Status Solidi B-Basic Solid State Physics, 2008. **245**(6): p. 1118-1124.

26. Nord, J., et al., *Modelling of compound semiconductors: analytical bond-order potential for gallium, nitrogen and gallium nitride.* Journal of Physics-Condensed Matter, 2003. **15**(32): p. 5649-5662.

27. Do, E.C., Y.H. Shin, and B.J. Lee, *Atomistic modeling of III-V nitrides: modified embedded-atom method interatomic potentials for GaN, InN and Ga1-xInxN.* Journal of Physics-Condensed Matter, 2009. **21**(32).

28. Regner, K.T., et al., *Broadband phonon mean free path contributions to thermal conductivity measured using frequency domain thermoreflectance.* Nature Communications, 2013(DOI: 10.1038/ncomms2630).

29. J. P. Freedman, J.H.L., E. A. Preble, Z. Sitar, R. F Davis, J. A. Malen, *Universal phonon mean free path spectra in crystalline semiconductors at high temperature.* 2013.

30. Su, Z., et al., *Layer-by-layer thermal conductivities of the Group III nitride films in blue/green light emitting diodes.* Applied Physics Letters, 2012. **100**(20).

31. Su, Z., et al., *The impact of film thickness and substrate surface roughness on the thermal resistance of aluminum nitride nucleation layers.* Journal of Applied Physics, 2013. **113**(213502).

32. Cho, J.W., et al., *Low Thermal Resistances at GaN-SiC Interfaces for HEMT Technology.* Ieee Electron Device Letters, 2012. **33**(3): p. 378-380.

33. Manoi, A., et al., *Benchmarking of Thermal Boundary Resistance in AlGaN/GaN HEMTs on SiC Substrates: Implications of the Nucleation Layer Microstructure.* Ieee Electron Device Letters, 2010. **31**(12): p. 1395-1397.

34. Koh, Y.K., et al., *Heat-Transport Mechanisms in Superlattices.* Advanced Functional Materials, 2009. **19**(4): p. 610-615.

35. Kotchetkov, D., et al., *Effect of dislocations on thermal conductivity of GaN layers.* Applied Physics Letters, 2001. **79**(26): p. 4316-4318.

36. Zou, J., et al., *Thermal conductivity of GaN films: Effects of impurities and dislocations.* Journal of Applied Physics, 2002. **92**(5): p. 2534-2539.

37. Asnin, V.M., et al., *High spatial resolution thermal conductivity of lateral epitaxial overgrown GaN/sapphire (0001) using a scanning thermal microscope.* Applied Physics Letters, 1999. **75**(9): p. 1240-1242.

38. Florescu, D.I., et al., *Thermal conductivity of fully and partially coalesced lateral epitaxial overgrown GaN/sapphire (0001) by scanning thermal microscopy.* Applied Physics Letters, 2000. **77**(10): p. 1464-1466.

39. Witek, A., *Some aspects of thermal conductivity of isotopically pure diamond - a comparison with nitrides.* Diamond and Related Materials, 1998. **7**(7): p. 962-964.

ECS Transactions, 58 (4) 351-363 (2013)
10.1149/05804.0351ecst ©The Electrochemical Society

High Voltage InAlN/GaN HFETs Achieved by Schottky-Contact Technology for Power Applications

Qi Zhou[a], Wanjun Chen[a], Shenghou Liu[b], Bo Zhang[a], Zhihong Feng[c], Shujun Cai[c] and Kevin J. Chen[b]

[a] State Key Laboratory of Electronic Thin Films and Integrated Devices, University of Electronic Science and Technology of China, Chengdu, China
[b] Department of Electronic and Computer Engineering, Hong Kong University of Science and Technology, Hong Kong
[c] The Science and Technology on ASIC Lab., Hebei Semiconductor Research Institute, Shijiazhuang, China

In this paper, a novel approach of Schottky-contact technology for high voltage InAlN/GaN power transistors is demonstrated. The improved breakdown voltage (BV) attributes to the effectively suppressed source carrier injection achieved by the Schottky source metallization featuring excellent metal morphology. Without using field-plate structure and buffer engineering technique, the record three-terminal off-state BV of 650 V is obtained for InAlN/GaN HFET featuring a Schottky source. The corresponding specific on-resistance is as low as 3.4 m$\Omega \cdot$cm^2. Moreover, the proposed Schottky-contact technology enables the removal of conventional alloyed Ohmic contact which is beneficial to device scaling. The BV of 58 V is measured in an InAlN/GaN HFET with L_{GD} of 250 nm, which is the highest BV reported on GaN-based HFETs with such a short drift region. With simple fabrication process as well as the device structure while achieving a significant breakdown voltage enhancement, the proposed technology opens up new perspectives for designing high voltage InAlN/GaN HFET for power applications.

Introduction

The InAlN/GaN heterojunction has been predicted to be capable of delivering record device performance owing to the superior material properties of wide band-gap, ultra-high spontaneous polarization induced charge density (1). Moreover, the In$_x$Al$_{1-x}$N ternary film can be grown to be lattice-matched to GaN templates for an Indium mole-fraction $x \approx 0.17$. Without the tensile strain as that presented in AlGaN barrier the lattice-matched InAlN/GaN heterojunction features reduced elastic energy in the InAlN barrier, which leads to important benefits with regard to better device reliability for high voltage applications (2,3). Stable operation up to 1000 °C of InAlN/GaN HFET was also reported suggesting prominent temperature stability and reliability of InAlN/GaN heterojunction (4,5). Furthermore, the InAlN barrier (e.g. <10 nm) in InAlN/GaN is much thinner than the AlGaN barrier (e.g. ~20 nm) in a conventional AlGaN/GaN heterojunction. The reduced gate-to-channel distance is beneficial to enhance the gate-to-channel modulation thus improved device performance. The native material properties and these already

351

demonstrated device characteristics also imply that the InAlN/GaN HFETs is a promising candidate for high-power switching devices with low R_{on} and high power density. However, reports on BV, one of the most important parameters for providing reliable performance at high voltage and high-power density operation of power devices, is still lacking up to date. In previous work, un-passivated InAlN/GaN HFETs with BV of 350 V at $L_{GD}=10$ μm was demonstrated (6), in which the breakdown was dominated by the source-drain leakage. In addition, a BV of 400 V at $L_{GD}=12$ μm for an SiN$_x$ passivated InAlN/GaN HFETs with a field-plated gate was reported (7). However, these data are well below the reported values in the AlGaN/GaN HFETs counterparts (8,9,10). The poor breakdown voltage of InAlN/GaN HFETs originates from the relatively large gate leakage current as a result of the conductive leakage path formed by Indium segregation related screw- and mixed-type threading dislocation (11,12), which indicates an immature stage of the InAlN/GaN heterojunction growth and device processing techniques. Besides, the ultra-high electron density induced in InAlN/GaN may also reduce the breakdown voltage.

In this paper, as a proof-of-concept, we present a new Schottky-Contact technology to improve the off-state BV of the InAlN/GaN HFETs. Based on this concept, the Schottky-Source/Drain (SSD) and Schottky-Source (SS) InAlN/GaN HFETs are proposed (13). The proposed devices with a $L_{GD}=15$ μm showed a BV more than 600 V, while conventional devices of the same geometry showed a maximum BV of 184 V. Without using any field-plate the BV of 650 V at drain leakage of 1 mA/mm was obtained in the SS HFET with $L_{GD}=15$ μm, which is the highest BV ever achieved on an InAlN/GaN HFET. The corresponding specific on-resistance ($R_{on,sp}$) is as low as 3.4 m$\Omega \cdot$cm^2. In order to further reduce the off-state leakage, a 12-nm high-k Al$_2$O$_3$ deposited by ALD was introduced as the gate dielectric in the SSD HFETs, leading to the SSD MISHFETs. The SSD MISHFET with $L_{GD}=10$ μm achieves a BV of 460 V at drain leakage current as low as 10 μA/mm, realizing 170 % improvement in BV compared with the conventional Al$_2$O$_3$/InAlN/GaN MISHFET (14). The remarkably improved BV in the proposed devices originates from the effective suppression of source carrier injection achieved by the smooth Schottky metallization which leads to a uniform electric-field distribution in GaN buffer.

Device Design and Fabrication

The detail of InAlN/GaN epilayer structure can be found in our previous work (13,14). The device fabrication commenced with mesa isolation by BCl$_3$/Cl$_2$ based plasma etching. The Ti/Al/Ni/Au (20/150/50/80 nm) Ohmic metal stack were deposited in the area of source/drain and drain for the conventional HFETs and SS HFETs, respectively. The Ohmic contact was annealed at 850 °C in N$_2$ ambient results in a contact resistance (R_c) of 0.6 $\Omega \cdot$mm obtained from linear transfer-length method (TLM). The Ni/Au gate was deposited by e-beam evaporation. The non-alloyed Ti/Au Schottky contact was formed on the InAlN/GaN heterostructure in the area of source/drain and source for the SSD HFETs/MISHFETs and SS HFETs, respectively. The 12-nm high-k Al$_2$O$_3$ in the SSD MISHFETs was deposited by ALD at 300 °C. On the same wafer, the conventional InAlN/GaN HFETs and MISHFETs featuring Ohmic source and drain were also fabricated for reference.

The absence of Ohmic contact indicates a low temperature process and rules out the reliability issue referring to alloyed Ohmic contacts both in the SSD HFETs and SSD MISHFETs. The proposed devices feature an identical L_{GS} of 0.25 μm. The ultra-short L_{GS} of proposed devices is achievable because of the elimination of lateral metal overflow that normally observed in thermally annealed Ohmic contacts, which is beneficial to the reduction of source-gate access resistance then the overall on-resistance. On the contrary, the L_{GS} of the conventional HFETs is 0.75 μm. Figure 1 shows the scanning electron microscope (SEM) images of the Ti/Au Schottky contacts (see Fig. 1(a)) and the Ti/Al/Ni/Au Ohmic contacts after high temperature annealing (see Fig. 1(b)), which confirm that the Schottky contact features much better surface metal morphology than the Ohmic contacts.

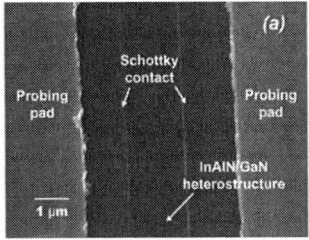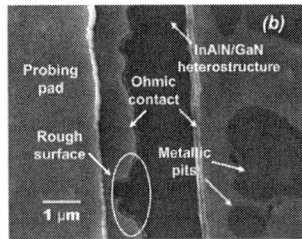

Figure 1. Top view SEM images of (a) Ti/Au Schottky contact and (b) conventional Ti/Al/Ni/Au Ohmic contact after high temperature annealing.

Measurements and Results

Schottky-Source/Drain (SSD) InAlN/GaN HFETs

The three-terminal off-state breakdown measurements were carried out with the samples immersed in Fluorinert to prevent surface flashover. We define BV as the voltage at which the off-state drain leakage current ($I_{D,leakage}$) increases to 1 mA/mm, while the gate is biased in pinchoff (-3.5 V). Figure 2 shows a comparison of the three-terminal BV for an InAlN/GaN SSD HFETs and conventional HFETs. The maximum BV for the SSD HFETs and the conventional HFETs were 605 V and 184 V, respectively, for an L_{GD}=15 μm. Compared with the conventional HFETs, the BV is substantially improved by 229% on the SSD HFETs, which well validates the effectiveness of the Schottky contact technology in BV improvement for InAlN/GaN HFETs. The breakdown of the SSD HFETs is dominated and induced by the gate leakage (see Fig. 2) instead of the source leakage as reported in (6). The source leakage of the SSD HFETs exhibits negligible increase during the high voltage measurement, which is a result of the effective suppression of the source carrier injection realized by using the Schottky contact in the source area. Furthermore, the leakage current of the SSD HFETs is consistently one order lower than that in the conventional HFETs prior to the device breakdown revealing better off-state leakage suppression capability of the Schottky contact over the conventional Ohmic contact. The on-resistance (R_{on}) of the SSD HFETs and conventional HFETs is calculated from the I-V curves, as shown in the inset of Fig. 2. The static R_{on} of the SSD HFETs and conventional HFETs were 26 and 13 $\Omega \cdot$mm, respectively. The SSD HFETs exhibits a drain forward-voltage drop of 3.2 V at I_D=120 mA/mm ($I_{D,max}/2$ of the SSD HFETs) at V_G=1.5 V, which is ~1.6 V higher than that obtained in the conventional HFETs.

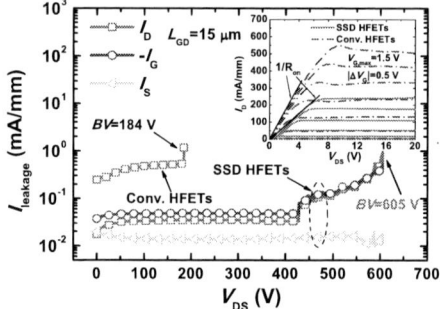

Figure 2. Three-terminal off-state BV while V_{GS}=-3.5 V for an SSD InAlN/GaN HFET and conventional HFET with L_{GD}=15 µm. The inset shows the output characteristics of the SSD HFET and conventional HFET.

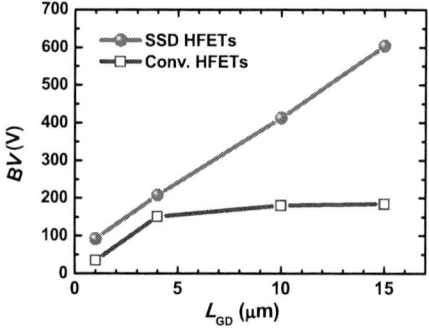

Figure 3. Three-terminal off-state BV versus L_{GD} of the SSD InAlN/GaN HFETs and conventional InAlN/GaN HFETs.

The BV versus L_{GD} of the SSD HFETs and conventional HFETs are depicted in Fig. 3. The BV of the SSD HFETs increases linearly with L_{GD} and shows a significant improvement compared with the conventional HFETs. On the contrary, the BV of the conventional HFETs saturates at ~180 V at a gate-drain distance $L_{GD}\geq10$ µm. The linear dependence on L_{GD} of BV for the SSD HFETs indicates that higher BVs can be achieved in the SSD HFETs with larger L_{GD}. The static specific on-resistance ($R_{on,sp}$) of the SSD and conventional HFETs plotted in Fig. 4 was extracted from the I-V curves (see the inset of Fig. 2), including a 3-µm transfer length from the contact pads. The SSD technology enables 605 V breakdown voltage for InAlN/GaN HFETs with $R_{on,sp}$ of 5.2 mΩ·cm^2.

With the similar $R_{on,sp}$ as shown in Fig. 4, the SSD HFETs can deliver much higher BV than the conventional HFETs while BV within the range of ~200-400 V. Compared with the dramatic improvement in BV the penalty paid in $R_{on,sp}$ is moderate especially for the SSD HFETs with BV of ~200-400 V, validating the feasibility of the SSD technology for breakdown voltage enhancement for InAlN/GaN HFETs. Moreover, the excellent surface metal morphology because of the absence of metallic overflow during the thermal annealing features decent scaling capability of the SSD HFETs that favors the applications of high-speed/high-power RF power amplification in which the high breakdown voltage and scaled device dimensions are simultaneously required. In order to study the BV in the scaled SSD HEMTs, device with short L_{GS} and L_{GD} of 0.25 μm was fabricated by using contact optical photolithography and lift-off process. The SSD HFETs features a three-terminal off-state BV of 58 V (not shown), which is the highest BV reported for GaN-based HFETs by linear scaling of L_{GD} to 0.25 μm. On the contrary, the conventional HFETs with such short L_{GS} and L_{GD} can not be obtained by using the same fabrication process due to the overflow of ohmic metal during high temperature annealing.

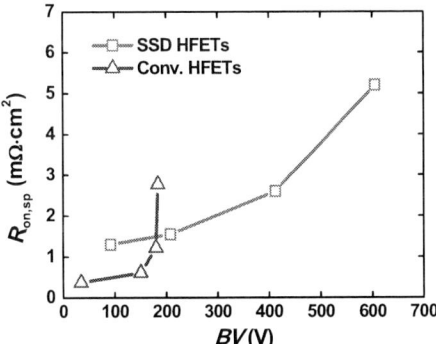

Figure 4. The specific on-resistance $R_{on,sp}$ versus three-terminal off-state BV of the SSD InAlN/GaN HFETs and conventional InAlN/GaN HFETs.

Schottky-Source InAlN/GaN HFETs

As aforementioned, in the SSD HFETs, the BV enhancement originates from the suppressed carrier injection by the Schottky source, however, at the cost of $R_{on,sp}$. In order to further reduce the $R_{on,sp}$ and simultaneously sustain the high BV, the Schottky-Source InAlN/GaN HFETs featuring an Schottky-Source and an Ohmic-Drain was further proposed. In the Schottky-Source HFETs, the Ti/Au Schottky source is responsible for the BV improvement while the ohmic drain is beneficial to the reduction of $R_{on,sp}$. Such an optimized device structure has been confirmed by the measurement results. The DC characteristics of the Schottky-Source InAlN/GaN HFETs with the dimensions of $L_G/L_{GS}/L_{GD}/W_G$=2.5/0.25/15/10 μm are plotted in Fig. 5. The conventional HFETs is also given in Fig. 5 for reference. It can be seen that the Schottky-Source HFETs features respectable drive current. The peak transconductance $G_{m,peak}$ is 158 mS/mm and the

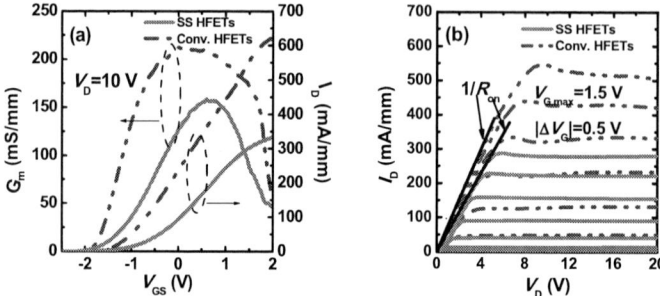

Figure 5. Measured dc characteristics of Schottky-Source HFET and conventional HFET. (a) Transfer curves measured at V_{DS}=10 V and (b) output curves. The device dimensions of the SS HFETs and conventional HFETs are $L_G/L_{GS}/L_{GD}/W_G$=2.5/0.25/15/10 μm and $L_G/L_{GS}/L_{GD}/W_G$=2.5/0.75/15/10 μm, respectively.

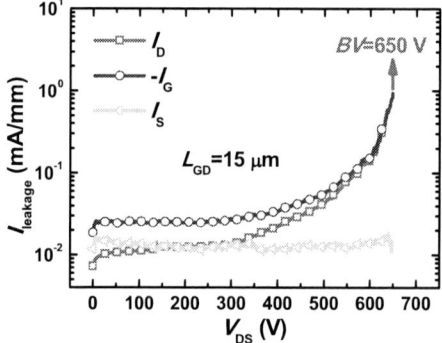

Figure 6. Gate, drain and source current (I_G, I_D and I_S) versus drain bias V_{DS} in the off-state BV measurement of SS InAlN/GaN HFETs.

maximum drain current density $I_{D,max}$ is 334 mA/mm at V_{GS}=2 V. The R_{on} of the Schottky-Source HFETs is 17 Ω·mm. Compared with the SSD HFETs, the R_{on} was reduced by 35 % in the SS HFETs. At I_D=120 mA/mm, the forward voltage drop of the SS HFETs is 2.3 V which is 28 % lower than that obtained in the SSD HFETs.

In an Schottky-Source HFETs with an L_{GD} of 15 μm, the BV of 650 V is achieved as shown in Fig. 6, which is the highest BV ever reported on InAlN/GaN HFETs so far. Compared with the conventional HFETs featuring conventional ohmic source/drain, the BV was improved by 253 %. The corresponding static $R_{on,sp}$ is 3.4 mΩ·cm^2, realizing 33 % reduction compared with the SSD HFETs with identical device dimensions. Similar to the SSD HFETs, the breakdown of the Schottky-Source HFETs is dominated by the gate

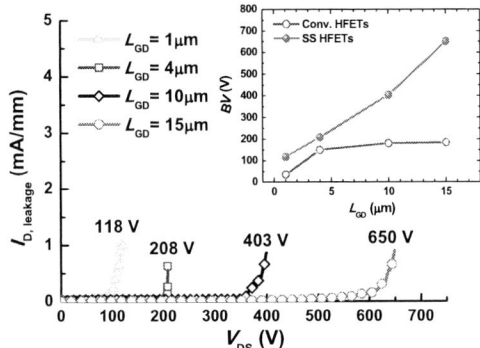

Fig. 7. Measured BV versus L_{GD}. The inset is the BV versus L_{GD} of the SS HFETs and conventional HFETs.

leakage. During the high voltage measurement the source leakage is effectively restrained and maintained at a relatively low level compared with the gate leakage current, which confirms that the Schottky source is able to provide effective prevention of source carrier injection and consequently leads to a significant improvement of BV. Fig. 7 shows the BV as a function of L_{GD} of the Schottky-Source InAlN/GaN HFETs, the inset plots the comparison of the BV versus L_{GD} for the SS HFETs and conventional HFETs. The BV of the SS HFETs was remarkably enhanced compared with the conventional HFETs. Different from the conventional HFETs that saturate at ~180 V, the BV of the SS HFETs exhibits a linear dependence on L_{GD}. In an SS HFETs with an L_{GD}=1 μm, the BV of 118 V is measured which is about fourfold higher than the result reported in (6). Without using field-plate (7) and buffer engineering technique (15) this value is already among the best reported results in GaN-based devices even compared with state-of-the-art AlGaN/GaN HFETs by linear scaling of L_{GD} to 1 μm (8,9,10).

Figure 8 plots the $R_{on,sp}$ versus BV of the SSD HFETs, SS HFETs and conventional HFETs for comparison. By replacing the Schottky drain with Ohmic contact in the SSD HFETs, the $R_{on,sp}$ was substantially reduced without sacrificing the breakdown voltage in the SS HFETs as shown in Fig. 8. In the devices with L_{GD} of 15 μm the $R_{on,sp}$ is 5.2, 3.4, 2.77 m$\Omega\cdot$cm^2 for the SSD HFETs, SS HFETs and conventional HFETs, respectively. The SS InAlN/GaN HFETs achieves 253 % improvement in BV with only a low cost of 23% in $R_{on,sp}$. In addition, the $R_{on,sp}$ exhibits a linear dependence on BV in the SS HFETs, which is different from that obtained in the SSD HFETs and conventional HFETs. By delivering high BV and low $R_{on,sp}$ simultaneously, the proposed Schottky-Source InAlN/GaN HFETs achieve the optimized balance between BV and $R_{on,sp}$ for devices with BV beyond 200 V.

Schottky-Source/Drain (SSD) Al$_2$O$_3$/InAlN/GaN MISHFETs

In order to improve the performance of InAlN/GaN power devices, the gate leakage current (16) need to be reduced, which will significantly reduce the power consumption.

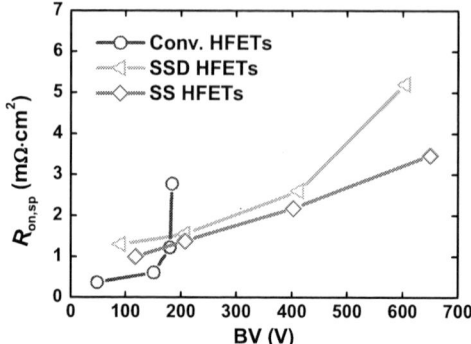

Figure 8. The $R_{on,sp}$ versus BV of the conventional HFETs, SSD HFETs and SS HFETs.

The use of gate dielectric showed a reduction of gate leakage current in HFETs (16). In this paper, the Al_2O_3/InAlN/GaN MISHFETs with Schottky-Source/Drain (SSD) was further demonstrated.

DC characterization of an SSD MISHFET with dimensions of $L_G/L_{GS}/L_{GD}/W$=1.0/0.25/0.25/10 μm is shown in Fig. 9. The maximum drain current density at V_{GS}=3 V is 416 mA/mm in the SSD MISHFET compared to 828 mA/mm in the conventional device. The peak transconductance is 113 mS/mm. The threshold voltage V_{th} of SSD MISHFETs is -2 V from the linear extrapolation of I_D. We observe an onset voltage of 0.16 V (at a drain current of 1 mA/mm) which is significantly lower than that reported in Schottky-drain AlGaN/GaN HFETs (17,18,19). The SSD MISHFET exhibits a drain forward-voltage drop of 1.9 at I_{DS}=200 mA/mm with V_{GS}=3 V, which is ~1 V higher than that obtained in the conventional MISHEMT. Comparing with the

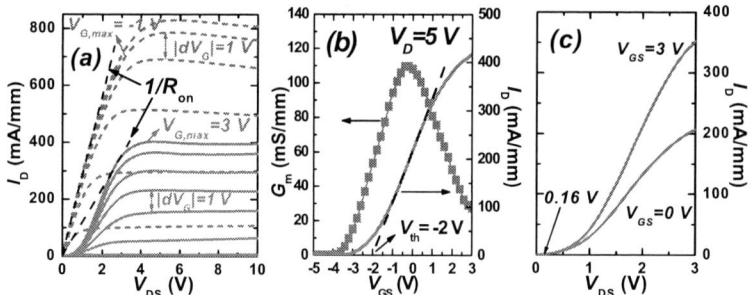

Figure 9. DC I-V of SSD MISHFETs. (a) Output I-V curves (dashed line—a conventional MISHFET). (b) Linear-scale transfer curves and (c) details in the onset of the drain current flow when forward biased.

Schottky gate (Ni/Au) directly made on the InAlN barrier , the 12 nm Al_2O_3 dielectric layer in the MISHFETs deliver a leakage reduction of ~10^2 at reverse bias and ~10^3 at forward gate bias (not shown). By directly extracting from the output curves in Fig. 9(a), the R_{on} of the SSD MISHFET and conventional device are 8.5 and 4.2 $\Omega\cdot$mm, respectively.

The high voltage measurement of the SSD MISHFETs was shown in Fig. 10. The criteria of the BV is an off-state drain current of 10 μA/mm which is 2-order lower than that used for SSD/SS HFETs. The BV is 55 V with an ultra-short L_{GD}=0.25 μm, and increases to 460 V with L_{GD}=10 μm. The corresponding $R_{on,sp}$ is as low as 2.27 m$\Omega\cdot$cm^2. The comparison of the BV of the SSD MISHFETs and conventional MISHFETs is given in Fig. 11. The conventional MISHFET with L_{GD} =10 μm delivers a BV of 170 V, significantly lower than that in the SSD MISHFETs.

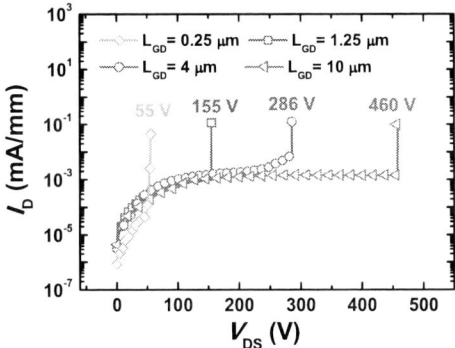

Figure 10. Measured BV of the SSD MISHFETs as a function of L_{GD}. The gate bias V_{GS} is set at -5 V.

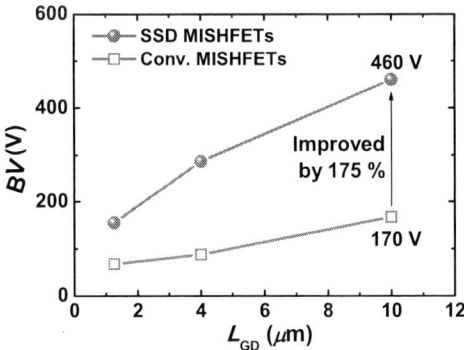

Figure 11. BV dependence on L_{GD} of SSD MISHFETs and conventional MISHFETs.

The $R_{on,sp}$-BV relations of the fabricated InAlN/GaN HFETs were plotted in Fig. 12, also given are the reported InAlN/GaN HFETs (6,7) and state-of-the-art AlGaN/GaN HFETs (8-10,17) reported in literatures for comparison. It is obvious that the BV of the InAlN/GaN HFETs were significantly improved by using the Schottky contact technology while paying a moderate cost in $R_{on,sp}$ compared with the conventional HFETs. The performance of the proposed devices is much better than the reported InAlN/GaN HFETs (6,7). The Schottky-Source HFETs exhibit almost a linear $R_{on,sp}$-BV dependence as shown in Fig. 12, by linear extrapolation the performance of the Schottky-Source InAlN/GaN HFETs is comparable with AlGaN/GaN HFETs. It is also noteworthy that the BVs in the proposed devices exhibit a good linear dependence on gate-drain distance, which implies that higher BV can be expected in the proposed devices with larger L_{GD}. The results validate the feasibility of Schottky contact technology for breakdown voltage improvement on InAlN/GaN HFETs, suggesting a new degree of freedom for designing high voltage InAlN/GaN power transistors with simplified device fabrication process as well as buffer layer structure.

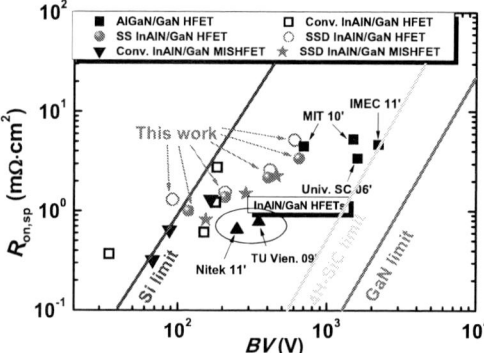

Figure 12. Comparison of $R_{on,sp}$-BV of proposed devices and reported results.

Discussion

In the proposed devices, the breakdown of the devices is dominated and induced by the gate-drain leakage instead of the source-drain leakage which has been observed in conventional InAlN/GaN HFETs (6). During the three-terminal off-state breakdown measurement the source leakage ($I_{S,leakage}$) is consistently lower than the drain leakage ($I_{D,leakage}$) for the proposed devices. In addition, the $I_{S,leakage}$ shows negligible increase over the entire drain bias range. By comparing the device structure of the conventional devices with the Schottky-Source HFETs it can be confirmed that the lower off-state leakage current and enhanced BVs of the proposed devices originate from the Schottky contact in the source area. In the case of conventional HFETs, it has been observed that very deep metallic spikes could be formed either to the physical downward spiking of the alloyed metal or due to a change in the crystalline material after high temperature anneal for ohmic contact formation (17,20). As the spikes are small in dimensions (tens of nanometers), at sufficient high drain bias, these deep spikes may cause the electric-field lines concentrate at the spikes and form a high electric-field region in GaN buffer where the electrons could be injected into GaN buffer. Since the existence of the leakage path formed by the background impurity doping (e.g. O or Si) in GaN buffer the injected electrons would then drift to the peak electric-field region beneath the drain side gate edge where they can be accelerated to high energies to trigger intraband or interband impact ionization and subsequently induce gate breakdown. The atomic force microscopy (AFM) measurements confirmed that deep pits were formed in the area of Ohmic contact after the removal of alloyed Ohmic metal as shown in Fig. 13(a). The alloyed Ohmic metal and non-alloyed Schottky metal were removed by dipping the sample in HCl:HNO$_3$=3:1 and BOE alternatively until all the metal has been etched away, in which the alloy-etch acid can etch away the alloyed Ohmic metal into semiconductor leading to the spiky features in the Ohmic area. On the contrary, a smooth surface morphology is obtained in the Schottky region after the removal of non-alloyed Schottky metal as shown in Fig. 13(b). Taking the advantage of smooth Schottky source a uniform electric-field distribution can be obtained in the proposed devices leading to effective alleviation of electron injection at source, which is responsible for the breakdown voltage enhancement.

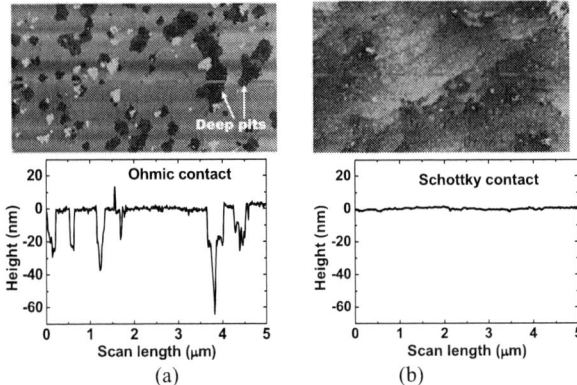

(a)   (b)

Fig. 13. AFM measurements of Ohmic and Schottky region after the removal of alloyed Ohmic and non-alloyed Schottky metal. (a) Deep pits up to 64 nm were found in the area of Ohmic after the removal of alloyed metal. (b) In the Schottky area the surface roughness is less than 3 nm yielding a smooth surface. The AFM scanning routes are highlighted in the upper pictures by dash-dot line.

On the other hand, the proposed devices exhibit decent drive current and low $R_{on,sp}$ in spite of the Schottky contact is used for the source and/or drain electrodes. In our previous report the TLM measurements reveal that the Ti/Au Schottky contact is able to deliver considerable reverse current (13), which ensures the current drive capability of the proposed devices featuring a Schottky source in this work. The good electrical conduction between the Schottky metal and the 2DEG channel is a collective effects of several factors. First, the threading dislocations present in InAlN barrier provide additional conducting path between the Schottky contact and the 2DEG channel (11,12). Second, low work function metal Titanium was used to form the Schottky contact. The low work function Ti can pull down the conduction band reducing the Schottky barrier height because of the Fermi level pinning at the metal/semiconductor interface. Compared with Ni/Au Schottky contact, the Ti/Au Schottky contact is able to deliver 3-order higher reverse current while featuring similar I-V characteristics in forward bias. Third, the 2DEG concentration of the InAlN/GaN sample used in this work is much higher than that in a normal AlGaN/GaN heterostructure ($1.85 \times 10^{13} cm^{-2}$ versus $1.2 \times 10^{13} cm^{-2}$). The high 2DEG concentration together with the small InAlN barrier thickness (7 nm in this work) results in a lower effective barrier to confine the 2DEG. Therefore, the proposed devices are capable of delivering respectable drain current and low on-resistance.

Conclusion

The superior material properties of lattice-matched InAlN/GaN heterojunction suggest that the InAlN/GaN HFETs is potential for power transistors. In this paper, the novel Schottky-contact technology in InAlN/GaN HFETs is experimentally demonstrated for breakdown voltage improvement. The respectable BV enhancement stems from the suppressed source carrier injection by the Schottky source contact. Based on the concept, several devices including SSD/SS HFETs and SSD MISHFETs were developed. The measurement results validate the proposed approach, which is promising for designing high voltage InAlN/GaN HFETs with simple fabrication process.

Acknowledgments

The devices in this work were fabricated in Nanoelectronics Fabrication Facility (NFF) of Hong Kong University of Science and Technology (HKUST).

References

1. J. Kuzmik, *IEEE Electron Device Lett.*, **22**, 11, p. 510–512, (2001).
2. J. Joh, and J. A del Alamo, *IEEE Electron Device Lett.*, **29**, 4, p. 287-289, (2008).
3. J. Kuzmik, G. Pozzovivo, C. Ostermaier, G. Strasser, D. Pogany, E. Gornik, J. F. Carlin, M. Gonschorek, E. Feltin, and N. Grandjean, *J. Appl. Phys.*, **106**, 12, p. 124503, (2009).
4. F. Medjdoub, J. F. Carlin, M. Gonschorek, E. Feltin, M. A. Py, D. Ducatteau, C. Gaquiere, N. Grandjean and E. Koln, *Intl. Electron Devices Meeting (IEDM)*, Tech. Dig., p. 927-930, (2006).
5. D. Maier, M. Alomari, N. Grandjean, J.-F. Carlin, M.-A. Diforte-Poisson, C. Dua, S.L. Delage, and E. Kohn, *IEEE Device Research Conf.*, Tech. Dig., p. 73-74, (2011).
6. J. Kuzmik, G. Pozzovivo, J. F. Carlin, M. Gonschorek, E. Feltin, N. Grandjean, G. Strasser, D. Pogany, and E. Gornik, *Phys. Status Solid-C*, s2, p. s925-s928, (2009).
7. Q. Fareed, A. Tarakji, J. Dion, M. Islam, V. Adivarahan, and A. Khan, *Phys. Stat. Solid-C*, **8**, 7-8, p. 2454-2456, (2011).
8. N. Tipirneni, A. Koudymov, V. Adivarahan, J. Yang, G. Simin, and M. Asif Khan, *IEEE Electron Device Lett.*, **27**, 9, p. 716-718, (2006).
9. B. Lu, and T. Palacios, *IEEE Electron Device Lett.*, **31**, 9, p.951-953, (2010).
10. P. Srivastava, J. Das, D. Visalli, M. V. Hove, P. E. Malinowski, D. Marcon, S. Lenci, K. Geens, K. Cheng, M. Leys, S. Decoutere, R. P. Mertens, and G. Borghs, *IEEE Electron Device Lett.*, **32**, 1, p. 30-32, (2011).

11. Th. Kehagias, G. P. Dimitrakopulos, J. Kioseoglou, and H. Kirmse, C. Giesen, M. Heuken, A. Georgakilas, W. Neumann, Th. Karakostas, and Ph. Komninou, *Appl. Phys. Lett.* **95**, 071905 (2009).

12. J. Song, F. J. Xu, X. D. Yan, F. Lin, C. C. Huang, L. P. You, T. J. Yu, X. Q. Wang, B. Shen, K. Wei, and X. Y. Liu, *Appl. Phys. Lett.* **97**, 232106, (2010).

13. Qi Zhou, Wanjun Chen, Shenghou Liu, Bo Zhang, Zhihong Feng, Shujun Cai, and Kevin J. Chen, *IEEE Trans. on Electron Devices*, **60**, 3, p. 1075-1081, (2013).

14. Qi Zhou, Hongwei Chen, Chunhua Zhou, Z. H. Feng, S. J. Cai, and Kevin J. Chen, *IEEE Electron Device Lett.*, **33**, 1, p. 38–40, (2012).

15. H.S. Lee, D. Piedra, M. Sun, X. Gao, S. Guo, and T. Palacios, *IEEE Electron Device Lett.*, **33**, 7, (2012).

16. M. A. Khan, X. Hu, G. Sumin, A. Lunev, J. Yang, R. Gaska, and M .S. Shur, *IEEE Electron Device Lett.*, **21**, 2, p. 63–65, (2000).

17. B. Lu, E. L. Piner, and T. Palacios, *IEEE Electron Device Lett.*, **31**, 4, p. 302–304, (2010).

18. E. B. Treidel, R. Lossy, J. Wurfl, and G. Trankle, *IEEE Electron Device Lett.*, **30**, 9, p. 901–903, (2009).

19. C. Zhou, W. Chen, E. L. Piner, and K. J. Chen, *IEEE Electron Device Lett.*, **31**, 7, p. 668–670, (2010).

20. Y. Dora, A. Chakraborty, S. Heikman, L. McCarthy, S. Keller, S. P. DenBaars, and U. K. Mishra, *IEEE Electron Device Lett.*, **27**, 7, p. 529–531, (2006).

ECS Transactions, 58 (4) 365-374 (2013)
10.1149/05804.0365ecst ©The Electrochemical Society

Interaction of Defects with Quantum Well States:
Electrostatic-Dependent Response Time for Traps in AlGaN/GaN HEMTs

M. J. Marinella[1], S. DasGupta[1], R. J. Kaplar[1], M. Sun[2], S. Atcitty[1], and T. Palacios[2]

[1]Sandia National Laboratories, PO Box 5800, MS 1084, Albuquerque, NM 87185-1084
[2]Massachusetts Institute of Technology, Cambridge, MA 02139

Recovery transients in high-voltage AlGaN/GaN HEMTs following both blocking voltage and ON-state stress show strong dependence on stress time. The defect time constant spectra exhibit a temperature-dependant component (TG1, E_a = 0.6 eV) and a temperature-independent component (TG2). With increased stress time and larger current collapse, the time constants for TG2 become progressively longer. The stress-time-dependent behavior of TG2 is shown to be consistent with the capture of trapped carriers in the AlGaN barrier directly by quantum well states, originating from the same defect giving rise to the 0.6 eV behavior. By modulating this alternate recombination pathway, the electric field normal to the AlGaN/GaN interface is shown to have a strong effect on reducing the response time of the trap to several orders of magnitude below its bulk response time. This demonstrates that barrier design may be utilized to tune the recovery characteristics of the HEMT.

Introduction

Recently, AlGaN/GaN High Electron Mobility Transistors (HEMTs) have seen widespread application in RF electronics. However, the ease of achieving a low on-state resistance resulting from high channel mobility (μ_{ch}) coupled with high breakdown field due to the wide bandgap of GaN (E_g = 3.4 eV) has led to significant advancements in developing the AlGaN/GaN HEMT as a device for the next generation of high-voltage power electronics [1-4]. Understanding the characteristics of defects in the AlGaN/GaN material system is a key factor in developing power devices with improved performance and reliability. Most techniques for characterizing defects use a bias condition to fill the traps (like the filling pulse in Deep Level Transient Spectroscopy). Following this, optical or thermal excitation is used to study emission of the trapped carriers. However, either in a simple material stack (like a Schottky diode) or in a device such as the HEMT under realistic operation conditions, trapped electrons are assumed to be thermally or optically emitted to the conduction band. Interaction of trapped carriers with the quantum well levels is usually not considered to influence the transient characteristics.

In switching applications, traps that are much slower than the switching time lead to parametric shifts, such as threshold voltage (V_{th}) shift or increase in on-state resistance (R_{ds}) over time. Faster traps with time constants on the order of the switching time lead to distortions in the switching waveform, affecting output power quality [5,6]. Electric field management is generally understood to be important for breakdown voltage enhancement in power devices.

In this study, we demonstrate strong stress-time-dependent behavior for detrapping time constants from deep defect levels near the AlGaN/GaN heterointerface in the HEMT. A temperature-independent detrapping component becomes progressively slower

365

as the stress time is increased. This behavior is shown to be consistent with the capture of trapped carriers in the AlGaN barrier directly by quantum well states. We show that the presence of the quantum well reduces the response time of the trap by several orders of magnitude below its thermal emission time in the bulk material, and the effect is influenced strongly by the field normal to the AlGaN/GaN interface. The strong influence of electric field on the trap response time indicates that electric field management likley has an important role in the switching performance of the AlGaN/GaN HEMT.

Device Details

The tested devices were fabricated at Massachusetts Institute of Technology on silicon (111) substrates. The devices were designed to achieve a maximum breakdown voltage of 1800V. The devices had gate-to-drain spacing (L_{gd}) ranging from 1.5 to 40 μm, gate-to-source spacing $L_{gs} = 1.5$ μm, and gate length $L_g = 2$ μm. These devices used $Al_{0.15}Ga_{0.85}N$ for the 50-nm thick barrier and had a threshold voltage (V_{th}) of ~ −4.1 V. The HEMTs used 4 μm epilayers with a carbon-doped 2.4 μm buffer and 1.4 μm of i-GaN, and the channel GaN was 200 nm thick. The surface passivation was an $Al_2O_3/SiO_2/Al_2O_3$ stack deposited by atomic layer deposition, and was deposited after the gate. A few monolayers (< 2 nm) of gallium oxide, resulting from oxygen plasma treatment before gate pattering, act as the gate dielectric [7]. No field plates were used in the structure.

Bias Time Dependence of Detrapping Transients

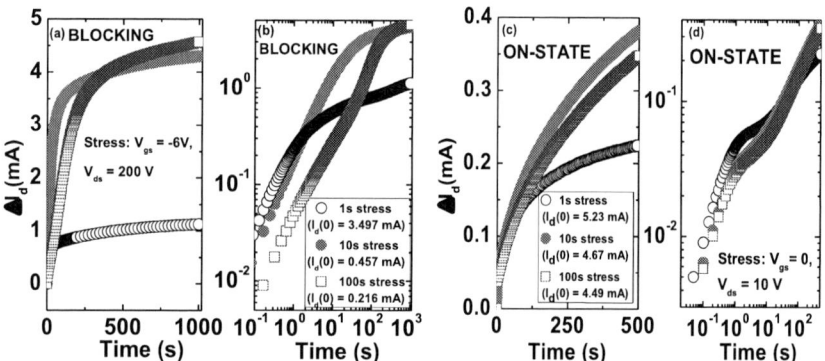

Fig. 1. Detrapping transients ($V_{ds} = 1V$, $V_{gs} = 0$) in a passivated $Al_{0.15}Ga_{0.85}N/GaN$ HEMT ($L_g = 2$μm, $W_g = 100$μm, $L_{gd} = 10$μm) following blocking voltage stress ($V_{gs} = -6$ V, $V_{ds} = 200$ V) for 1s, 10s and 100s on (a) linear and (b) log scales. Pre-stress I_d ($V_{ds} = 1$ V, $V_{gs} = 0$) = 4.91 mA. Detrapping transients ($V_{ds} = 1V$, $V_{gs} = 0$) following ON-state stress ($V_{gs} = 0$, $V_{ds} = 10$ V) for 1s, 10s and 100s on (c) linear and (d) log scales. Pre-stress I_d ($V_{ds} = 1$ V, $V_{gs} = 0$) = 5.51 mA.

The devices were stressed in complete darkness with V_{gs} ~ -6 V at $V_{ds} = 200$ V (blocking voltage state) or $V_{ds} = 10$ V at $V_{gs} = 0$ (ON-state). The stress voltage was applied for variable periods of time, following which the detrapping transient was recorded at $V_{ds} = 1$ V, $V_{gs} = 0$. Prior to every stress, the drain current (I_d) was completely recovered to the pre-stress value by shining the probestation microscope light (a halogen lamp) on the sample. Thus, all the effects to be described in subsequent sections are related to variations in occupancy of pre-stress traps. There was no indication of permanent degradation or reduction in I_d at any stage of the experiments. The current transient method (described in [8]) was used to characterize the trapping components.

Fig. 1 shows the detrapping transients following blocking voltage stress ($V_{gs} = -6$ V, $V_{ds} = 200$ V) for 1s, 10s and 100s stress times. The recovery transients were measured at $V_{gs} = 0$, $V_{ds} = 1$ V. Fig 1a clearly shows that the absolute amount of drain current recovery remains smaller for detrapping following the 100s stress than for detrapping following 10s stress for the first ~100s. Fig. 1b (showing the same data on a log scale) shows that ΔI_d is greater for detrapping following 10s stress than detrapping following 1s stress for the first ~2s.

Figs. 1c and 1d show similar detrapping transients following ON-state voltage stress ($V_{gs} = 0$, $V_{ds} = 10$V) for 1s, 10s, and 100s duration. Although less pronounced than in Fig. 1a, the reduced ΔI_d with increased stress time is clearly visible, similar to detrapping following the blocking voltage stress.

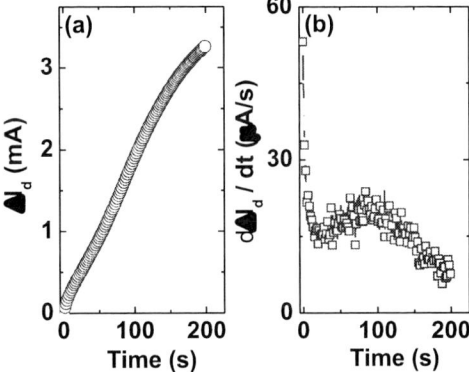

Fig. 2. (a) First 200s of the detrapping transient ($V_{ds} = 1$V, $V_{gs} = 0$) shown in Fig. 1 following blocking voltage stress ($V_{gs} = -6$V, $V_{ds} = 200$V) for 100s and (b) time derivative of the detrapping transient.

The results in Fig. 1 are not possible for a trap or a combination of traps with fixed emission time constants for the following reason. If we compare the transients following 10s and 100s stress, the stress conditions are the same for the first 10s. Therefore, at least the component of trapped charge emitted following the 10s stress should also appear in the transient following the 100s stress. Thus for all times, $\Delta I_d(t)$ following 100s stress should be the same or greater, but never smaller than $\Delta I_d(t)$ following 10s stress. The fact that Fig. 1 show the exact opposite of this is clear evidence that the trap response time slows down as the stress time is increased.

Also, a careful observation of the first 200s of the detrapping transient following the 100s blocking voltage stress reveals that the transient has an increasing time derivative between ~25 and ~75s (Fig. 2). If we consider the detrapping transient to be composed of a number of exponential detrapping transients with amplitudes A_i and time constants τ_i, then

$$\Delta I_d(t) = \Sigma a_i(1 - \exp(-t/\tau_i)), \text{ or } \partial \Delta I_d(t)/\partial t = \Sigma(a_i/\tau_i)\exp(-t/\tau_i) \qquad (1)$$

Clearly, for a fixed set of τ_i, the time derivative in eqn. 1 can never increase with time. Thus, Fig. 2 conclusively shows that the trap response times are long at the start of the detrapping process ($t = 0$), and decrease as detrapping proceeds.

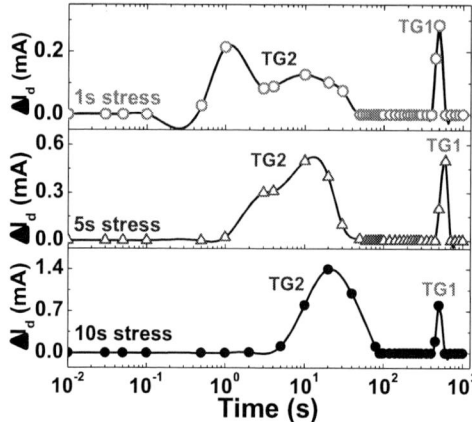

Fig. 3. Extracted time constant spectra from room-temperature detrapping transients for blocking voltage stress (V_{gs} = -6 V, V_{ds} = 200 V) for 1s, 5s, and 10s. Stress time dependence of transients can be seen to come from TG2.

The transients for longer stress times do not fit a sum of decaying exponentials with fixed time constants very well (Fig. 2) for reasons described in the last paragraph, and hence the analysis was done for stress times ranging from 1 to 10 sec. Fig. 3 shows the extracted time constant spectra from detrapping transients after blocking voltage stress (V_{gs} = -6 V, V_{ds} = 200 V) for 1s, 5s and 10s at room temperature. The time constant spectra shows two major components: TG1 at ~500s, and TG2, which is more distributed in time with almost a continuum of time constants spanning over an order of magnitude. The amplitudes of both components increase with increasing stress time. As the stress

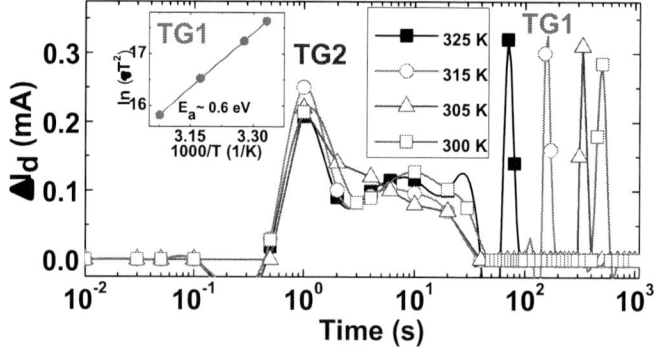

Fig. 4. Extracted time constant spectra from detrapping transients for blocking voltage stress (V_{gs} = -6 V, V_{ds} = 200 V) for 1s at 300K, 305K, 315K and 325K. TG1 shows temperature dependence, but TG2 does not.

time is increased from 1s to 10s, TG2 progressively shifts to higher time constants, but TG1 is invariant in time. Thus, TG2 is the component of trapping which is responsible for the stress time dependence of the detrapping transients.

Fig. 4 shows the temperature dependence of the two trapping components TG1 and TG2. Extracted time constant spectra for detrapping transients following 1s blocking voltage stress are shown over the range 300 to 325K (Fig. 4). TG1 shows a clear temperature dependence with activation energy $E_a \sim 0.6$ eV, whereas TG2 shows negligible temperature dependence (Fig. 4). The component with $E_a \sim 0.6$ eV has been observed in multiple studies of charge trapping in the AlGaN/GaN device literature [8-13].

Increase in TG2 time constant: Thermal Emission, Electron Capture and Tunneling

The increase in the detrapping time constant of TG2 with increasing stress time can be due to (a) reduced thermal emission, (b) an increase in electron capture by the defect level during the recovery bias ($V_{ds} = 1$ V, $V_{gs} = 0$), or (c) reduced tunneling from the trap level. Hole emission or capture can be ignored in this wide-bandgap majority-carrier device.

(a): Reduced Thermal Emission: At a given temperature, decreased thermal emission rate can be caused by a decrease in the defect capture cross section or a change in the defect energy level to a value further from the conduction band. Such physical changes to the defect or emergence of new defects constitute permanent degradation. It should be remembered that no sign of permanent degradation was observed during any stress experiment. As stated at the beginning of Section III, any current collapse was completely recoverable in less than 1s by shining the probestation microscope lamp on the sample. Thus, the change in the time constant of TG2 is unrelated to creation of new defects, as observed in hot-carrier degradation or the inverse piezoelectric effect [9, 14-16]. Without creation of new defects or physical changes to existing defects, reduced thermal emission can be ruled out as the reason for the increase in the time constant of TG2.

The other factor to consider within the context of reduced thermal emission is the variation of channel temperature with channel current and its influence on the emission time constant of the traps [17-19]. As Fig. 1 shows, the biggest change in detrapping rate with stress time is observed between 10s and 100s for blocking voltage stress, where the post-stress current $I_d(0)$ changes from ~0.45 mA to ~0.22 mA. However, in spite of a much bigger difference in $I_d(0)$ from ~3.5 mA to ~ 0.45 mA, the change in detrapping rate is much smaller between 1s and 10s stress than between 10s and 100s stress. Thus it is unlikely that the decreased detrapping rate with increased bias time is related to variations in channel temperature or channel current. In any case, most of the bias-dependant variation is shown to come from a trapping component with negligible temperature dependence. The temperature-independent nature of TG2 also rules out combined thermal and electrostatic effects like field-enhanced emission, which shows a clear temperature dependence that grows weaker with increasing electric field.

(b): Increase in electron capture during recovery bias ($V_{ds} = 1$ V, $V_{gs} = 0$): Enhanced electron trapping due to increased stress time can significantly alter the conduction band profile and hence the electron distribution in the device. If the change in the band profile causes the free electron density to locally increase at the location of TG2, there can be an overall slowdown in the detrapping of TG2.

An increase in trapped electron density tends to increase the conduction band energy and reduce electron current in a semiconductor. However, in the GaN HEMT, injection of electrons from the gate electrode influences the electron density in the device. An increase in the level of gate injection from the altered conduction band profile post-stress could explain enhanced electron capture at TG2 during recovery bias, and hence an increase in time constant for TG2. However I_g decreases after both ON-state and blocking voltage stress for all V_{gs} from OFF-state to recovery bias ($V_{gs} = 0$). This rules out the increase in electron capture during recovery bias as the reason for an increase in detrapping time for TG2.

(c): Reduced Tunneling from the Trap Level: Since both reduced thermal emission and enhanced electron trapping have been shown to be inconsistent with the experimental observations, reduced tunneling from the trap level is the only possible explanation for the stress-time dependant behavior of TG2.

To understand the dependence of detrapping time constants on stress time, we perform self-consistent 1D solutions of the Schrödinger and Poisson equations on the HEMT structure for different trapping magnitudes. The two cases shown in Fig. 5a are (a) Electron trapping in the AlGaN with a density of 10^{17}cm^{-3}, ranging in spatial extent from the AlGaN/GaN interface to a distance 7.5 nm from the interface, emulating the trapping for shorter stress time, such as 10s; and (b) Electron trapping in the AlGaN with a density of 10^{17}cm^{-3}, ranging in extent from the AlGaN/GaN interface to a distance 10 nm from the interface, emulating trapping for higher stress time, such as 100s. The trap energy level is shown for each trapping situation with a dashed line ($E_{t,10s}$ and $E_{t,100s}$ respectively). Since a trap with $E_a \sim 0.6$ eV is observed experimentally, we choose this trap to demonstrate the effect of stress time on trap response time. Although the results are discussed for traps in AlGaN, qualitatively the same arguments and conclusions hold for traps in the GaN buffer.

For an electron trapped in a defect level in the AlGaN barrier, there are two possible detrapping mechanisms. Thermal emission to the conduction band (shown by the vertical solid arrow in Fig. 5a is possible, and is the conventional view. However, another parallel avenue is possible. The two-dimensional electron gas (2DEG) penetrates for some distance into the AlGaN, and as Fig. 5b shows, the 2DEG wavefunction decays rapidly in

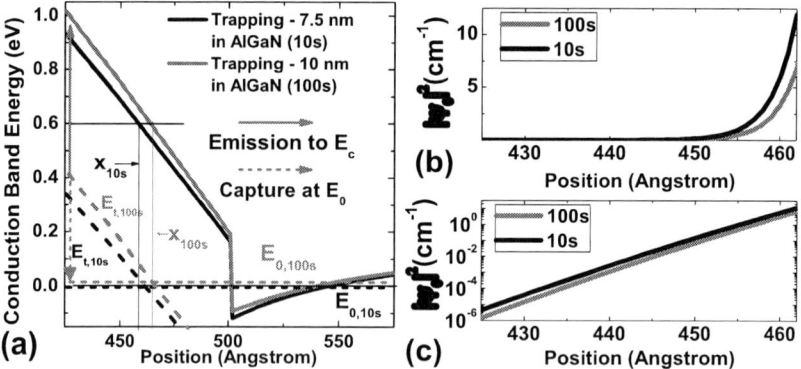

Fig. 5. (a) Simulated conduction band profiles, ground state energy E_0, and trap position $E_t \approx E_c - 0.6$ eV for trapping in AlGaN from the heterointerface to a physical depth 7.5 nm and 10 nm away from the interface, emulating short and long duration stresses respectively. Simulated ground-state wavefunction in the AlGaN barrier for the two cases are shown on linear (b) and log (c) scales.

the AlGaN barrier. Thus, at position x, if the wavefunction for the nth 2DEG state with energy E_n is given by $\psi_n(x)$, and the effective density of this state is given by $|\psi_n(x)|^2 D$, where D is the density of states of the two-dimensional band, given by:

$$D = (8\pi m k_B T \ln 2)/h^2 \tag{2) [20]}$$

Thus, for a defect at position x close to the heterointerface, once stress is discontinued the carriers trapped in the defect level can be captured directly by the 2DEG states with local density $|\psi_n(x)|^2 D$. This process is shown by the vertical dashed arrow in Fig. 5a. Figs. 5b and 5c show the wavefunction profile for the ground state (with energy E_0) for the two stress times 10s and 100s on linear and log scales.

From Figs. 5b and 5c, we find that the penetration of the wavefunction into the AlGaN barrier decreases as we increase the stress time. This is because as more electrons are trapped in the AlGaN, the conduction band edge rises and increases the barrier height between the GaN and the AlGaN. Thus, the process of capturing electrons trapped in defect states directly by 2DEG states potentially explains both the temperature independence and the stress-time-dependant nature of TG2.

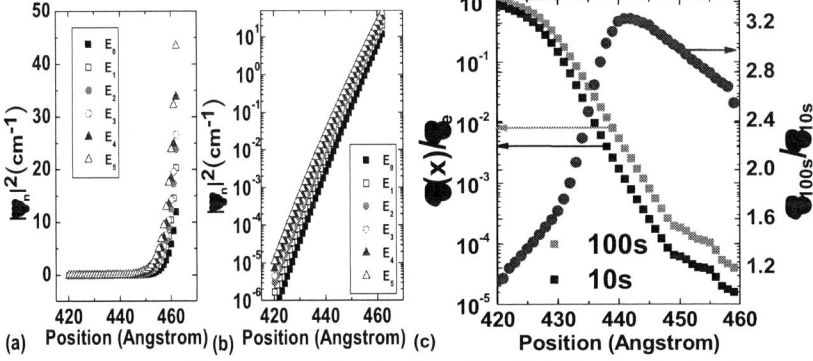

(a) Position (Angstrom) (b) Position (Angstrom) (c) Position (Angstrom)

Fig. 6. Decaying tail of the wavefunction into the AlGaN barrier for the six bound states on a a) linear and b) log scales. c) Effective detrapping time constant $\tau(x)$ as a function of position in the AlGaN barrier relative to the bulk emission time constant τ_e for the two trapping magnitudes in Fig. 6, representative of long (100s) and short (10s) stress times. The longer stress time can lead to detrapping time contants that are higher by a factor of ~3.2.

For both the 10s and 100s stress scenarios simulated in Fig. 5, there are six bound states associated with the 2DEG. The higher-energy states show greater penetration into the AlGaN barrier and have a greater effect on the stress-time-dependant behavior (Figs. 6a,b). The capture time constant at an empty defect state in a bulk material is given by $\tau_c = (N_t \sigma v_{th})^{-1}$, where N_t is the density of the level at which capture takes place, σ is the capture cross-section, and v_{th} is the thermal velocity [21]. The captured carriers do not increase the fill fraction of the quantum well states (which are delocalized in the plane of the 2DEG and free to conduct), but rather become part of the drain current. Thus the 2DEG states may be considered to be empty throughout the capture process, and the capture time constant at each of the 2DEG states is given by:

$$\tau_n(x) = (|\psi_n(x)|^2 D\sigma v_{th})^{-1} \tag{3},$$

The thermal emission time constant to the AlGaN band edge is given by $\tau_e = (N_{c,AlGaN}\exp(E_t - E_c)/k_BT)\sigma v_{th})^{-1}$ [21]. At a given position, the sum of all of these processes gives an effective time constant

$$\tau(x) = (\Sigma 1/\tau_n(x) + 1/\tau_e)^{-1} \tag{4}$$

Fig. 6c shows the change in the effective time constant $\tau(x)$ for 10s and 100s stress. E_t is below the Fermi level for the first ~4 nm from the AlGaN/GaN interface, and traps in this region will be filled pre-stress. The Fermi level crossover points for the $E_c - 0.6$ eV trap are denoted as x_{10s} and x_{100s} in Fig. 5a. Direct capture to 2DEG states strongly dominates $\tau(x)$ for the next 3nm, and the response time is less than the thermal emission time constant by several orders of magnitude. Beyond ~7 nm from the heterointerface, $\tau(x)$ is dominated by thermal emission to the conduction band, since the wavefunction rapidly decays into the AlGaN barrier. The enhanced barrier from GaN to AlGaN caused by increased stress time (shown in Fig. 5) causes an overall increase in $\tau(x)$ by a factor that is position dependent and peaks at a value of ~3.2 (Fig. 6c). It is important to note that the increase in time constants between long and short time stresses is most pronounced in a region where $\tau(x)$ shows continuous variation over several orders of magnitude. Figs. 3 and 4 show that the stress-dependant component TG2 spans a much wider range of time constants compared to the thermally activated TG1. This provides another point of consistency between the observed results and the proposed detrapping model.

One might consider the possibility of tunneling from the trap level to the AlGaN barrier conduction band to explain the stress time dependence of TG2. However, this explanation is not consistent with the experimental data. The wavefunction of conduction band states is independent of electric field. As Fig. 5a shows, the increased electron trapping increases the electric field for the electron in the barrier-to channel direction. This reduces the tunneling distance from the trap to the AlGaN conduction band, making the tunneling rate increase with stress time (similar to an increase in the tunneling rate from an oxide trap to the semiconductor with increased electric field from oxide to channel in an MOS structure). Therefore, trap to band-edge tunneling also has to be ruled out as a possible explanation for the stress-time dependence of TG2.

The long- and short-time stresses have been emulated in this study by considering two regions of trapping with different thicknesses in the AlGaN barrier. Another alternative to this is to consider trapping over the same spatial location, but with different concentrations. This is plausible, since either increasing the density of trapped charge or the spatial extent of trapping results in an enhanced barrier to electrons from the channel to AlGaN. While quantitative differences will result from different charge distributions in the barrier, increased barrier height due to charging of the barrier reduces the penetration of the wavefunction into the AlGaN, reducing capture into the 2DEG and increasing the overall time constant. For very small penetration, the overall time constant will be close to the bulk emission time constant. Thus, the qualitative nature of the evolution in detrapping time constants with increased stress and current collapse stays the same, independent of variations in trapping location or density.

Design-Dependent Trap Response Time

The variations in the band profile (shown in Fig. 5a) in the direction normal to the 2DEG can be created by design and processing steps such as varying the AlGaN thickness, molefraction, or surface passivation. Thus, the spatial distribution of detrapping time constants (as shown in Fig. 6b), which is an important factor in the switching performance of the HEMT, becomes a function of the process and design parameters controlling the potential barrier profile normal to the 2DEG. For example, a higher AlGaN surface charge density results in a larger electric field in the AlGaN barrier, reducing the coupling between the trap levels and the 2DEG and hence reducing the response time of the traps. Conversely, a higher 2DEG density results in reduced net charge at the heterointerface, since the polarization charge is positive. Therefore, systems with higher 2DEG densities can be expected to have higher coupling between barrier traps and 2DEG levels, reducing the trap response times, all other factors remaining constant. Our results suggest that by using appropriate barrier design techniques, it should be possible to tune the time constant of the recovery of the device following stress, which may be optimized for performance or reliability.

Conclusions

We have demonstrated strong stress-time-dependant behavior of detrapping time constants for near-heterointerface defects in AlGaN/GaN HEMTs under both blocking voltage stress and ON-state conditions. The stress-time-dependant behavior is shown to be consistent with the capture of trapped carriers in the AlGaN barrier directly by quantum well states. The proximity of the quantum well and the penetration of the bound-state wavefunctions into the barrier are shown to reduce the response time of a barrier trap 0.6 eV below the conduction band edge by several orders of magnitude compared to its bulk response time. The electric field normal to the 2DEG is shown to have a strong effect on the trap response time.

Acknowledgments

This work was supported by the GaN Initiative for Grid Applications (GIGA) program managed by Dr. G. Bindewald of the U.S. Department of Energy's Office of Electricity Sandia National Laboratories is a multi-program laboratory managed and operated by Sandia Corporation, a wholly-owned subsidiary of Lockheed Martin Corporation, for the U.S. Department of Energy's National Nuclear Security Administration under contract DE-AC0494AL85000.

References

1. B. Lu and T. Palacios, IEEE Electron Device Letters, **31**(9), 951 (2010).
2. N. Tipirneni, A. Koudymov, V. Adivarahan, J. Yang, G. Simin, and M. Asif Khan, **27**(9), 716 (2006).
3. J. Das, J. Everts, J.Van Den Keybus, M. Van Hove, D. Visalli, P. Srivastava, D. Marcon, Kai Cheng, M. Leys, S. Decoutere, J. Driesen, G. Borghs, Electron Device Letters, IEEE , **32**(10), 1370 (2011).
4. Y.-F. Wu, M. J. Mitos, M. L. Moore, and S. Heikman, IEEE Electron Device Lett., **29**(8), 824 (2008).

5. D. Pavlidis, P. Valizadeh, and S. H. Hsu, in Proc. GaAs Symp., 265 (2005).

6. R. Vetury, N. Q. Zhang, S. Keller, and U. K. Mishra, IEEE Trans. Electron Devices, **48**(3), 560 (2001).

7. J. W. Chung, J. C. Roberts, E. L. Piner, and T. Palacios, IEEE Electron Device Lett., **29**(11), 1196 (2008).

8. Jungwoo Joh and J.A. del Alamo, IEEE Trans. Electron Devices, **58**(1), 132 (2011).

9. A. Sozza, C. Dua, E. Morvan, M. A. diForte-Poisson, S. Delage, F. Rampazzo, A. Tazzoli, F. Danesin, G. Meneghesso, E. Zanoni, A. Curutchet, N. Malbert, N. Labat, B. Grimbert, and J.-C. De Jaeger, in IEDM Tech. Dig., 590 (2005).

10. T. Mizutani, T. Okino, K. Kawada, Y. Ohno, S. Kishimoto, and K. Mzezawa, Phys. Stat. Sol. (A), **200**(1), 195 (2003).

11. A. Chini, F. Fantini, V. Di Lecce, M. Esposto, A. Stocco, N. Ronchi, F. Zanon, G. Meneghesso, and E. Zanoni, in IEDM Tech. Dig., 1 (2009).

12. M. Tapajna, R. J. T. Simms, Y. Pei, U. K. Mishra, and M. Kuball, IEEE Electron Device Lett., **31**(7), 662 (2010).

13. M. Meneghini, C. de Santi, T. Ueda, T. Tanaka, D. Ueda, E. Zanoni, and G. Meneghesso, IEEE Electron Device Letters, **33**(3), 375 (2012).

14. J. Joh and J. A. del Alamo, in IEDM Tech. Dig., 461 (2008).

15. G. Meneghesso, G. Verzellesi, F. Danesin, F. Rampazzo, F. Zanon, A. Tazzoli, M. Meneghini, and E. Zanoni, IEEE Transacions on Device and Materials Reliability , **8**(2), 332 (2008).

16. J. A. del Alamo and J. Joh, Microelectronics Reliability , **49**(5), 1200 (2009).

17. J. Joh, U. Chowdhury, T. M. Chou, H. Q. Tserng, and J. L. Jimenez, "Method for estimation of the channel temperature of GaN high electron mobility transistors," in Proc. ROCS Workshop , 87 (2007).

18. A. Sarua, H. Ji, M. Kuball, M. J. Uren, T. Martin, K. P. Hilton, and R. S. Balmer, IEEE Trans. Electron Devices , **53**(10), 2438, (2006).

19. J. Kuzmik, P. Javorka, A. Alam, M. Marso, M. Heuken, and P. Kordos, IEEE Trans. Electron Devices, **49**(8), 1496, (2002).

20. John H. Davies, The Physics of Low Dimensional Semiconductors, 6th ed. NY: Cambridge University Press, 2006.

21. D. K. Schroder, Semiconductor Material and Device Characterization, 3rd ed. Hoboken, NJ: Wiley, 2006.

Monolithic Integration of High Temperature Silicon Carbide Integrated Circuits

Mihaela Alexandru, Viorel Banu, Josep Montserrat, Philippe Godignon, José Millán

Centro Nacional de Microelectrónica, Instituto de Microelectrónica de Barcelona, CNM-IMB-CSIC, Campus UAB, 08193 Bellaterra, Barcelona, Spain.

This paper deals with the design, the fabrication and the characterization of SiC ICs able to work at high temperature. The key SiC device is a planar MESFET specially designed for high temperature and high integration density. The other element required for integrating the basic cells library is a SiC resistor built in the same epilayer than the MESFET. Our preference for the MESFET is due to the already proved temperature stability of its Tungsten-Schottky barrier. The SPICE models of these two components are based on experimental measurements in the 25°C-300°C temperature range. The process technology setup is also analyzed, which contains three metal levels fully compatible with a standard CMOS technology. The digital library allows implementing multi-stage logic embedded in power management circuitry. The fabrication and the testing of the elementary logic gates library are also analyzed at high temperature and high frequencies. Furthermore, the standard CMOS topologies can be transferred to various Flip-Flops based on 4H-SiC MESFET basic logic gates. The multi-stage SiC ICs show a similar behavior as the logic gates at room and high temperatures, and at high frequency. The functionality of multi-stage SiC ICs has been proved.

Introduction

The current trends for innovative electronic applications require high frequency, high temperature, high efficiency and even high radiation operation stability. Although the most dominant semiconductor material for commercial devices and integrated circuits (ICs) is Silicon (Si), it has been shown several limitations for RF and harsh environment applications, due to its physical properties, such as its relative narrow bandgap and low thermal conductivity. Therefore, there is a strong motivation for switching to wide bandgap (WBG) materials, which can offer a viable alternative solution for proper device performance in such extreme environments. From the WBG semiconductors that have been studied so far, Silicon Carbide (SiC) is the most suitable semiconductor that can overcome the previous limitations. It has become particularly interesting thanks to its material improvement and to its nature process technology. Hence, providing a real opportunity for producing dedicated devices and ICs for harsh environment applications.

Currently there is no major technical obstacle for SiC to penetrate the power electronics market, except the material cost. However, because of the great benefits from SiC technology, this could offset part of the total costs for systems. Furthermore, the SiC process technology maturity makes also realistic to switch to SiC devices and ICs.

Aside from the practical advantages, some of them still to be proved on SiC devices, accurate models are required for designing and optimizing SiC based devices, as well as for developing SiC ICs. Therefore, the attention that SiC research has received in the last decade has resulted also in a great industrial interest. The improvements of the epitaxial growth and their commercial availability have led SiC devices into a vast variety of electronic applications, even in the consumer application field. It is also predicted that with the SiC power device replacement in the power generation–conversion–consumption chain, the power systems' losses will be considerably reduced. Therefore, one of the global environmental problems, the reduction of energy usage, can be a solvable problem. Great benefits are foreseen in the case of automotive and aerospace engines. Improved electronic telemetry and control from high-temperature engines are necessary to more precisely control the combustion process, thus improving fuel efficiency while reducing polluting emissions. High-temperature capability eliminates performance, reliability, and weight obstacles associated with liquid cooling, fans, thermal shielding and longer wire runs, typically needed to realize similar functions in engines using conventional Si electronics.

Even though SiC devices are commercially available, the development of ICs on SiC is still in its early stage. However, it is predicted that SiC ICs can become a competitor of the smart Si power technology in certain application niches requiring high temperature, high voltage and/or high switching speed. There are many applications in which SiC ICs seems to have a remarkable effect, such as sensors and control electronics for jet or rocket engines that must be removed from the target area, carefully shielded, protected and cooled.

The 4H-SiC Device Library

Either analog or digital applications require components whose parameters are closely matched and, consequently, on the same die, as in an IC. Therefore, this section deals with the description of the 4H-SiC device library that has been further used in the SiC ICs development.

The 4H-SiC Normally-*ON* Planar-MESFET

The Metal Semiconductor Field Effect Transistor (MESFET) has quite a similar structure as the Junction Field Effect Transistor (JFET). The main difference between the JFET and MESFET is that the latter has a Schottky junction (metal-semiconductor junction) instead of a pn-junction for the gate terminal. The device architecture is the lateral one, the MESFET channel is built in a moderately doped layer at the surface of the wafer. Our preference for MESFETs instead of the nMOSFET/JFET is due to the already proved stability of the Tungsten-Schottky barrier, technology that has been successfully used in the fabrication of stable SiC Schottky diodes for the European Space Mission BepiColombo [1]. Moreover, the MESFET shows a very stable pinch-off voltage (V_P) behavior with temperature.

A new planar-MESFET has been developed for developing SiC ICs. As the purpose of the new structure is to be integrated in future circuit development containing a large device number, it is necessity to accomplish some of the most important ICs requirements, generally applicable to Si-ICs; i.e., i) planar topology – it assures good interconnections between different metal levels of the circuit, especially when multiple metals are necessary for devices interconnection; ii) device scalability – it offers important advantages concerning ICs design flexibility and chip area reduction; and iii) multi-level interconnection – it highly increases the device integration density.

Therefore, some important and novel design and fabrication features have been considered in the development of the new structure: a) the P^+-implanted walls isolation technique which, to our knowledge, has been utilized in SiC for the first time; b) the transistor layout approach (finger-gate), which is typically used in the Si-CMOS ICs; and c) the process flow is fully compatible with CMOS processes. The groundwork of the new planar-MESFET mainly relies on the electrical and geometrical analysis of the already fabricated mesa-MESFET [2].

a) The Isolation Technique – The P^+-Implanted Walls Isolation

A significant aspect considered in developing the new structure is the isolation technique of the devices. With the mesa isolation technique, the 4H-SiC MESFETs have shown a proper behavior up to 300°C [2]. This method has proved to assure a good isolation between devices and it is widely used for individual device definition due to its simplicity [3]. However, as planarity is an important standpoint in developing ICs, the mesa isolation brings some problems concerning devices interconnection in ICs.

The isolation technique widely used in Si ICs is the junction isolation based on a deep P^+-implant. This technique was typically used for the early fabrication technology of Si ICs based on bipolar transistors [4]. The P^+-type impurities are deeply implanted into the N-type epitaxial layer so that it reaches the P-type substrate, forming N-type islands. This method generates N-type wells surrounded by P-type moats in which individual components are implemented. A comparison between the mesa and junction isolation techniques on SiC is reported in [5]. The schematic cross-section of the new proposed 4H-SiC MESFET structure using the P+-implant technique as a device isolation method is show in Figure 1.

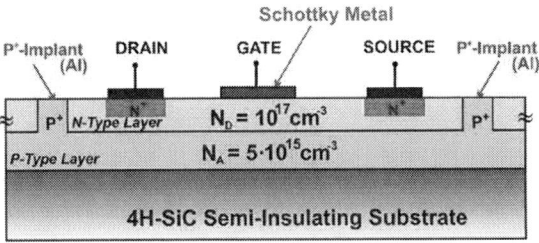

Figure 1. Schematic cross-section of the 4H-SiC MESFET with P^+-implant isolation

The benefits of using the P^+-implanted isolation are the improvement of the device integration density and the planar topology. Since a certain number of devices are fabricated on the same IC chip, it becomes necessary to provide good isolation between the various components and their interconnections. This technique offers not only a good isolation between devices, but also a better planarization of the wafer surface, making the device scalability easier and also the interconnection between various devices or circuits. Therefore, this typical Si-CMOS isolation technique is our choice for 4H-SiC individual device definition.

b) The Transistor Layout Approach – Device Scalability

One of the most important aspects in developing ICs is related to the device scalability. Device scalability brings important advantages, such as the improvement of both circuit speed and integration density [6]. Concerning scalability, the technique widely used in CMOS ICs is the finger-gate technique [7].

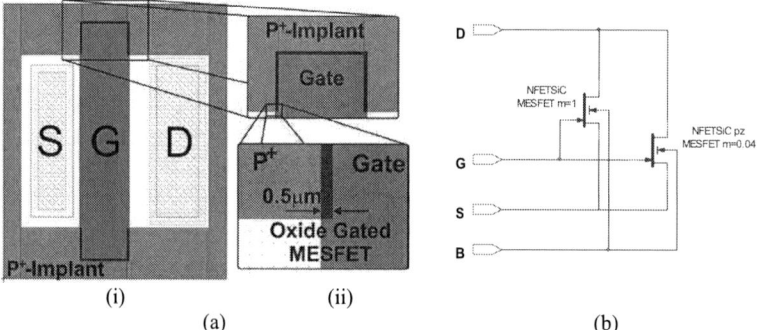

Figure 2. (a) The finger-gate layout for a single MESFET (m=1) using P$^+$-implant isolation walls, (i) the general view and (ii) geometry trick. (b) The SPICE equivalent circuit of the 4H-SiC MESFET

The embraced finger-gate technique allows applying the concept of transistor multiplicity, which accounts for the scalability requirement. The proposed transistor layout is shown in Figure 2(a-i) where the P$^+$-implanted isolation ring can be observed (the red ring). The gate electrode W_G/L_G ratio for the novel device is considered to be roughly 10 times smaller than that of the mesa-MESFET [2]; hence, it is 48μm × 8μm. However, due to the special fabrication processes on SiC, the embraced geometry presents a built-in drain-source residual current through the gate terminal. In order to minimize it, a geometry trick has been adopted on the lateral edges of the gates (Figure 2(a-ii)). An additional oxide gated MESFET, 25 times smaller than the main MESFET, is implemented under the prolonged gate metallization (0.5μm), in order to block any residual drain-to-source leakage current. This oxide gated MESFET shows a slightly higher V_P than the main planar-MESFET.

In order to get accurate and realistic simulations, the oxide gated MESFET has been considered in the SPICE model. Hence, the SPICE model of the new planar MESFET structure is more complex. As one can observe from Figure 2(b) the extrapolated SPICE model for a single planar MESFET involves two devices in parallel. It has been considered V_P=-8V for the main planar-MESFET, which implies the N-epilayer of 0.35μm thick. The main reason why V_P of the planar MESFET has been considered lower than that of the mesa-MESFET [2] is for decreasing the power consumption and the logic levels of the digital circuits. For a proper functionality, the voltage level of the power supplies should be roughly 30% higher than $2xV_P$. Thus, the power supply voltage could be decreased significantly just lowering the V_P value.

c) The Process Flow – The Planar-MESFET Fabrication

The new planar-MESFETs have been fabricated on 4H-SiC wafer supplied by CREE Research Inc. The P-layer grown on a semi-insulating substrate is 5µm thick with $5 \cdot 10^{15} cm^{-3}$ of doping concentration, and the N-layer is 0.5µm thick with a doping concentration of $10^{17} cm^{-3}$. Even though one of the challenges in designing a new device lies on the reduction of the mask counting, the fabrication of the planar-MESFET, requires 10 photolithographic masks. One mask is especially dedicated for the N-type tubs formation. An additional etch of the N-epilayer is also performed in order to reduce the N-epilayer to a thickness of 0.35µm. The last four masks are required for forming the interconnection metal strips between devices and circuits (two metals and two vias). Figure 3a shows the cross-section of the new planar MESFET with the three metal levels. Figure 3b shows the fabricated planar-MESFET with one gate finger.

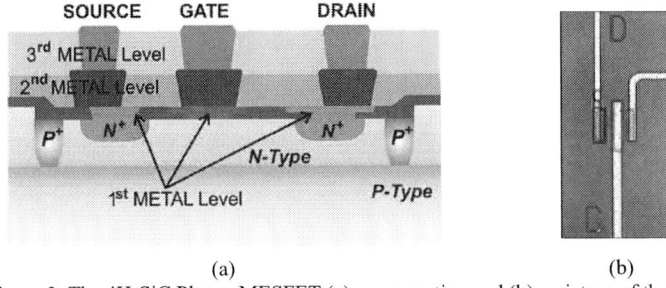

| (a) | (b) |

Figure 3. The 4H-SiC Planar-MESFET (a) cross-section and (b) a picture of the fabricated planar-MESFET

Figure 4 shows the experimental I-V curves for MESFETs with 1, 2 and 4 finger-gates, respectively, at room temperature. Figure 4(a) shows the experimental forward I-V curves of the planar-MESFET. Note that the I_{Dsat} (the saturation drain current) is proportional to the number of fingers. One can observe from Figure 4(b) that V_P remains constant for the three devices, being slightly lower than -8V. Hence, the electrical scalability at room temperature is demonstrated. At high temperatures the scalability of the devices is also proven and reported in [8].

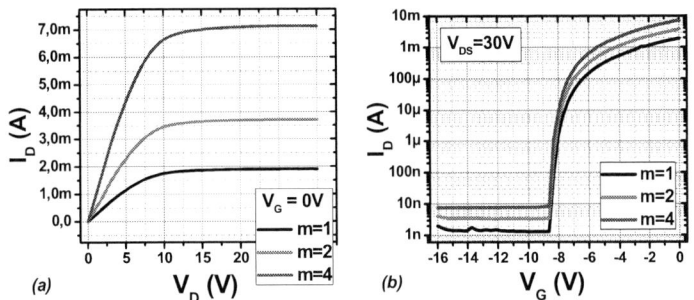

Figure 4. The I-V characteristics of the 4H-SiC planar-MESFETs with 1, 2 and 4 gate-fingers at room temperature

The 4H-SiC Epitaxial Resistors

For high temperature ICs operation the temperature matching between various devices is an important aspect concerning circuit's functionality. Ref. [9] shows that using 4H-SiC epitaxial resistors, the temperature matching between the various components is accomplished. Thus, because the epitaxial resistors are implemented on the same N-layer as the planar-MESFET, its value can be controlled by either varying the effective cross-section of the resistor or by the doping concentration [8]. The resistors values are chosen as a function of the MESFET's DC parameters. Therefore, we have fabricated 4H-SiC epitaxial resistors on the same N-type epitaxial layer.

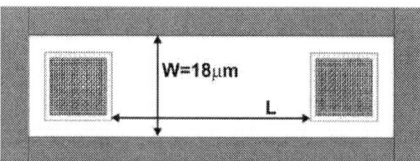

Figure 5. The 4H-SiC Epitaxial resistance layout

The proposed 4H-SiC resistor layout is shown on Figure 5. They are implemented in N-type islands, isolated by the same P^+-implanted walls. Thus, for an N-layer uniformly doped with $N_D=10^{17}cm^{-3}$ and 0.35μm thick, the theoretical sheet resistance value at room temperature is $R_S = 3.4k\Omega$/square.

Eq. (1) describes the modeling of the epitaxial resistor:

$$R = \rho \cdot L/A = \rho \cdot L/(t_{epi} \cdot W) = R_S \cdot L/W \quad \Rightarrow \quad R=R_S \cdot N_{sq} \qquad (1)$$

where L and W are the length and width of the resistor (Figure 5), t_{epi} is the epitaxial thickness, R_S is the sheet resistance and N_{sq} the number of square units.

N_{sq} is chosen depending on the desired resistance value. From technological limitations we have chosen W=18μm as the minimum width for the resistor. To obtain resistors with different values, the parameter that has to be varied is the length L. Moreover, in order to obtain accurate simulation results of the ICs, a SPICE model has been defined for the epitaxial resistors. The effect of temperature on the resistors value is:

$$R(T) = R(T_0)[1+TC_1(T-T_0)+TC_2(T-T_0)^2] \qquad (2)$$

where T is the circuit temperature, T_0 is the nominal temperature (25°C for SPICE modeling) and TC_1 and TC_2 are the first and the second order temperature coefficients, parameters required for the SPICE model.

In order to demonstrate the resistors scalability, Figure 6(a) shows the linear I-V characteristics of 4 resistors with 2, 4, 6 and 8 squares units. Figure 6(b) shows their experimental values extracted from a linear fitting of the I-V curves.

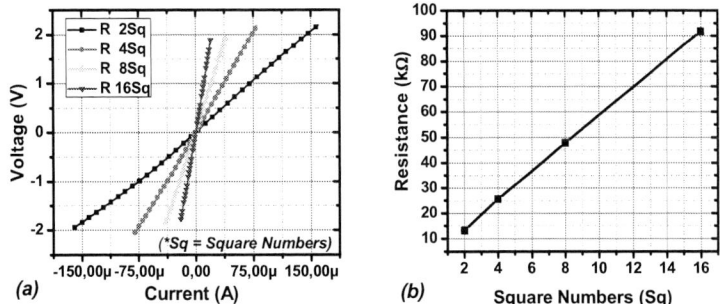

Figure 6. (a) V-I characteristics of the epitaxial resistors and (b) Resistance vs. Square Numbers for the 4 different epitaxial resistors

The experimental sheet resistance R_S (in the range of 6.6kΩ - 5.7kΩ per square) extracted from the V-I curves is ×2 the theoretical one. Several causes may determine this difference, such as the non-uniformity of the epilayer doping concentration, the incomplete ionization of the N dopants at room temperature, or even the non-uniformity of the SiC etch rate. Moreover, high temperature measurements have been carried out on the fabricated resistors, demonstrating that the experimental epitaxial resistors evolution with temperature is consistent with the theory.

Figure 7 shows the epitaxial sheet resistance evolution with temperature. The extracted values of the temperature coefficients resulting from a polynomial fitting of Figure 7 are TC_1=2.118·10^{-3}K and TC_2=5.367·10^{-5}K. These values are considered in the SPICE epitaxial resistance model, thus defining its SPICE model for high temperature operation.

The ICs on 4H-SiC can be modeled and simulated more accurately with the extracted SPICE models for the epitaxial resistor and for the planar-MESFET.

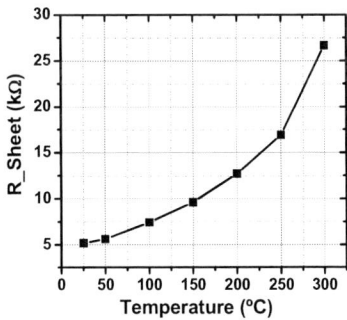

Figure 7. Temperature dependence of the sheet resistance

The 4H-SiC MESFET Basic Logic Gates Library

The proposed 4H-SiC digital circuits monolithically integrate the planar-MESFETs with the epitaxial resistors on the same chip. In order to demonstrate the functionality of more complex digital circuits using standard CMOS topologies, the basic logic gates (Inverter, NAND and NOR gates) have been designed and simulated with a SPICE model extracted for high temperature operation [9].

Design of the 4H-SiC MESFET Basic Logic Gates

Figure 8 shows the circuit's schematics for the basic logic gates built with 4H-SiC MESFETs and epitaxial resistors, adopting the normally-on JFET topology [9, 12].

The inverting logic gates have a first stage with a positive voltage output and the MESFETs have to be controlled with a negative voltage. Therefore, to accommodate the output voltage of the logic gates to the input control voltage, a level shifter is necessary. This function is performed by a voltage follower (the NFETSiC2 MESFET) and a two-resistor voltage divider (R_1 and R_2). This level shifter is designed to obtain 0V for the High logic level at the B output, and a voltage close to V_P, (in our case -8V) for the Low logic level.

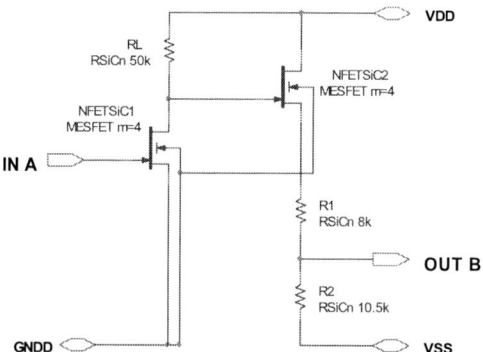

Figure 8. Circuit configuration of the Inverter basic logic gate

The NAND and NOR gates (Figure 9) result from the addition of a supplementary transistor in series and in parallel, respectively, with the first stage of the Inverter gate. The negative voltage control requires two power supplies; one positive (V_{DD}) and another negative (V_{SS}), both referred to ground. However, the Low and High logic levels can be easily tuned by adjusting properly the supply voltages.

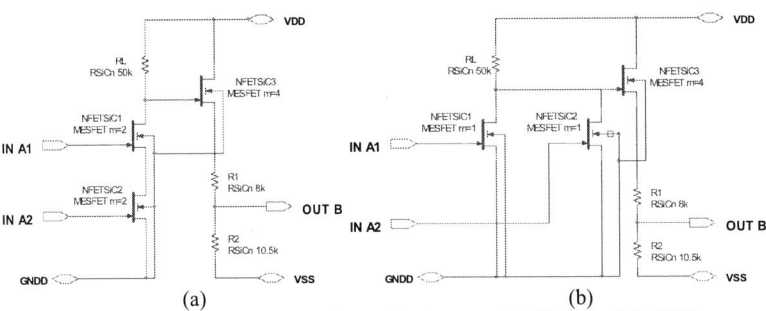

(a)   (b)

Figure 9. Schematics of the universal logic gates (a) NAND and (b) NOR

Experimental Results of the 4H-SiC MESFET Basic Logic Gates at Room and High Temperature

In accordance with the elementary logic gates design and due to complexity of the circuits, the ICs fabrication require 10 masks, four of which are needed for the metal and via levels, thus resulting in a three metal technology. The last two metal levels are required for the circuit's components interconnection and to make high multiplicity planar-MESFETs, hence obtaining high integration density ICs. To our knowledge, such a complex technology with three metal levels for designing SiC ICs has not been reported so far.

This process technology has allowed the fabrication of digital multi-stage ICs. In order to demonstrate the functionality of complex digital circuits based on standard CMOS topologies with the elementary logic gates, the Inverter, NAND and NOR gates have been simulated, fabricated and finally tested.

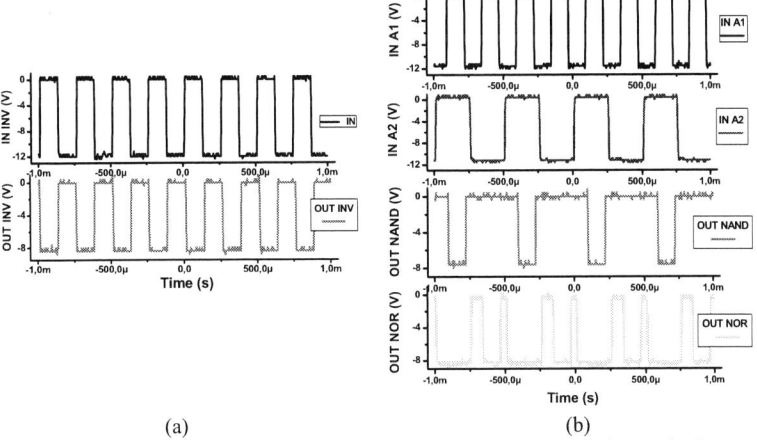

(a)   (b)

Figure 10. The experimental waveforms of the (a) INV, (b) NAND and NOR logic gates at room temperature

Figure 10 shows the experimental waveforms of the 4H-SiC elementary logic gates at room temperature. One can observe that the Input signal level is swept between 0V and -12V, ensuring the turn-off of the planar-MESFET. From the experimental Inverter waveforms, one can easily see that the Output signal shows opposite logic levels with respect to the Input level. Hence, the proper Inverter operation at room temperature is performed. The NAND and NOR gates also perform properly their respective logic functions. The Output responses for the elementary logic gates have the amplitude of 8V approximately.

Figure 11 shows the experimental waveforms of the Inverter up to 300°C. The Output responses show quite a similar behavior in the whole temperature range (25°C-300°C), performing its proper logic function. One can notice that up to 200°C no changes of the Output waveforms are observed with respect to those at room temperature. However, for the two highest temperatures (250°C and 300°C), the High and the Low levels drift negatively. This temperature evolution of the gates is mainly explained by the level shifter temperature behavior.

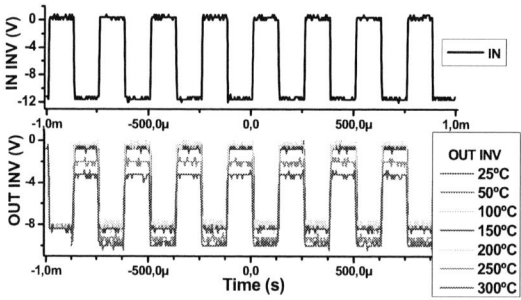

Figure 11. The Inverter experimental waveforms vs. temperature (25°C-300°C)

Concerning the epitaxial resistors, it is shown that the experimental sheet resistance is double than expected. Nevertheless, it is important to mention that the resistor divider ratio (R_1/R_2) is maintained constant to the same value since the individual resistors show the same increasing rate. Moreover, the two-resistor divider shows the same temperature coefficients, thus the voltage divider maintains the same ratio in the whole temperature range, not affecting the Output response.

The observed changes are mainly caused by the follower transistor temperature behavior. As the voltage follower transistor works in saturation and because of the difference between the theoretical and experimental values of the epitaxial sheet resistance, the voltage drop across the follower transistor increases considerably at high temperature. The follower drain-gate leakage current increases significantly, adding a supplementary voltage drop, thus affecting the gate Output logic levels. The high temperature behavior of the NAND and NOR logic gates show similar evolution as the Inverter.

Nevertheless, the 4H-SiC MESFET logic gates output negative drift at high temperatures can be minimized using a higher Schottky barrier for the MESFET gate contact. However, it is important to mention that the 4H-SiC MESFET logic gates perform their respective logic functions acceptably in the whole temperature range. The temperature measurements have been limited to 300°C due to the high reverse leakage of the Schottky gate contact.

4H-SiC MESFET Logic Gates High Frequency Experimental Results

As our main interest is to confirm the functionality of the digital SiC ICs, the previous measurements were performed with a 4 kHz Input signal. Generally, low frequency and high frequency inverters perform the same function. In order to check the behavior of the SiC MESFET logic family in frequency, the Inverter's Output evolution up to 300 kHz has been investigated.

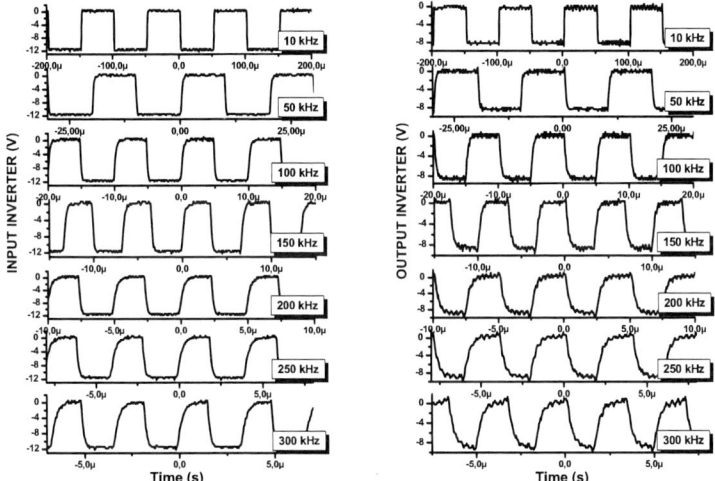

Figure 12. The Inverter's Input and Output experimental signals in the 10kHz-300kHz frequency range

These measurements are performed using a Test Interface circuit initially built for verifying the logic functions of the fabricated circuits, not being optimized for high frequency measurements. Therefore, the generated Input signals deteriorate at high frequencies and will impact the output signal of the gates. However, our results still provide important information regarding the SiC circuit's performance in frequency.

The 4H-SiC MESFET Multi-Stage Digital Integrated Circuits

From the above analyzed 4H-SiC MESFET Basic Logic Gates one has built different complex circuits, such as pulse or edge triggered latches and Flip-Flops. The basic logic gates can be used to build other dynamic blocks like ring oscillators, voltage controlled oscillators, phase detectors, digital counters, phase locked loop circuits (PLL) and other typical digital logic schematics able to operate at HT. As Flip-flops are fundamental building blocks of digital electronic systems used in computers, communications, and many other types of systems, [13] reports the fabricated circuit and the experimental waveforms of the standard CMOS topology of the D Flip-Flop designed with 4H-SiC MESFET elementary logic gates.

Moreover, a high temperature compensated analog voltage reference ICs implemented with the previous 4H-SiC planar-MESFET has been designed and fabricated, as our knowledge, for the first time on SiC [14].

In CMOS technology many circuits can be implemented at transistor level, but also with different gates combinations. If a limited number of specific gates is available in a certain topology, digital ICs can be constructed with these gates. Although the Exclusive OR (XOR) is considered a single digital logic gate, we have considered it as a complex IC since its implementation requires more elementary logic gates. The NAND and NOR are so-called "universal gates" because any logical function can be constructed with them.

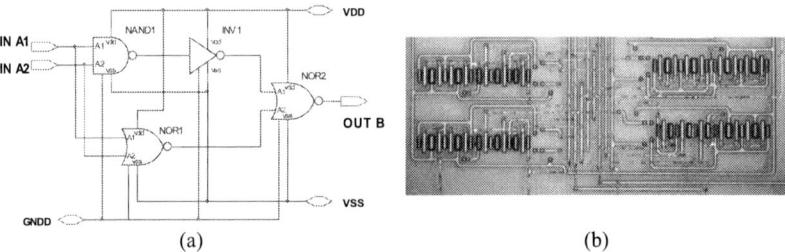

(a)   (b)

Figure 13. XOR circuit architecture based on 4H-SiC MESFET universal gates

Therefore, a XOR circuit can be easily configured from NAND or NOR gates. Figure 13(a) shows the XOR circuit configuration with a NAND, an Inverter and two NOR basic gates. In order to show the fabrication complexity, Figure 13(b) shows the XOR fabricated circuits.

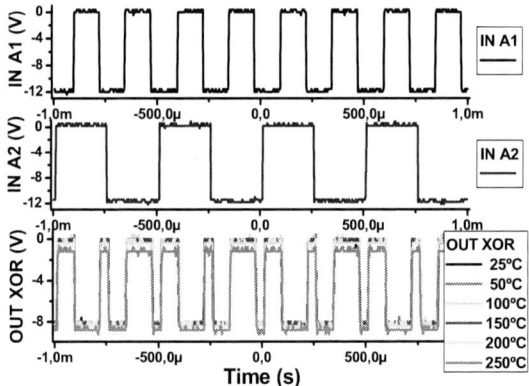

Figure 14. The XOR experimental waveforms at room temperature

From the experimental waveforms of the XOR gate (Figure 14) we can see that the Output is at High level whenever one of the Inputs is High, and is at Low level when both Inputs are Low. The Output amplitude is around 8V. Therefore, we can confirm that the

SiC XOR logic gate performs properly its natural logic function for the whole temperature range. The temperature evolution of the XOR is fairly similar to the elementary logic gates temperature behavior.

Therefore, the functionality of multi-stage digital ICs fabricated with 4H-SiC MESFET logic gates at room and high temperatures is ensured by the functionality of the individual basic logic gates at room and high temperature operation, and limited by the defects density of the starting material or process technology. The circuits before studied can be further used for other logic application, like for the implementation of frequency dividers, counters, latches among others. More complex circuits can be realized using these elementary building blocks.

Conclusions

This paper presents the functionality of the 4H-SiC MESFET elementary logic gates library: Inverter, NAND and NOR gates. Due of the difference between the designed and experimental epitaxial resistors at room temperature, the High and Low logic levels of the gates show a negative voltage shift mainly caused by the voltage follower transistor. This drop is easily eliminated by adjusting properly the circuits supply voltages. The gates functionality is also demonstrated up to 300°C. It has been observed that above 200°C the gate-drain leakage current of the follower increases considerably. Hence, affecting the Output response of the gates by drifting negatively their logic levels. However, this drift can be reduced using a different metal for the MESFET gate contact for a better high temperature operation. It has been also seen that the Inverter's output fairly reproduces the reversed Input slopes from the characterization at frequencies up to 300 kHz, performing a proper commutation on each High-to-Low and Low-to-High front.

We have also demonstrated experimentally that the standard CMOS topologies can be transferred to the Data Flip-Flops based on 4H-SiC MESFET basic logic gates. The multi-stage SiC ICs show a similar behavior as the logic gates at room and high temperatures and at high frequency. The functionality of multi-stage SiC ICs has been also proved, provided by the elementary logic gates library operation.

Acknowledgments

This work was supported by the Spanish Ministry of Science and Innovation under the grants "Advanced Wide Band Gap Semiconductor Devices for Rational Use of Energy" (CSD 2009-00046) and TRENCH-SIC (TEC2011-22607).

References

1. P. Godignon et al. *"Long term stability of packaged SiC Schottky diodes in the -170°C/+280°C temperature range"*, Proc. of 22nd ISPSD, pp. 9-4, 2010.
2. A. Devie et al. *"Fabrication and Testing of 4H-SiC MESFETs for Analog Functions Circuits"*, Materials Science Forum Vols. 645-648, pp. 1159-1162, 2010.
3. J. P. Colinge, *"Silicon-on-Insulator Technology Materials to VLSI"*. Boston, MA. Kluwer, 1991

4. Kurt Lehovec, *"Multiple Semiconductor Assembly"*, U.S. Patent 3, 029, 366, 1962.
5. M. Alexandru et al., *"Comparison between Mesa Isolation and P^+ Implantation Isolation for 4H-SiC MESFET Transistors"*, IEEE, International Semiconductor Conference Vol. 2, pp.317-320, 2011
6. G.V. Ram and M.I. Elmasry, *"On the Scaling MESFET"*, IEEE Electron Device Letter, Vol. EDL-1, No.12, pp. 259 - 262, 1980.
7. J.P. Colinge, *"FinFETs and Other Multi-Gate Transistors"*, Ed. Springer, pp.8, 2008.
8. M. Alexandru et al., *"4H-SiC MESFET Specially Designed and Fabricated for High Temperature Integrated Circuits"*, to be presented at ESSDERC 2013.
9. Michael J. Krasowski, *"Logic Gates Made of N-Channel JFETS and Epitaxial Resistors"*, NASA Tech Briefs, LEW-18256-1, 2007.
10. Wai-Kai Chen, *"The Circuits and Filters Handbook"*, Ed. CRC Press, pp.1554, 2003.
11. M. Alexandru et al., *"Design of Logic Gates for High Temperature and Harsh Environment made of 4H-SiC MESFET"*, IEEE, International Semiconductor Conference, Vol. 2, pp. 413-416, 2010.
12. P. Neudeck et al., *"Extreme Temperature 6H-SiC JFET Integrated Circuits Technology"*, Phys. Status Solidi A 206, No. 10, pp. 2329–2345, 2009.
13. M. Alexandru et al., *"4H-SiC Digital Logic Circuitry Based on P^+ Implanted Isolation Walls MESFET Technology"*, Mater. Sci. Forum, Vols.740–742, pp.1048–1051, 2012.
14. V. Banu et al., *"High Temperature-Low Temperature Coefficient Analog Voltage Reference Integrated Circuit Implemented with SiC MESFETs"*, to be presented at ESSCIRC 2013.

CHAPTER 10

POWER ELECTRONICS CURRICULUM

390

ECS Transactions, 58 (4) 391-398 (2013)

Power Semiconductor Device Modeling and Simulation

H. A. Mantooth, S. Ahmed, S. S. Ang

National Center for Reliable Electric Power Transmission, Department of Electrical
Engineering, University of Arkansas, Fayetteville, Arkansas 72701, USA

Power semiconductor devices continue to be an active area of
research and development. With the emergence of wide bandgap
technologies this area has blossomed over the past decade beyond
silicon boundaries. A brief survey of courses available around the
world on the topic of power semiconductor device modeling and
simulation is provided. An exemplar course at the University of
Arkansas is described to demonstrate its applicability to power
electronics students interested in circuit design, controls, device
design, electronic packaging, and compact modeling of devices.
This course is built on model-based engineering principles that are
applicable to any area of electronic design.

I.   Introduction

With the advent of wide bandgap semiconductors (e.g., SiC and GaN), it is now possible
to fabricate power devices with high voltage and current ratings and extremely fast dv/dt
and di/dt. This is why, even in the presence of minimal parasitics, a significant amount of
oscillation is observed in the switching characteristics of the circuits that employ wide
bandgap power devices. Additionally, these wide bandgap devices can be effectively
incorporated in numerous high temperature applications. In order to predict these
oscillations and high temperature behavior as well as estimate rise and fall times, voltage
breakdown phenomena, leakage current, and conduction and switching losses accurately,
physics-based compact models of power devices are a must for power electronics circuits
and systems simulation. The phrase 'compact models' refers to the device models created
for use in circuit simulation. The 'compactness' is achieved through reasonable
mathematical approximations to the two and three dimensional charge distribution and
movement within the actual device as is reflected more precisely through finite-element
analysis. Compact models are necessary in order to simulate circuitry that consists of tens,
hundreds and even thousands of devices.

At the University of Arkansas (UA), educational efforts in power semiconductor
device modeling and simulation are undertaken through course work and research
projects. The primary objective of the course (ELEG 5313 – Power Semiconductor
Devices) is to convey the physics of operation, performance capabilities and limitations
to the students. These descriptions and derivations are accompanied by discussions on the
applicability of devices to certain circuit applications. Common approaches applied to
compact modeling of the various kinds of power devices are surveyed and class projects
involve creating first level device models. The UA actively pursues research projects in
the area of compact modeling of power devices. As such, this course lays the groundwork
for subsequent graduate level research into compact modeling. The UA's research
program has developed new compact models for a number of Si and SiC devices. For

example, models are available for Si and SiC power MOSFETs, diodes, IGBTs, GTOs, SITs, JFETs (1)-(8).

Although many universities teach concepts of power devices in power electronics courses, some offer separate courses on power semiconductor devices with the goal of educating the students on the advanced features of power devices. Apart from teaching DC and transient characteristics of the devices, these courses help students learn the trade-offs between on–state resistance and breakdown voltage, ambipolar diffusion, vertical structures, reverse recovery, effects of interface states and temperature on mobility, leakage current, theory of super junction etc. In TABLE I, courses offered by different universities around the world are listed as an illustration of these. An attempt at a complete listing has been made. Any omissions are simply an oversight.

TABLE I. Power Semiconductor Device Courses Offered by Universities Worldwide.

Course No.	Course Name	Institute	Course Level
ELEG 5313	Power Semiconductor Devices	University of Arkansas	Graduate
ECE 5204	Power Semiconductor Devices	Virginia Polytechnic Institute and State University	Graduate
EEL6317	Power Semiconductor Devices	University of Central Florida	Graduate
EE666	High Power Semiconductor Devices	Indian Institutes of Technology, Bombay	Graduate
EECS498	Solid State Power Devices	University of Michigan at Ann Arbor	Senior
ECE553	Semiconductor Power Devices	North Carolina State University	Graduate
IH2661	Power Semiconductor Devices	KTH Royal Institute of Technology, Sweden	Graduate
ECE5340	Power Semiconductor Devices	Cornell University	Graduate
CE00261-7	Power Semiconductor Devices	Staffordshire University, UK	Graduate
EIE583	Advanced Power Semiconductor Devices and Design Criteria for Applications	The Hong Kong Polytechnic University	Graduate
ECE 442	Power Semiconductor Devices and Integrated Circuits	University of Illinois at Chicago	Senior
EE4505	Power Semiconductor Devices & ICs	National University of Singapore	Senior
ECE 5354	Power Semiconductor Devices	Texas Tech University	Graduate
ECSE6260	Semiconductor Power Devices	Rensselaer Polytechnic Institute	Graduate
80220252	Principles and Applications of Power Electronic Devices	Tsinghua University, China	Graduate
ECE 423	Power Semiconductor Devices	University of Rochester	Senior

The courses that have the title 'Power Semiconductor Devices', 'High Power Semiconductor Devices', 'Solid State Power Devices' or 'Semiconductor Power Devices' primarily focus on teaching design concepts and fabrication processes of different power device structures. Finite element numerical simulation is used in the design process for semiconductor devices in different material systems, and optimization is performed in order to achieve higher breakdown voltage, lower on-state resistance, higher switching speed, reliability, etc. The courses with 'Advanced Power Semiconductor Devices and Design Criteria for Applications', 'Power Semiconductor Devices and Integrated Circuits', 'Power Semiconductor Devices and ICs' or 'Principles and Applications of

Power Electronic Devices' mostly deal with the application side of power devices in power electronics systems. These courses put emphasis on circuit versus device parametric tradeoffs, safe operating area, circuit modeling, parasitic analysis, etc. The goal of the UA course, in contrast, is to establish a strong base for device modeling research and convey detailed concepts of device operation to power electronics students whether they are circuits, packaging or controls oriented. The course is principally concerned with the physical descriptions of the device structure and their practical behavior as a switch with the appropriate mathematical governing equations and their culmination into forms suitable for compact modeling in circuit simulators.

The remaining sections of the paper describe the UA power semiconductor devices course and how it is intertwined with the research program as part of the UA's power electronics program. This program consists of a research agenda that includes transportation, grid-connected power electronics, energy exploration, and space exploration topics. Section II describes the overall pedagogical approach behind our power semiconductor device education and places it in context with other coursework at the UA. Section III focuses on the details of the course itself and is followed by a section describing the methods used to evaluate student performance. The paper is concluded with a description of how research activities are fed by the course, but also returned into the course to keep it current and relevant to the latest emerging topics in power electronics.

II.   Overview of Pedagogical Approach

The UA power electronics course sequence at the graduate level builds off of expectations of having basic fundamentals of electronics, energy conversion, and control systems found in most ABET-accredited engineering programs. There are many options for the sequence of courses at the graduate level, but the most popular is to have taken the first semiconductor devices course as either a senior undergraduate or first semester graduate student. The UA teaches this course from Streetman (9). From this starting point, students take design courses, advanced controls courses, power distribution and transmission courses, and electronic packaging courses in addition to the power semiconductor devices course.

The expectation from students who enroll for the power semiconductor devices course is that they have taken the Streetman course, have some level of mastery with circuit simulators such as PSpice, Spectre, or Saber, and have had basic courses in circuits and electronics. Without this pre-requisite knowledge, students tend to fall behind because they either do not have the proper context for understanding the description of device characteristics such as mobility, minority/majority carriers, semiconductor doping, and band diagrams or an ability to conceptualize and analyze the circuits within which the devices are used.

The primary philosophy behind this power semiconductor devices course is to teach the main silicon and silicon carbide devices (GaN is a coming attraction) in terms of basic structure, device characteristics, and tradeoffs between devices. This information is taught at a level that forms the basis for understanding and deeper research by those students involved in device modeling while being fundamentally useful to circuit designers and electronic packaging students that will use such devices as part of their research. Modeling of each device, a secondary philosophy underlying the course, is discussed as a part of a model-based engineering concept that all electronic design is performed with models, and therefore it is vital that circuit designers learn what models to apply at various points in the design process (10). The various levels of device model are described and existing device models are used in simulations of basic circuit power electronic topologies as homework assignments (11)-(12).

The UA course is a traditional lecture-style course taking students through the fundamentals of power devices, silicon versus silicon carbide, some laboratory experimentation, the basic principles of compact modeling, and use of device models to simulate power electronic circuit topologies of a standard textbook nature. An entire course is devoted to model-based engineering techniques, but a single module from that course is reused here on compact modeling. This is vastly accelerated through the use of the integrated modeling environment ModLyng^{TM} (13). ModLyng, previously known in the literature as Paragon, is a graphical modeling tool that enables the conceptual study (and implementation) of compact models without the usual tethers of: a) having to learn a modeling or programming language and b) having to understand circuit simulator data structures. To summarize, the students are instructed on the theory of devices, practical issues in testing those devices, methods on how to model device behavior, and experience in using existing compact models to analyze circuits.

III. Course Flow

The syllabus for the UA course is summarized in this section. While the course is taught from course notes, five books form the foundation for these notes (10), (14)-(17). The course is described to the students in the syllabus as follows:

"This course on power semiconductor devices will introduce the students to basic material properties of silicon and silicon carbide – the two most prevalent power device technologies. Some of the silicon material may be a repeat of material in ELEG 4203 – Semiconductor Devices, but will serve as a solid baseline for the study of device structures. All of the devices will be presented in silicon technology first and then compared and contrasted with those or similar devices in silicon carbide technology. The devices lectures will begin with diodes, proceed to MOSFETs and JFETs, and then proceed to bipolar devices. It is always helpful to provide a context for device work, so some power electronic application circuits will be described. This class is recommended for all students considering employment involving power electronic circuit design, power electronic packaging, power systems, and, of course, power semiconductor device modeling."

The primary overall learning objectives for this course are:

- Learn the physics and principles of operation of several commonly used power semiconductor devices
- Learn how power devices differ from low-voltage, microelectronic devices
- Learn both silicon and silicon carbide material properties and power device technologies associated with each material
- Learn the various ways these devices are modeled for circuit simulation
- Explore the characteristics of the devices through lab measurements and circuit simulation

The following is a rough outline of the power semiconductor device course along with the approximate number of lectures (80 min/lecture) spent on each topic (26 lectures total).

I. *Power Electronic Application Considerations (1-2 lectures)*
An overview of power electronic applications and where semiconductor devices fit into this realm are covered. This will motivate the attributes desired in power semiconductor devices. The course returns to this material repeatedly when devices are being described.

II. *Material Properties for Power Semiconductor Devices (1-2 lectures)*
The basic material properties for silicon and silicon carbide are described. This includes lattice structures, mobilities, band gap considerations, thermal properties, typical doping characteristics, etc. Those properties that are attractive for power device performance will be emphasized. Other alternative materials may be introduced, but will be discussed more in-depth towards the end of the course as a topic of future device technology.

III. *Compact Modeling & Model-Based Engineering (2-3 lectures)*
An overview of model-based engineering as it pertains to power electronics circuit design is given. The importance of compact semiconductor device models in the design process is demonstrated. Example models are described so that models of many of the upcoming devices to be described can be more quickly summarized.

IV. *Computer-Aided Device Analysis (2 lectures)*

An important aspect of the course will be to learn the primary tools for the analysis and design of new semiconductor devices. Commercial tools such as Taurus-Device from Synopsys will be used for teaching the basics of device analysis. Later homework assignments will involve the analyses of devices under study.

V. *Semiconductor Devices (17 lectures)*
  A. *Diodes (p-i-n, Schottky, Si-FRD) (5)*
  B. *Field Effect Devices: MOSFETs/SiC JFETs (5)*
  C. *Bipolar Devices: Si BJTs/Thyristor devices (GTO, MCT, S-GTO) (4)*
  D. *IGBTs (could also include IGCT) (3)*
Generally, each device will be covered starting with the silicon device and then describing the silicon carbide counterparts. The description of each device will cover the principles of operation with reference to representative cross-sectional structures as well as the governing relationships that describe current, voltage and charge/capacitance relationships within the device. Some laboratories outside of class will be performed involving device measurements on the high power probe station and associated instrumentation. Compact models for some devices will be summarized and homework assignments given using these models in circuit analysis.

VI. *Alternative Materials for Power Devices (2 lectures)*
The last topic to be covered in this course will be on other materials that are either under investigation or in use in special applications for power circuitry. Among these are gallium nitride (GaN), silicon germanium (SiGe) and diamond.

A future extension is to bring GaN devices into the main body of the course (part V).

At particular points during the course, experts in a particular device technology may give lectures associated with their areas of expertise. This provides added insight into the course material for the students, gives them access to leaders in the field, and generally makes the course more intellectually stimulating.

IV. Assessment and Evaluative Methods

The assessment and evaluative methods are standard. As with most graduate level courses, homework assignments are not given as frequently as in undergraduate courses. However, the nature of this material requires a degree of consistency in assessment and evaluation of student comprehension. The course description given in the previous section indicates a top-down flow where the device physics and descriptions are motivated first by the power electronic applications and circuits and then by the need for models to analyze these circuits before delving into the details of various devices.

Homework and/or in-class quizzes are given after each of parts I through V. Sometime after the diodes are described, a mid-term examination is given. This exam covers everything to that point. Later, a final exam covering the latter half of the course material along with selected topics from the first half of the course is given. Along the way, homework assignments and quizzes provide feedback to the instructor as to student understanding. Some homework assignments are "mini-projects" in that they require more substantial analysis and reporting such as a lab assignment.

V. Relation to Research

As mentioned previously, the UA's research program has been producing compact models of power semiconductor devices for over a decade. These models form the basis for student interaction with these devices and serve to focus attention on modeling, parameter extraction, and simulation. Clearly, this is a choice that could be taken in another direction. The course could involve a more comprehensive laboratory where more devices are physically tested using standard test configurations. Because of the desire to cover so many devices, and to make the course more appealing to a broad range of graduate students, the choice was made to focus more on the modeling and model-based engineering aspects with a limited exposure to device testing.

As new device models are created by the UA research group, these students create modules for this course that teach students about the device that they have modeled. The model(s) are available to future classes. This completes the feedback loop of the course providing a foundation for the graduate student to build their knowledge upon and then the student re-injecting expertise that they have gained through their research back into the course for future students to learn from. This course was actually created from such an approach. Several modeling students in the research program compiled their notes into the starting point of this course and augmented those notes with the textbook information needed to complete the flow of the course.

The knowledge rendered by this course may also be utilized in active research of device design and performance optimization. University of Arkansas has now industry standard finite element simulators like Atlas from Silvaco and Sentaurus TCAD from

Synopsys. Students who are interested in design of power semiconductor devices may use these simulators and apply the concepts learnt in the course to invent innovative device structure with enhanced performance capability.

VI. Conclusions

The UA power semiconductor devices course has been taught since 2008 and has been shown very effective at accelerating student learning in the field. It has undergone two major revisions in that time to reflect the fast-changing nature of the power devices field – particularly given the advent of wide bandgap devices. Device modeling students have benefitted greatly from the course as anticipated. However, other power electronics students have demonstrated a greater knowledge and insight when designing their packages and circuitry and the overall power electronics program has benefitted through more team-based research projects. A typical project might include a device modeling student, a power packaging student, and a power electronic design student all focused on the research and development of a new integrated power module.

Future modifications to the course will certainly include GaN devices. These devices have emerged commercially over the past few years and the UA has begun accumulating experience with them in circuit design, electronic packaging, and in compact modeling. Others have already developed some GaN device models that may enable more rapid introduction of this material into the course (18).

Acknowledgments

The first author would like to thank the many fine graduate students that have participated in the development of the UA's power semiconductor devices course through their research, the creation of course notes, and the creation of sample homework problems and simulation assignments.

References

1. T. McNutt, A. Hefner, A. Mantooth, D. Berning, R. Singh, *Journal of Solid-State Electronics*, Elsevier, vol. 48, no. 10-11, pp. 1757-1762, (2004).
2. T. R. McNutt, A. R. Hefner, H. A. Mantooth, J. L. Duliere, D. Berning, R. Singh, *IEEE Trans. on Power Electronics*, vol. 19, no. 3, pp. 573-581, (2004).
3. T. R. McNutt, A. R. Hefner, H. A. Mantooth, D. Berning, S.-H. Ryu, *IEEE Trans. on Power Electronics*, vol. 22, no. 2, pp. 353-363, (2007).
4. M. Saadeh, H. A. Mantooth, J. L. Hudgins, E. Santi, A. Agarwal, S.-H. Ryu, *IEEE Applied Power Electronics Conference (APEC)*, pp. 1728-1733, Orlando, Florida, (2012).
5. O. S. Saadeh, H. A. Mantooth, J. C. Balda, *Energy Conversion Congress and Exposition (ECCE), 2012 IEEE*, pp. 3589-3594, Raleigh, North Carolina (2012).
6. A. S. Kashyap, T. R. McNutt, T. Funaki and H. A. Mantooth, *India International Conference on Power Electronics (IICPE'04)*, 8 pgs., Mumbai, India, (2004).
7. K. Speer, T. R. McNutt, A. Lostetter, H. A. Mantooth, D. Berning, A. Hefner, K. J. Olejniczak, *European Power Electronics Conference*, 6 pgs., paper 1092, Toulouse, France, (2003).

8. M. Mudholkar, M. Saadeh, H.A. Mantooth, *Power Electronics and Applications (EPE 2011), Proceedings of the 2011-14th European Conference on* , pp.1-10, Birmingham, England, (2011).
9. B. Streetman, S. Banerjee, *Solid-State Electronic Devices*, Pearson Prentice Hall, New Jersey (2005).
10. P. R. Wilson, H. A. Mantooth, *Model-based Engineering of Complex Electronic Systems*, Elsevier, London (2013).
11. E. Santi, J. L. Hudgins, H. A. Mantooth, *2007 Summer Computer Simulation Conference (SCSC'07)*, 4 pgs., (2007).
12. I. K. Budihardjo, P. O. Lauritzen, K. Y. Wong, R. B. Darling, H. A. Mantooth, *IEEE Industry Appl. Soc. meeting*, pp. 1084-1090, vol. 2, (1995).
13. H. A. Mantooth, A. M. Francis, Y. Feng, W. Zheng, *IET Proc. of Computer and Digital Techniques*, pp. 519-527, (2007).
14. S. K. Ghandi, *Semiconductor Power Devices: Physics of Operation and Fabrication Technology*, John Wiley & Sons, New York (1977).
15. B. J. Baliga, *Modern Power Devices*, John Wiley & Sons, New York (1987).
16. B. J. Baliga, *Silicon Carbide Power Devices*, World Scientific Publishing Company, New Jersey (2006).
17. B. J. Baliga, *Advanced High Voltage Power Device Concepts*, Springer Scientific, New York (2012).
18. K. Shah and K. Shenai, *IEEE Trans. on Electron Devices*, vol. 59, no. 10, pp. 2735-2741, (2012).

ECS Transactions, 58 (4) 399-411 (2013)
10.1149/05804.0399ecst ©The Electrochemical Society

Application Engineering of Wide Bandgap Semiconductors
(Invited Paper)

B. Sarlioglu, D. Han, J. Noppakunkajorn, A. Ogale

Wisconsin Electric Machines and Power Electronic Consortium,
University of Wisconsin-Madison, Madison, Wisconsin 53705, USA

This paper will address the importance and benefits of wide
bandgap (WBG) semiconductors and provide example research
results on utilization of these new devices in power electronic
circuits. The paper will elaborate on device and circuit level
opportunities, challenges, and potential solutions for various power
electronic applications using WBG semiconductors. The paper will
also present the current power electronic curriculum and education
for undergraduate, graduate, and outreach education. Furthermore,
suggestions are included on changes that can be made to include
and promote WBG devices in power engineering education to
develop a new workforce.

Introduction to Wide Bandgap Semiconductor Devices

Wide bandgap semiconductors have the capabilities to block higher voltage, switch faster,
and dissipate heat more efficiently than their silicon (Si) counterparts (1). Traditionally,
Si devices are divided into two subgroups. First, the majority carrier device group, which
includes MOSFET and Schottky barrier diode (SBD), has fast switching capabilities.
However, majority carrier devices are limited by the high conduction loss for high
blocking voltage above 600 V. Second, minority carrier device group, which includes
IGBTs and PN diodes, has high voltage blocking capabilities and approximately constant
on-state voltage. In particular, every time an IGBT and PN diode switched from an on to
an off state, the tail and reverse recovery current phenomenon occurs and produces a
significant amount of switching loss. This is why typically a minority carrier device's
hard switching frequency is limited to around 20 kHz. Currently, silicon carbine (SiC)
and gallium nitride (GaN) devices are being widely researched. Due to the significantly
higher breakdown electric field than Si, SiC and GaN offers potential to manufacture
majority carrier devices with the same blocking voltages as minority carrier devices,
while still have low conduction and minimal hard switching loss (2). From a system-level
perspective, this will allow the power electronic converters to operate at a higher
switching frequency, reduce the size of the passive filtering components, reduce the size
of thermal management system, and increase the overall power density.

For SiC devices, MOSFETs and JFETs are currently available. Due to the early
maturity of manufacturing techniques, there are many literatures on SiC JFET (3)-(8). To
many power electronics engineers, MOSFET is still a preferred choice, due to an ease of
gate driver design. SiC MOSFET requires Vgs of 20V for a turn-on with minimized
conduction loss, which is similar to the traditional Si MOSFET that requires Vgs in the
range of 10-15V. With SiC JFETs, there are two different types. They are normally-on

399

and normally-off JFETs. Normally-on JFET requires negative Vgs to turn off the device and requires an additional protection circuit to ensure that the device is in the commanded state. However, the normally-on JFET can be cascaded in series with a low-voltage Si MOSFET to allow the switch to be controlled like a traditional MOSFET (8). This technique adds additional complexity and loss to the overall design. For normally-off JFET, the gate current has to be controlled.

Similarly to SiC, GaN has a wider bandgap, higher electron saturation velocity, and higher breakdown electric field than Si. There are currently two types of GaN device available, which are enhancement and depletion mode. The enhancement GaN transistor is a normally-off device, while the depletion GaN transistor is a normally-on device (9).

Application of Wide Bandgap Semiconductor Devices in Power Converters

To better illustrate and quantify the benefits that WBG devices may bring about on the power converter level, two research example cases are provided in this section, which covers the application of SiC MOSFETs and SBDs in a 20 kW DC/DC converter and a 12 kVA inverter.

Case Study 1: SiC-based 20 kW Buck/Boost Converter

In a previous study of authors, a 20 kW bidirectional buck/boost converter (shown in Figure 1), which is suitable for the hybrid/electric vehicle application, has been analyzed (10). A comparison has been made among all Si (Si IGBTs and Si PN diodes), hybrid (Si IGBTs and SiC Schottky diodes), and all SiC (SiC MOSFETs and SiC Schottky diodes) converters. The devices used are listed in the Table I below. The converter specifications are shown in Table II.

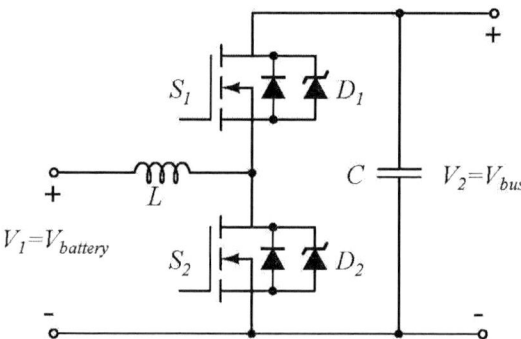

Figure 1. Bidirectional buck/boost converter (with SiC MOSFETs and SiC Schottky diodes as switches)

TABLE I. Devices Under Comparison

Part Number	Manufacturer	Description	Max Ratings	No. Paralleled
IRGP20B120U-EP	International Rectifier	Si IGBT	1200V, 20A	3
DSEP 29-12A	IXYS	Si Diode	1200V, 30A	2
CMF20120D	CREE	SiC MOSFET	1200V, 20A	3
C4D40120D	CREE	SiC SBD	1200V, 54A	2
GA35XCP12-247	GeneSiC	Si IGBT with SiC SBD co-pack	1200V, 35A	2

The loss components and the overall efficiency under different load conditions are calculated for each converter. The results are shown in Table III. The efficiencies are also plotted in Figure 2 for better comparison.

It is shown that, via the use of SiC devices, both conduction and switching losses are significantly reduced, which is consistent with the prediction. The all-SiC converter has a total switching loss (P_{sw}) of only 36, 27, and 17 Watts at full load, 75% load, and 50% load, respectively, which are approximately one-sixth of the total switching losses of the all-Si converter. The conduction loss (P_c) of all-SiC design is significantly reduced as well, which is around 40%-50% less than conduction loss of the all-Si design.

Inductor and capacitor loss (P_r) of the three converters under consideration are almost always the same because the rms current values and component ESRs are the same. The all-SiC design has an overall efficiency of 98.98%, 99.17%, and 99.37% for the full load, 75% load, and 50% load, respectively, and is about 1.4% higher than that of the all-Si design, while the performance of the hybrid converter is consistently at some level between the performances of other two converters.

TABLE II. Specifications of the Converter

Specifications	Value
Low side voltage V1 (V)	300
High side voltage V2 (V)	600
Power rating P (kW)	20
Switching frequency f (kHz)	20
Junction temperature Tj (C°)	125
Chock inductor L1 (mH)	2.5
DC bus capacitor C (mF)	100

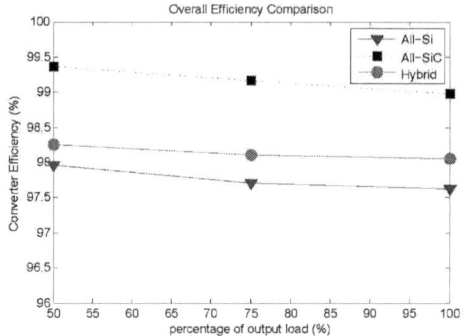

Figure 2. Efficiency comparison of three converters under different loads

TABLE III. Calculated Losses and Efficiencies

	All-Si	All-SiC	Hybrid Device
		Full load	
Conduction loss Pc (W)	211.8	131.8	158.0
Switching loss Psw (W)	222.7	36	193.5
Inductor and capacitor loss Pr (W)	34.7	34.9	34.8
Total loss Pt (W)	469.2	202.7	386.3
Converter efficiency h (%)	97.62	98.98	98.05
		75% load	
Conduction loss Pc (W)	165	97.1	134.3
Switching loss Psw (W)	174.4	27.1	147
Inductor and capacitor loss Pr (W)	19.5	19.6	19.5
Total loss Pt (W)	358.9	143.8	300.8
Converter efficiency h (%)	97.71	99.17	98.11
		50% load	
Conduction loss Pc (W)	91	45.9	71.7
Switching loss Psw (W)	111.3	17	100.5
Inductor and capacitor loss Pr (W)	8.7	8.7	8.7
Total loss Pt (W)	211	71.6	180.9
Converter efficiency h (%)	97.96	99.37	98.26

Case Study 2: SiC-based Voltage Source Inverter

A three-phase voltage source inverter (as shown in Figure 3) is also analyzed and simulated based on the commercially available SiC devices (CMF20120D and C4D20120D) and conventional Si devices of the same rating (IXGQ20N120B and DSEP29-12A). Some specifications of the inverter are listed in Table IV.

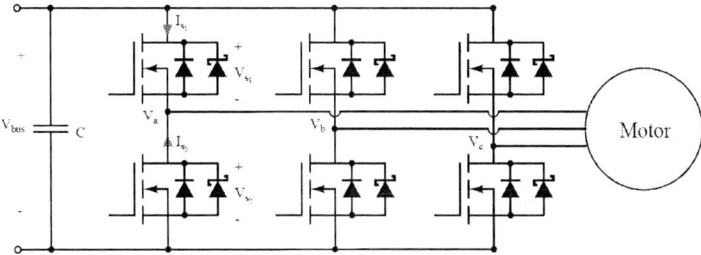

Figure 3. Three phase voltage source inverter used to drive a motor (with SiC MOSFETs and SiC SBDs as switches)

TABLE IV. Specifications of the Inverter

Specifications	Value
DC bus voltage V_{bus} (V)	800
Max output voltage line-to-line V_{ll} (Vrms)	490
Power rating S (kVA)	12
Max load current I_{load} (Arms)	15
Motor stator resistance R (Ω)	0.1
Motor synchronous inductor L (mH)	40
Rated back EMF of motor line-to-neutral (Vrms)	339

It is assumed that inverter works at full load with unity power factor. Standard SPWM method is used for generating gate signal and the carrier frequency is chosen to be 20 kHz. The simulation results for the SiC inverter and Si inverter under given condition are listed in Table V below.

As shown in Table V, the SiC inverter outperforms Si inverter. Due to the larger voltage drop across Si IGBT and Si diodes during on-state and the longer dead-time used in Si inverter, the output voltage of the Si inverter is lower than the SiC inverter.

On the loss aspect, the power dissipated on Si IGBTs (343 W) is 4.5 times the power loss on SiC MOSFETs (76 W). This is caused by the larger on-state collector-emitter voltage drop of Si IGBT (which results in larger conduction loss) as well as the presence of current tailing in IGBT and reverse recovery effect in PN diode (which results in larger switching loss). Similarly, the loss on Si PN diode is also much larger than (9.5 times) the loss on SiC SBD. As a result, the total loss on the inverter is reduced by 78% (348.1W to 76.56W) when Si devices are replaced by SiC devices. It is impressive that the overall efficiency of the SiC inverter is as high as 99.4%, which is about 2.3% higher than the Si inverter.

TABLE V. Simulation Results under Given Conditions

Full load, unity PF, 20 kHz	SiC MOSFET Inverter	Si IGBT Inverter
Power in (W)	12604	11944
Loss on Main Switch Device (W)	76.02	342.97
Loss on Diodes (W)	0.54	5.14
Total Loss (W)	76.56	348.12
Output Power (W)	12528	11595
Efficiency (%)	99.40	97.08

Issues with Wide Bandgap Semiconductor Devices on Circuit Level

Stray Inductance and Ringing

In order to maximize the power density of the power converter design with WBG devices, the design needs to take full advantage of its fast switching capabilities. With the fast switching capabilities and the absence of minority charge carrier removal, SiC and GaN majority carrier are more sensitive to parasitic oscillations. Previously with traditional Si devices, the device's resistance, capacitance, reverse recovery, or tail current damped any oscillation created by the packaging and printed circuit board (PCB) inductances during switching transients. Figure 4 shows a MOSFET schematic with the device's capacitances and packaging inductances. Before the MOSFET can be in turn-off condition, the C_{DG} and C_{DS} capacitors have to charge. Similarly for turn-on condition, the C_{DG} and C_{DS} capacitors discharge to allow the current to flow into the drain terminal of the MOSFET (11). There have been a significant amount of studies on the performance impact introduced by the parasitic components in Figure 4 (12)-(15). Reference (12) investigated the cause of the parasitic oscillations and proposed potential methods to reduce the ringing with snubber circuits and ferrite beads. In reference (13), the performance comparison was made between SiC SBD and Si PN diode. Due to SiC SBD's sensitivity to ringing, the Si PN diode outperformed the SiC SBD between -40°C and 0°C (13). Reference (14) derived an analytical expression to estimate the overshoot voltage during MOSFET turn off transient. Reference (15) used a finite element tool to compute the inductances and capacitance of the device packaging.

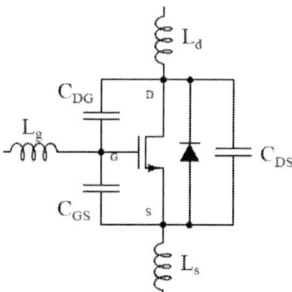

Figure 4. MOSFET schematic with parasitic elements shown

Figure 5 shows the SiC and Si MOSFET body diode turn off current comparison. The SiC MOSFET is CMF20120D from CREE and Si MOSFET is APT28M120B2 from Microsemi. Figure 6 shows the schematic of the experimental setup to obtain the waveforms in Figure 5. The current was measured at the drain terminal of S1. For both tests, S1 is CMF20120D. For D1, both SiC and Si MOSFET's gate and source terminals were shorted together for their respective test.

By looking at Figure 5, the SiC MOSFET body diode demonstrated sensitivity to ringing with a peak overshoot current of 3.22 A. For Si MOSFET body diode, it did not show any ringing because the transient was damped by the reverse recovery with a peak overshoot current of 15 A. Although the SiC MOSFET body diode showed significantly lower overshoot current at 1A load, for higher voltage and current operations, the ringing may be more severe and cause an overshoot value over the device's rating. Furthermore, if the packaging and PCB parasitic are not analyzed carefully, the ringing transient can last longer than the switching cycle. For any high switching frequency operation, the parasitic within the packaging and PCB have to be considered. Especially for the higher-voltage application, the electromagnetic interference and reliability becomes a major concern.

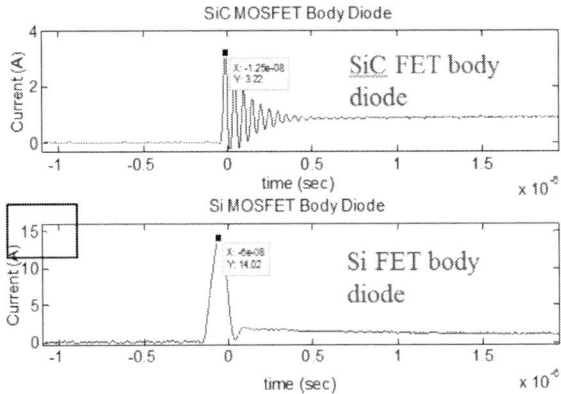

Figure 5. Current waveforms comparison of SiC and Si MOSFETs during body diode turn-off transients

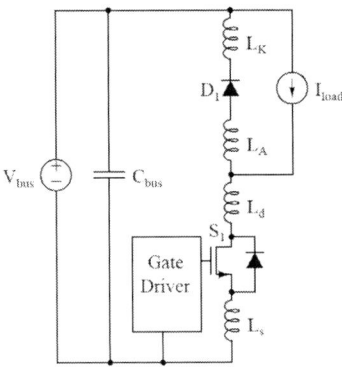

Figure 6. Schematic of experimental setup used to obtain the waveforms of Figure 5

Thermal Design

High Temperature Operation

The main device physics considerations that limit of operating Si semiconductor devices at high temperature, which can be superseded by WBG devices are (16):

1. Intrinsic Carrier Concentration: Sufficient control of the local free carrier concentration is vital to operation of any semiconductor device and is primarily accomplished during device fabrication through the intentional introduction of doping impurities into various desired regions of the device. As ambient temperature increases, the intrinsic carrier concentration becomes greater than or comparable to dopant concentration and thus free carrier concentration in the device no longer remains equal to the dopant concentration. Intrinsic carrier concentration unwantedly influences conductivity of the region. In WBG devices, due to much lower intrinsic carrier concentrations than Si (which is a result of wider bandgap), the upper ambient temperature limit is much higher (600°C as opposed to 300°C for Si).

2. P-N junction leakage: In almost all circuit applications, leakage currents of P-N junctions in power electronic devices should be kept negligible with respect to the desired signal currents. For negative i.e., reverse bias greater than a few tenths of a volt at temperatures below 1000°C, $I_{reverse}$ depends chiefly on intrinsic carrier concentration and is thus much lower for WBG than Si at given temperature. Further, it allows for safe operation of WBG devices at temperatures much higher than those for Si devices.

3. Thermionic Leakage: Thermionic emissions and reverse leakage currents originating from the same, increase with temperature, as carriers get more and more energy to cross energy barrier height. Thus, much larger (almost twice) barrier heights in WBG result in much lower leakage currents and allow operation at much higher voltages and temperatures.

Thermal Management

In all three situations, namely high power, high ambient temperature or high-frequency operation, controlling heat in power electronic circuit design becomes one of the major concerns. With the advent of wide bandgap devices, a significant reduction in the required die size of MOSFET for given voltage rating has become feasible. It also results in lower forward on-state resistance of MOSFET and the associated conduction losses. However, if the reduction in die size exceeds the corresponding reduction in the power loss to be dissipated in the device in the form of heat, thermal management becomes more critical as the power dissipation density increases.

Fortunately, with its higher value of thermal conductivity (5 Watts/cm^2 K), SiC can supersede conventional Si devices in both thermal as well as electrical performance aspects. For GaN devices however, the lower thermal conductivity of GaN (1.3 Watts/cm^2 K) as compared to Si (1.5 Watts/cm^2 K) and SiC forces one to monitor thermal limitations more cautiously. This is one of the reasons why the use of GaN devices is still chiefly limited to high-frequency, low-power applications. In case of GaN-based converters, whether there is a net decrease in the size of cooling requirements depends on a relative decrease in total losses due to better switching performance and an increase in heat sink size requirement due to poorer thermal conductivity.

Wide Bandgap Device Power Electronics in Educational System

Incorporation of WBG Device Technology into Undergraduate and Graduate Curriculum

Various courses exist in electrical engineering curriculum where the fundamental semiconductor device physics and electronic circuits are typically covered. It is also possible that power electronics courses can be part of the curriculum if a particular university has faculty for power electronics expertise. For graduate-level curriculum, advanced power electronics, power electronics lab, and some other application courses related to power electronics may exist.

Table VI shows the proposed enhanced curriculum to include the WBG device technology. The new topics that can be covered for WBD devices are device physics, device structures, reliability, thermal characteristics, packaging, switching speed capabilities, gate drives, stray inductance and capacitance calculations and measurement techniques, electromagnetic interference (EMI) considerations due to WBG operation, and required design solutions including EMI filters and printed circuit board design guidelines. In terms of the application of WBG devices to power electronics, major subjects to cover will include the efficiency and reliability calculations, high-temperature operation and thermal management, high switching operations of WBG devices in power electronic topologies, filter sizing, and new power electronic topologies that can utilize WBG device characteristics.

TABLE VI. Enhancement of Undergraduate, Graduate, and Outreach Education for WBG Device Technology

Current Curriculum (Example)	Enhanced Curriculum for WBG Device
Undergraduate Courses	
Introduction to Solid State Electronics	Incorporate WBG device technology into existing courses
Power Electronic Circuits	
Semiconductor Physics and Devices	
Electronic Circuits	
	Suggested New Courses
	Materials and Fabrication Technology
	Power Semiconductor Devices
Graduate Courses	
Advanced Power Electronics	Incorporate WBG device technology into existing courses
Power Electronics Laboratory	
Power Electronic Systems for Specific Applications (Sustainable, EV, Aerospace)	
	Suggested New Courses
	Advanced Power Semiconductor Devices
	Power Semiconductor Device Thermal Packaging and Management
	Reliability Engineering for Power Devices and Power Electronics
	WBG Power Devices and Power Electronic Applications

As a first attempt, it is easier to include the WBG device technology into existing courses. In addition, new courses can be developed to increase the depth and breadth of the subject. For example, materials and fabrication technology and power semiconductor devices can be added for undergraduate education if they are not already part of the curriculum. Similarly, advanced power semiconductor, thermal packaging and management, and reliability engineering courses can be developed to advance the education which includes the new WBG device technology. It is also possible to create a unique course which fully covers the WBG device technology and its application to power electronics.

Incorporation of WBG Device Technology to Outreach Curriculum for Practicing Engineers

There are outreach programs at various universities that are geared towards training practicing engineers in electrical engineering. Due to the time delay associated with newly educated undergraduate and graduate students getting into the workforce, it is particularly important to educate practicing engineers to accelerate the adaptation and incorporation of WBG devices and technology into products and systems.

Various short courses, degree courses, and credit courses are available to engineering professionals at the University of Wisconsin-Madison by the Department of Engineering Professional Development (EPD). Short courses already exist in the area of power electronics, motor drives, electromagnetic interference and compatibility (EMI/EMC), and energy storage devices and systems as shown in Table VII. Each short course runs for three or four days. There could be up to five lecturers for each course. In order to achieve an effective educational experience, the course is divided into small sections wherein each section is about one hour long. The learning objects for WBG devices and technologies can be produced in few sections that can be configured to support various short courses in a modular way. In this way, it is possible to minimize the effort and optimize the accessibility and applicability of the curricular objects into many short courses. The curriculum developed for undergraduate and graduate studies and faculty expertise will be immensely helpful to achieve re-use of the material in a succinct way for the short courses. Experts from the industry and national labs are typically included as lecturers. In this way, a strong team of lecturers from academia and industry can cover both academic and practical perspective of the subjects in a more effective way.

TABLE VII. Enhancement of Outreach Short Courses for WBG Device Technology

Outreach/Certificate Courses	
Current Curriculum (Example)	Enhanced Curriculum for WBG
Introduction to Power Electronics	Incorporate WBG device technology into existing courses
Dynamics and Control of AC Drives	
Permanent Magnet Machines and Drives	
Introduction to Electric Machines and Drives	
Introduction to EMI/EMC and Best Practices	
Introduction to Energy Storage Devices and Systems	
	Suggested New Course
	WBG Power Devices and Power Electronic Applications

The Industries that Benefit from WBG Device Technology

The reference (17) identified new trends in key areas such as power semiconductors, power converter circuits, component level simulation, and modeling and mentioned that the new era of power electronics will begin the millennium by commercialization of SiC power devices. As Case 1 and Case 2 examples illustrated, the power electronic technology will be immensely improved because of the significant benefits of the WBG device technology. There are many different power electronic circuits for converting one power type (ac or dc) to another power type (ac or dc). All these circuit topologies strive for efficiency and reliability improvements. These power electronic circuits have critical functions in the systems that they serve. The weight, volume, reliability, and cost have utmost importance for consideration during design of these power electronic circuits. Improvements due to the adoption of WBG device technology will allow reduced weight and volume, reduced thermal cooling requirements, increased reliability, and increased functionality and performance. The cost of WBG technology will come down due to wide adaptation of WBG devices and scale of economics projected. Additional system-level savings, for example, due to reduced cooling or reduced packaging, will also offset some of the initial cost disadvantage of WBG devices compared to silicon counterparts.

TABLE VIII.

Industries Benefiting from WBG technology	Power Electronic Products
Electric Vehicles	Inverters, dc/dc converters, battery chargers, air conditioner compressor, fan, pump controllers
Aerospace	Compressor, fan, pump controllers, actuators, main engine and auxiliary power unit starter units
Wind Power	Power Conversion Units (AC-DC and DC-AC conversion) and actuator controllers for pitch controller
Solar Power	Power Conversion Units (DC-AC)
Ocean Power	Power Conversion Units (AC-DC and DC-AC conversion)
Heating Ventilation Air Conditioners (HVAC)	Motor drives to run compressors, fans, and pumps
Appliances (Refrigerators, Washer/Dryers, etc.)	Motor drives
Uninterruptable Power Supplies (UPS)	Power Converters (AC-DC and DC-AC conversion)
Industrial and Servo Industries	Motor drives, and dc/dc converters, active rectifiers, active filters
Oil and Mining Industry	High temperature power electronics for motor drives, power supplies, actuator controllers
High-Temperature Applications	Power electronic converters placed in jet engine compartments, automobile engine compartments, space applications

The major industries that will benefit WBG device technology are electric vehicles, aerospace, renewable energy, HVAC units, appliances, uninterruptable power supplies. These industries and their products are listed in Table VIII. Certainly, industries that have large volume such as electric vehicles, appliances, and industrial applications will enable the large number of WBG device utilization and help out the scale of economics to work in favor of WBG devices. The aerospace industry has high interest in WBG device technology due to the weight and volume savings and can become early adopters of the

WBG device technology. Renewable and sustainable applications also require efficient systems to deliver energy from the harvesting source to the grid. Hence, efficiency improvements that can be provided by WBG technology will be beneficial to these technologies. Finally, there are applications such as mining that require high-temperature operating conditions. For applications such as power converters in jet engine compartments, automobile engine compartments, and space applications where the high-temperature capability of power electronics is required, certainly SiC technology will be very important and useful.

Conclusion

This paper described the benefits of the WBG devices and provided research results on two application examples of these devices using a dc-dc converter and an inverter. Both of these examples showed significant efficiency increases which translates into reduced thermal requirements. The reduced thermal requirements will achieve cost, weight, and volume savings along with better reliability and performance of the future power electronic circuits and systems designed using WBG devices. The paper proposed a new power electronic curriculum model for undergraduate, graduate, and outreach education and devised a strategy to include the WBG device technology into the curriculum. Education of future engineers and current workforce is very important for the successful dissemination, adaptation and implementation of WBG devices into power electronic industries. Finally, a brief discussion is provided on the power electronic industries and products that WBG devices can be used in, and the overall benefits of these new devices.

References

1. A. Elasser, T. Chow, "Silicon Carbide Benefits and Advantages for Power Electronics Circuits and Systems," *Proceedings of the IEEE*, vol.90, no.6, pp. 969- 986, Jun 2002.
2. B. Jayant Baliga, "Fundamentals of Power Semiconductor Devices," Springer, 2008.
3. J. Biela. M. Schweizer, S. Waffler, J.W. Kolar, "SiC versus Si—Evaluation of Potentials for Performance Improvement of Inverter and DC–DC Converter Systems by SiC Power Semiconductors," *IEEE Transactions on Industrial Electronics*, Vol.58, no.7, pp.2872-2882, July 2011
4. J. Dong, R. Burgos, F. Wang, D. Boroyevich, "Temperature-Dependent Characteristics of SiC Devices: Performance Evaluation and Loss Calculation," *IEEE Transactions on Power Electronics*, Vol.27, no.2, pp.1013-1024, Feb. 2012
5. R.Wang, D. Boroyevich, P. Ning, Z. Wang, F. Wang, P. Mattavelli, K. Ngo, K. Rajashekara, "A High-Temperature SiC Three-Phase AC–DC Converter Design for $> 100^\circ C$ Ambient Temperature," *IEEE Transactions on Power Electronics*, Vol.28, no.1, pp.555-572, Jan. 2013
6. S. Waffler, S. D. Round, and J. W. Kolar, "High Temperature ($> 200 \circ C$) Isolated Gate Drive Topologies for Silicon Carbide (SiC) JFET," *Proc. 34th Annu. Conf. IEEE Ind. Electron.*, 2008, pp. 2867–2872.

7. Z. Chen, "Characterization and modeling of high-switching-speed behavior of SiC active devices," M.S. thesis, Dept. Elect. Eng., Virginia Polytechnic Institute and State University, Blacksburg, 2009.
8. J. Biela, D. Aggeler, D. Bortis, and J. W. Kolar, "5 kV/100 ns Pulsed Power Switch Based on SiC-JFET Super Cascode," *Proc. IEEE Int. Pulsed Power Conf.*, Washington, DC, Jun. 2009, pp. 358–361.
9. S. Ji, D. Reusch, F.C. Lee, "High-Frequency High Power Density 3-D Integrated Gallium-Nitride-Based Point of Load Module Design," *IEEE Transactions on Power Electronics*, Vol.28, no.9, pp.4216-4226, Sept. 2013
10. D. Han, J. Noppakunkajorn, B. Sarlioglu, "Efficiency Comparison of SiC and Si-Based Bidirectional DC-DC Converters," *IEEE Transportation Electrification Conference and Expo (ITEC), 2013 IEEE*, pp.1-7, 18-20 June 2012
11. Y. Ren, M. Xu, J. Zhou, F.C. Lee, "Analytical Loss Model of Power MOSFET," *IEEE Transactions on Power Electronics*, Vol.21, no.2, pp.310-319, March 2006
12. Josifovic, J. Popovic-Gerber; J.A. Ferreira, "Improving SiC JFET Switching Behavior Under Influence of Circuit Parasitics," *IEEE Transactions on Power Electronics* , vol.27, no.8, pp.3843-3854, Aug. 2012
13. O. Alatise, N. Parker-Allotey, D. Hamilton, P. Mawby, "The Impact of Parasitic Inductance on the Performance of Silicon–Carbide Schottky Barrier Diodes," *IEEE Transactions on Power Electronics*, vol.27, no.8, pp.3826-3833, Aug. 2012
14. W. Teulings, J. L. Schanen, J. Roudet, "MOSFET Switching Behavior Under Influence of PCB Stray Inductance," *Proc. IEEE IAS 1996*, Vol. 3, pp. 1449-1453, 1996.
15. J. Z. Chen, L. Yang, D. Boroyevich, and W. G. Odendaal, "Modeling and Measurements of Parasitic Parameters for Integrated Power Electronics Modules," *Proc. IEEE APEC 2004*, vol. 1, pp. 522-525, 2004.
16. P. Neudeck, R. Okojie and L. Chen, *Proceedings of the IEEE,* **90**(6), 1065 (2002).
17. S. Abedipur and K. Shenai, "Power Electronics Technologies for the New Millennium," *Proc. IEEE Devices, Circuits, Systems Conference*, pp. 111/1-111/9, 2000

412

ECS Transactions, 58 (4) 413-423 (2013)
10.1149/05804.0413ecst ©The Electrochemical Society

Studying the Performance of Series-Connected GaN FETs in Higher Voltage Switching Applications

A. Hasanzadeh and A. Khaligh

Power Electronics, Energy Harvesting and Renewable Energies Laboratory
Electrical and Computer Engineering Department and the Institute for Systems Research
2347 A.V. Williams Building, University of Maryland, College Park, MD 20742
Email: Khaligh@ece.umd.edu; Url: www.ece.umd.edu/~khaligh

This manuscript investigates cascading of gallium-nitride (GaN) field-effect transistors (FETs) to accommodate for higher voltage switching operation, beyond the ratings of commercially available single switches. The current GaN FETs present substantial advantages over corresponding silicon (Si) and silicon-carbide (SiC) FETs in terms of switching speed, conduction and switching loss, and thermal capability. However, currently GaN based switching semiconductor devices are not available in high voltage ratings, which could be a limiting factor in their use for high voltage applications. This paper presents the study, design, simulation, and implementation of cascading enhancement-mode GaN (eGaN) devices to demonstrate their superior performance capabilities in comparison to their Si-counter parts. The presented study in this manuscript paves out a way to utilize future high-voltage GaN semiconductor devices in plasma technologies, pulsed microwave tubes, and high voltage direct current transmission applications, which need very high voltages.

Introduction

The Gallium Nitride (GaN) based high-electron-mobility transistors (HEMTs) have been proven to be the most promising switching devices for future power electronics application. The eGaN devices offer lower conduction resistance, smaller gate charge, and faster switching capability than comparable metal–oxide–semiconductor FETs (MOSFET) [1]. In addition, the developed models for eGaN devices have shown the accuracy and simplicity of this power transistor in high-speed power electronic converters [2, 3]. Currently as of July 2013, the EPC Corporation is the only manufacturer, which offers commercially available GaN switching power devices with 200V drain-to-source voltage V_{ds}. Various research groups are investigating GaN semiconductor devices with higher voltage ratings [4-6]. A non-commercial 900V GaN HEMT on SiC substrate device was used to design a 350V boost converter with 97.8% efficiency. In this design even though the voltage rating of switches is 900V; however, they are used for much less rating voltage [7]. In another study, a boost converter with 140V output voltage is designed, which achieves 96% efficiency. In this study, the eGaN switches are fabricated on Si and preliminary designed for 550V operating voltage [8]. In addition to above applications, which operate under 1 MHz switching frequency, a 16V

413

to 34V and 15W boost converter was designed for radio frequency (RF) applications, which is being switched at 10MHz switching frequency, using 100V GaN-on-Si HEMT devices [9]. Based on abovementioned designs and studies it is observed that majority of eGaN devices have been used in designs with much lower voltages than their rated drain-to-source voltages.

On the other hand, a higher blocking voltage is necessary for variety of applications such DC/AC inverters in uninterruptible power supplies, AC/DC and DC/DC converters in electric vehicles battery chargers, and multilevel inverters in motor drive systems, which could benefit significantly from GaN technology [10-12]. In order to accommodate for higher voltages, beyond the ratings of commercially available single GaN FETs, there are three different approaches: (a) employing transformers, (b) adopting multilevel topologies, and (c) connecting devices in series. Utilizing transformers would severely increase the volume and cost of converter. Pursuing multilevel topologies would not only increase the size of the system, but also complicate the structure and control of converter [13-15]. Hence, connecting power semiconductor devices in series to account for higher voltage ratings than a single device is the most direct, effective, and low cost contrivance [16].

From aforementioned discussion, it is obvious that developing power electronic circuits and systems using wide band gap GaN devices will still need various research and manufacturing challenges in particular in high voltage and high current applications [1, 17]. Even though research is being conducted to develop switches with extended voltage ratings and current carrying capability, more time is needed to develop and introduce high voltage GaN switchs like their Si and SiC based counterparts [18]. In the meantime, the series connection of the GaN switches is promising to account for high voltages. Challenges of series connection of GaN FET devices include: high-side switch grounding and clearance, series connection balancing with minimum circuit complexity, low- and high-side switches gate-driving.

In this manuscript, series connection of eGaN switches is discussed and analyzed. The gate- and load-side operation and topologies are explored. The analytical and simulation results are validated through experimental set-ups. The fabricated set-up is a test circuit composed of two series connected 200V EPC2012 eGaN devices. Also, in order to show superior performance capabilities of eGaN series connection, the results is compared with their Si-counter parts. The gate-drive and load-balancing circuits for this set-up are outlined. Finally, the paper conclusions are presented.

Cascading FETs

Figure 1 shows a circuit composed of series connection of GaN switches. This topology can be used for different applications when a higher switching voltage device is required [10-12]. In this structure, the maximum V_{DS} of the each series-connected FET must be kept less than its absolute maximum rating in both transient and steady-state operations. The reference [14] overviews several concepts for series connection of FETs; such as balancing network on load-side, gate drive circuits on gate-side and influence of parasitic capacitances of FETs in their series connection. The voltage rise- and fall-time is a critical parameter to select a proper approach for cascading FETs with appropriate balancing network on the load-side and suitable gate drive circuit in the gate-side [13,14,16,18].

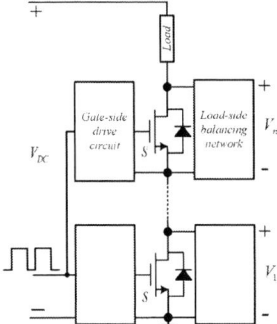

Figure 1 Schematic of DC to pulse converter circuit of cascaded FETs.

Load-side voltage balancing

The load-side voltage balancing methods include impedance symmetrisation, clamp circuits and snubber circuits. Figure 2 shows a DC to pulse converter circuit of cascaded FETs to achieve a higher switching voltage. The balancing network limits V_{DS} to a maximum value, which is less than nominal V_{DS} of FET (approximately 80%-90% of nominal V_{DS}). Also the overall DC bus voltage V_{DC} is evenly distributed between FETs during on- and off-states of switching operation.

$$V_1 = \cdots = V_i = \cdots = V_n = V_{DC}/n \tag{1}$$

where, V_i is the voltage across each FET.

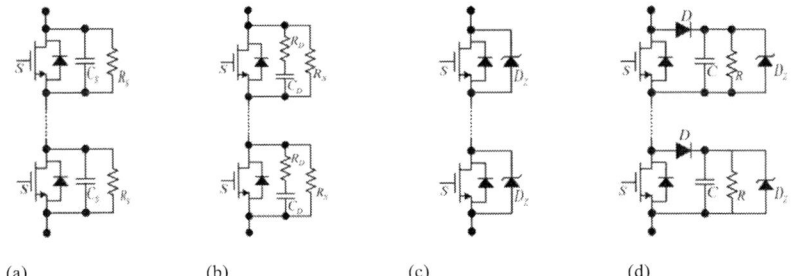

(a)   (b)   (c)   (d)

Figure 2 Load-side voltage balancing configurations for cascading FETs with (a) static parallel $R_S C_S$ network (b) static parallel-series $R_S C_D R_D$ network (c) clamping zener diode network (d) combination of (b) and (c) configuration using a decoupling diode.

The balancing network should not have any influence on switching behaviour of FETs. Nevertheless, in figure 2(a-b) schemes, the voltage balancing circuit slightly alters the rise- and fall-time of pulse voltage due to addition of passive capacitive components. This alteration increases the switching losses. In addition, the resistive elements contribute to some additional conductive losses. In figure 2(c), the zener diode D_Z is an overvoltage protection element and should have lower voltage rating than the rated V_{DS} of FET. The zener diode should be able to limit the voltage across each FET by dissipating overvoltage losses, and consequently it must be a high power rated zener diode. The voltage balancing circuit demonstrated in Figure 2(d) is a combination of figure 2(a-c) configurations, with its inherent advantages and disadvantages. The capacitor C, similar to C_S and $R_D - C_D$ in figures 2(a-b), is used to prevent transient overvoltages and the resistance R, similar to R_S in figure 2(a-b), is used to guarantee static balancing. The diode D, in figure 2(d), is used to decouple the balancing network from the FETs when switches are on. Therefore, the balancing network has no negative influence on the switching behaviour and switching losses of the FETs. The disadvantage of the figure 2(d) scheme, which shows up here at the first glance, is higher number of components and cost. In this study, to simplify series connection of eGaN devices, the balancing network of figure 2(c) is chosen.

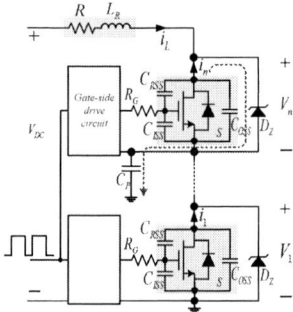

Figure 3 Leakage capacitances in cascaded eGaN FETs.

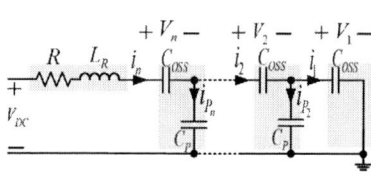

Figure 4 Equivalent circuit for series connected eGaN FET output capacitors and leakage capacitor of each cascaded stage to ground.

The non-synchronous switching and/or parasitic capacitances of FETs will cause different i_i ($i \in 1 \ldots n$) transient currents from FETs, shown in Fig. 3 in blue trace, during turn-off switching instants. The capacitance C_P is the parasitic capacitance between each FET and the system ground. Also, input capacitance C_{ISS}, output capacitance C_{OSS}, and reverse transfer capacitance C_{RSS} are the FET's internal capacitances as shown in figure 3. To simplify the analysis of this transient voltage distribution behaviour, the balancing networks are neglected. The circuit of figure 4 represents the equivalent circuit at turn-off instant of all switches. The Kirchhoff law leads to the following equations.

$$V_{i-1}(s) = \begin{cases} -V_{DC}\dfrac{C_p}{C_{OSS}} + V_i(s)\left(1 + \dfrac{C_p}{C_{OSS}}\right) + V_n(s)(Ls + R)sC_p & ; i = n \\[4mm] -V_{DC}\dfrac{C_p}{C_{OSS}} + V_i(s) + \dfrac{C_p}{C_{OSS}}\displaystyle\sum_{k=i}^{n} V_k(s) + V_n(s)(Ls + R)sC_p & ; 2 \leq i < n \end{cases} \tag{2}$$

where

$$V_{DC} = \mathcal{L}^{-1}\left\{\sum_{k=1}^{n} V_k(s)\right\} \tag{3}$$

By solving equations (3), the transient voltage can be calculated. However, it is apparent that the smaller C_p in comparison to C_{OSS} will lead to smaller deviations of the voltages V_i. Consequently, If absolutely synchronous switching occurs and the parasitic capacitances to ground are zero, which is the ideal condition, the currents i_i will be zero or negligible, and consequently drain-source voltages of FETs V_i (figure 1) will be equal. If these two conditions are not fulfilled, the switching frequency of circuit will be restricted to smaller frequencies than switching frequency of each FET.

Gate Drive

The gate-side voltage balancing methods consist of dv/dt- and di/dt -control, active overvoltage-protection by dynamic clamp, high precision gate drive timing, cascaded synchronization and time delay compensation [14]. Figure 5 shows three different gate driving methods for series-connected FETs. The Figure 5(a) shows the conventional method, in which each gate-side driver is composed of a gate-drive integrated circuit (IC), an optocoupler IC, and an isolated dc-dc converter to switch FETs. This method requires high number of components; however, it is very reliable. In this method, implementing the isolated dc-dc converter for upper series-connected FET stages becomes very challenging in particular in high-voltage applications.

The second method (figure 5(b)) is a capacitively coupled gate drive circuit [14]. This method is easy to implement; however, it has non-synchronous switching behaviour. Thus, this restricts the switching frequency. The third method (figure 5(c)) is a transformer isolated gate drive [14]. Each FET is derived through a transformer with toroid core. The primary of these transformers is a common single turn winding. Thus, synchronous switching of all FETs in one series-connected stack is achievable. The primary winding transfers the switching command as well as the gate drive energy. The single-turn primary winding can be realized by a high voltage cable to ensure high voltage isolation. However, such cable would make the design expensive in comparison to other approaches.

In summary, there is a compromise between first and third approaches pertaining to components counts, synchronization, cost, and implementation complexity to choose a gate-drive circuit. In this study, the first method is employed for sake of cost and implementation effort for cascading eGaN FETs. Although minimal differences in the values of the driver-ICs and optocouplers parameters could be a reason for non-synchronous switching in the first method in comparison to third one, this influence on the balancing of the drain-source voltages is negligible compared to the effect of the parasitic capacitance from each gate drive circuit to ground.

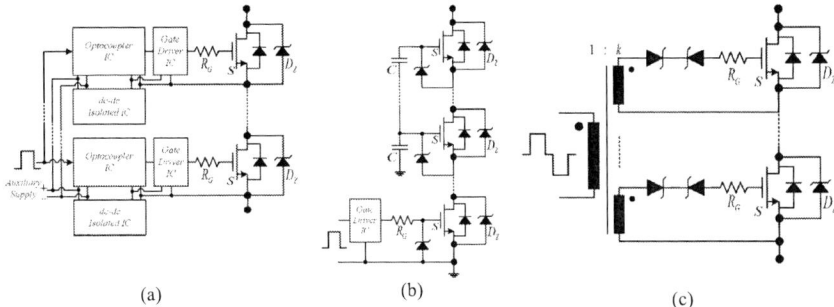

(a)   (b)   (c)

Figure 5 Gate drive circuits: (a) conventional, (b) capacitively coupled, (c) transformer isolated.

Cascading GaN Schematic

The figure 6 shows the schematic of cascading two EPC2012 eGaN FETs, including gate drive IC, optocoupler IC, and isolated dc-dc converter. The schematic and printed circuit board (PCB) have designed by free EAGLE PCB design software. A brief description of gate- and load-side circuits will be described in continued text.

Gate-side circuit

Being configured as enhancement mode, allows eGaN devices to operate similar to power MOSFETs. However, driving eGaN devices requires special considerations due to the low threshold voltage, fast switching speed and 6V gate-to-source maximum voltage rating [19]. Fortunately, newly optimized gate drivers like LM5114 are now available to resolve the challenges of driving eGaN devices. The eGaN FETs have a typical gate-to-source threshold voltage of 1.4V, which is low in comparison to many MOSFETs used in power electronics converters, particularly higher voltage rated devices.

The optocoupler, ADuM11001, is a digital isolator based on combining high speed CMOS and monolithic air core transformer technology. The ADuM1100 supports data rates as high as 25 Mbps. To isolate high-side eGaN device ground, which is drain of low-side eGaN, from low-side ground, the gate signal is transmitted via an ADuM1100 to upper EPC2012 gate driver. To have same propagation delay on transmission of gate signal to both LM5114 drivers, they receive the gate signal through two ADuM1100 although the low-side ADuM1100 does not need isolation for its input and output.

The isolated dc-dc converter, RI-0505S, is a 2W converter with efficiencies up to 87% and very suitable for applications where board space is a premium. With an input-output voltage isolation of 1kV DC, it is used to provide isolated +5V supply for receiver (secondary) side of ADUM1100 and LM5114 of the high-side EPC2012.

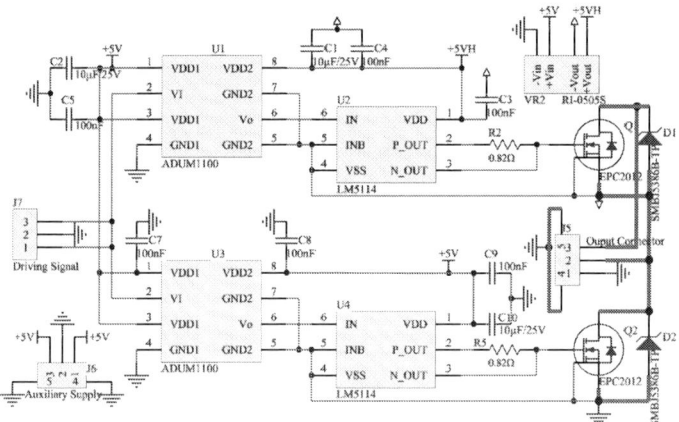

Figure 6 The schematic of cascaded EPC2012 eGaN devices including FET driver, optocoupler and isolated dc-dc converter ICs.

Load-side Circuit

The series switches have to be switched simultaneously. A small delay can reflect all DC bus voltage across one switch, which could be more than rated voltage of that switch. As can be found in the schematic of figure 6, the load-side balancing networks are composed of two zener diodes, SMBJ5386B, to protect each GaN device from extra blocking voltage. The SMBJ5386B is rated for 180V and 5W, which limits the EPC2012 V_{Ds} to 180V and dissipates up to 5W over voltage energy. The EPC2012 is a eGaN device with nominal $V_{DS} = 200V$, continuous $I_D = 3A$, and rated drain-source ON resistance $R_{Ds(ON)} = 70m\Omega$.

Simulation Results

Before presentation of the PCB and experimental results, the eGaN series-connected switches are simulated in a realistic scenario using ORCAD CADENCE 16.6 software. The simulated circuit only encompasses the series modeled eGaN devices, load equivalent impedance, and gate-drive equivalent Thevenin circuit.

As can be seen in figure 7, the eGaN devices has been modeled using an ideal switch combined with $C_{ISS} = 128pF$, $C_{OSS} = 73pF$, $C_{RSS} = 3.3pF$, $R_{DS(ON)} = 0.07\Omega$, and gate threshold voltage $V_{GS(TH)} = 1.4V$. The equivalent load, measured by GW INSTEK LCR-821 (LCR meter), is composed of resistive and inductive portions of 469.3Ω and 15.2µH, respectively. As mentioned in introduction, most of fabricated eGaN devices have been utilized in circuits with much lower peak voltages than their originally rated V_{DS}. Typically, as a rule of thumb, the blocking voltage of a FET device is considered around 70%-75% of its nominal V_{DS}. In this study, the maximum voltage on V_{DS} should not be more than 140V. Nonetheless, the applied voltage in this simulation across two series eGaN is 175V. Around 50% overvoltage (spike) will be acceptable in this hard switching test for the sake of resonance/oscillations pertaining to the equivalent series inductance (ESL) of load and the eGaNs' cascading equivalent capacitance (figure 4). The gate equivalent Thevenin circuit, as shown in figure 7, has 0.82Ω resistance. Hence, the LM5114 gate-driver is able to source around 6A ($I_G = 5V/0.82\Omega$). The leakage inductance of gate circuit is assumed to be around 10nH, which is more likely acceptable value for this application.

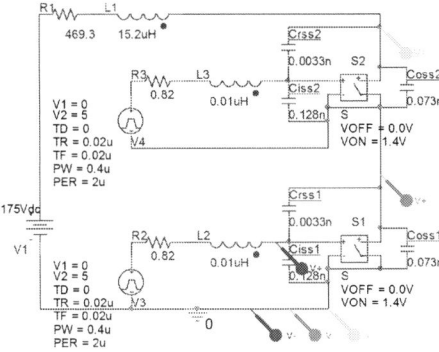

Figure 7 The simulation schematic of two modeled series EPC2012 in ORCAD CADENCE 16.6 environment.

The simulation result in figure 8 shows the low-side gate voltage and drain voltages of two cascaded FETs to ground. Based on the previous paragraph discussion, the high-side drain voltage has passed the 280V, which is more than each eGaN drain-to-source nominal voltage. The hard switching frequency for this test has been considered 500kHz with 20% duty cycle on-time gate signal. The oscillations on gate and drain voltages are the consequence of the resonance between the eGaN capacitances and the gate and load leakage inductances, successively.

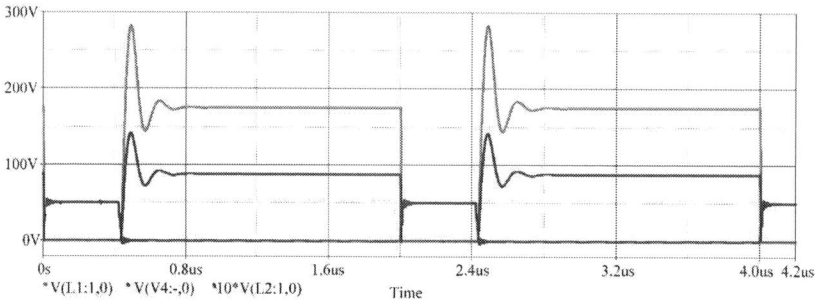

Figure 8 The simulation result of series connected two EPC2012; low-side gate voltage (blue trace) and low- (red trace) and high-side (green trace) drain voltage to ground.

Experimental Verifications

The eGaN FET has very low ratio of gate to drain charge (Q_{GD}) to gate to source charge (Q_{GS}) in comparison to Si MOSFETs. Therefore the PCB pattern for converters employing eGaN devices and an optimized driver like LM5114 can be laid out such that common source inductance can become very small, so even under the most extreme dv/dt conditions, the driver will hold the devices OFF. As shown in figure 9(a), the PCB was designed, built, and assembled using the same parameters and components values, which described in schematic of figure 6. Similar to simulation analyses, the set-up is being switched with 500kHz switching frequency and 20% duty cycle to demonstrate the appropriate performance of cascaded FETs. The high-side switch grounding and clearance considerations can also be seen from figure 9.

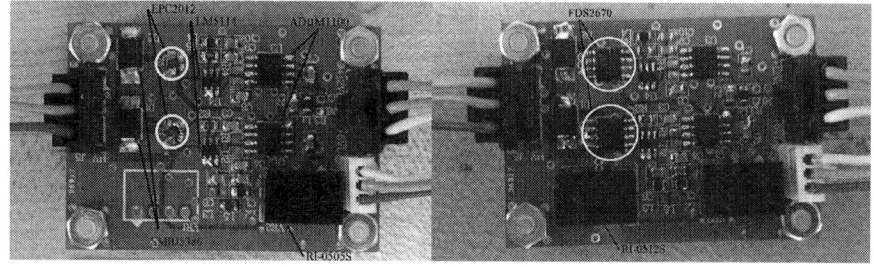

(a)   (b)

Figure 9 The PCB picture of two cascaded: (a) EPC2012 eGaN FETs; (b) FDS2670 Si MOSFETs.

Also as can be seen in figure 9(b), an exactly similar PCB using a Si MOSFET with almost similar parameters is designed, built, and tested to accomplish the study of switching performance of series connected eGaN devices in comparison to Si MOSFETs. Table I represents the EPC2012 eGaN FET and FDS2670 Si MOSFET parameters whose deal with switching functions.

TABLE I. eGaN FET and Si MOSFET Parameters.

Parameters	Symbol	EPC2012 eGaN FET	FDS2670 MOSFET	Unit
Drain-to-Source Voltage	V_{DS}	200	200	V
Drain-to-Source Continuous/pulsed Current	I_D	3/15	3/20	A
Gate-to-Source Voltage	V_{GS}	+6/-5	±20	V
Gate Threshold Voltage	$V_{GS(TH)}$	1.4	4	V
Drain-Source On Resistance	$R_{DS(ON)}$	0.07	0.1	Ω
Input Capacitance	C_{ISS}	0.128n	1.228n	F
Output Capacitance	C_{OSS}	0.073n	0.112n	F
Reverse Transfer Capacitance	C_{RSS}	0.0033n	0.017n	F

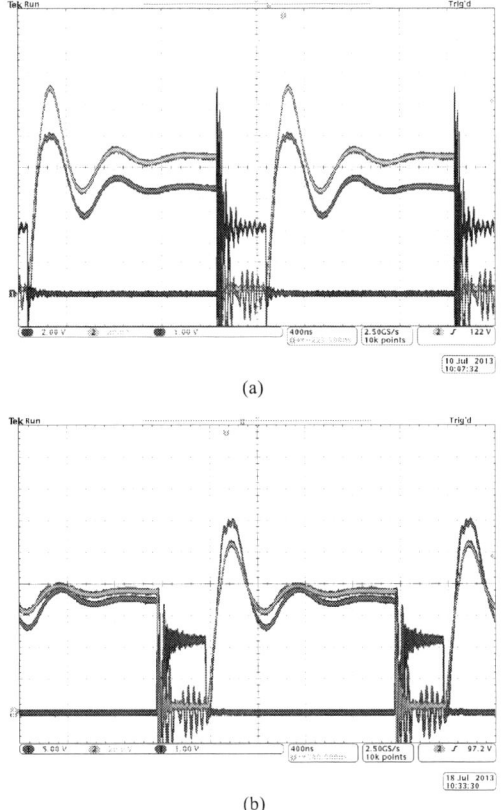

(a)

(b)

Figure 10 The experimental results of DC to pulse converter circuit for series connected: (a) EPC2012 eGaN FETs (b) FDS2670 Si MOSFETs. X-axis is time (400nsec per division). Y-axis consists of low-side gate voltage as well as low- and high-side drain voltages to ground in blue (5V per division), cyan (20V per division) and pink (50V per division), respectively.

The experimental results in figure 10(a-b) show the operation of two cascaded EPC2012 eGaNs FETs and two cascaded FDS2670 Si MOSFETs. The low-side gate voltage, low- and high-side drain voltages are shown in figure 10 (a-b). There are lots of ringing/resonance on gate voltage (blue) and low-side drain voltage (pink), which part of them is due to leakage inductances and eGaN corresponding capacitances; however, some other parts is due to measuring and grounding noise. The Tektronix MSO 4034B oscilloscope and TPP0540 probes are used to capture aforementioned blue and pink waveforms. Also, P5200A differential high voltage probe is used to measure high-side drain to ground (cyan) voltage, which shows that some parts of the ringing and oscillations are due to noise.

The peak voltage (transient) of high-side drain voltages reach to 250V and 300V for EPC2012 and FDS2670, respectively, which are 63% and 75% of the 400V nominal voltage of cascaded FETs. The extra overshoot in the case of Si FDS2670 MOSFETs yields more undesirable resonance. In the case of EPC2012, the maximum voltage of low-side drain voltage reaches to 132V, which is almost equal to half of the high-side drain voltage 250V and shows an acceptable voltage sharing between two eGaN devices. However, for FDS2670, this maximum low-side drain voltage is 106V, which is equal to approximately 1/3 of the high-side drain voltage 300V and reveals serious unbalance behavior in the Si MOSFETs series connection. The steady-state low-/high-side drain voltages are 88V/175V and 76V/175V for EPC2012 and FDS2670, respectively. To solve this unbalance problem for series connected Si MOFETs, a 68kΩ/0.5W static resistance R_S (see figure 2) was added to each FDS2670 and the results are presented in figure 9(b); however, still the voltage distribution is not completely balanced. The switching speed difference is not very considerable in these two case studies; however, the figures 10(a-b) show slightly faster switching operation in the case of series connected eGaN devices in comparison to Si FETs series operation. In addition, it should be noted that, as shown in figure 9(b), the Si MOSFET gate drive circuit needs additional +5 to +12V isolated DC-DC converter (RI-0512S in figure 9(b)) since MOSFET gate drive cannot be driven by +5V like eGaN.

Conclusion

This paper demonstrates the series connation of two EPC2012 eGaN devices to investigate the voltage switching operation of them; beyond the ratings of commercially available single eGaN switches. However, the demonstration of this work will not be limited to cascading just two EPC2012 200V eGaN devices and all the discussion, simulation, and experiment can be extended for future purposes to implement high voltage switches using stacking the lower V_{DS} eGaN devices. The paper also addresses the challenges of series connection of GaN FET devices including high-side switch grounding and clearance, series connection balancing with minimum circuit complexity, low- and high-side switches gate-driving. The switching speeds and voltages waveforms of cascaded eGaN devices demonstrate superior performance of eGaN switching devices in comparison to Si MOSFET with the same specifications. It is important to note that in future in the event of introduction of new GaN semiconductor devices with voltage-ratings potentially up to even 5kV, the presented approach in this manuscript paves out a way to utilize these high-performance switching devices in very high-power and high-voltage converters in the utility systems such as unified power flow control (UPFC) and high voltage direct current transmission (HVDC); in the plasma technologies such as dielectric barrier discharge (DBD); and in driving high-power pulsed microwave tubes such as klystron and magnetron.

References

1. F. C. Lee and Q. Li, *IEEE Transactions on Power Electronics*, **28** (2013) 4127.
2. K. Shah and K. Shenai, *IEEE Transactions on Electron Devices*, **59** (2012) 2735.
3. A. Nakajima, K. Takao, and H. Ohashi, *IEEE Transactions on Electron Devices*, **60** (2013) 646.
4. HRL Laboratories, LLC; http://www.hrl.com/.
5. Redefining Energy Efficiency | Transphorm, Inc.; http://www.transphormusa.com/.
6. International Rectifier; http://www.irf.com/product-info/ganpowir/.
7. Y. Wu, M. Jacob-Mitos, M. L. Moore, and S. Heikman, *IEEE Electron Device Letters*, **29** (2008) 824.
8. J. Das, J. Everts, J. Van den Keybus, M. Van Hove, D. Visalli, P. Srivastava, D. Marcon, K. Cheng, M. Leys, S. Decoutere, J. Driesen, and G. Borghs, *IEEE Electron Device Letters*, **32** (2011) 1370.
9. F. Gamand, M. D. Li, and C. Gaquiere, *IEEE Transactions on Circuits and Systems II: Express Briefs*, **59** (2012) 776.
10. A. Hasanzadeh, O.C. Onar, H. Mokhtari, A. Khaligh, *IEEE Transactions on Power Delivery*, **25** (2010) 468.
11. L. Young-Joo, A. Khaligh, A. Emadi, IEEE Transactions on Vehicular Technology, **58** (2009) 3970.
12. A. Dey, P.P. Rajeevan, R. Ramchand, K. Mathew, K. Gopakumar, IEEE Transactions on Industrial Electronics, **60** (2013) 1989.
13. T. W. Rasmussen, *European Conference on Power Electronics and Applications*, (2005) 1.
14. D. Tastekin, Q. K. Nguyen, A. Lunk, and J. Roth-Stielow, *IEEE 8th International Conference on Power Electronics and ECCE Asia*, (2011) 1558.
15. D. Okula, *IEEE Twenty-Seventh International Power Modulator Symposium*, (2006) 256.
16. T. Lu, Z. Zhao, S. Ji, H. Yu, L. Yuan, *7th International Power Electronics and Motion Control Conference (IPEMC)*, **2** (2012) 1502.
17. M. Araghchini, J. Chen, V. Doan-Nguyen, D. V. Harburg, D. Jin, J. Kim, M. S. Kim, S. Lim, B. Lu, D. Piedra, J. Qiu, J. Ranson, M. Sun, X. Yu, H. Yun, M. G. Allen, J. A. del Alamo, G. DesGroseilliers, F. Herrault, J. H. Lang, C. G. Levey, C. B. Murray, D. Otten, T. Palacios, D. J. Perreault, and C. R. Sullivan, *IEEE Transactions on Power Electronics,* **28** (2013) 4182.
18. S. Castagno, R. D. Curry, and E. Loree, *IEEE Transactions on Plasma Science*, **34** (2006) 1692.
19. B. Wang, N. Tipirneni, M. Riva, A. Monti, G. Simin, and E. Santi, *IEEE Transactions on Industry Applications*, **45** (2009) 843.

CHAPTER 11

MATERIAL SYNTHESIS AND PROCESSING

426

Materials Issues for Vertical Gallium Nitride Power Devices

Adrian D. Williams and Theodore D. Moustakas

Department of Electrical & Computer Engineering and Materials Science and Engineering Division, Photonics Center, Boston University, 8 Saint Mary's St., Boston, MA 02215, USA

In this paper we are addressing some of the fundamental materials issues for the development of vertical GaN-based power devices. Major components of such device are the n+ GaN freestanding substrate on which a thick (~50 µm), low defect density and low carrier concentration ($<10^{16}$ cm^{-3}) n-GaN drift region is grown homoepitaxially. We show that the hydride-vapor-phase-epitaxy (HVPE) is a method capable of producing economically free standing n+ GaN substrates as well as the required thick and low defect and carrier concentration n-GaN drift region. The formation of freestanding GaN substrates by a natural separation mechanism effectively eliminates the need for post-growth processes such as laser liftoff, chemical etching or mechanical lapping to form freestanding GaN substrates. A number of GaN thick films were grown onto sapphire substrates by the Hydride Vapor Phase Epitaxy (HVPE) method with thickness varying from 150µm to 3.8mm using either a low-temperature GaN or an AlN buffer as the nucleation step. We have found that samples grown on a low temperature GaN buffer naturally delaminate from the sapphire substrate post-growth over the entire thickness range studied. However, the GaN films grown on AlN buffers did not delaminate. These results were accounted for by calculating the thermal stresses in the GaN film and substrate as a function of film thickness using Stoney's equation and assuming that the GaN buffer undergoes decomposition at the growth temperature. The structure of these films was determined by x-ray diffraction and the dislocation density was measured to be as low as $5x10^6$ cm^{-2}. The lowest carrier concentration in these heteroepitaxially grown films was found to be 10^{17} cm^{-3}. Furthermore, we have identified the origin of this n-type auto-doping and proposed method to reduce the carrier concentration to values 10^{16} cm^{-3} or lower.

Introduction

Because of the low on-resistance, high breakdown field strength, and good thermal conductivity, GaN power devices are very promising to address a number of current day power electronic applications. The most GaN power devices reported in the literature are based on lateral geometry; however, the development of vertical power devices will lead to higher blocking voltage, lower parasitic inductance and current collapse-free operation. Although there are a number of reports on vertical GaN power devices they generally suffer from a number of problems, the most important of which is high leakage current due to the high concentration of extended defects, such as threading dislocations and basal plane stacking faults.

One basic materials requirement to form a GaN power device rated up to 5 kV, is the ability to form n+ GaN / n- GaN (~50 µm thick, low defect density, and doping level 10^{16}cm^{-3} or lower). The requirement that the drift region of such vertical power devices to be approximately 50 µm thick, rules out as uneconomical the traditional epitaxial methods of MBE and MOCVD for its growth because of their relatively low growth rates of 1-2 µm / h. Furthermore, the currently commercially available GaN substrates for the growth of the GaN drift region are prohibitively expensive. In this paper we are addressing the development of both an n+ GaN free standing substrate as well the n- GaN drift region using the HVPE method.

Development of free standing n+ GaN substrates

The progression of nitride semiconductor technology has been rapid with a plethora of devices having been demonstrated to date. The majority of work in this field has been carried out on non-native substrates, predominantly C-plane sapphire and 6H silicon carbide due to the commercial unavailability of GaN substrates [1]. Such substrates are not ideal for the growth of GaN due to large mismatches in lattice constants and thermal expansion coefficients leading to high dislocation and crack densities in the epitaxial layers. Thus far, the increase in the performance metrics of nitride-based devices has relied on advancements in the hetero-epitaxial growth of such materials including the use of nitridation (in the case of growth on sapphire) and optimized buffer layers [2,3], as well as more exotic techniques such as lateral epitaxial overgrowth [4] and dislocation filtering layers [5]. While these techniques allow for the growth of device quality GaN, it is commonly believed that the successful development of the next generation of GaN based devices relies inexorably on the availability of high quality GaN substrates. Such substrates would allow device growth to take place homoepitaxially, which has a number of benefits. Of paramount importance is the markedly lower dislocation densities that can be realized in device structures grown homoepitaxially onto low dislocation bulk GaN substrates, allowing for higher carrier mobilities [6], increased radiative recombination efficiency [7], longer device lifetime [8], and lower leakage current [9,10]. The matching of thermal expansion between epitaxial layer and substrate is also realized and thus eliminating the associated thermal stress, which is a significant source of cracking. The availability of such substrates yields several additional advantages in the arena of device processing. Since GaN substrates can be made electrically conductive, the processing of certain device structures would be made simpler by enabling the deposition of metal contacts onto the substrate, eliminating the need for an etching step to contact underlying device layers. GaN's relatively high thermal conductivity also allows for efficient heat sinking of devices grown on such substrates without the need for flip-chip bonding. And lastly, homoepitaxial layers generally follow the same orientation as the substrate, allowing for the matching of cleavage planes, which becomes important when one desires to cleave facets for lasers. Given the many virtues of homoepitaxial growth, it is abundantly clear why there is a great deal of motivation to develop GaN substrates.

Currently, there are a number of articles that address the formation of freestanding GaN substrates. Owing to the technological difficulties of GaN growth from the liquid phase, the hydride vapor phase epitaxy (HVPE) method has emerged as the front-running candidate for the commercialization of such substrates due to its high growth rates and relatively low cost. By this method, the formation of freestanding bulk GaN is preceded by the heteroepitaxial deposition of a thick layer of GaN onto an appropriate substrate and a separation step to free the bulk GaN from its substrate. With regards to separation of film and substrate, researchers have employed a number of techniques including mechanical lapping [11], chemical etching [12], laser liftoff [13], and void assisted separation [14]. Each of these methods, however, requires ex-situ processing of the substrate either before or after the growth.

Our group has developed the growth of crack-free bulk freestanding GaN by the HVPE method, whereby separation of GaN from the substrate occur naturally post-growth and all required steps are performed within the same reactor, effectively streamlining the process [15]. In this section we describe this method and address the issues related to the suppression of crack formation, as well as the mechanism that allows for natural separation to occur.

The hydride vapor phase epitaxy method was used to grow the GaN in this study. The reactor employed is a home-built cylindrical quartz system operating horizontally within a three-zone furnace. Gallium chloride (GaCl) is synthesized in-situ by flowing hydrogen chloride gas (HCl) over a quartz boat containing liquid gallium held at 900°C in a central tube. Ammonia is flowed in an outer tube, and is appropriately mixed via convective and diffusive means with the GaCl flow prior to impinging onto the substrate where GaN growth takes place. Nitrogen is used as the carrier gas and the reactor is operated at atmospheric pressure. Under the system's current configuration (geometry and flow conditions), growth rates up to 700μm/hr have been established.

A number of GaN films with thickness ranging between 150μm and 3.8mm were grown onto C-plane (0001) sapphire utilizing a two-step growth method. Prior to growth, each wafer was exposed to a flow of GaCl as a surface pretreatment at 1000°C [16]. The exact influence of this substrate pretreatment step is not well understood. We hypothesize that upon the decomposition of GaCl on the substrate the Ga reacts with physibsorbed oxygen to form volatile GaO and Cl reacts with surface carbon to form volatile chlorocarbons. Both of these impurities are undesirable for epitaxial growth. The substrates were then cooled and a low-temperature GaN buffer layer, approximately 30-100nm thick, was deposited at low temperatures (400-600° C) [2, 3]. The substrate was then ramped up to a temperature of 1020° C at which the bulk GaN film was grown. Post-growth, the cooling rate was dictated by the natural loss of heat from our furnace, with an initial rate of approximately 60°C/min that gradually declined with the temperature. Some samples were also grown on an AlN buffer, pre-deposited on a (0001) sapphire substrate by the MBE method.

The samples were examined using optical microscopy in reflection mode to image crack densities on the surface of the grown films. While additional cracks may exist in the bulk, it is the surface cracks that we have concerned ourselves with since it is those that will affect further epitaxy. Cracks in a material are initiated and propagated due to stress. The causes of stress are numerous during the heteroepitaxial growth of GaN on sapphire with significant contributions originating from lattice mismatch and island coalescence [17]. These stresses are relieved, to a large extent, through the formation of dislocations, grain boundaries and cracks during the first few microns of growth [18]. Cracks that are generated during the growth may be overgrown and therefore buried within the bulk, resulting in a relaxed epitaxial layer at the growth temperature [18].

Post-growth, the stress that is typically observed is dominated by thermal stress. The act of cooling the sample from the growth temperature to room temperature imparts stress in both the sapphire substrate as well as the GaN layer due to differences in the thermal expansion coefficient between the two materials. Specifically, the average in-plane thermal expansion coefficient in the temperature range 300 to 1000 K for sapphire (7.5×10^{-6} K^{-1}) is greater than that of GaN (5.6×10^{-6} K^{-1}) causing the sapphire to contract at a faster rate than GaN upon cooling [19]. However, the sapphire is mechanically constrained by the GaN epi-layer, and therefore held at a length that is elongated compared to the length it would have achieved had it not been constrained. This results in a tensile stress in the sapphire substrate. The stress analysis of the GaN layer follows analogous logic. GaN would like to contract at a certain rate but is coerced to contract faster than normal under action of the sapphire substrate. The final length of the GaN layer is shorter than it would be had the layer contracted independently and is thus under compressive stress. Such stresses can be quantified utilizing the Stoney Equation (1), which expresses the thermal stress in the film as a function of fundamental parameters of the film and substrate [20].

$$\sigma_f = \frac{(\alpha_s - \alpha_f)\Delta T}{[(1-v_f)/d_f E_f) + ((1-v_s)/d_s E_s)](d_f)} \tag{1}$$

According to Eq. 1, the epitaxial film stress (σ_f) depends on the thermal expansion coefficients (α), Poisson's ratios (v), Young's moduli (E), thickness (d), and the temperature change (ΔT). The subscripts "f" and "s" denote values for the film and substrate respectively.

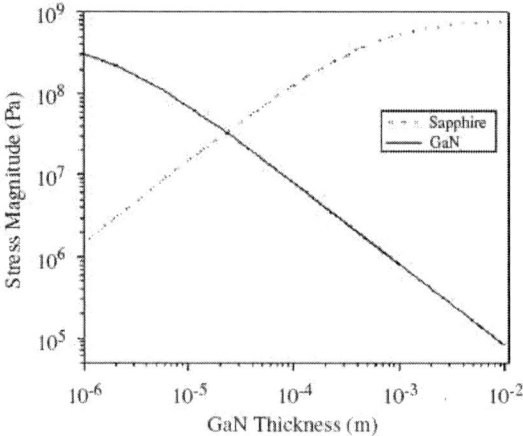

Figure 1: Calculated thermal stress in sapphire and GaN as a function of GaN deposition thickness.

Table I: Input Parameters to the Stoney equation for thermal stress calculation.

Stoney Eqn. Parameter	GaN	Sapphire
Thermal Expansion Coeff.	$5.6\times10^{-6}\,K^{-1}$ $\perp$c [19]	$7.5\times10^{-6}\,K^{-1}$ $\perp$c [19]
Poisson Ratio	0.17 [21]	0.25 [1]
Young's Modulus	210Gpa [22]	352Gpa [1]
Thickness	Variable	330μm
$\Delta T = 1000K$		

Figure 1 shows the theoretical thermal stress generated as a result of cooling GaN on sapphire from 1025°C (GaN growth temperature) to 25°C (room temperature) as a function of the GaN thickness. Table I shows the associated material properties used in the calculation of the data of Figure 1. The values of the thermal expansion coefficient listed in this Table are the average values determined from X-ray measurements as a function of temperature [19]. It should be noted that what is plotted in Figure 1, is the magnitude of the stress and that the nature of the stress in the GaN layer is compressive and that of sapphire, tensile. It should also be noted that this equation does not take relaxation mechanisms such as cracking into account, and thus the actual stress measured for a film will be lower when the predicted stress exceeds the threshold for activating a stress relieving mechanism in either material. This limitation aside, the relevant feature to note in this plot is that the compressive stress of GaN is reduced by over two orders of magnitude as the film thickness increases to 1mm. Correspondingly, the tensile stress of sapphire increases by over two orders of magnitude in the same range of film thickness.

The mechanisms of post growth cracking of GaN on sapphire substrates are still not very well understood to this day. However, empirical observations in the literature have shown that cracking in the sapphire substrate may initiate cracks in GaN at the interface. It is also shown that cracks, which are overgrown during the growth process, may leave voids within the material as demonstrated from SEM images in Ref [17]. In the case of the sapphire substrate, one expects it to develop cracks due to the increasing post-growth tensile stress with GaN thickness. In the case of GaN, the large compressive stress presents a problem as well. While cracking is not typically associated with compressive stress failure, compressive stress acts as a driving force for crack propagation in a material with preexisting flaws such as cracks or voids [23]. Uncontrolled buckling and rupture are also unwanted by-products of large compressive stresses. The reduction of such stress in the GaN layer, by means of growing the GaN film very thick, should effectively reduce the propagation of cracks initiated at the GaN/Sapphire interface.

Figure 2, shows the crack density in three GaN films with thickness of 200μm, 800μm and 1300μm, grown on low temperature GaN buffers. These data show that the density of surface cracks is reduced as the film thickness increases and specifically the 1300μm film is free of cracks.

The preceding discussion on crack formation has yielded conditions under which GaN can be grown crack-free. However, there remains the task of separating the GaN film from the substrate. Post-growth separation techniques such as laser lift-off or mechanical lapping may be employed at this point, but one would ideally like to eliminate such procedures if at all possible to streamline the process.

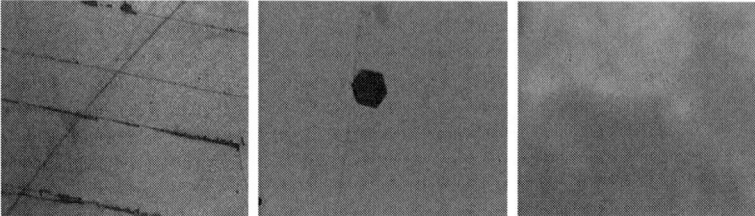

Figure 2: 800μm x 1000μm optical images of surface crack density for GaN films grown to various thicknesses. (a) 200μm film suffers from dense network of cracks separated by 200-300μm. (b) 800μm film is sparsely populated with cracks separated by several mm. (c) 1300μm film is crack-free.

We have empirically observed that thick GaN films (200μm+) grown at high temperature (1020°C) onto a low temperature GaN buffer naturally separate from the sapphire substrate upon cooling. This is achieved in a single HVPE reactor without any special processing of the substrate either before or after the growth procedure. We attribute the observed separation to the low temperature GaN buffer. The GaN buffer layer undergoes decomposition during the high temperature bulk GaN growth. Such decomposition acts to appreciably weaken adhesion at the hetero-interface. Mechanical failure ensues in the buffer layer during the cool down phase on account of thermal stress resulting in the liberation of GaN from the sapphire substrate. Such a process is akin to the void assisted separation process [14]. In this case, the voids originate from the dissociation of nitrogen in the buffer layer.

The facilitation of separation through the buffer layer is seen as viable explanation to the observed separation based on observations of gallium metal on both the sapphire and GaN side of the once hetero-interface post-separation. Koleske et al. has also demonstrated a stark difference in the decomposition rate of GaN nucleation layers grown at low temperatures (550°C) and GaN films grown at high temperatures, with measured rates for nucleation layers seen as high as an order of magnitude greater under the same experimental conditions [24]. Further evidence supporting such a mechanism is the failure of GaN films grown under similar conditions onto AlN buffers or directly onto sapphire (without a LT GaN buffer layer) to separate upon cooling. This is due to the higher onset temperature for decomposition of AlN for the former case and the lack of a weakened interfacial structure for the latter.

Properties of GaN films grown by HVPE: Development of the n- GaN drift region

A. Structural Properties

The surface morphology of the GaN films, produced by the HVPE method, was determined by atomic-force-microscopy (AFM). Under optimized growth conditions the surface is atomically smooth. Figure 3 shows a typical AFM image of a GaN film produced by this method. The rms roughness from this image was found to be 1.2 nm.

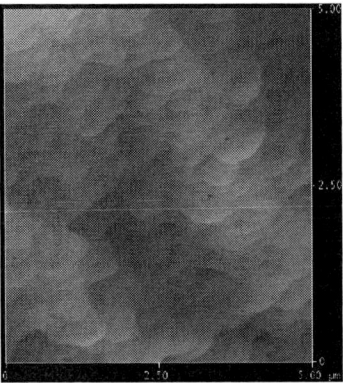

Figure 3: AFM image of a typical as-grown GaN surface. Image dimension is 5µm×5µm. Rms roughness = 1.2nm.

XRD measurements were performed to assess the crystalline quality of HVPE-grown GaN films in this work. Owing to the initial Volmer-Weber growth mode during nucleation of highly lattice mismatched GaN on sapphire, epitaxial GaN layers have a mosaic structure consisting of single-crystal sub-grains with slight misorientations with respect to one another. Misorientation of these grains with respect to the C-plane is termed the tilt and the in-plane misorientation of the grains with respect to each other is termed the twist [25].

The on-axis (002) rocking curve as well as two off axis rocking curves corresponding to (102) and (201) planes were examined in this study. From a qualitative standpoint, an assessment of the relative structural quality between two samples can be made by judging the narrowness of the rocking curves of various planes in each. Both on- and off-axis planes were chosen due to the fact that different defects present themselves in different planes. An example of this is the lack of contribution of edge dislocations to the broadening of on-axis (002) rocking curves, but yield contribution to the broadening of off-axis peaks [26]. The distribution of tilt is associated with screw dislocations and is measured directly by (002) rocking curves. The distribution of twist is associated with edge dislocations and is measured indirectly by extrapolating the FWHM of planes at an inclination angle of 90° with respect to the C-plane using planes of intermediate inclination angles [27]. In this work, the (102) and (201) planes were observed, with inclination angles of 43.2° and 75.1° respectively, and used to extrapolate the twist component.

Figure 4 shows the FWHM of X-ray rocking curves of three planes for a 1 µm thick sample, grown by MBE on sapphire, a 5µm HVPE sample on sapphire, and a 400µm freestanding HVPE GaN sample. Table II summarizes this data. A progressive reduction of the twist and tilt distribution is observed. The difference in the structural quality of GaN grown by MBE versus HVPE samples can readily be attributed to the difference in the growth temperature employed by these methods. The higher temperature growth by HVPE leads to larger single crystal grain size, which translates into lower dislocation densities by virtue of the reduced number of grain boundaries. The difference between the quality of the 5μm and 400μm samples is attributed to dislocation annihilation processes that occur as GaN films are grown thicker.

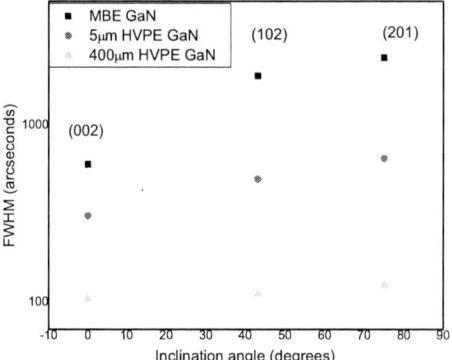

Figure 4: FWHM of X-ray rocking curves for on-axis and off-axis planes for the GaN samples described in the figure legend.

Table II: Summary of XRD rocking curve FWHMs displayed in Figure 4

	(002) - Tilt	**(102)**	**(201)**	**Twist**
1μm MBE	600	1900	2400	2570
5μm HVPE	306	495	647	718
400μm HVPE	104	111	126	132

There are a variety of sources that contribute to the broadening of the rocking curves. These include wafer curvature broadening, instrumentation broadening, dislocation broadening due to strain fields, and intrinsic crystal broadening [28]. Theoretically, it is possible to extract dislocation densities from the broadening of rocking curves using a model proposed by Ayers taking these various components into account [29]. However, there remain some issues in the application of this model to GaN due to analytical shortcomings such as the lack of account for planar defects such as stacking faults in the theory, which skew the results [28]. There tends to be a systematic under-counting of dislocations by this method; it was therefore not employed to determine dislocation densities in these studies.

A rapid characterization method for determining dislocation densities in GaN films is by measuring surface etch pit densities. Chemically etching a GaN surface under appropriate conditions can delineate dislocations due to the different electrical activity in the vicinity of dislocations that modify the local etch rate. Researchers have explored etching GaN in various wet chemistries in an effort to derive a quick method to accurately count dislocations without using a time and labor intensive method such as TEM. One such method that has been correlated well to TEM results is etching in phosphoric acid at 160°C for six minutes [30].

Various HVPE-grown GaN samples, ranging in thickness from 2.75μm to 1100μm, were etched in H_3PO_4 under conditions just described. Figure 5 shows the resulting etch pit density measurements as a function of film thickness. A reduction of the etch pit density by two orders of magnitude is observed in the 1000μm film, with a value of 5×10^6pits/cm^2 compared to 5×10^8pits/cm^2 for the 2.75μm film.

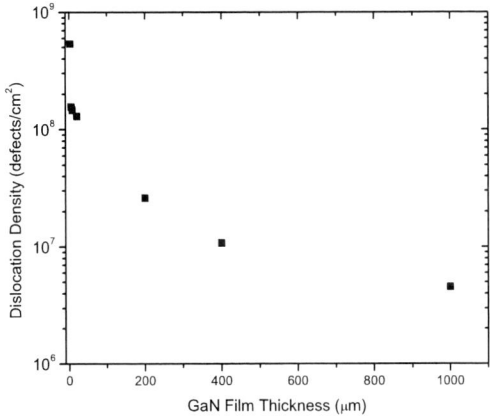

Figure 5: Dislocation densities, determined from EPD measurements, as function of GaN film thickness

B. Transport Properties

Hall effect measurements were carried out on a variety of HVPE GaN samples and their mobilities were studied as a function of their carrier concentration. Such measurements are plotted in Figure 6. The majority of thin films grown for this work were found to be degenerate with n-type conductivity. The most likely cause for this is the presence of unintentional dopants that enter the reactor with the process gasses, the outgasing of chamber walls, and/or potential atmospheric leaks in the system. The presence of such dopants (e.g. oxygen and silicon) within the material leads to a large degree of impurity scattering, which limits the mobility of carriers. The freestanding samples were found to have lower carrier concentration, potential due to the longer growth times, which leads to the coating of the reactor walls and limits outgasing or leaching of impurities from the quartz walls. The decomposition of the buffer layer by the separation method employed also gets rid of the degenerate buffer layer that acts to skew the measurement. The sample exhibiting the highest mobility is a freestanding sample with a carrier concentration of 1×10^{17}/cm^3 with a mobility of 550 cm^{2}V^{-1}s^{-1}.

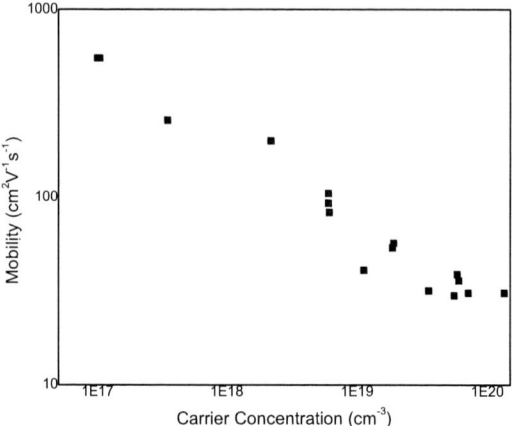

Figure 6: Mobility versus the carrier concentration of for a number of GaN films grown by HVPE

The data presented in Figures 5 and 6 were measured on films grown on sapphire substrates. Thus, the homoepitaxial growth of GaN on GaN substrates by the HVPE method has the potential to produce GaN films with concentration of defects 10^6 cm^{-2} or lower. Furthermore, the employment of appropriate gas purifiers as well as appropriate liners to prevent the leaching of impurities from the quartz tube and their incorporation in the films should lead to GaN films with carrier concentration 10^{16} cm^{-3} or lower and electron mobilities in excess 1000 cm^{2}V^{-1}s^{-1}. Such films will be appropriate to be used as the drift region in vertical GaN power devices.

Conclusions

In conclusion, we reviewed a novel method of generating free standing GaN substrates developed in our laboratory using the HVPE method. We have found that separation of epitaxially grown GaN from sapphire substrates can occur naturally when one employs a LT GaN buffer layer prior to high temperature growth. The enhanced thermal decomposition of the highly defective GaN buffer layer leading to a severely compromised interfacial structure upon cool-down was argued to be the key to this natural separation mechanism. Additionally, a first order analysis of the residual thermal stress in the GaN epilayer and sapphire substrate as a function of the GaN deposition thickness was discussed. Such analysis has shown that the post-growth tensile stress in the sapphire substrate increases by about two orders of magnitude for films 1mm thick, causing the substrate to develop cracks. On the other hand, the compressive thermal stress in the GaN epilayer is reduced by over two orders of magnitude, thus preventing the propagation of cracks generated at the GaN/sapphire interface. These GaN films, grown on sapphire, were found to have a dislocation density as low as 5x10^6 cm^{-2} and carrier concentration as low as 10^{17} cm^{-3}. We proposed that the homoepitaxial growth of such GaN films on GaN substrates by this method should lead to lower dislocation density, and that the employment of gas purifiers and appropriate liners in the HVPE reactor should lead to carrier concentration 10^{16} cm^{-3} or lower. Thus, the HVPE method, as described in this paper, is capable of producing n+ GaN / n- GaN, a basic component for vertical GaN power devices.

References

[1] L. Liu, J.H. Edgar, Material Science and Engineering R 37 (2002) 61-127

[2] T.D. Moustakas, R.J. Molnar, T. Lei, G. Menon, C.R. Eddy Jr., Materials Research Society Symposium Proceedings, 242 (1992) 427

[3] T.D. Moustakas, T. Lei, R.J. Molnar, Physica B 185 (1993) 36

[4] Ok-Hyun Nam, Michael D. Bremser, Tsvetanka S. Zheleva, Robert F. Davis, Applied Physics Letters 71 (1997) 2638

[5] Y. Fu et al., Journal of Applied Physics 99 (2006) 033518

[6] H.M. Ng, D. Doppalapudi, T.D. Moustakas, N.G. Weimann, L.F. Eastman, Applied Physics Letters 73 (1998) 821.

[7] T. Miyajima et al., Physica Status Solidi B 228 (2001) 395.

[8] S. Nakamura et al., Applied Physics Letters 72 (1998) 211.

[9] P. Kozodoy, J.P. Ibbetson, H. Marchard, P.T. Fini, S. Keller, J.S. Speck, S.P. DenBaars, U.K. Mishra, Applied Physics Letters 73 (1998) 975.

[10] X.A. Cao, J.A. Teetsov, F. Shahedipour-Sandvik, S.D. Arthur, Journal of Crystal Growth 264 (2004) 172

[11] H.M. Kim, J.E. Oh, T.W.Kang, Materials Letters 47 (2001) 276

[12] Kensaku Motoki et al., Material Science and Engineering B 93 (2002) 123

[13] R.P. Vaudo, G.R. Brandes, J.S. Flynn, X. Xu, M.F. Chriss, C.S. Christos, D.M. Keogh, F.D. Tamweber, Proceedings of the IWN2000, IPAP Conference Series C1, Nagoya, Japan 2000, p15

[14] Y Oshima, T Eri, M Shibata, H Sunakawa, K Kobayashi, T Ichihashi, A Usui, Japanese Journal of Applied Physics Part 2-Letters 42 (1A-B): L1-L3 (2003)

[15] Adrian D. Williams and Theodore D. Moustakas, Journal of Crystal Growth, **300** (2007) 37

[16] R.J. Molnar, W. Gotz, L.T. Romano, N.M. Johnson, Journal of Crystal Growth 178 (1997) 147

[17] E.V. Etzkorn, D.R. Clarke, Journal of Applied Physics 89 (2001) 1025

[18] B. Monemar, H. Larson, C. Hemmingsson, I.G. Ivanov, D. Gogova, Journal of Crystal Growth 281 (2005) 17

[19] M. Leszczynski, T. Suski, H. Teisseyre, P. Perlin, I. Grzegory, J. Jun, S. Porowski, T. D. Moustakas, Journal of Applied Physics 76 (1994) 4909

[20] Milton Ohring, The Materials Science of Thin Films, Academic Press, London, 1992, p. 419

[21] S. Stepanov, W.N.Wang, B.S.Yavich, V. Bougrov, Y.T.Rebane, Y.G.Shreter, MRS Internet J. Nitride Semicond. Res. 6, 6(2001)

[22] S.O. Kucheyev, J.E. Bradby, J.S. Williams, C. Jagasish, M. Toth, M.R. Phillips, M.V. Swain, Applied Physics Letters 77 (2000) 3373

[23] Siavouche Nemat-Nasser, Muneo Hori, Journal of Applied Physics 62 (1987) 2746

[24] D. D. Koleske, M. E. Coltrin, A. A. Allerman, K. C. Cross, C. C. Mitchell, J. J. Figiel, Applied Physics Letters 82 (2003) 8
[25] H. Heinke, V. Kirchner, S. Eifeldt, D. Hommel, Appl. Phys. Lett. **77** (2000) 2145
[26] B. Heying, X. H. Wu, S. Keller, Y. Li, D. Kapolnek, B. P. Keller, S. P. Denbaars, J. S. Speck, Appl. Phys. Lett. **68** (1996) 643
[27] V. Srikant, J. S. Speck, D. R. Clarke, J. Appl. Phys. **82** (1997) 4286
[28] A. Pelzmann et al., MRS Internet Journal of Nitride Semiconductors **1** (1997) Art 40
[29] J. E. Ayers, Journal of Crystal Growth, **135** (1994) 71
[30] P. Visconti et al., Materials Science and Engineering **B, 93** (2002) 229

ECS Transactions, 58 (4) 439-445 (2013)
10.1149/05804.0439ecst ©The Electrochemical Society

Electrochemical Hydrogenation of Dimensional Carbon

K.M. Daniels[a], S.Shetu[a], J. Staser[b], J. Weidner[b], C. Williams[b], T. S. Sudarshan[a], MVS Chandrashekhar[a]

[a] Department of Electrical Engineering, University of South Carolina, Columbia, SC, 29208, USA
[b] Department of Chemical Engineering, University of South Carolina, Columbia, SC, 29208, USA

The hydrogen adsorption and electrochemical properties of epitaxial graphene (EG) grown on silicon carbide (SiC) and bulk graphite in dilute sulfuric acid (H_2SO_4) and perchloric acid ($HClO_4$) solutions were investigated by cyclic voltammetry (CV) and electrochemical impedance spectroscopy (EIS). According to EIS, electrochemical activity is dependent on starting I(D)/I(G) ratio suggesting a dependence on point defects in the material for successful hydrogenation. Electrochemical hydrogen loading and unloading on epitaxial graphene was also demonstrated and confirmed by Raman spectroscopy and CV. These results show that hydrogenation of graphene is electrochemically reversible and history dependent impacting H loading. This process demonstrates a new pathway to hydrocarbon bond formation for synthesis of advanced organic/inorganic carbon-based compounds.

Introduction

Hydrogenated graphene, graphane, was theoretically predicted in 2007 by Sofo (1) and have since been synthesized by a number of means (2, 3, 4). The difficulty with functionalizing graphene with hydrogen is the need for atomic hydrogen, since hydrogen gas H_2 does not directly react with carbon. Techniques implemented by other groups involve *in situ* development of atomic hydrogen by plasma-assistance which can cause damage due to energetic ions (3) or hot filaments (4), as the H-H bond in hydrogen gas requires high energy/temperature to break (3). With the limitations of current techniques to functionalize graphene, notably the need to form it *in situ*, an alternative, *ex situ* electrochemical (5) means was demonstrated.

Electrochemistry offers the most controlled route to systematic functionalization, as the extent of the hydrogenation of graphene can be precisely controlled by changing the current level (or voltage) and time. Such tunability is not easily achievable using the other techniques described above. Furthermore, through electrochemistry, reactions can be conducted at ambient conditions, opposed to *in situ* as in other techniques. The convenience and controllability of electrochemical hydrogenation of graphene provides a realistic approach for a tunable bandgap in graphene, though an observed dependence on the underlying SiC substrate using this technique has been previously shown (5).

439

Like other hydrogenation techniques, electrochemical hydrogenation of EG was found to be thermally reversible by thermal anneal (5). Hydrogen removal may also be possible electrochemically by reversing the polarity of the electrodes allowing for new possibilities in fine tuning hydrogenation. In this paper, we investigate electrochemical reversibility, hydrogen loading and unloading of epitaxial graphene, confirmed by cyclic voltammetry (CV) and Raman spectroscopy. Material dependence was also investigated and observed by CV and electrochemical impedance spectroscopy (EIS) to study the hydrogenation mechanisms of carbon.

Experimental Details

Epitaxial graphene was grown on semi-insulating, on axis (0001), 6H SiC obtained from II-IV, Inc. The Si-face was chemical-mechanical polished (CMP) and the C-face optically polished. The wafer was diced into 10 mm^2 samples and cleaned using standard RCA clean (Trichloroethylene [TCE], acetone, methanol) and HF used to remove native oxide. Growth of EG was done on both faces by thermal decomposition of SiC *in vacuo* (6), $< 10^{-5}$ Torr, using an RF furnace. Raman was used to check the quality of EG.

Raman spectroscopy was performed using a micro-Raman setup with laser excitation wavelength at 632nm and a spot size of ~2 μm. The Raman system was calibrated using the known Si peak at 520.7 cm^{-1}. Reference blank SiC substrate spectra were scaled appropriately and subtracted from the EG/SiC spectra to show only the graphene and functionalized graphene peaks (3, 7). All the spectra shown are difference Raman spectra obtained in this manner. There are three peaks associated with graphene: The D peak (~1345 cm^{-1}) corresponds to the disorder and diamond like sp^3 content in the material, The G peak (~ 1595 cm^{-1}) corresponds to the in plane vibration of the graphene lattice and the 2D peak due to double resonance (3). Graphene functionalized with hydrogen shows an increase in I(D)/I(G) ratio indicating an increase in sp^3 content. A fluorescence background in the Raman spectra was used to distinguish functionalized graphene from damage. Hydrogen functionalized graphene is a direct gap material (8), resulting in a fluorescence background being seen in the Raman spectra. This can also be used to measure what fraction of sp^3 content is present in the material and therefore estimate of the extent of EG functionalization (8).

To investigate the electrochemical reversibility of EG, hydrogen was loaded and unloaded on EG in 10 minute bursts to maintain material conductivity as fully hydrogenated graphene is an insulator. This experiment was performed using a home-built electrochemical setup, using Hg/HgSO$_4$ reference (0.67V vs. NHE) and dilute H$_2$SO$_4$ as source of hydrogen as shown in Figure 1. A 99.6% Pt wire and exposed EG (approximately a 4mm diameter circular area) were used as the anode and cathode respectively. A potentiostat (Series G^{TM}300, Gamry Instruments, Warminster, PA, USA) was used for potential holds and CV measurements. CV was performed to determine oxidation and reduction potentials at a scan rate of 25mV s^{-1}. The sample was then cycled between potentials, ensuring hydrogen loading and unloading between cycles by Raman spectroscopy.

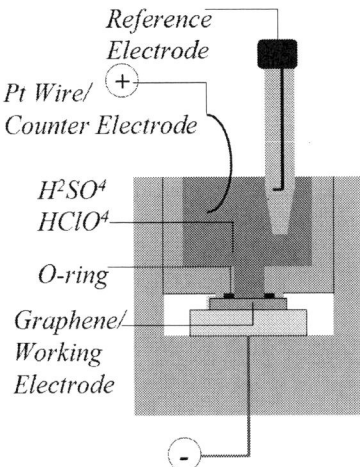

Figure 1. Schematic of home-built electrochemical cell for hydrogenation of epitaxial graphene.

To determine the material dependence on electrochemical functionalization three different EG samples (Si-face EG [~2ML], C-face EG [~10ML] and defective Si-face EG [>50ML]) and a smooth graphite disk [>1000ML] were characterized using electrochemical cell using dilute $HClO_4$, Ag/AgCl reference (0.198 V vs. NHE) and potentiostat. CV and EIS measurements, recording the impedance spectrum of the carbon electrodes between 300 kHz and 0.1 Hz with a perturbation signal of 10 mV were taken along with Raman to compare electrode reactivity with material properties.

Electrochemical Reversibility

In the electrochemical hydrogen reversal experiment oxidation (0.217 V vs. $Hg/HgSO_4$ [0.64 vs. NHE]) and reduction (-0.212 V vs. $Hg/HgSO_4$ [0.64 V vs. NHE]) potentials were obtained by CV. Using these potentials, samples were first held above the hydrogen loading potential observed from CV at -0.3V versus reference to ensure hydrogenation, confirmed with Raman. Through Raman, as shown in Figure 2, an increase in I(D)/I(G), from 0.15 to 0.37 and blue shifts in peak positions was observed after the first hydrogenation. A small fluorescence background was also observed after functionalization. The samples were then held above the hydrogen unloading potential at 0.3V and checked again, Raman confirming removal of hydrogen.

Dehydrogenation of graphene showed decrease in I(D)/I(G), 0.167 close to that of pristine EG at 0.146, though G and 2D peaks were significantly blue shifted beyond pristine EG suggesting bonding with another functional group. This is further supported by the addition of a fluorescence differing from that observed with hydrogenated EG and similar to spectra observed with graphene oxide. Sulfur groups from the acid are believed to be responsible for this behavior, as SO_4^{-2} ions are known to passivate C-surfaces in hydrocarbon polymers. The 2nd hydrogenation of the sample supports this claim with increase in I(D)/I(G) to 0.326, less than that of the first hydrogenation, suggesting energy was required to remove functional groups from the oxidation cycle before hydrogenation of the sample could take place.

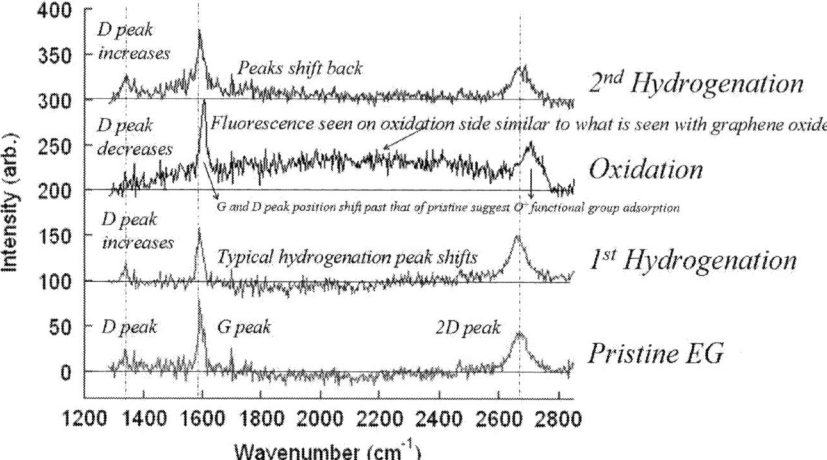

Figure 2. Raman spectra of electrochemical cycling of hydrogen on EG.

Detailed values obtained with Raman are shown in Table I. G peak width was observed to increase after each cycle, suggesting damage/strain in the lattice caused by the changing C-C bond length after each cycle. Changes in I(D)/I(G) ratio shows clear hydrogen adsorption and desorption with a decrease in functionalization shown after hydrogen unloading due to the addition of non-hydrogen functional groups. These results show that graphene hydrogenation is electrochemically reversible and history dependent. This process demonstrates a new pathway to hydrocarbon bond formation for synthesis of advanced organic/inorganic carbon-based compounds.

TABLE I. Raman Spectra Data from Hydrogen Cycling Experiment

Cycles	D peak position (cm^{-1})	G peak position (cm^{-1})	2D peak position (cm^{-1})	I(D)/I(G)	D peak width (cm^{-1})	G peak width (cm^{-1})	2D peak width (cm^{-1})
Pristine EG	1339	1593	2671	0.15	41.8	18.7	67.1
1st Hydrogen Loading	1335	1591	2663	0.37	25.3	21.9	63.3
Hydrogen Unloading	1345	1604	2705	0.17	86.2	22.8	83.7
2nd Hydrogen Loading	1339	1590	2669	0.33	45.6	26.6	81.3

Carbon Reactivity

Figures 3-6 show CV and EIS data obtained from various carbon materials, bulk graphite, C-face graphene, Si-face graphene and defective Si-face graphene. CV was used to determine the reactivity of the carbon electrodes and EIS was used to investigate the charge transfer mechanisms involved. Oxidation and hydrogenation peaks observed in the CV of the graphite disk, Figure 3 were not well defined, most likely due to its poor surface area and layer inaccessibility to the solution. This is further supported by the EIS data which shows a relatively straight line, a system dominated by mass transfer. The system is diffusion limited due to inaccessibility to reaction sites.

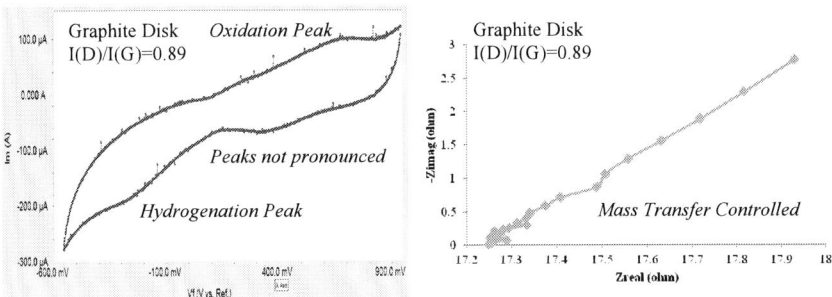

Figure 3 . CV and EIS data of a smooth graphite disk.

C-face EG, Figure 4, showed an absence of hydrogenation peak most likely due to the lack of point defects in the material as I(D)/I(G)~0. This suggests that grain boundaries in the material are not electrochemically active and that functionalization is dependent instead on the presences of point defects.

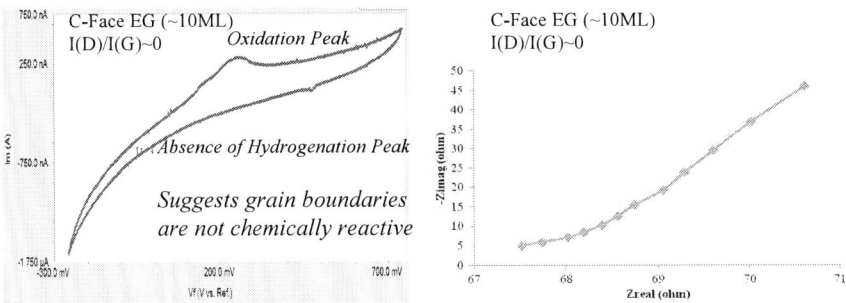

Figure 4. CV and EIS data of C-face EG.

Si-face EG, Figure 5, showed a sharp hydrogenation peak most likely due to point defects present in the material, I(D)/I(G)~0.08. This is further supported by the plateau observed in the high frequency regime in the EIS nyquist plot. The large charge transfer resistance shown in the EIS however, means that the system is kinetically slow.

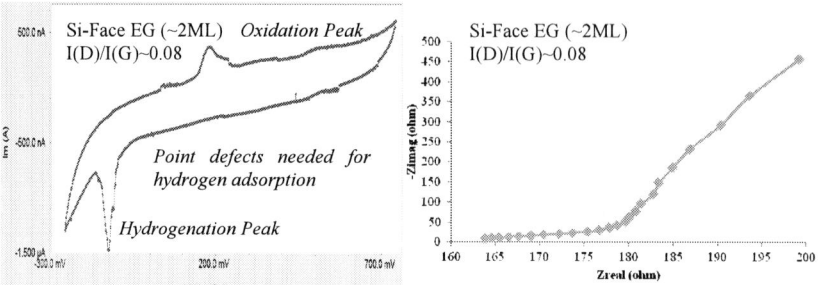

Figure 5. CV and EIS data of Si-face EG.

Defective Si-face EG, Figure 6, further supports dependence on the presence of point defects present in the material, I(D)/I(G)~1.70, as the system revealed a semicircle in the high frequency regime. The diameter of the semicircle revealed a kinetically fast system with ample sites for hydrogen to adsorb. This data suggests that there may not be a substrate to substrate dependence in electrochemical functionalization but a dependence on the quality of the EG film grown on it, with more defective EG ideal for electrochemical functionalization.

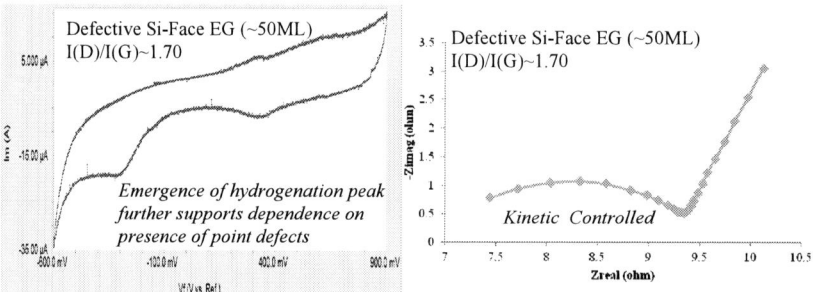

Figure 6. CV and EIS data of defective Si-face EG.

Summary

The hydrogen adsorption and electrochemical properties of epitaxial graphene (EG) grown on silicon carbide (SiC) were shown. Electrochemical hydrogen loading and unloading on EG showed that electrochemical hydrogenation is electrochemically reversible and history dependent impacting H loading. According to EIS, electrochemical activity is dependent on starting I(D)/I(G) ratio suggesting a dependence on point defects in the material for successful hydrogenation. Electrochemical hydrogenation of EG demonstrates a new pathway to hydrocarbon bond formation for synthesis of advanced organic/inorganic carbon-based compounds

Acknowledgments

Kevin M. Daniels acknowledges support from the South East Alliance for Graduate Education and the Professoriate, Southern Regional Education Board, and African American Professors Program. The authors also acknowledge the Southeastern Center for Electrical Engineering Education (SCEEE) and the National Science Foundation (NSF) ECCS-EPDT Grant #1029346, under the supervision of program director Rajinder Khosla, for funding this work.

References

1. J. O. Sofo, A. S. Chaudhari, G. D. Barber., *Physical Review B,* **75**, 153401 (2007)
2. S. Ryu, M. Y. Han, J. Maultzsch, T. F. Heinz, P. Kim, M. L. Steigerwald, L. E. Brus., *Nano Lett.,* Vol.
3. D. C. Elias, R. R. Nair, T. M. G. Mohiuddin, S. V. Morozov, P. Blake, M. P. Halsall, A. C. Ferrari, D. W. Boukhvalov, M. I. Katsnelson, A. K. Geim, K. S. Novoselov., *Science* **30**, Vol. 323, No. 5914, 610-613 (2009)
4. C. Riedl, C. Coletti, T. Iwasaki, A. A. Zakharov, U. Starke., *PRL* **103**, 246804 (2009)
5. K. M. Daniels, B. K. Daas, N. Srivastava, C. Williams, R. M. Feenstra, T. S. Sudarshan, MVS Chandrashekhar., *J. Appl. Phys.* **111**, 114306 (2012)
6. B. K. Daas, K. Daniels, S. Shetu, T. S. Sudarshan, MVS Chandrashekhar., *Mater. Sci. Forum* **717-720**, 633-636 (2012)
7. C. Srinivasan and R. Saraswathi., *Current Science* **97**, No. 3 (2009)
8. A. C. Ferrari and J. Robertson., *Phys. Rev. B* **61**, 14095-14107 (2000)

446

ECS Transactions, 58 (4) 447-453 (2013)
10.1149/05804.0447ecst ©The Electrochemical Society

Abrasive-Free Polishing of SiC Wafer Utilizing Catalyst Surface Reaction

Y. Sano [a], K. Arima [a], and K. Yamauchi [a,b]

[a] Department of Precision Science and Technology, Graduate School of Engineering, Osaka University, 2-1 Yamadaoka, Suisa Osaka 565-0871, Japan
[b] Research Center for Ultra-Precision Science and Technology, Graduate School of Engineering, Osaka University, 2-1 Yamadaoka, Suisa Osaka 565-0871, Japan

Catalyst-referred etching (CARE) is an abrasive-free chemical polishing method that utilizes a catalytic chemical reaction at the contact points of a wafer and a catalyst plate surface. Using a platinum plate catalyst and hydrogen fluoride aqueous solution, an atomically flat 4H–SiC (0001) wafer was obtained. The removal rate was found to be proportional to the rotational velocity and processing pressure, and it reached approximately 500 nm/h, which is comparable to chemical mechanical polishing methods. Furthermore, it was found that an atomically flat SiC surface can be obtained by CARE using only deionized water and a platinum catalyst, which is significant from the viewpoint of the widespread use of CARE.

Introduction

Although silicon (Si) power devices are currently used to regulate the frequency and impedance of electric power for their efficient usage, silicon carbide (SiC) power devices have received much attention in recent years because of superior properties of SiC as compared to Si. These properties include a large band gap and high breakdown electric field, which result in devices with lower power consumption. Owing to recent progress in the development of crystal growth technologies, the quality of bulk SiC crystal continues to improve. In addition to bulk quality, surface quality is important. SiC wafers are finished by chemical mechanical polishing (CMP), which can achieve atomically smooth SiC surfaces with atomic steps (1-3). However, because CMP uses a slurry of abrasives and chemical solutions, it may leave subsurface damage or subsurface scratches on polished surfaces, affecting the quality of subsequent epitaxial growth. We have been developing a novel, abrasive-free polishing method, catalyst-referred etching (CARE) (4-6), which does not, in principle, leave subsurface damage or subsurface scratches on polished surfaces. In this paper, we describe a brief overview of the CARE process and recent progress of the process.

CARE Process

Concept

If a catalyst generates reactive species that are chemically active only on the catalyst surface, and/or if only the contact points of the work surface with the catalyst are chemically etched, the work surface will be chemically removed from the topmost parts

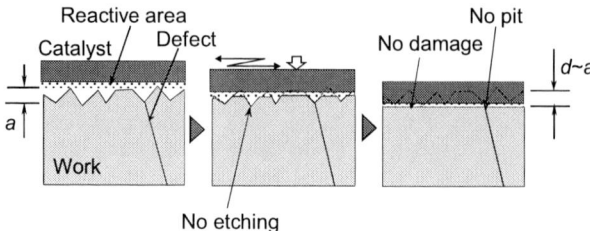

Figure 1. Schematic of catalyst-referred etching (CARE)

and will be efficiently planarized without damage (Fig. 1). We considered that F and OH radicals would be effective reactive species because such radicals are commonly known to be among the most reactive chemical species. We selected platinum (Pt) as the catalyst material and hydrogen fluoride (HF) aqueous solution as the solution because Pt has catalytic properties, such as the ability to dissociate various molecules, and F and OH radicals can be generated by the catalytic dissociation of HF and H_2O, respectively.

Apparatus

Figure 2 shows the schematic of the CARE apparatus, which is quite similar to the conventional CMP apparatus. The catalyst plate and solution are used instead of the polishing pad and slurry of CMP, respectively. To ensure the safe use of dangerous chemicals, our homemade CARE apparatus was built using chemically inert materials, piping with coaxial tubes, closed containers with local ventilation, and so on. The catalyst plate made of Pt was finished by turning and had narrow grooves on its surface for solution delivery and the prevention of adhesion of the work surface. For wafer-level polishing, a catalyst pad consisting of a Pt thin film deposited on a fluorine-containing elastomer was used instead of a bulk solid Pt plate to avoid nonuniform processing due to the undulation of the wafer surface.

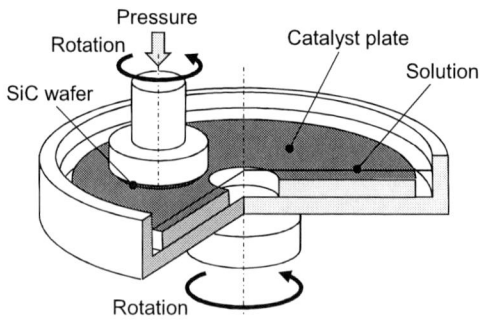

Figure 2. Schematic of CARE apparatus

Processing properties

From the results of CARE of a lapped 4H–SiC (0001) substrate (on-axis, n-type, 0.02–0.03 Ω•cm) whose surface had many scratches and microcracks, we found that CARE preferentially removed the topmost sites of the wafer surface in contact with the catalyst surface, and that the frictional force between the catalyst and wafer surface was sufficiently small to preempt the introduction of mechanical scratches or cracks (4). An evaluation of the CARE-processed surface by atomic force microscopy (AFM) revealed the surface's step-and-terrace structure, with a step height of a single bilayer of Si and C (5). This showed the CARE-processed surface to be atomically flat. Such a surface structure strongly indicates that the removal mechanism is based on chemical phenomena and that a type of step-flow removal occurs on the (0001) Si surface of SiC. Thus, we can consider the removal rate to be dependent on the step density of the work surface. From the results of the experiment using several SiC wafers with different off-angles, it was shown that the removal rate was proportional to the step density of the substrate surface (7). Of course, the catalyst surface was not atomically flat, unlike the flattened SiC surface. However, the topmost regions on the SiC surface came into contact with the catalyst surface more frequently than the other sites through the averaging effect caused by the relative motion of the catalyst plate and SiC wafer, which allowed the formation of an atomically flat SiC surface.

A mechanically polished 2-inch 4H–SiC (0001) 8° off-axis wafer was used for wafer-level processing by CARE (8). Figures 3(a) and 3(b), respectively, show typical images of a 4H–SiC wafer before and after CARE, observed by optical interferometry (Zygo

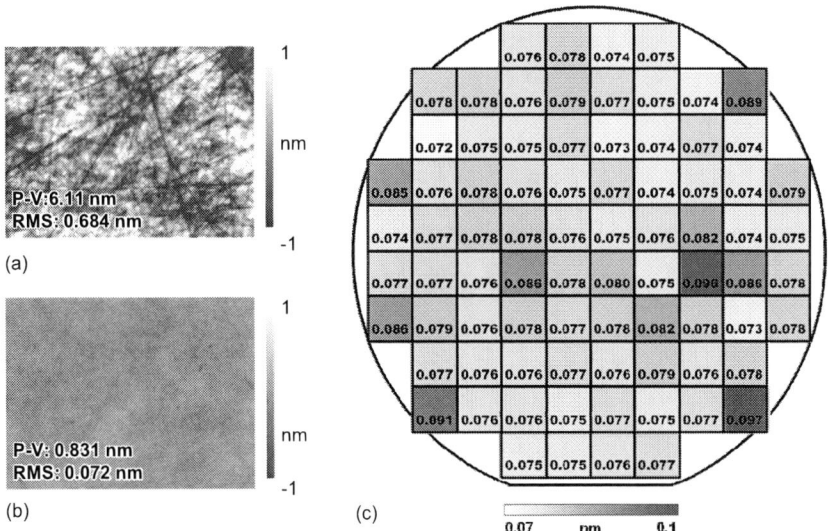

Figure 3. Surface roughness of a 2-inch 4H–SiC (0001) 8° off-axis wafer measured by an optical interferometer (64×48 mm^2): (a) image of an as-received wafer (PV:6.11 nm, RMS: 0.684 nm), (b) image of CARE-processed wafer (PV: 0.831 nm, RMS: 0.072 nm), and (c) RMS value map at 5 mm intervals over the entire processed wafer.

NewView200). Figure 3(c) shows the RMS values of the entire processed wafer. The roughness is markedly reduced to less than 0.1 nm at all points of the wafer. Also, we tried a commercially available 3-inch n-type 4H–SiC (0001) 4° off-axis substrate. The RMS values of the as-received surface and the 2-inch wafer were successfully reduced to less than 0.1 nm (9). It is noteworthy that the removal rate of the 3-inch wafer was 37 nm/h, which was nearly equal to that of the 2-inch 4° off-axis wafer (33 nm/h). In general, when polishing using an abrasive, such as by CMP, the removal rate may decrease with increasing wafer diameter because it becomes more difficult for the abrasive to flow into the center of the wafer. Because CARE does not use abrasives, there should not be an issue regarding abrasive transportation, which is a major advantage of CARE over CMP.

Recent Progress

Improvement of removal rate

An issue of CARE with regard to its practical use is its relatively low removal rate. In an attempt to improve the removal rate, we investigated its dependence on the rotational velocity and processing pressure using a 2-inch n-type 4H–SiC (0001) 8° off-axis substrate. From the results, we found the removal rate to be nearly proportional to both the rotational velocity and processing pressure. Increasing the rotational velocity (25 rpm) and processing pressure (980 hPa) to the maximum values of the apparatus, we could obtain a removal rate of 492 nm/h, which is comparable to that of conventional CMP processes (10). Figure 4 shows the proportional relationship between the removal rate and the product of rotational velocity and processing pressure. A further increase in the removal rate would be possible with an increase in the rotational velocity and/or processing pressure.

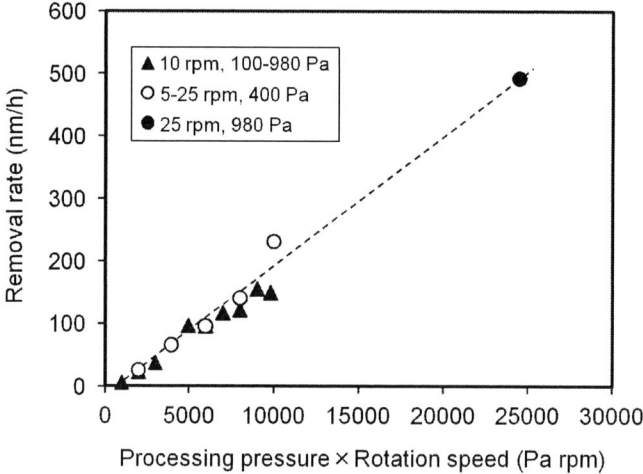

Figure 4. Relationship between removal rate and product of rotational velocity and processing pressure.

In CARE, to remove surface atoms, alternating contact between the catalyst surface and solution is required. Thus, the removal rate was considered to be proportional to the rotational velocity because the frequency at which the SiC surface alternately makes contact with the catalyst and solution increases with the rotational velocity. This proportional increase in the removal rate with the processing pressure is considered to be due to the increase in the contact area between the wafer surface and catalyst pad with increasing pressure.

In the case of mechanical polishing, it is well known that the amount of processing is proportional to the polishing time, polishing pressure, and relative velocity between the work surface and polishing pad (11). However, during CARE, although the amount of processing is proportional to the polishing pressure and relative velocity, it is often not proportional to the processing time and depends highly on the solution species and catalyst surface conditions such as the presence of adsorbing species and contaminants.

Decreasing use of HF

HF is a well-known toxic chemical and is not preferred for widespread use of CARE. For this purpose, it is very important to know the role of HF in CARE. We first considered that the mechanism is a direct etching of SiC by F radicals or a combination of the oxidation of SiC on the catalyst surface and dissolution of the oxide to an HF solution. However, it was found that the removal rate became extremely slow when using a dilute HF solution, which was still sufficiently concentrated for the dissolution of SiO_2 (12). Thus, we believe that the role of the HF solution is not limited to the source of an F radical or the dissolution of SiO_2. Currently, we assume that the main chemical reaction is the dissociative adsorption of HF molecules to the backbond of the step-edge Si atom in the presence of the catalyst surface (13). If so, there may be dissociative adsorption of H_2O molecules to the backbond of the step-edge Si atom in the presence of the catalyst surface. Thus, we attempted to use only deionized (DI) water. The sample used was a commercially available 4H–SiC (0001) on-axis wafer. After processing with a rotational velocity of 10 rpm and an applied pressure of 400 hPa, we observed a 1-bilayer-high step-and-terrace structure, which is the same structure as that processed with HF (Fig. 5). The removal rate as calculated by the weight loss was 1.2 nm/h. Although this is extremely slow, the removal rate can be increased by increasing the rotational velocity, processing pressure, and step density (using an off-axis wafer).

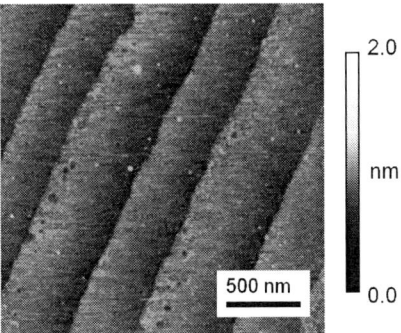

Figure 5. AFM image of 4H–SiC (0001) on-axis surface ($2 \times 2 \ \mu m^2$) processed by CARE with Pt catalyst and DI water.

Conclusion

CARE is an abrasive-free chemical polishing method that utilizes a catalytic chemical reaction at the contact points of the work surface and the surface of a catalyst plate. An atomically flat SiC surface can be obtained by CARE using not only a Pt catalyst and HF solution but also a Pt catalyst and DI water. Although the removal rate using DI water is very slow, it can be improved by increasing the rotational velocity and/or processing pressure. Even though the removal rate is not significantly high, it is still very useful, especially for the use of smoothing the slightly roughened surface during device processing such as a high-temperature annealing and an epitaxial growth.

Acknowledgments

This work was partially supported by grants for the 21st Century COE Program and the Global COE Program from the Ministry of Education, Culture, Sports, Science and Technology (MEXT), Japan, and the Industrial Technology Research Grant Program in 2005 from the New Energy and Industrial Technology Development Organization (NEDO), Japan.

References

1. P. Vicente, E. Pernot, D. Chaussende, and J. Camassel, *Mater. Sci. Forum*, **389-393**, 729 (2002).
2. T. Kato, K. Wada, E. Hozomi, H. Taniguchi, T. Miura, S. Nishizawa, and K. Arai, *Mater. Sci. Forum*, **556-557**, 753 (2007).
3. K. Hotta, K. Hirose, Y. Tanaka, K. Kawata, and O. Eryu, *Mater. Sci. Forum*, **600-603**, 823 (2009).
4. H. Hara, Y. Sano, H. Mimura, K. Arima, A. Kubota, K. Yagi, J. Murata, and K. Yamauchi, *J. Electron. Mater.*, **35**, L11 (2006).
5. K. Arima, H. Hara, J. Murata, T. Ishida, R. Okamoto, K. Yagi, Y. Sano, H. Mimura, and K. Yamauchi, *Appl. Phys. Lett.*, **90**, 202106 (2007).
6. Y. Sano, K. Arima and K. Yamauchi, *ECS J. Solid State Sci. Technol.*, **2**, N3028 (2013).
7. T. Okamoto, Y. Sano, K. Tachibana, K. Arima, A.N. Hattori, K. Yagi, J. Murata, S. Sadakuni, and K. Yamauchi, *J. Nanosci. Nanotechnol.*, **11**, 2928 (2011).
8. T. Okamoto, Y. Sano, H. Hara, T. Hatayama, K. Arima, K. Yagi, J. Murata, S. Sadakuni, K. Tachibana, Y. Shirasawa, H. Mimura, T. Fuyuki, and K. Yamauchi, *Mater. Sci. Forum*, **645-648**, 775 (2010).
9. T. Okamoto, Y. Sano, K. Tachibana, K. Arima, A.N. Hattori, K. Yagi, J. Murata, S. Sadakuni, and K. Yamauchi, *Mater. Sci. Forum*, **679-680**, 493 (2011).
10. T. Okamoto, Y. Sano, K. Tachibana, B.V. Pho, K. Arima, K. Inagaki, K. Yagi, J. Murata, S. Sadakuni, H. Asano, A. Isohashi, and K. Yamauchi, *Jpn. J. Appl. Phys.*, **51**, 046501 (2012).
11. F.W. Preston, *J. Soc. Glass Tech.*, **11**, 214 (1927).

12. A. Isohashi, Y. Sano, T. Okamoto, K. Tachibana, K. Arima, K. Inagaki, K. Yagi, S. Sadakuni, Y. Morikawa, and K. Yamauchi, *Mater. Sci. Forum*, **740-742**, 847 (2013).
13. P.V. Bui, K. Inagaki, Y. Sano, K. Yamauchi, and Y. Morikawa, *Current Appl. Phys.*, **12**, S42 (2012).

454

ECS Transactions, 58 (4) 455-461 (2013)
10.1149/05804.0455ecst ©The Electrochemical Society

Growth of GaN by MOCVD on Rare Earth Oxide on Si(111)

F. Erdem Arkun[a], Rytis Dargis[a], Andrew Clark[a], Robin S. Smith[a], Michael Lebby[a]
Jeffrey M. Leathersich[b], F. Shahedipour-Sandvik[b]

a. Translucent Inc., 952 Commercial St. Palo Alto, CA 94303
b. College of Nanoscale Science and Engineering, State University of New York,
Albany, NY 12203

Growth of GaN on rare earth oxide (REO) buffers grown on
silicon (111) was performed. A novel low temperature buffer layer
was developed in order to suppress the decomposition of the REO
buffer under hydrogen flow during MOCVD. GaN films, grown by
this technique, exhibit smooth surface morphologies. XRD studies
reveal the formation of an ErN layer during growth.

Introduction

Increased global power demands and penetration of new technologies such as solar
energy, hybrid and electric vehicles (EV) require efficient power conversion systems
capable of switching large currents at high voltages with minimal loss. The current
technology based on silicon LDMOS devices are reaching their limit in terms of speed
and efficiency of power conversion.
One of the emerging devices for efficient power conversion is based on GaN, which
combines high breakdown voltage and low specific on resistance R_{onsp}, making this
material an ideal candidate for such applications. The possibility of growing an
AlGaN/GaN heterostructure and formation of a two dimensional electron gas (2DEG) in
this material system creates high electron density, high electron mobility channel, and
results in HEMT devices with enhanced performance. The high breakdown voltage of
GaN allows large blocking voltages and makes this material system ideal for power
conversion applications. Growth of GaN on silicon is very attractive for power devices
due to the cost and size benefit of silicon along with the high thermal conductivity of
silicon. Many different buffer schemes based on AlN/Si have been implemented for
growth of GaN on silicon. All of these schemes make use of an AlN buffer layer grown at
high temperature on silicon wafers which results in the formation of a conductive layer at
the interface due to the diffusion of silicon into the growing GaN layer at high growth
temperatures. This conductive layer in turn limits the performance of AlGaN/GaN
HEMT devices at high drain biases by causing the device to breakdown at the epilayer
substrate interface. Growth of REO epitaxial films on Si(111) substrates and the
subsequent growth of GaN on this layer make this material system especially attractive as
a templates for growth of GaN. Growth of GaN by this method directly onto an oxide
layer will eliminate the formation of a conductive layer formed by unintentional doping
of the silicon substrate.

Experiment

Growth of REO materials on silicon has been reported by several researchers in the
field and used as buffer layers for GaN growth as well as high-k dielectric layers on

455

silicon substrates [1-7]. Further insight about the growth of REOs can be found in the references.

Growth of Er$_2$O$_3$ on Silicon (111) by MBE

Growth of Er$_2$O$_3$ single crystal films with excellent crystal quality is possible on Si(111) surfaces due to a unique alignment between the lattice parameter of silicon and most of the rare earth oxides ($a_{REO} \approx 2a_{Si}$). Samples for this work were grown in a Veeco Gen 20 reactor using elemental rare earth metals and molecular oxygen. Si(111) wafers were dipped in a 1% HF solution in water to remove the native oxide and passivate the surface of the wafer. Cleaned wafers were introduced into the growth chamber and heated up to 750°C to obtain a clean 7x7 reconstructed surface as determined by reflection high energy electron diffraction (RHEED). Growth of Er$_2$O$_3$ was commenced on this surface and the morphology of the growth surface was monitored using RHEED; reflectometry at 650nm was used to determine the growth rate of the Er$_2$O$_3$ films. Details of the growth of Er$_2$O$_3$ on Si(111) is given in previous publications [8]

Breakdown Voltage Measurements on Gd$_2$O$_3$

Breakdown voltage measurements were performed on Gd$_2$O$_3$ films on silicon to determine the average breakdown voltage. Gd$_2$O$_3$ on silicon films were metallized through a shadow mask with circular pads ranging from 100 μm to 1mm in size and were measured until breakdown. The measurements were performed under vacuum conditions to avoid parasitic breakdown. It was found that the majority of the samples tested had a breakdown voltage around 4MV/cm as shown in Figure 1. There was no direct link between the pad size and the breakdown voltage observed in these samples. Breakdown voltage measurements were not performed for Er$_2$O$_3$ films on silicon at the time of the publication, however physical and electrical properties of Er$_2$O$_3$ and Gd$_2$O$_3$ are similar.

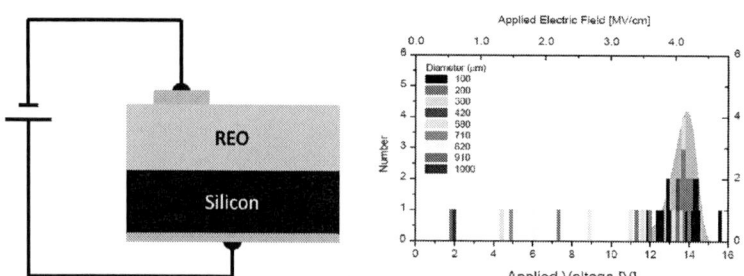

Figure 1 breakdown voltage distribution of Gd2O3 films grown on Si(111) substrates. Gd2O3 was found to have a breakdown voltage around 4MV/cm

Growth of GaN on Si/Er$_2$O$_3$/Silicon (cSOI^{TM}) templates by MOCVD

Growth of GaN was performed on a single crystal SOI platform (cSOI^{TM}) with an Er$_2$O$_3$ oxide layer in order to test the potential of the REO materials as a buffer for GaN.
This approach provides a route to de-convolute the nucleation of GaN and the stability of the oxide during the growth of GaN. cSOI^{TM} templates comprising of a Si/Er$_2$O$_3$/Si(111) structure were prepared by MBE and transferred into an Veeco D-180 reactor in a "6x2"

configuration for further growth of GaN. Silicon wafers were also loaded alongside these templates as monitor wafers for the experiment. Wafers were heated up to 1060°C and cleaned under a H_2 flow in order to remove the native oxide and prepare the surface for subsequent growth. An AlN buffer was then deposited on the wafers at a temperature of 1050°C to a thickness of ~200nm. Following the AlN, two 300nm stress mitigating layers of Al_6Ga_4N and Al_4Ga_6N were deposited at 1050°C. This buffer stack was immediately followed by growth of a 1μm thick GaN film as measured by an *in-situ* reflectometer. An AlGaN cap with 25% Al content and 30nm was deposited at the end of the experiment.

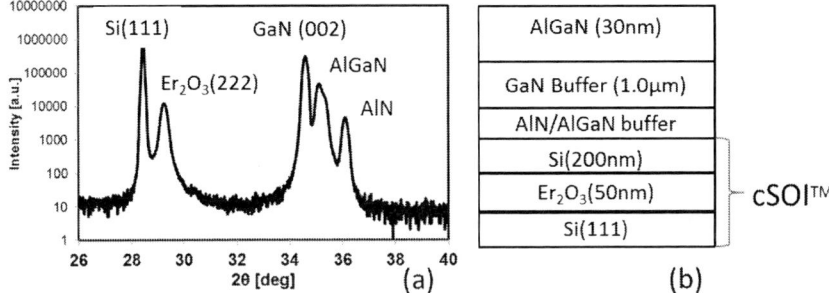

Figure 2(a) Symmetrical XRD spectra showing reflections from all the layers in the multilayer stack formed by MBE and MOCVD. (b) Schematic representation of the multilayer stack

Figure 2(a) shows the symmetrical X-ray diffraction pattern of the structure grown by MOCVD on cSOI™ templates. The layer structure is given in Figure 2(b). The layers in the structure were identified clearly and the absence of secondary phases (i.e. metal-silicides, metal–nitrides) in the spectra confirms the thermal stability of the oxide under growth conditions used in the process. Moreover it was also concluded that the top silicon layer shields the oxide layer against the reducing H_2 environment employed to clean the surface of the wafers at 1060°C.

Surface morphology of this sample was characterized by AFM and compared to that of the surface morphology of GaN on the bare silicon wafer. Figure 3 (a) and (b) shows the 5μm x 5μm AFM scan of the surfaces. Both samples showed smooth surfaces with terraces indicating a two dimensional growth process. The growth on cSOI™ sample looked slightly more undulated which was attributed to the rough silicon starting surface.

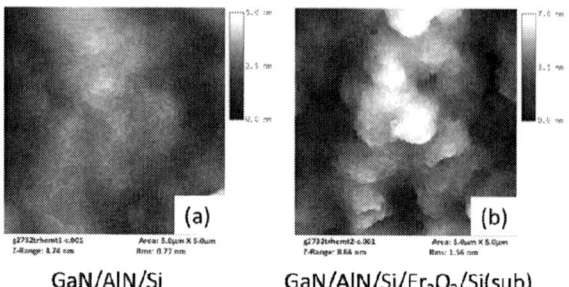

GaN/AlN/Si   GaN/AlN/Si/Er₂O₃/Si(sub)

Figure 3 (a) 5µmx5µm AFM scan of GaN on silicon using an AlN buffer layer showing a RMS roughness of 0.77nm. (b) 5µmx5µm AFM scan of GaN on Si/Er₂O₃/Si(sub) using an AlN buffer layer showing a RMS roughness of 1.56nm.

Growth of GaN on Er₂O₃/Silicon (111) templates by MOCVD

Growth of GaN on a cSOI™ sample showed that Er₂O₃ is stable in contact with the silicon substrate and capping Si layer under MOCVD growth conditions. The growth of GaN directly on an Er₂O₃ layer was found to be more complicated due to the presence of hydrogen and its reducing effect on the Er₂O₃ layer during high temperature growth.

The oxide template used in this part of the growth campaign was Er₂O₃ (150nm)/Gd₂O₃(10nm)/Si(sub). It was found that growth of a thin (10nm) Gd₂O₃ layer before the growth of a thick Er₂O₃ (150nm) layer helped bridge the lattice mismatch between the silicon substrate and Er₂O₃ and resulted in smoother films. In order to circumvent the reduction of the Er₂O₃ layer in the presence of H₂, a low temperature (LT) GaN buffer layer using TEGa and N₂ carrier gas (no H₂) was employed. This buffer layer serves as a nucleation layer for the high temperature (HT) GaN layer, while protecting the underlying Er₂O₃ layer from reduction by hydrogen during the HT- GaN growth step.

Optimization of the LT GaN buffer layer was performed empirically by growing two thicknesses of buffers on Er₂O₃. LT GaN buffers of 100nm (buffer A) and 250 nm (buffer B) thick were grown on Er₂O₃ using TEGa and nitrogen exclusively as a carrier gas at temperatures ranging from 500-700°C. Both of the buffers resulted in relatively rough films with root mean square roughness of 8 and 20nm respectively on a 2µmx2µm AFM scan. Optical microscopy reveals that buffer A shows a continuous layer without cracks whereas buffer B shows severe cracking. This is attributed to the difference in the thermal expansion coefficient of the substrate and the GaN layer.

Growth of HT GaN on these buffers was used as a tool to evaluate the effectiveness of the LT buffers on Er₂O₃ with respect to their ability to protect the underlying oxide against reducing effects of H₂. Buffer A and buffer B were paired with a previously grown AlN/Si wafer to be used as a monitor wafer and loaded in the growth reactor. Substrate temperature was raised under an ammonia flow up to the HT GaN growth temperature (1030°C); TMGa and H₂ were turned on to commence the growth. The best growth conditions for HT GaN on both of the buffer layers was found to be a two part growth in which the first part was performed under a low (3000sccm) H_2 flow followed by a GaN layer with full H_2 (9000sccm) flow. Both buffer A and buffer B showed an increase in the reflectivity signal as the temperature of the wafers was raised to the growth temperature. As the growth commenced the reflectivity signal for buffer A underwent a sharp drop indicating a sudden roughening of the surface, whereas no such drop was observed for the reflectivity signal of buffer B or for the AlN/Si monitor wafer. This indicates that buffer A was either roughened due to nucleation of GaN on the surface, or the Er_2O_3 layer was consumed by the presence of hydrogen and ammonia in the ambient. Growth of GaN was performed under reduced H_2 flow until a thickness of 50nm of GaN was reached, after which the H_2 flow was gradually ramped to the full flow as the N_2 flow was reduced. Table I summarizes the samples presented in this study along with growth performed on these samples.

TABLE I. Summary of samples used in this study.

Sample	Growth1	Growth 2	Comments
cSOI^{TM}	AlN/GaN/AlGaN	N/A	Er_2O_3 thermally stable
Er_2O_3/Gd_2O_3/Si(111)	100nm GaN Buf. A	250nm HT GaN 2-Step growth	ErN formation observed
Er_2O_3/Gd_2O_3/Si(111)	250nm GaN Buf. B	250nm HT GaN 2-Step growth	ErN formation suppressed

The surfaces of the wafers looked shiny to the naked eye, with some haze around the edges of the wafers, which is believed to be due to process gasses affecting the sidewall of the epi structure on the wafers. The morphology of the surfaces measured by AFM is shown in Figure 4 a-c. AFM shows two dimensional growth on both buffer A and B, and a roughness between 1-2nm RMS. The surface morphology of GaN grown on the AlN/Si monitor wafer appeared to be similar to that on buffer A and buffer B. It is believed that growth on the oxides can be further improved by further optimizing the GaN growth conditions.

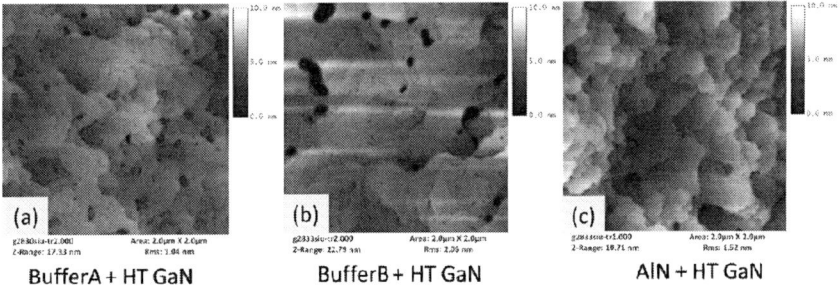

Figure 4 2μmx2μm AFM scan of GaN on buffer A showing a RMS roughness of 1.04nm. (b) 2μmx2μm AFM scan of GaN on buffer B showing a RMS roughness of 2.05nm. (c) 2μmx2μm AFM scan of GaN on AlN/Si showing a RMS roughness of 1.52nm

Further characterization of the GaN layers grown on these buffers were performed by symmetrical X-ray diffraction and a secondary phase was observed. The XRD spectra of the GaN layers grown on buffer A and buffer B are shown in Figure 5(a) and (b). The secondary phase reflections were identified as ErN (111) d=0.2789nm, 2theta=32.05° and ErN(200) d=0.2437nm 2theta=36.85° which is in agreement with d-spacings reported in the literature for ErN $d(111)$=0.279nm and ErN $d(200)$=0.242 nm [9]. The intensity of the X-ray signal at the ErN(200) position was found to be quite weak to conclusively identify the presence of this particular reflection. Moreover, the presence of an ErN (200) reflection in a single crystal symmetrical scan would indicate the presence of a polycrystalline layer. Therefore further investigation of the layer structure by additional methods is required to confirm the existence of a polycrystalline ErN layer in the post-growth structure of the films.

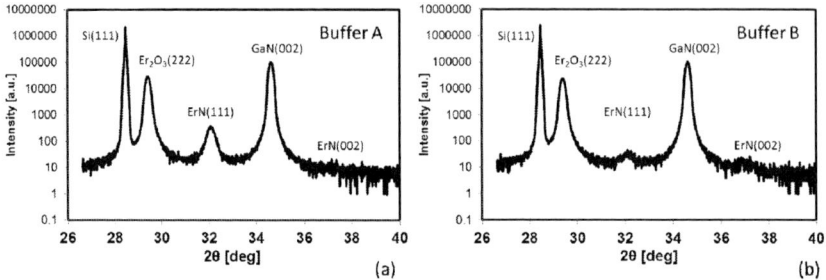

Figure 5 (a) show the XRD spectra of GaN on buffer A. ErN (111) reflection is identified as a secondary phase formed during growth of GaN (b) XRD spectra of GaN on Buffer B. ErN(111) reflection is much weaker indicating a thinner layer.

The presence of an ErN layer is detected in both of the films, however, intensity of the ErN(111) reflection was much lower in the film grown on buffer B indicating a thinner layer. It is believed that ErN was formed as a result of nitrogen diffusion through the GaN buffer layer and conversion of Er_2O_3 to ErN during growth. It can further be concluded that buffer B which was 250 nm thick, was more effective in preventing nitrogen diffusion to the Er_2O_3 layer. It was assumed that as nitrogen diffuses through the growing GaN layer during growth, Er_2O_3 gradually transforms to ErN and the amount of ErN formed is limited by the amount of nitrogen diffusing through the GaN layer. The quality of the GaN layers was also measured by using the FWHM of the rocking curves of the symmetrical (002) reflections. The films grown on buffer A and buffer B have 0.378° and 0.5° FWHM of the GaN (002) peak respectively.

Conclusion

Growth of GaN on cSOI™ and Er_2O_3/Gd_2O_3/Si(sub) REO structures were performed to evaluate the thermal stability of REO layers and feasibility of growing GaN directly on the Er_2O_3 layers. Successful growth of GaN was performed on the cSOI™ structure using a typical GaN on silicon recipe. In the case of growth of GaN directly on Er_2O_3/Gd_2O_3/Si(sub) structure, a LT GaN buffer layer was developed to avoid reduction of Er_2O_3 by hydrogen during HT GaN growth. A two-step growth on this buffer was found to be successful in avoiding reduction of Er_2O_3 by hydrogen, however, an ErN secondary phase was identified. In the case of the thicker buffer layer (buffer B) ErN was found to be thinner, which might be due to the reduced amount of nitrogen diffusing through the GaN layer to the oxide layer. Further characterization of the grown layers by TEM is underway to determine the exact location and thickness of the ErN layer formed in these samples. Optimization of growth will include growth of thicker GaN buffers in an effort to completely inhibit the formation of ErN during HT GaN growth. Additional process improvements to remedy cracking in the layers and refine surface morphology will also be necessary to achieve layers conducive to device structures. Growth of GaN on REO may potentially increase the breakdown voltage of HEMT devices in this material system.

Acknowledgements

Authors would like to acknowledge with much appreciation, Dr. Gilberto A. Umana Membreno of University of Western Australia for performing breakdown voltage measurements on Gd_2O_3 samples reported in this study.

References

1. P. Zaumseil, L. Tarnawska, P. Storck and T. Schroeder, *J. Phys. D: Appl. Phys.* **44** (2011) 315403
2. O. Seifarth, B. Dietrich, P. Zaumseil, A. Giussani, P. Storck and T. Schroeder, *Journal of Applied Physics*, **108**, 073526 (2010)
3. A. Laha, E. Bugiel and H. J. Osten, *Applied Physics Letters*, **88**, 172107 (2006)
4. A. Laha, H. J. Osten, A. Fissel, *Applied Physics Letters*, **89**, 143514, (2006)
5. R. Dargis, A. Fissel, D. Schwendt, E. Bugiel, J. Krügener, T. Wietler, A. Laha and H. J. Osten, *Proceedings of the 4th Symposium on Vacuum based Science and Technology*, **85**, 4, pages 523-526, (2010)
6. J. Kwo, M. Hong, A. R. Kortan, K. T. Queeney, Y. J. Chabal, J. P. Mannaerts, T. Boone, J. J. Krajewski, A. M. Sergent and J. M. Rosamilla, Applied Physics Letters, **77**, 1, pages 130-132 (2000).
7. L. Tarnawska, A. Giussani, P. Zaumseil, M. A. Schubert, R. Paszkiewicz, O.Brandt, P. Storck and T. Schroeder, *Journal of Applied Physics*, **108**, 063502 (2010).E. Gaura and R. M. Newman, *ECS Trans.*, **4**(1), 3 (2006).
8. R. Dargis, D. Williams, R. Smith, E. Arkun, R. Roucka, A. Clark and M. Lebby, *ECS J. Solid State Sci. Technol.*, (2012), **1**, 2, N24-N28
9. Standard X-ray Diffraction Powder Patterns: US Department of Commerce, National Bureau of Standards, *NBS Monograph 25 Section 4, page 54*, (1966)

Author Index

Abdel-Motaleb, I. M.	245	D'Evelyn, M. P.	287
Ahmed, S.	391	Daniels, K. M.	439
Aktas, O.	295	Dargis, R.	455
Alexandru, M.	375	DasGupta, S.	211, 365
Alvarez, B.	295	Detchphrom, T.	261
Ancona, M. G.	279	Dhar, S.	51
Anderson, T. J.	221, 279	Diduck, Q.	295
Ang, S. S.	253, 391	Disney, D.	295
Arima, K.	447	Dobrinsky, A.	129
Arkun, F. E.	455	Downey, B. C.	287
Atcitty, S.	211, 365	Dudley, M.	315
		Dupuis, R. D.	261
Bahat-Treidel, E.	145, 187		
Bakowski, M.	61	Eddy, C. R. Jr.	279
Balda, J. C.	253	Edwards, A.	295
Banu, V.	375	Ehrentraut, D.	287
Baranyai, R.	279	Erlbacher, T.	71
Bauer, A.	71, 81		
Bisi, D.	187	Feng, Z.	351
Borowicz, P.	61	Feygelson, T. I.	279
Bour, D.	295	Flicker, J. D.	211
Brosselard, P.	331	Frewin, C.	119
Brunner, F.	145, 187	Frey, L.	71
Brylinski, C.	331	Fujimoto, T.	3
Burenkov, A.	71, 81		
Byrappa, S.	315	Gaska, R.	129
		Godignon, P.	375
Cai, S.	351	Goto, H.	25
Chandrashekhar, M. V. S.	97, 439	Goto, H.	25
Charles, M.	269	Green, R.	87
Chen, K. J.	351	Gutt, T.	61
Chen, W.	351		
Chung, G.	315	Habersat, D.	87
Clark, A.	455	Han, D.	399

Hansen, D.	315	Lebby, M.	455
Harada, K.	155	Lee, D. S.	311
Hasanzadeh, A.	413	Lee, Y. C.	261
Hasebe, K.	155	Lelis, A.	87
Hilt, O.	145, 187	Liu, S.	351
Hirano, H.	3	Loboda, M.	315
Hite, J. K.	221, 279	Lochner, Z.	261
Hobart, K. D.	221, 279		
Hughart, D. R.	211	MacElwee, T.	167
		Mahadik, N. A.	9, 325
Ikeda, N.	155	Malen, J. A.	343
Imhoff, E. A.	325	Mantooth, H. A.	253, 391
Ito, S.	111	Marinella, M. J.	211, 365
		Mastro, M. A.	279
Ji, M. H.	261	Masumoto, K.	111
Jiang, W.	287	Matsueda, T.	25
		Matsui, K.	33
Kambayashi, H.	155	Meneghesso, G.	187
Kamber, D. S.	287	Meneghini, M.	187
Kao, T. T.	261	Meunier, R.	269
Kaplar, R. J.	211, 365	Meyer, D. J.	279
Katsuno, M.	3	Millin, J.	375
Khaligh, A.	413	Mizuno, J.	25
Kim, J.	261	Mohan, N.	237
Kitabatake, M.	17	Montserrat, J.	375
Kizilyalli, I. C.	295	Mooro, E.	87
Klowak, G.	167	Morancho, F.	269
Knauer, A.	187	Morozumi, Y.	155
Knauer, A.	145	Morvan, E.	269
Koehler, A. D.	221, 279	Moustakas, T. D.	427
Kojima, K.	111	Mueller, S.	315
Kotara, P.	145		
Krishnan, B.	311	Nagata, A.	111
Kub, F. J.	221, 279	Nakagawa, T.	25
Kuball, M.	279	Nath, A.	325
		Nepal, N.	279
Lafontaine, H.	167	Nie, H.	295
Leathersich, J. M.	455	Nipoti, R.	325

Nomura, T.	155	Shahedipour-Sandvik, F.	455
Noppakunkajorn, J.	399	Shen, S. C.	261
		Shenai, K.	179, 199, 229, 301
Ogale, A.	399	Shetu, S.	439
Ohmi, T.	155	Shinohara, H.	25
Okada, A.	25	Shur, M.	129
Okumura, H.	111	Silvestri, R.	187
Omar, S. U.	97	Simin, G.	129
		Smith, R. S.	455
Pakalapati, R. T.	287	Song, H.	97
Palacios, T.	365	Specht, P.	221
Papasouliotis, G. D.	311	Stahlbush, R. E.	9, 325
Paranjpe, A.	311	Staser, J.	439
Pate, B. B.	279	Strenger, C.	71, 81
Phung, L. V.	331	Su, J.	311
Pichler, P.	71, 81	Su, Z.	343
Piskorski, K.	61	Sudarshan, T. S.	97, 439
Planson, D.	331	Sugawa, S.	155
Plissonier, M.	269	Sun, M.	365
Pomeroy, J. W.	279	Sunakawa, H.	25
Porter, M.	221		
Przewlocki, H. M.	61	Tadjer, M. J.	221, 279
		Tani, K.	3
Raghothamachar, B.	315	Tanimoto, S.	33
Rana, T.	97	Tanisawa, H.	33
Reyes, M.	119	Teramoto, A.	155
Robbins, W. P.	237	Torres, A.	269
Roberts, J.	167	Tournier, D.	331
Ryou, J. H.	261	Tsuge, H.	3
Ryssel, H.	71		
		Ueda, H.	155
Saddow, S. E.	119	Uhnevionak, V.	71, 81
Sanchez, E. K.	315	Ushio, S.	3
Sano, Y.	447	Usui, A.	25
Sarlioglu, B.	399		
Sato, S.	33	Wang, H.	315
Sato, S.	3	Ward, P. J.	119
Scott, I.	167	Watanabe, K.	33

Weatherford, T. R.	221
Weaver, B.	221
Weidner, J. W.	439
Williams, A. D.	427
Williams, C.	439
Wu, F.	315
Wuerfl, J.	145, 187
Yamaguchi, A. A.	25
Yamauchi, K.	447
Yano, T.	3
Yashiro, H.	3
Yoo, H. D.	287
Yushyna, L.	167
Zanandrea, A.	187
Zanoni, E.	187
Zhang, B.	351
Zhou, Q.	351
Zhytnytska, R.	145